# Global Arctic

Matthias Finger • Gunnar Rekvig
Editors

# Global Arctic

## An Introduction to the Multifaceted Dynamics of the Arctic

*Editors*
Matthias Finger
Northern Arctic Federal
University (NArFU)
Arkhangelsk, Russia

Gunnar Rekvig
UiT The Arctic University of Norway
Tromsø, Norway

ISBN 978-3-030-81252-2     ISBN 978-3-030-81253-9  (eBook)
https://doi.org/10.1007/978-3-030-81253-9

© The Editor(s) (if applicable) and The Author(s), under exclusive license to Springer Nature Switzerland AG 2022
This work is subject to copyright. All rights are solely and exclusively licensed by the Publisher, whether the whole or part of the material is concerned, specifically the rights of translation, reprinting, reuse of illustrations, recitation, broadcasting, reproduction on microfilms or in any other physical way, and transmission or information storage and retrieval, electronic adaptation, computer software, or by similar or dissimilar methodology now known or hereafter developed.
The use of general descriptive names, registered names, trademarks, service marks, etc. in this publication does not imply, even in the absence of a specific statement, that such names are exempt from the relevant protective laws and regulations and therefore free for general use.
The publisher, the authors and the editors are safe to assume that the advice and information in this book are believed to be true and accurate at the date of publication. Neither the publisher nor the authors or the editors give a warranty, expressed or implied, with respect to the material contained herein or for any errors or omissions that may have been made. The publisher remains neutral with regard to jurisdictional claims in published maps and institutional affiliations.

This Springer imprint is published by the registered company Springer Nature Switzerland AG
The registered company address is: Gewerbestrasse 11, 6330 Cham, Switzerland

*Gunnar Rekvig dedicates this book to his wife, Minako Kikkawa, and son, Arn Kikkawa Rekvig. Their patience, love, and support were invaluable.*

# Acknowledgment

Gunnar Rekvig would like to acknowledge the Nansen Professorship at the University of Akureyri for having supported the production of this book.

# Contents

| | | |
|---|---|---|
| 1 | **Introduction**........................................................<br>Matthias Finger and Gunnar Rekvig | 1 |

**Part I   Evolving Images and Perceptions**

| | | |
|---|---|---|
| 2 | **Indigenous Peoples of the Circumpolar North**...................<br>Yvon Csonka | 21 |
| 3 | **Arctic Cities-Pioneers of Industrialization:**<br>**More than "Mining Towns"**....................................<br>Nadezhda Zamyatina | 41 |
| 4 | **Arctic as a New Strategic Region in the Soviet Union**<br>**in the 1920s–1930s and Transformation of the Arctic Science**......<br>Alexander Saburov | 63 |
| 5 | **Norway in the Cold War – A Contemporary Case for Today**.......<br>Lars Saunes | 91 |
| 6 | **The Post-Cold War Arctic**....................................<br>Lassi Heininen | 109 |
| 7 | **The Arctic as the Last Frontier: Tourism**......................<br>Edward H. Huijbens | 129 |
| 8 | **A Primary Node of the Global Economy: China and the Arctic**....<br>Liisa Kauppila | 147 |

**Part II   Geography, Environment, and Climate**

| | | |
|---|---|---|
| 9 | **Climate Effects of Other Pollutants – Short-Lived**<br>**Climate Forcers and the Arctic**................................<br>Kaarle Kupiainen, Mark Flanner, and Sabine Eckhardt | 171 |

| 10 | **Permafrost Climate Feedbacks** .............................. | 189 |
|---|---|---|
|  | Benjamin W. Abbott |  |
| 11 | **Impacts of Global Warming on Arctic Biota** .................... | 211 |
|  | Mathilde Le Moullec and Morgan Lizabeth Bender |  |
| 12 | **Pollution and Monitoring in the Arctic**......................... | 229 |
|  | Tatiana Yu. Sorokina |  |

**Part III Economics and Geopolitics**

| 13 | **The Quest for the Ultimate Resources: Oil, Gas, and Coal**......... | 257 |
|---|---|---|
|  | Andrey Krivorotov |  |
| 14 | **Arctic Fisheries in a Changing Climate** ........................ | 279 |
|  | Franz J. Mueter |  |
| 15 | **Infrastructure Projects in the Global Arctic** ..................... | 297 |
|  | Alexander Pilyasov |  |
| 16 | **New Arctic Seaways and the Role of China in Regime Formation** .......................................... | 315 |
|  | Mariia Kobzeva |  |
| 17 | **Sustainable Development of the Arctic?**........................ | 331 |
|  | Matthias Finger |  |

**Part IV Governance**

| 18 | **Between Resource Frontier and Self-Determination: Colonial and Postcolonial Developments in the Arctic** ............. | 351 |
|---|---|---|
|  | Peter Schweitzer |  |
| 19 | **Understanding Cold War Trust-Building Between Norway and the Soviet Union. Norwegian-Russian Relations – A History of Peaceful Coexistence**................... | 369 |
|  | Gunnar Rekvig |  |
| 20 | **Regional Governance: The Case of the Barents Region** ........... | 389 |
|  | Florian Vidal |  |
| 21 | **The Arctic Council at 25: Incremental Building of a More Ambitious Inter-governmental Forum** ................. | 407 |
|  | Timo Koivurova and Malgorzata Smieszek |  |
| 22 | **The European Union and Arctic Security Governance**............ | 425 |
|  | Andreas Raspotnik and Andreas Østhagen |  |

| 23 | **Global Conventions and Regional Cooperation: The Multifaceted Dynamics of Arctic Governance** .............. 443 |
|---|---|
| | Cécile Pelaudeix and Christoph Humrich |
| 24 | **Arctic Order(s) Under Sino-American Bipolarity** ................ 463 |
| | Rasmus Gjedssø Bertelsen |

**Index**........................................................ 483

# Abbreviations

| | |
|---|---|
| 2D | Two-Dimensional |
| 3D | Three-Dimensional |
| A5 | Five Arctic Ocean Rim States |
| AC | Arctic Council |
| ACAP | Arctic Contaminants Action Program |
| ACIA | Arctic Climate Impact Assessment Report |
| AD | Anno Domini |
| AEPS | Arctic Environmental Protection Strategy |
| AFN | Alaska Federation of Natives |
| AFNORTH | Allied Forces Northern Europe |
| AHDR | Arctic Human Development Report |
| AMAP | Arctic Monitoring and Assessment Program |
| AMSA | Arctic Marine Shipping Assessment |
| ANCSA | Alaska Native Claims Settlement Act |
| ANWR | Alaska National Wildlife Refuge |
| ANWR | Arctic National Wildlife Refuge, Alaska |
| AZRF | Arctic Zone of the Russian Federation |
| B/D | Barrels a Day |
| BAEC | Barents Euro-Arctic Council |
| Bbl | Billion Barrel |
| Bcm | Billion Cubic Meters |
| Bcm/Y | Billion Cubic Meters a Year |
| BEAC | Barents Euro-Arctic Council |
| Bfrs | Brominated Flame Retardants |
| Bn | Billion |
| BRC | Barents Regional Council |
| BRI | Belt and Road Initiative |
| Bt | Billion (Metric) Tons |
| CAFF | Conservation of Arctic Flora and Fauna (Working Group of the Arctic Council) |
| CAO | Central Arctic Ocean |

| | |
|---|---|
| CAS | Chinese Academy of Sciences |
| CAST | Canadian Air-Sea-Transportable Brigade Groupe |
| CBD | Convention on Biological Diversity |
| CCP | Chinese Communist Party |
| CIAO | China Icelandic Arctic Observatory |
| CINCNORTH | Commander-in-Chief North |
| CNARC | China-Nordic Arctic Research Centre |
| CNOOC | China National Offshore Oil Corporation |
| CNPC | China National Petroleum Corporation |
| CNPGS | China Remote Sensing Satellite North Polar Ground Station |
| COPE | Committee for Original People's Entitlement, Canada |
| COSCO | China Ocean Shipping Company, Limited |
| COVID-19 | Coronavirus Disease Pandemic |
| CP | Central Passage |
| CPAN | Circumpolar Protected Area Network Plan |
| CPC | Communist Party of China |
| CSDP | Common Security and Defense Policy |
| EEA | European Economic Area |
| EEZ | Exclusive Economic Zone EU European Union |
| HR | High Representative of The Union for Foreign Affairs and Security Policy |
| DDT | Dichlorodiphenyltrichloroethane |
| DEW | Distant Early Warning |
| DNA | Deoxyribonucleic Acid |
| $CO_2$ | Carbon Dioxide |
| EAEU | Eurasian Economic Union |
| EEC | European Economic Community |
| EEZ | Exclusive Economic Zone |
| EIA | Environmental Impact Assessment |
| EPPR | Emergency Prevention, Preparedness and Response |
| ESPO | East Siberia-Pacific Ocean |
| EU | European Union |
| GIUK | Greenland-Iceland-United Kingdom Gap |
| FPSO | Floating Production, Storage, and Offloading Unit |
| GDP | Gross Domestic Product |
| GIUK | Greenland-Iceland-United-Kingdom Gap |
| GUSMP | Glavsevmorput |
| Gosplan | The State Planning Committee |
| GPS | Global Positioning System |
| HSE | Health, Safety, and Environment |
| ICSU | International Council of Scientific Unions |
| IGBP | International Geosphere Biosphere Program |
| IIASA | International Institute for Applied Systems Analysis |
| ILO 169 | ILO's Convention on the Indigenous and Tribal Peoples |
| ILO | International Labour Organization |

| | |
|---|---|
| IMO | International Maritime Organization |
| Ipos | Indigenous Peoples' Organizations |
| IPTRN | International Polar Tourism Research Network |
| IPY-2 | Second International Polar Year |
| IR | International Relations |
| ISR | Ice Silk Road |
| ITK | Inuit Tapiriit Kanatami |
| IUCN | International Union for the Conservation of Nature |
| Jogmec | Japan Oil, Gas and Metals National Corporation |
| JV | Joint Venture |
| KGH | Kongelige Grønlandske Handel (Royal Greenlandic Trading Company) |
| Km | Kilometer |
| LKAB | Luossavaara-Kiirunavaara Aktiebolag |
| LNG | Liquefied Natural Gas |
| LPG | Liquefied Petroleum Gas |
| MAD | Mutual Assured Destruction |
| Mbl | Million Barrels |
| Mbl/D | Million Barrels a Day |
| Mcm | Million Cubic Meters |
| MOSPA | Marine Oil Pollution Preparedness and Response in the Arctic |
| MoU | Memorandum of Understanding |
| Mt | Million (Metric) Tons |
| Mt/Y | Million Tons a Year |
| Mtoe | Million Tons of Oil Equivalent |
| Narkompros | The People's Commissariat of Education |
| NATO | North Atlantic Treaty Organization |
| ND | Northern Dimension |
| NEP | Northeast Passage |
| NGO | Nongovernmental Organization |
| NORDEFCO | Nordic Defense Cooperation |
| NPR-A | National Petroleum Reserve-Alaska |
| NSR | Northern Sea Route |
| NWP | Northwest Passage |
| OECD | Organisation for Economic Co-operation and Development |
| PAG | People's Action Group, Norway |
| PAME | Protection of the Arctic Marine Environment |
| Pbdes | Polybrominated Diphenyl Ethers |
| PCB | Polychlorinated Biphenyls |
| Plavmornin | The Floating Marine Research Institute |
| Pops | Persistent Organic Pollutants |
| PRC | People's Republic of China |
| PRIC | Polar Research Institute of China |
| RAC | Russian-American Company, Russia |
| Rannís | Icelandic Centre for Research |

| | |
|---|---|
| REE | Rare Earth Elements |
| ROV | Remotely Operated Vehicle |
| RSFSR | Russian Soviet Federative Socialist Republic |
| SACEUR | Supreme Allied Commander Europe |
| SACLANT | Supreme Allied Commander Atlantic |
| SAG | Sámi Action Group |
| SAR | Search and Rescue in the Arctic |
| Scm | Standard Cubic Meters |
| Sdgs | Sustainable Development Goals |
| SDWG | The Sustainable Development Working Group |
| SLBM | Submarine-Launched Ballistic Missile |
| Slcfs | Short-Lived Climate Forcers |
| SLOC | Sea Lines of Communication |
| SNK | Sovnarkom, Council of People's Commissars |
| SPAR | Single Point Anchor Reservoir |
| SSBN | Sub-surface Ballistic Nuclear |
| STRIKEFLT | Striking Fleet Atlantic |
| TALC | Tourist Area Life Cycle |
| Tcf | Trillion Cubic Feet |
| Tcm | Trillion Cubic Meters |
| TFN | Tungavik Federation of Natives, Canada |
| TPP | Trans-Pacific Partnership |
| TPP | Transpolar Passage |
| TRC | Truth and Reconciliation Commission, Canada |
| Trn | Trillion |
| TTIP | Transatlantic Trade and Investment Partnership |
| UN | United Nations |
| UNCED | United Nations Conference on Environment and Development |
| UNCLOS | United Nations Convention on the Law of the Sea |
| UNDRIP | United Nations Declaration on the Rights of Indigenous Peoples |
| UNEP | United Nations Environment Program |
| UNFCCC | United Nations Framework Convention on Climate Change |
| UNPFII | United Nations Permanent Forum on Indigenous Issues |
| UNWTO | United Nations World Tourism Organization |
| US | United States |
| USA | United States of America |
| USGS US | Geological Survey |
| USSR | Union of Soviet Socialist Republics |
| VAI | All-Union Arctic Institute |
| VSNKH | Supreme Council of the National Economy |
| WHO | World Health Organization |
| WSSD | World Summit on Sustainable Development |

# List of Figures

| | | |
|---|---|---|
| Fig. 1.1 | The initial Global Arctic matrix. (Source: Matthias Finger, first draft for the GlobalArctic website, April 2014) | 3 |
| Fig. 2.1 | Share of indigenous population in the Arctic regions. (Source: R. Rasmussen and J. Sterling, © Nordregio. https://archive.nordregio.se/en/Maps/01-Population-and-demography/Indigenous-population-in-the-Arctic-regions/) | 25 |
| Fig. 2.2 | Arctic peoples subdivided according to language families. (Source: Dallmann and Schweitzer (2004)) | 28 |
| Fig. 3.1 | *The upper picture*: The suburbs of the city of Vorkuta, which was once one the biggest Arctic city in 1980th. *The lower picture*: A part of the city of Vorkuta, the place from which Vorkura ran it's developing (houses in the left are now abandoned). (*Source*: Photos by Zamyatina) | 42 |
| Fig. 10.1 | Permafrost organic carbon stocks compared to other global pools. Citations for the estimates for each pool are as follows: subsea permafrost (Sayedi et al., 2020), peatlands and shallow and deep permafrost (Hugelius et al., 2014; Mishra et al., 2021), vegetation (Abbott, Jones, et al., 2016b), cumulative human emissions since the Industrial Revolution (Raupach & Canadell, 2010), living biomass (Bar-On et al., 2018), atmosphere and other soil (Chapin et al., 2012), and marine sediments (Atwood et al., 2020) | 193 |

Fig. 11.1   Schematic representation of (**a**) multi-year sea-ice Arctic ecosystems above shallow seas with the Arctic wildlife of polar bear and walrus close to the sea-ice edge; sea-ice algae, amphipods *G. wilkitzkii*, *A. glacialis*, and polar cod embryos under the ice; and ice-associated seals, narwhals, polar cod, and large-sized Arctic copepods species in the cold water; (**b**) under a possible climate change scenario, with thin first-year ice above the deep Arctic Ocean (accessible to cargo ships), incoming boreal species (Atlantic cod, capelin, and small-sized copepods) and altered primary productivity patterns with warmer waters, and few remaining Arctic species (*A. glacialis* and polar cod) ............ 216

Fig. 11.2   Schematic representation of Arctic tundra ecosystems (**a**) when sea-ice still reaches land supporting polar bears and Arctic foxes and cooling coastal terrestrial climates where Arctic herbivores like reindeer/caribou and lemmings dominate; and (**b**) under a climate change scenario where the sea-ice loss isolates populations, drives walruses ashore, and warms coastal terrestrial weather. The tundra landscape is greening with shrub expansion, but also shows patches of browning, while boreal species like red fox and raven predate/scavenge on Arctic herbivores ............ 219

Fig. 13.1   Investor-government interaction in resource extraction under commercially viable (**a**) and subsidized (**b**) projects. (Source: Author) ............ 259

Fig. 14.1   Bathymetry of the Arctic Ocean with major basins and shelf seas. Arrows denote major inflows and dashed ovals highlight the three inflow shelves ............ 280

Fig. 14.2   Daily sea ice concentrations on the two major Arctic inflow shelves (Fig. 14.1) based on data from the NOAA National Snow and Ice Data Center (Meier et al., 2017). Thin black line and grey bands denote the mean and range of observed sea-ice concentrations from 1988–2010. Daily concentration for the most recent years shown as separate lines ............ 282

Fig. 14.3   (**a**) Average latitudinal trends in the density of fish from the southeast Bering Sea to the northern Chukchi Sea based on trawl surveys conducted during the cool period 2010–2012 (dashed line) and during recent warm years (solid line) with 95 percent confidence bands. The location of Bering Strait is indicated by a vertical dashed line. No samples were available for the Chukchi Sea during the recent warm period. Dotted vertical line denotes northern extent of the historical survey area. Data from the Alaska Fisheries Science Center,

|  |  | NOAA. (**b**) Station locations sampled during trawl surveys in the standard survey area (triangles), the northern Bering Sea (grey crosses) and the Chukchi Sea (circles) ................ 284 |
| --- | --- | --- |
| Fig. 17.1 | Conceptualization of the Earth System dynamics. (Source: *Global Change*, Issue 84. November 2015, page 10) ............................................................... 334 |
| Fig. 17.2 | The Great Acceleration. (Source: https://en.wikipedia.org/wiki/Great_Acceleration, accessed 5.5.2021) ........................... 336 |
| Fig. 17.3 | Existing, prospective, and potential Arctic sites of non-renewable resources. (Source: https://nordregio.org/maps/resources-in-the-arctic-2019/) ......................................... 341 |
| Fig. 19.1 | Map of Barents Region. (Source: Arctic Center, University of Lapland) ....................................................................... 384 |

# List of Tables

| | | |
|---|---|---|
| Table 7.1 | Arctic tourism indicators | 132 |
| Table 7.2 | Inbound tourism in the Arctic regions, 2015 | 133 |
| Table 7.3 | Advancing new thematic areas for polar tourism research | 140 |
| Table 10.1 | Characteristics of greenhouse gases associated with permafrost climate feedbacks | 196 |
| Table 13.1 | Key figures on Norwegian continental shelf (end 2020) | 267 |
| Table 13.2 | Resources production goals, Russian Arctic and Energy Strategies through 2035 | 270 |
| Table 15.1 | Comparison of the features of the Arctic transport megaproject in the industrial and post-industrial eras | 304 |
| Table 22.1 | The European Union's Arctic Policy Milestones (2008–2021) | 431 |
| Table 23.1 | The nexus between regional cooperation and global conventions | 445 |

# Chapter 1
# Introduction

**Matthias Finger and Gunnar Rekvig**

When the GlobalArctic project was launched back in 2014, together with Professor Lassi Heininen, we made the claim that the Arctic should be looked at as a global region as it had during the Cold War. After the end of the Cold War, however, with the now-famous Murmansk speech by Mikhail Gorbachev in 1987 calling for the Arctic to become a "Zone of Peace", the Arctic's global dimension declined somewhat, at least among politicians and academics. While it was true that the Arctic Council, the main governance body in Arctic matters, had been admitting "observers" since its inception in 1996, they had only remained as observers, until today. As for academia, one could indeed note an increased interest in Arctic research, but almost no researchers looked at the big picture; that is, at how the Arctic was transformed by global forces and in turn affected the globe, whether ecologically, economically, or in terms of geopolitics.

At around the same time (2013–2014) the Arctic Circle Conference was established, and the GlobalArctic project has been present there ever since. But the Arctic Circle is more of a trade show, channeling all the new interests in Arctic matters, especially the business interests. With the changing Arctic, the business interests became ever more global. The Arctic Frontiers Conference was a precursor to this (starting in 2006), but had more of a niche focus. Nevertheless, our GlobalArctic project was not created as a reaction to these conferences; rather, it was geared toward the academic community. We wanted to help it to conceptualize what we saw unfolding, namely the melting of the Arctic ice, the ensuing environmental problems, especially the global ones, and simultaneously the new gold rush to

M. Finger (✉)
Northern Arctic Federal University (NArFU), Arkhangelsk, Russia
e-mail: matthias.finger@epfl.ch

G. Rekvig
UiT The Arctic University of Norway, Tromsø, Norway
e-mail: gunnar.rekvig@uit.no

extract whatever resources one could find in the Arctic, as well as the ensuing new geopolitics. The geopolitical situation has worsened in recent years, especially since the 2014 Ukraine Crisis.

This introductory chapter has three main aims. In the first section, we present the GlobalArctic project and how it has unfolded since its inception. This section serves as the context and *raison d'être* of this book. The second section stresses our substantive conceptualization of the GlobalArctic, as well as how this conceptualization has (somewhat) evolved over time. In the third section we summarize the main contributions to this book and explain what we want to achieve with it. At the end of the chapter, we will offer some considerations about the future perspectives of the GobalArctic project.

## The GlobalArctic Project

In January 2014, the GlobalArctic was simply an idea born in a brainstorming meeting between Rasmus Bertelsen, now professor at UiT the Arctic University of Norway, Lassi Heininen, now professor emeritus from the University of Lapland, and Matthias Finger, now professor emeritus from Ecole Polytechnique Fédérale and a research professor at the Northern Arctic Federal University. The idea rapidly materialized in a website, www.globalarctic.org, which we have developed ever since. The website featured in particular our initial matrix containing the following eight cells (see Fig. 1.1).

The basic structure and even content of this initial matrix has not really changed since then. However, the conceptualization of the GlobalArctic has become much more sophisticated and has continued to evolve ever since.

Based on this matrix, Lassi Heininen and Matthias Finger launched an open call for a first edited book on the GlobalArctic in Fall 2015 and subsequently gathered 16 highly qualified authors. As a first tangible outcome of this project, the (first) Global Arctic Handbook was published by Springer (Finger & Heininen, 2018).

In parallel, in November of 2016 Matthias Finger submitted a proposal to EPFL for a Massive Open Online Course (MOOC) along the GlobalArctic matrix featuring over 20 experts on various Arctic issues to cover the entire above GlobalArctic matrix. The MOOC, managed by Florian Vidal, was finally recorded in early 2018 and put online on Coursera later that year.

Considering the success of the MOOC, Matthias Finger teamed up with Gunnar Rekvig and asked the MOOC lecturers to contribute to a book. Many agreed to do so and some new experts were enlisted, resulting in this (second) Global Arctic book.

This process over the past 6 years has illustrated how the GlobalArctic has evolved to become an increasingly multidisciplinary endeavor, as well as an increasingly collective effort. Over time, however, it became apparent that such an interdisciplinary and collective effort required further framing to avoid simply turning into a collection of individual efforts, and this is what we are trying to do in the next section.

|  | Global > Arctic | Arctic > Global |
|---|---|---|
| Resources, Infrastructures, Technology | Pressure to extract oil, gas, minerals, fisheries; development of corresponding infrastructures; land-grabbing; expansion of industries and their TNCs and SOEs | Significant oil, gas, and minerals reserves for further industrial development; shortening of sea routes facilitating global trade; strengthening of SOEs and TNCs |
| Earth System Dynamics | Global warming affecting the Arctic disproportionately (permafrost, biodiversity loss); pollution (military, shipping, oil spills, Arctic haze, ocean acidification) | Climate forcing (albedo effect, methane release); the Arctic as a laboratory of the Anthropocene |
| (Geo)Politics, Security, Governance | Arctic resource geopolitics (and conflicts); weakening of the states' ability to protect their sovereignty; TNC-State alliance against indigenous peoples | The Arctic as a laboratory of governance (e.g., Arctic as a commons; governance innovations; regionalism) |
| Peoples, Society and Cultures | McDonaldization and urbanization of the Arctic; threats to health, food security and well-being; resource curse (inequitable development) | The Arctic as a laboratory of transculturalization; the role of indigenous peoples and self-determination in resource governance and sustainable development |

Source: Matthias Finger, first draft for the GlobalArctic website, April 2014.

**Fig. 1.1** The initial Global Arctic matrix. (Source: Matthias Finger, first draft for the GlobalArctic website, April 2014)

## What Is the GlobalArctic?

The point of departure for such framing is still the above matrix, which, in its basic structure, remains valid today. The matrix has two axes: a vertical one and a horizontal one; especially the vertical axis, with its different categories, can be discussed at length. In its most simple version, we think that the matrix must contain a geo-physical and societal/economic dimension. However, it also makes sense to further distinguish between an economic and a (geo-)political dimension, and a social/cultural dimension should probably also be considered.

However, the core of the matrix is undoubtedly its horizontal axis, which is the interaction between the Arctic and the Global. The "Global" in/of the Arctic begins with the Cold War, even though the colonization of the Arctic predates the Cold War. In fact, it is not the Arctic that is global at that time; rather, it is the Cold War that is global and happens to take place also in the Arctic. Furthermore, the "war" (conflict) was not about the harsh lands of the Arctic, which were not considered to be of much value. At that time, the effects of global warming were not yet being taken seriously, and even though there was knowledge of the energy and mineral resources under the ice, these were considered to be difficult to extract. Similarly, the trade routes were not seen as being practicable either, at least not in the foreseeable future.

The Cold War has never really ended and is being revived as we publish this book. Also, the militarization of the Arctic never really stopped and is actually increasing with the deteriorating global security situation. So, with the Arctic now becoming global because of other reasons (see below), the militarization of the Arctic and the Cold War, at least in the Arctic, take on a new, and we think even more dangerous dimension. Furthermore, the Cold War is so important in the Arctic that most of the other issues that have emerged since are now framed in Cold War terms. If this is not the case yet, they will almost certainly be framed in such terms in the future.

On the other hand, the GlobalArctic is really the result of global warming and of Earth System dynamics more generally. Of course, there was the colonization of the Arctic by the Southern powers, mainly for reasons of minerals extraction in Alaska, Norway, Sweden and Russia. But this did not make the Arctic global. We now know that global warming is affecting the Arctic disproportionately more severely than any other geographical part of the world. Also, global warming is the result of the fossil fuel-based economy in the industrialized countries and of the "Great Acceleration" of industrial development since the 1960s, especially in the emerging economies (Thomas et al., 2020).

As a result of global warming, the Arctic's mineral and energy resources have become increasingly accessible for exploration and exploitation. Furthermore, this process is facilitated because of more accessible maritime and land transport routes in the Arctic. On the maritime transport side, this increased accessibility of Arctic resources is facilitated because shipping through Arctic waters leads to significant competitive trade advantages.

However, the exploitation of Arctic minerals and energy resources would not have taken place without the planet's hunger for these very resources; that is, because of the pressure from the industrialized and emerging economies. The population in the Arctic and not even the Arctic countries would not need to exploit all these resources. What happens in the Arctic, at least in economic terms, is mostly the result of a "resource curse", a development that is similar to what happens in other resources-rich countries in the world, such as Saudi Arabia, the Congo, etc. (Murshed, 2018). In this sense, the GlobalArctic simply means some form of "resource colonization", albeit on a much more massive scale than in the early twentieth century.

Because of this massive scale, the exploitation of the Arctic's resources is accompanied by equally massive infrastructure developments in the form of roads, ports, railways, airports, and entire cities (such as mining towns and drilling platforms). Such infrastructure developments add still other dimensions to the GlobalArctic: not only does it make the transformation of the Arctic much more permanent, it also has cultural consequences, as the Arctic's indigenous peoples are increasingly also becoming minorities; this is something that goes far beyond simple colonization. In other words, the Arctic is now global.

The development of these infrastructures must be financed and built, which leads to the entry of new global actors. There are, of course, the companies, especially transnational corporations (TNCs) and state-owned enterprises (SOEs), but moreover there are the investors behind these companies, such as private equity funds (for example, Guggenheim Investments) or entire countries (such as China). In particular, the role of China as an investor and builder of Arctic infrastructures adds a new aspect to geopolitics and complexifies the existing Cold War in the Arctic.

More generally, the entry of TNCs, SOEs, investors, and China and other States (South Korea, Japan, etc.) within the broader context of the global race for resources, along with the new Cold War, raises the question of governance: governance of the Arctic, but more precisely the governance of the race for what is left. This, we believe, is what the GlobalArctic is all about.

## Structure of the Book and Summary of the Chapters

Against this backdrop, we have assembled 25 book chapters that illustrate the various aspects, dimensions, and dynamics of the GlobalArctic. We have structured these chapters into four parts, which together present a holistic and multidisciplinary view of the Arctic. The first part contains chapters about the evolving "images" and perception of the Arctic. Part II covers the Arctic's geography, environment, and climate. Part III deals with economics and geopolitics and Part IV covers governance. Below is a short summary of each of the chapters.

### *Evolving Images and Perceptions*

In the first substantive chapter of this book, entitled *Indigenous peoples of the Circumpolar North* (Chap. 2), Yvon Csonka provides an overview of the diversity of the cultures and societies of indigenous peoples, as well as of their dynamics under rapidly changing circumstances. Indigenous identities make reference to pre-colonial and colonial pasts, and to privileged links with the Arctic environments through activities such as hunting, herding, fishing, and gathering. The indigenous peoples' identities are shaped by their interactions with other indigenous peoples, with the rest of the nations they are part of, and increasingly with the globalizing world. Indigenous identities currently rest on different combinations of foundations, which include their relationship with the natural environment, language, spirituality, system of knowledge, and the arts. In most regions, their situation is underprivileged compared with the rest of the country they are a part of. Health, education, unemployment, poverty, access to renewable resources, food security, and environmental change are some of the challenges that many of them face today. Csonka argues that the intrusion of globality, both virtually through the Internet and physically through the emerging rush to exploit the non-renewable resources of the

Arctic, represents new potential pitfalls that indigenous residents will have to deal with.

In Chap. 3, entitled *Arctic cities – pioneers of industrialization: More than "mining towns"*, Nadezhda Zamyatina takes a historical look at the early industrialization of the Arctic, which was marked by the rapid development of transport infrastructures and the growth of new cities in areas of large mineral deposits. The area of gold mining in Alaska and the Yukon is highlighted. Other major projects were the development of iron ores and the founding of Kiruna in Sweden; the development of a transport network in the Murmansk region of the USSR; and the development of gold in the Kolyma basin, the Norilsk industrial region, and others. In a number of cases, road construction "opened" up the development of mineral deposits (Kiruna, Kirovsk, Monchegorsk), while in other regions (mainly gold mining ones), railway construction followed the initial deployment of the mining industry. The development of river shipping also played an important role, as did the "opening" of the Northern Sea Route along the coast of Eurasia. The period of initial industrialization spans from roughly the 1880s until the outbreak of the Second World War. The development of different regions of the Arctic at this time can be characterized by several general trends, namely population growth due to mass migration, and the formation of large companies (often engaged in the extraction of certain types of raw materials, but also in the construction of infrastructures, including cities, as well as the management of the territory generally).

In his chapter on the *Arctic as a new strategic region in the Soviet Union in the 1920s–1930s and transformation of the Arctic science* (Chap. 4), Alexander Saburov provides an overview of the Soviet Arctic science transformation processes that took place in the interwar period (1920s–1930s). Corresponding research activities are considered in the wider context of Soviet Arctic policy development and new approaches to the Arctic as a specific object of governance. Based on archival sources and published materials, this chapter reviews several aspects of Arctic research: major factors that have determined the development of Arctic science, the role of the state and the scientific community in the organization of research activities, the interplay of research activities and challenges of domestic and foreign Arctic policy, and the development of research institutions. Saburov concludes that Arctic research in the RSFSR–USSR between 1920 and 1940 underwent significant developments as a result of the prospective annexation of previous no-man's lands and the development of shipping and extractive industries. These endeavors laid the foundation for the creation of scientific organizations specializing in Arctic research during this period.

Chapter 5 by Lars Saunes is entitled *Norway in the Cold War – a contemporary case for today*. Saunes argues that while the importance of Norway in shaping interactions between great powers during the Cold War has been studied at length, its application in this era of great power competition in the Arctic has not. Indeed, Norwegian defense and security policy had to adapt to the ideological and strategic conflict between the Soviet Union and United States. The author also states that Norwegian leadership succeeded in adapting to the great powers' strategies during the Cold War and says that the return of global great power competition has brought a need to revisit the previous policies developed for the Cold War. Saunes identifies the central dilemma confronting Norwegian leadership from 1947–1991 and explores its effects on decision-making and implementation related to its military strategy and military capabilities in the Arctic region. The chapter builds on existing historical sources, declassified NATO documents, and the author's own experience from the Cold War and great power competition in the Arctic. Saunes concludes that Norway could play a role in bridging the rivalry between the great powers in the Arctic by initiating a circumpolar security dialogue.

In his chapter on *The Post-Cold War Arctic* (Chap. 6) Lassi Heininen states that, during the 1980s, the Arctic was defined as a homeland of indigenous peoples vis-à-vis a land for discovery, a storehouse of resource vis-à-vis a nature reserve, and a military theater of the Cold War vis-à-vis a "Zone of Peace". Indeed, the region faced the threat of a global nuclear war, as well as the final destination of long-range pollution. Now, in the 2020s, the Arctic faces a wicked problem consisting of growing risks of climate change and environmental degradation due to global warming, pollution, and the collapse of biodiversity, as well as globalization due to growing interests by Arctic stakeholders and new actors from outside the region. Among new images, narratives, and visions of the global Arctic are an environmental linchpin for, and a "climate archive" of, the planet, and an exceptional "political space" in world politics characterized by "high geopolitical stability and constructive cooperation". The chapter draws a holistic picture of the post-Cold War Arctic. It discusses and analyzes the changing dynamics of main themes/trends of Arctic geopolitics and governance, as well as common interests of the Arctic states and indigenous peoples, and special Arctic features that make it possible to restructure Arctic geopolitics and governance towards stability and cooperation.

Chapter 7, entitled *The Arctic as the last frontier: tourism*, by Edward H. Huijbens focuses on the development of the tourism industry in the Arctic from the perspective of transport and mobilities and how this links with the particular geography of the region and sustainability challenges. The Arctic is an amalgam of sovereign territories, nations, and peoples. This diversity is also reflected in the sporadic nature of tourism growth and development across the region. In order to understand tourism development from this perspective of diversity, Huijbens emphasizes tourism as premised on transport, accessibility, and mobility infrastructure in place or being developed. Beyond this, the emerging tourism geography of the Arctic relies on images of the region and particular allures. These revolve around climate change, "last chance to see type of tourism", cultural encounters, comprehensions of wilderness and protected areas, second-home tourism, and the role tourism plays in regional and economic development. Tourism tropes of the Arctic under any of these guises revolve around the region, defined loosely as a type of last frontier. These guises also throw into particularly sharp relief the challenges tourism faces if it is to be developed to the benefit of local peoples and ecosystems.

The last chapter on the evolving images of the Arctic by Liisa Kauppila is entitled *A Primary Node of the Global Economy: China and the Arctic*. In it, the author discusses the role of the Arctic in China's global vision, which seeks to reposition the country as a primary node of global flows of energy, goods, technology, data, and knowledge. The chapter begins by introducing the flow-centric lens through which China's global economic outreach and its region-based execution are interpreted. In particular, it proposes an alternative approach of China's functional economic regions, a novel conceptual framework that challenges the Euro-centric reading of space that dominates most analyses of China's global engagement. The chapter also provides an overview of the political goals of China's primary node vision, setting the scene for an empirical section that investigates the make-up of the China–Arctic functional economic region. This is done by tracing flows of (1) natural resources; (2) seaborne goods; and (3) technology, knowledge, and data, and by discussing their role in China's economic development and goals of the primary node vision. As China's Arctic engagement is as much about future potential as about actual developments, the empirical analysis also takes into account possible and probable developments that carry the potential to shape China–Arctic flows. Finally, in her conclusion Kauppila elaborates on the future prospects and implications of the emerging China–Arctic functional economic region.

## Geography, Environment, and Climate

In their chapter on the *Climate effects of other pollutants – Short-Lived Climate Forcers and the Arctic* (Chap. 9), Kaarle Kupiainen, Mark Flanner, and Sabine Eckhardt describe the evolution and current state of short-lived climate forcers (SLCFs) in the Arctic and more generally. These SLFCs directly or indirectly alter the Earth's radiative energy budget and have relatively short residence times in the atmosphere compared with the major climate driver, carbon dioxide. Gaseous species such as ozone, methane, and nitrous oxides, and aerosols such as black carbon, sulfate, and mineral dust all have those properties. SLCFs have both anthropogenic and natural emission sources, albeit with different proportions and different seasonal cycles. SLCFs influence climate in various ways. Methane and tropospheric ozone amplify Earth's greenhouse effect by absorbing infrared radiation. Aerosols mostly affect climate by altering the amount of solar energy absorbed by Earth. For instance, black carbon absorbs a high proportion of sunlight and warms the climate system. The authors argue that a relevant impact for the snow-covered Arctic is that, after depositing to snow and ice surfaces, dark particles like black carbon can hasten ice melt by increasing solar heating at the surface. Sulfate, in turn, cools climate by scattering solar radiation back to space.

In his chapter on *Permafrost and climate feedbacks* (Chap. 10), Benjamin W. Abbott presents and discusses the dynamics of permafrost in the Arctic, as well as more generally. Permafrost ecosystems have accumulated a vast pool of organic carbon, three times as large as the atmospheric carbon pool and five times as large as the carbon contained by all living things. The high elevations and high latitudes where permafrost occurs are experiencing some of the most extreme climate change on Earth. Consequently, the ecological reaction of the permafrost zone could influence the trajectory of the climate system for thousands of years to come. As permafrost regions warm, more carbon and nitrogen will be exposed to decomposition, combustion, and hydrologic export, increasing greenhouse gas production and release. At the same time, plants may take advantage of the extended growing season and nutrient release to take up more atmospheric carbon dioxide. In this chapter, Abbott lays out recent advances in understanding of permafrost climate feedbacks, focusing primarily on the production, uptake, and release of carbon dioxide, methane, and nitrous oxide. He attempts to answer the following questions: Why do permafrost regions contain so much organic matter? How sensitive is this organic matter to climatic perturbations? How important are permafrost feedbacks compared to anthropogenic greenhouse gas production? Current estimates of the permafrost climate feedback vary in magnitude and sign, representing an important unknown risk for local communities and ecosystems. However, compared to direct human emissions, potential greenhouse gas uptake or release from the permafrost zone is quite small. This emphasizes the importance of continued permafrost research and the imperative for rapid decarbonization of the global economy.

Chapter 11 by Mathilde Le Moullec and Morgan Lizabeth Bender on the *Impacts of global warming on Arctic biota* starts by highlighting the magnifying effects of climate change in the Arctic, which alter marine and terrestrial communities in profound ways. The authors argue that, together with alterations in resource availability, quality, and quantities, and increasing pressure by boreal competitors and predators invading from the South, Arctic organisms will be strained. The ongoing changes in the environment act across the range of ecological levels of arctic biota (that is, the wildlife). In particular, changes at the base of the food web can ripple throughout the ecosystem, affecting iconic top predators. For instance, earlier spring onset advances sea-ice break-up and vegetation green-up, creating a potential mismatch between trophic levels; "replacement prey" invading from the south may be of lower quality to Arctic predators; and winter warm spells modify snowpack properties that control vegetation available to herbivores. These examples have implications for the entire ecosystem and can modify abiotic factors through feedback loops such as vegetation cover affecting climate. The speed at which cold-adapted Arctic species can adapt to changing conditions will depend on their life history strategies, gene flow between populations, and hybridization with boreal species. Population isolation increases with habitat fragmentation from sea-ice loss and industrialization. The spatial extent to which climate change will synchronize population responses will determine population resiliency. The more heterogenous environmental gradients are, the more likely it is that Arctic species will find a suitable niche, maintaining biotic biodiversity.

Chapter 12, by Tatiana Yu. Sorokina, on *Pollution and monitoring in the Arctic* offers a broad understanding of what environmental monitoring in the Arctic is, what objects are being investigated, and who are the main actors in organizing and conducting monitoring activities. The issue of the availability of environmental information on the Arctic and monitoring data is also being considered. The author states that the main sources of pollution of the Arctic nature are local production facilities and settlements. Absolutely all environmental objects suffer from anthropogenic impact and related environmental pollution: water bodies, soils, atmospheric air, and bioresources. Toxic pollutants are detected in various samples. An analysis of peer-reviewed scientific articles and comprehensive reports of international organizations has shown that the problems of environmental pollution with persistent organic pollutants (POPs) and plastic (microplastics) are of the greatest concern. The concentrations of many POPs in various samples are gradually decreasing, which indicates the effectiveness of the international measures taken. At the same time, the academic literature notes the emergence of new substances that raise new concerns. Publications on plastics so far have simply stated the fact of the discovery of microplastic particles in various objects of nature, including in the Arctic fauna. Data on possible effects on human or animal health is not yet sufficient to draw unambiguous conclusions. The main stakeholders in environmental monitoring in the Arctic remain the Arctic states. At the same time, the positions of international entities are being strengthened, for example, within the framework of the Arctic Council or international environmental agreements. Sorokina concludes that not all environmental monitoring data in the Arctic are open to access. A significant revision of the existing regulation is required, so that the boundaries between open information and state secrets become clearer.

## *Economics and Geopolitics*

Andrey Krivorotov, in his chapter entitled *The quest for the ultimate resources: oil, gas, and coal* (Chap. 13), shows how the Arctic's fossil fuel potential attracts governments and investors with its political stability, short distances to key markets and good geological prospects. Arctic exploration and development is also of major political importance as a tool to ensure national energy security, establish a visible presence in the area, and give strong impetus to regional economies. However, there are also many challenges, like harsh natural conditions, insufficient or lacking technologies and infrastructure, and higher environmental risks. The political and industrial interest for the area has varied greatly over time, reflecting both long-term trends and current international developments. While Arctic coal production peaked in the twentieth century, the oil and gas industry has witnessed two periods of high activity, from the 1960s to the mid-1980s and in the last two decades. Krivorotov's country-by-country analysis shows that despite numerous optimistic statements, few actual Arctic projects have been implemented or have realistic chances to be, and all of them have been concentrated offshore Norway, onshore Russia, and in Alaska. Future industrial development will likely be very heterogeneous, unfolding under a strong environmentally motivated pressure, a broad variety of actors and national policy approaches, with a strong emphasis on innovative and sustainable technological solutions. The ongoing low-carbon transition will affect such development heavily in the longer run. The author concludes that while Arctic field developments are highly important for the respective countries and investors, the region as a whole will remain a marginal supplier of natural gas, oil, and especially coal at global scale.

In Chap. 14 on *Arctic fisheries in a changing climate,* Franz Mueter recalls that warming oceans, the loss of sea ice, and changes in advection drive changes in highly productive Subarctic marine ecosystems and the borealization of Arctic marine ecosystems. Borealization refers to the northward shift or expansion of marine organisms, including commercially important fish stocks, into Arctic waters. These shifts in distribution have been particularly pronounced on the major Arctic inflow shelves, which are important gateways from the Atlantic and Pacific oceans into the Central Arctic Ocean. Climate-driven changes in the abundance and distribution of fish stocks can pose significant challenges for fisheries stock assessment and management as some Subarctic fish stocks decline and others are displaced or expand into new areas. Shifting stocks are a challenge for resource surveys as fish move out of historically surveyed areas; for managers as changing stock and fishery interactions may lead to resource conflicts; and for international institutions as stocks cross national boundaries or expand into the central Arctic ocean. To meet these challenges, researchers will have to adopt new and cost-effective strategies for monitoring; managers will have to be flexible to adapt to rapidly changing conditions; and enhanced international cooperation will be required to address transboundary issues and ensure the conservation and sustainable management of living marine resources in the central Arctic ocean.

Chapter 15, by Alexander Pelyasov, on *Infrastructure projects in the global Arctic,* recalls that, for a significant part of the twentieth century, the Soviet Union was a leader in the deployment of large-scale and ambitious infrastructure megaprojects in the Arctic. It was here that the theoretical concepts of the territorial structures of the regions were elaborated, including *trassas* – land roads and railways of the Arctic frontier. An empirical generalization of large infrastructure megaprojects of the global Arctic, which have been developing over the past 100 years, made it possible to identify the main drivers: geopolitical circumstances; dramatic changes in comparative world prices for natural resources; and the ability of the state to carry out large-scale mobilization of material, financial, and labor resources. The specific national features of the infrastructure megaproject depend on the size of the country, its economic and geographical position within the continent (primarily, the number of neighbors), and the model of state economic policy. A specific technological era has an impact on the spatio-temporal organization of the construction of a transport megaproject and the organization of the system that supports it. In recent decades, numerous hopes that many Arctic megaprojects can be implemented by the forces of globalization have proven to be unfounded. A more powerful factor was global warming, which led to intensive melting of sea ice in the Arctic and allowed for transport projects (either in the form of linking to seaports, or directly, in the form of increased use of Arctic sea routes). All transport megaprojects, both those discussed and implemented, can be distinguished as latitudinal and meridian, each with its own institutional and physical and geographical features.

Mariia Kobzeva's chapter on *New trade routes* (Chap. 16) evaluates China's involvement in the Arctic regime formation regarding shipping and defines a pivotal moment that will turn China into an Arctic shipping nation. New Arctic seaways have proven to be an attractive goal for European and Asian businesses. Despite uncertainties regarding navigation in Northern waters, Arctic shipping is advancing impressively. However, this new level of connectivity fits into the deeply changing balance of power. New influential actors, such as China, have acquired the label of "challengers" to the international order, while the NATO and Russian military activities in Arctic waters are gradually heating up the security dilemma. The global shifts have stoked tensions in the Arctic region, emphasizing the deficiencies of the international regime. This situation has exacerbated the need to re-evaluate the perspectives of the development of new Arctic seaways, the established regime, and China's role in it. This chapter studies a new level of Arctic seaway connectivity, considering the active participation of China and determines the key nexus between Russian extracting industries, the European market, and China as a contributor to Arctic shipping development. Through the prism of regime theory, Kobzeva evaluates the impact of China's engagement in the Arctic navigation and answers the following key research question: What is the role of the new circumpolar seaways for China's involvement in Arctic regime-making?

In his chapter entitled *Sustainable development of the Arctic?* (Chap. 17), Matthias Finger adopts an Earth Systems dynamics perspective and argues that, from such a perspective, the "sustainable development of the Arctic" is illusionary. The by-now-global Arctic is in fact a case in point of unsustainable development – unsustainable for the Arctic and unsustainable for the planet. Global warming affects the poles and the Arctic disproportionately, to the point that the rapid warming of the Arctic may well lead to tipping points of the Earth System. This Earth System dynamics perspective is furthermore combined with an analysis of the dynamics of development, especially with an analysis of development's key actors and institutions. Finger concludes that the main actors in the Arctic – namely nation-states and state-owned enterprises – are unlikely to refrain from extracting resources from the Arctic and may well go to war over them. Finger argues that this is an institutional dynamic that will ultimately impede a realistic understanding of the detrimental effects of development on the Earth's life-sustaining biosphere. Could it be, he asks, that the concept of "sustainable development of the Arctic" serves other purposes?

## *Governance*

Chapter 18, entitled *Between Resource Frontier and Self-Determination: Colonial and Postcolonial Developments in the Arctic,* by Peter Schweitzer, reminds us that human habitation of the Arctic dates back millennia and is characterized by technological inventions necessary to survive and thrive in the High North. However, the main focus of this chapter is on the colonial entanglements in which the indigenous residents of the Arctic found themselves from the sixteenth century onwards (and even earlier in the case of northern Fennoscandia). The nineteenth and early twentieth centuries brought an intensification of these processes: the sale of Alaska to the USA, the founding of Canada, and the independence of Norway created conditions that led to a nationalization of colonial ideologies and to increased assimilation pressure on indigenous groups. The young Soviet Union presented itself as a counterweight to colonialism, while at the same time pushing for forced industrialization of the North. The chapter concludes with a review of self-determination movements after the Second World War. This process, which is neither complete nor has a uniform success rate across the Arctic, is also tied up with attempts at regional autonomy and political devolution. At the same time, the question arises as to whether there is still room for indigenous self-determination in times of the growing global economic and geopolitical significance of the Arctic.

In his chapter on *Understanding Cold War Trust-Building Between Norway and the Soviet Union* (Chap. 19), Gunnar Rekvig argues that the Cold War was a conflict of incompatible ideologies between the United States and the Soviet Union and their respective allies. The global order during this time was marked by bipolarity between the two superpowers, and the peace was paradoxically maintained by balancing the power with a nuclear arms race. Thus, the realpolitik of the Cold War was shaped by deterrence. In the Nordic Arctic, however, the Cold War would pan out differently. The very real confrontation between the Kingdom of Norway and the Soviet Union was placated by a series of trust-building initiatives that led to the ad hoc creation of a zone of low tensions during the height of the Cold War between Norway – a founding member of the North Atlantic Treaty Organization (NATO) – and the Soviet Union, the founder of the Warsaw Treaty Organization (WTO). From culture and science agreements, joint management of fisheries, a moratorium for hydrocarbon extraction in contested waters, to perhaps most importantly, Norwegian self-imposed military restrictions, the initiatives were groundbreaking. These activities did not appear in a vacuum, as there were long historical ties between Norwegians and Russians. Especially in the north, the relationship was strong, with the nineteenth century Pomor trade, to the liberation of Finnmark County from German occupation in the Second World War. The relations were marked by neighborliness, and it is against this historical backdrop that during the Cold War, the iron curtain began to open with a culture agreement in 1956.

In Chap. 20, on *Regional Governance: The Case of the Barents Region*, Florian Vidal shows that the Barents region has experienced a unique and complex governance model in the Arctic since the end of the Cold War. Recalling that this part of the Arctic has been connected to Europe since the Viking era, it stands as a singular case for the whole Polar region. For centuries, this subarctic region was under the influence of Norwegian–Russian relationships and their dynamics. The institutionalization process since the 1990s has largely focused on economic, social, and environmental matters. Diversity in stakeholders and institutional multilayers are key features of the region's governance. In the light of the Ukrainian crisis and the annexation of Crimea in 2014, the Barents region endured its most serious crisis after the Cold War. However, despite a significant increase in military activities, these new geopolitical conditions did not break long-standing cooperation. At a time when economic opportunities are soaring in the region and climate change and biodiversity losses are threatening socio-economic conditions, the Barents area must address extraordinary challenges. From oil and gas projects to tourism development, the Barents region presents itself as a showcase for socio-ecological balance in the Arctic. However, its institutional arrangements are clearly set to face growing tests of its resilience.

In Chap. 21, Timo Koivurova and Malgorzata Smieszek discuss the changes in the functioning of the *Arctic Council* during recent years. These changes are clearly due to the increased interest in Arctic affairs in general, at a time when climate change continues to progress in the region at a pace twice the global average and various economic possibilities for utilizing the region have become apparent. Even if the Council has clearly become a much more ambitious governance body, it also seems clear that much of its current form still stems from the early days of Arctic inter-governmental co-operation. The main goal of this chapter is to examine how much the changes in the Council have built on these foundational premises upon which the Arctic Environmental Protection Strategy and, later, the Arctic Council were founded. The 25th anniversary of the Arctic Council marks an appropriate point for reflection on these fundamental issues in the Council's performance. How has the Council reacted to new developments? How has it coped with the regional challenges? What do the past experiences say about the Council's ability to address future challenges? At the heart of this enquiry is an examination of how the Arctic Council has changed over the course of these last 25 years. The authors claim that the Council has retained its core functions whilst also changing other functions, some quite dramatically, in response to current needs. In the concluding section, the authors explore how the Council can meet the challenges of tomorrow.

Chapter 22 on *The European Union and Arctic Security Governance* by Andreas Raspotnik and Andreas Østhagen shows that, over the past decade, the European Union (EU) has established itself as a recognized Arctic actor with an obvious presence in the north in terms of geography, legal competence, market access, and its environmental footprint and contribution to Arctic science. However, the EU's Arctic policy is essentially only soft in nature, with a rather cautious approach to Arctic geopolitics and matters of hard security. Naturally, the European Union has its historical, legal, and competence-related problems with questions of (hard) security. This also becomes evident in the Arctic region. However, the global significance of the Arctic in terms of security and strategy is steadily increasing. In this chapter, the authors examine the EU's role in Arctic security governance. First, they discuss the Arctic's security environment and three levels of Arctic geopolitics. Second, they place these developments in an EU-Arctic context and discuss how a union of 27 states – only three of which are Arctic – will continue to engage with the Arctic region. How does the EU's Arctic endeavor fit with the emerging security concerns, related to great-power relations, in the Arctic? What role, if any, is there for the EU in such matters?

Cécile Pélaudeix and Christoph Humrich, in their chapter entitled *Global Conventions and Regional Cooperation: The Multifaceted Dynamics of Arctic Governance* (Chap. 23), show how the Arctic is a globally embedded space. This is as true for the impact of global climate change on the Arctic and the consequences of Arctic climate change for the rest of the world as it is for governance for and in the Arctic. This chapter analyzes the dynamics that structure and result from the coexistence of global and regional governance mechanisms in four issue areas: the governance of marine and maritime spaces, indigenous peoples, climate governance, and environmental protection and conservation. It assesses how global conventions impacted regional governance and, in turn, how regional cooperation influenced governance on the global level. In order to do this, the authors distinguish four ways in which the nexus between global governance and regional cooperation can be established: as harmonious, cooperative, conflictive, and indifferent. For each way, an outside-in (from the global to the regional level) and an inside-out (from the regional to the global level) perspective can be considered. This yields a typology of eight different kinds of links between regional cooperation and global conventions. Much of the previous research has focused on harmonious and cooperative links. By contrast, Pélaudeix and Humrich intend to show that as global interest in the Arctic grows, along with the need for regulatory governance in the region, the nexus might become increasingly conflictive. In order to retain control over the region and its governance, Arctic states' cooperation seeks to limit both their own global commitments and the influence of exogenous actors or institutions.

The book concludes with Rasmus Bertelsen's chapter on *Arctic Order(s) under Sino-American bi-polarity* (Chap. 24). The Arctic remains closely integrated into and reflects the international system and international order. Post-Cold War American unipolarity and Russia in crisis formed the basis for a liberal circumpolar Arctic order after the Cold War with the Arctic Council emphasizing liberal topics as environment, sustainable development, and indigenous rights. According to the Bertelsen, the emerging Sino-American bipolarity will create a realist, rather than a liberal, international order. Each superpower will create its own bounded order, as the USA had the Western bloc and the USSR the Eastern bloc. Russia is not a third pole in this Sino-American bipolarity, except in nuclear weapons. Russia will fall into either the bounded order of the USA or China. With Russian-Western miscalculations in the Caucasus and in Eastern Europe (Georgia 2008, Ukraine 2014, and perhaps Belarus 2021), Russia now appears in the Chinese bounded order, but perhaps not a natural member of that order. The Cold War Arctic order was bipolar and realist reflecting US-USSR bipolarity. The post-Cold War Arctic order was circumpolar and liberal, reflecting American unipolarity and hegemony and Russian crisis. The post-post-Cold War Arctic order will either be circumpolar and agnostic (downplaying human rights and environmental protection in the Russian Arctic) if Russia leaves the Chinese order and joins the US bounded order, or it will be bipolar, with a liberal Western bounded order and an agnostic Sino-Russian bounded order, Bertelsen concludes.

## Conclusion

Overall, this volume offers a comprehensive and multidisciplinary overview of an Arctic that is becoming increasingly global – or, should we say, entangled in global issues. It also shows how the Arctic gradually is losing the ability to govern itself, not just the Arctic's indigenous peoples, but also most of the Arctic States, driven as it is by different global forces, be they economic, ecological, or geo-political. While this overall picture of the direction of the Arctic emerges relatively clearly, the question of the future of our GlobalArctic project needs to be posed explicitly.

This is actually the question about what we want this GlobalArctic project to be. Over the past 7 years, both through publications and lectures, we hope that we have raised awareness of an ever more globally driven Arctic. However, this situation is now so obvious that not much more needs to be done. Therefore, we need to think about the next steps, and it seems to us that, from now on and for the next few years, we need to follow a dual strategy.

On one hand, it will become necessary to show how unsustainable the "development" of the Arctic is, imposed as it is by global economic (predatory resources extraction, tourism), environmental (conservation and "sustainable development", but also exploitation of rare minerals absolutely needed for the global green deal), and geo-political forces. If left unchecked, these forces and the ensuing conflicts will only increase. Indeed, many people, especially academics, still seem to believe that something like "sustainable development" of and in the Arctic is possible. However, given the global forces in place, including the ongoing and accelerating Earth System dynamics, the Arctic's sustainable development will emerge as only a very short-lived illusion.

On the other hand, the GlobalArctic project will have to focus more on the global systemic part and show how the profound mainly ecological transformation of the Arctic will equally profoundly affect the planet, mainly through what has become called "tipping points" (Lenton, 2012; Young, 2012). In this sense, we stated back in 2015 that the Arctic should be looked at in the context of the fate of the planet, namely as what we called a "laboratory of the Anthropocene" (Finger, 2015). This is a topic we will pursue in our GlobalArctic project more forcefully in the future.

## References

Finger, M. (2015). The Arctic, laboratory of the Anthropocene. In L. Heininen (Ed.), *Future security of the global Arctic* (pp. 121–137). Palgrave Macmillan.
Finger, M., & Heininen, L. (Eds.). (2018). *The global Arctic handbook*. Springer.
Lenton, T. (2012). Arctic climate tipping points. *Ambio, 41*, 10–22.
Murshed, S. M. (2018). *The resource curse*. Agenda Publishing.
Thomas, J. A., Williams, M., & Zalasiewicz, J. (2020). *The Anthropocene. A multidisciplinary approach*. Polity Press.
Young, O. (2012). Arctic tipping points: Governance in turbulent times. *Ambio, 41*, 75–84.

**Matthias Finger** is a Professor Emeritus from Ecole Polytechnique Fédérale in Lausanne, Switzerland (EPFL), a part-time professor at the European University Institute (EUI) in Florence, Italy, a full professor at Istanbul Technical University (ITÜ) and a research professor at Northern Arctic Federal University (NArFU), Arkhangelsk, Russia.

**Gunnar Rekvig** is an Associate Professor at UiT The Arctic University of Norway (UiT), Norway, and the 2019–2021 Nansen Professor at the University of Akureyri (UNAK), Iceland. He was an Assistant Professor at Tokyo University of Foreign Studies (TUFS), Tokyo, Japan (2018–2019).

# Part I
# Evolving Images and Perceptions

# Chapter 2
# Indigenous Peoples of the Circumpolar North

Yvon Csonka

## Baseline of Indigeneity: The Peopling of the Arctic

Indigenous to the Arctic[1] are those peoples whose ancestors settled the circumpolar North several millennia ago, well before the arrival of settlers with a European background. Their particular adaptations to the different types of Arctic environments remain the basis of their common status as indigenous.

Humans in what are now the temperate zones of Eurasia became equipped to cope with Arctic-like environments (with the exception of the protracted dark seasons) during the last Ice Age, which lasted for around 100,000 years and ended approximately 11 millennia ago. As the environmental zones started approximating those of today (roughly 8000 years ago), humans were pushing north of the Arctic circle in Eurasia, from Fennoscandia to the Pacific Coast, as well as in what is now Alaska. The Arctic coasts of North America and Greenland were first settled by humans between 5000 and 4000 years ago.

Over their long stretch of residence, these groups of humans have adapted to their challenging environments. Their subsistence was (and, for many, remains) based on hunting and fishing, with some plant gathering. Sophisticated technologies were invented and refined to hunt sea mammals on the sea ice or in open water. Dog or reindeer traction, and watercrafts, improved the mobility potential. Housing design and sewing techniques allowed comfortable living in the coldest temperatures. The domestication and herding of reindeer developed in the Eurasian North over the last millennium or so, but was only recently imported in a few regions in the North American Arctic, including, at one point, Greenland.

---

[1] In this chapter, the Arctic is defined as in *the Arctic Human Development Reports* I (2004: 17–18) and II (2014: 44–45). See boundary on Map 1.

Y. Csonka (✉)
Ilisimatusarfik, The University of Greenland, Nuuk, Greenland

© The Author(s), under exclusive license to Springer Nature Switzerland AG 2022
M. Finger, G. Rekvig (eds.), *Global Arctic*,
https://doi.org/10.1007/978-3-030-81253-9_2

The fate of these humans was tied to the availability of the animals upon which their survival depended, and access to their prey was periodically threatened by the sharp and often unpredictable changes in weather and natural conditions. Thus, settlement, after a number of generations, was sometimes followed by retreat or extinction through starvation. Archaeology documents some of these episodes, as well as the succession of prehistoric cultures, the evolution of technologies and changes in ways of life, such as settlement patterns, housing, tools and weapons for fishing and hunting, means of transportation, distribution of fish and game in the diet, domestic activities, etc. Spirituality and world views, as well as social organization, were shaped over the millennia by the interactions with the environment, particularly with other living beings.

It is often not possible to reconstruct the situation of specific indigenous peoples at the time of first contacts with incomers of European origin, since it was not documented until much later, by observers from the outside. In Russia, as well as in North America, the fur trade was the main motivation for early contacts, and the involvement of indigenous residents in trapping caused changes in their activities and geographical displacements. However, one may distinguish some broad types of aboriginal adaptations to the diverse environments of the circumpolar North. The Inuit, Yup'ik, and Aleut subsisted mainly by hunting marine mammals (various species of whales and seals, walruses, and polar bears) on the sea ice or in open water, and pursuing caribou inland on the tundra. The many Dene (Athabaskan) groups in Alaska and Canada lived entirely from the diverse resources of the boreal forest, or from a combination of forest and tundra, where they pursued the migratory caribou. In northern Eurasia, including Fennoscandia, adaptations to hunting and fishing in the boreal forest and, north of the treeline, to the tundra, predominate. Reindeer, the Old World equivalent of the North American caribou, were domesticated first as hunting decoys and auxiliaries in transportation, but since contact times, they became more and more intensively herded.

In northern Fennoscandia, the first contacts between Sámi and populations south of them date back a couple of millennia. Norse immigrants from Fennoscandia settled Iceland, which (except for a few Irish monks) was free of human presence, from 870 CE onwards. In 985 CE, a group of Icelanders went on to establish colonies in the southwest of Greenland. As for the ancestors of today's Inuit Greenlanders, bearers of the Thule culture arriving from what are now Alaska and the Canadian Arctic, they entered the island from the Northeast in the twelfth century CE. The first encounters between the two peoples occurred shortly thereafter, and must have been the first between people of European origin and indigenous people originating in North America. Apparently, contact was limited; the Norse did not survive in Greenland beyond the fifteenth century, whereas the Inuit thrived there. Greenland was colonized again in 1721, with the arrival of Lutheran priests from Norway. In Russia, regular contacts between incomers and indigenous communities occurred only in the last two to four centuries, as fur traders penetrated further and further east and north across the continent. The North American Arctic was colonized last, with some Inuit communities in Canada barely contacted before the beginning of the twentieth century.

## Recent History

Colonization had an impact on the ways of life of indigenous peoples that still reverberates today. These peoples were affected by contact with incomers at different times, different rhythms, and in different ways. Missionaries usually followed close behind the fur traders and did their utmost to replace local religions with Christianity. Up until the Second World War, however, many indigenous communities led a relatively autonomous way of life. Most were involved with colonizer societies through the fur trade. Along the Arctic coasts of Greenland, North America, and the Russian side of Bering Strait, they also interacted, worked for, or bartered with commercial whaling crews.

Over a period of many centuries, the Sámi experienced a gradual encroachment on their lands, interference in their way of life, and Christianization. In northernmost Fennoscandia, borders between States were established in the nineteenth century, separating Sámi groups from each other. Assimilation policies, particularly in Norway, culminated in the late nineteenth century and the first half of the twentieth century.

Starting in the 1920s, the Soviet government developed policies for dealing with its indigenous minorities, and in the 1930s it made massive efforts to exploit the renewable resources of the northern parts of the Soviet Union. The acceleration of changes affecting the circumpolar North was not synchronous across Arctic countries, but everywhere it reflected similar modernist ideologies regarding the development of society and economy.

The Second World War can be considered a watershed in the history of the Arctic, initiating an unprecedented rate of social and cultural change. Greenland was cut off from Denmark during the war and was de facto administered by the USA. Vast areas of Sápmi, (the Sámi homeland) in Norway and Finland, were burned by the retreating Germans. Following the Japanese invasion in the Aleutians, Alaska saw a rise in military activity. While the Canadian Arctic and the Asian part of the Russian Arctic were less directly affected, in all regions the war ended the Arctic's relative isolation from centers further south.

During the Cold War (1948–1988), military presence was increased in the Arctic. Among other developments, the Distant Early Warning (DEW) line of radars was built right across Inuit lands on the arctic shores of Alaska, Canada, and Greenland. The exploitation of the non-renewable resources of the North, such as mining, oil, and gas, and construction of hydroelectric dams increased, generally with little regard for environmental consequences or impacts on indigenous societies. In what was then the Soviet North, this was accompanied by an influx of immigrants such that, in some regions, the indigenous population was outnumbered by as much as 10 to one.

A major factor of change was the spread of welfare state policies which, among the indigenous peoples, were implemented in authoritarian and paternalistic ways. In the Arctic countries with indigenous populations, apart from the provision of housing, health care, and education, the goal was to assimilate these minorities into the mainstream population. A main instrument was the creation of boarding schools, where the native languages were forcibly replaced by the dominant language. Many young people lost fluency in their mother tongue and were alienated from their families and communities (Csonka & Schweitzer, 2004).

The still nomadic populations were encouraged – and in some cases forced – to move into settlements, in settings and types of housing that were unfamiliar to them. Throughout the Arctic, a number of indigenous settlements were forcibly closed and their residents moved without their consent (for examples in the Soviet Union, among the Yupik, see Krupnik and Chlenov (2007), and among the Sámi, see Allemann (2020); in Canada, see Damas (2002), Laugrand et al. (2011), Tester and Kulchyski (1994), Marcus (1995); in Greenland, see Robert (1971: 19), Walsøe (2003)). The explicit rationale given for these relocations was to improve living conditions (access to not-yet-depleted hunting or trapping grounds, easier provision of public services), but in a number of cases they served geopolitical motives, occupying lands that might have otherwise been claimed by other nations, or obeying strategic military considerations. These relocations, and the changes in ways of life they implied, had a profound impact, largely negative, on the populations involved.

With a few exceptions, such as the Nenets reindeer herders of the Yamal Peninsula in Russia, centralization of nomadic populations was achieved by 1970. Health services were vastly improved. However, some populations were decimated by epidemics, such as tuberculosis and poliomyelitis, which peaked among the Inuit population in the Canadian Arctic in the period 1940–1960. It is estimated that up to one-third of the population was infected, and many were taken to sanatoria in the south of the country for extended periods, without contacts with their relatives (Government of Canada, 2015).

In the newly created settlements, incomers from the south were hired to fill the positions as administrators, teachers, health professionals, construction workers, etc. Many of these professionals considered themselves superior to locals, at least implicitly. In these settlements, local people could not escape the impression that they were passively watching while things were being done around them and "for" them. As the late Greenland Home Rule Premier Jonathan Motzfeldt put it: "… things were administered by Danes, decisions were taken by Danes, and problems were solved by Danes. […] The common Greenlander had a feeling of standing outside, of being observer of an enormous development, which s/he did not have the necessary background to understand" (Motzfeldt, 1999: 7).

With the introduction of wage work and of social welfare, the relative importance of traditional subsistence activities decreased, and many young people who had been educated in boarding schools did not learn the skills sufficiently to practice them. This, together with the feeling of loss of control over their destiny, contributed to the social problems that emerged in those decades, such as alcohol and drug abuse, violence, suicide, and other harmful behaviors.

For most indigenous peoples of the Arctic, the decades around the middle of the twentieth century were marked by suddenly accelerating changes, most of which had been imposed from the outside, and which they could not control (Schweitzer et al., 2010). Any improvements in their physical well-being and food security offered by new housing, health care, Western-style education, welfare, wage work, and so on were offset by the feeling of despondency, resulting in a lack of possibility to shape their destiny. To this day, standards of living are still far inferior in indigenous communities compared with the rest of the country in which they live.

Notwithstanding the extreme pressure they endured, most indigenous cultures and identities did not vanish, nor were they absorbed in the mainstream culture of the rest of the nations they are part of. In the context of worldwide decolonization, some Western countries were willing to negotiate claims to the land and measures of autonomy pushed forward by indigenous residents of their own internal colonies. Political assertion came hand in hand with cultural reaffirmation. By the 1970s at the latest, most indigenous residents were living in settled communities, surrounded by accoutrements of modern life in terms of household goods, modes of transportation, community administration, schools, etc. Thus, cultural reaffirmation cannot consist of a return to earlier ways of life. It simply testifies – there, as everywhere else – that the traditions of all human societies are in a constant state of evolution (Fig. 2.1).

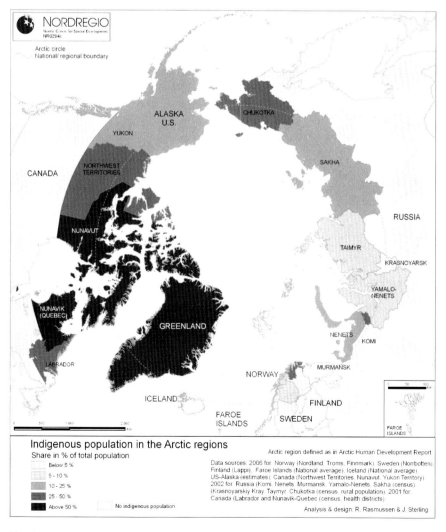

**Fig. 2.1** Share of indigenous population in the Arctic regions. (Source: R. Rasmussen and J. Sterling, © Nordregio. https://archive.nordregio.se/en/Maps/01-Population-and-demography/Indigenous-population-in-the-Arctic-regions/)

## Self-Identification, Recognition by Arctic States, and Demography

Although they differ from each other in many ways, the indigenous peoples of the Arctic share origins in ways of life shaped by the Arctic environments, and histories of having been colonized, subjected to rapid social and cultural changes, and marginalized within nation states. Today, he historical continuity of indigenous peoples with their (pre-)colonial past expresses itself in one or more characteristics, such as strong links to their territories and surrounding natural resources, a will to pursue traditional activities linked to their environment, distinct languages, beliefs, customs and artistic expressions, and distinct social, economic, or political systems (see next section). However, the main criterion, at the individual level, is self-identification as a member of one of the indigenous peoples, together with the acceptance by the community as their member (UNPFII, n.d.).

Not all people residing in the Arctic today fit the dichotomic categorization of indigenous and non-indigenous. There are also peoples with mixed ancestries dating back centuries, with distinct identities and cultures, such as the Métis in Canada and the Kamchadals in Kamchatka. Mixed or plural identities have also become increasingly common in recent times, resulting from ethnically mixed unions. In Alaska for instance, where more than one answer was allowed to the question on "race" in the 2010 census, 18% of those who declared themselves Iñupiat reported one or several more "races" (US Census, 2010). Also, some of the indigenous peoples straddle the boundary between Arctic and Subarctic, such as the Evenks in Russia and some of the Dene in Canada. Many Native individuals have also settled outside the Arctic, either temporarily or permanently, but most of them continue to identify with, and be recognized by, their home community.

Countries with Arctic territories have different ways of identifying, (failing to) recognize, and (aside from language differences) collectively naming, among their citizens, the peoples who were already present at the time when colonizers with a European background arrived. Consequently, population figures are not comparable from country to country, do not always differentiate between the different ethnic groups, and in some cases are very rough estimates. Also, the names of some of the peoples in the official documents (such as census data) differ from the self-designations (for example, the Russian Yupik, who are still referred to as "Eskimosy").

In 2010, there were 138,000 "American Indians" and "Alaska Natives" in Alaska. Alaska Natives comprise Iñupiat, Yup'ik, Aleut, Athabaskan, Tlingit-Haida, and Tsimshian.[2]

---

[2] In Alaska, in 2010, 30,900 persons identified as Yup'ik, 25,700 as Iñupiat, 16,600 as Alaskan Athabaskan, 11,200 as Aleut, 13,200 as Tlingit-Haida, and 1900 as Tsimshian. The total exceeds the number of individuals, as some of them declared that they identify with more than one "race" (US Census 2010).

Canada recognizes First Nations (Indians), Inuit, and Métis as "Aboriginal Peoples". Of the 65,000 Canadian Inuit, 47,000 (72%) live in Inuit Nunangat (Inuit land), the majority of them in Nunavut. The remaining 28% reside in other parts of Canada (Census, 2016). The Inuit population living outside Inuit lands in Canada has grown by 62% between 2006 and 2016 (ITK, 2018). In the Canadian Arctic, west of Hudson Bay, also dwell several groups of Athabaskan languages speaking Dene First Nations.

In Kalaallit Nunaat (the land of the Kalaallit, Greenland), population statistics only distinguish among those "born in Greenland", other Danish citizens born outside Greenland, and foreigners. The first category corresponds roughly to Greenlandic Inuit. In 2020, there were 56,081 residents in Greenland, 50,189 (89%) of whom were born there (Grønlands Statistik, 2020b). In addition, 16,770 people born in Greenland were living in Denmark, a figure that is related to a strong migration out of Greenland, particularly of young women (Danmarks Statistik, 2020).

In the Fennoscandian countries, censuses do not record ethnicity. Estimates of the Sámi population in these three countries vary widely, between 50,000 and 100,000, with a majority in Norway, followed by Sweden and then Finland. To escape stigmatization, many Sámi had given up on claiming their ethnic identity. This has been reversed in recent decades, with renewed pride in "Sámihood". About 1600 Sámi live on the Kola Peninsula in nearby Russia.

Since the mid-1920s, the Soviet and then Russian governments have recognized "indigenous small-numbered peoples of the North, Siberia and the Far East". The main objective criterion is that they each number fewer than 50,000; other criteria include living on their historical territory, preserving their traditional way of life, and identifying themselves as a distinct ethnicity. The original list comprised 26 peoples; it now lists 46, with a total population of 250,000 (Bogoyavlenskii, 2012). Nineteen of these peoples, with a total population of fewer than 200,000, live predominantly in the Arctic, as defined in this chapter (Heleniak 2014: 86). These people are, from West to East, the Sámi, Nenets, Mansi, Selkup, Enets, Khanty, Ket, Nganasan, Dolgan, Evenk, Even, Yukaghir, Chuvants, Chukchi, Kerek, Alyutor, Koryak, Aleut, and Yupik. Among other non-Russian peoples with long residence in the Arctic are the Komi and the Sakha (Yakuts), who by virtue of their numbers (circa half a million each) are not granted special status in the Russian Federation.

Indigenous populations of the Arctic are characterized by generally higher fertility rates than the rest of the countries they live in. In Nunavut, as well as among Alaska Natives, the fertility rate is as high as almost three children per woman (Heleniak, 2014: 60). In the Russian North, the fertility rate of the indigenous population is also significantly higher than the national average.

The age structure of the indigenous population is generally younger than that of the rest of the countries they are a part of. Accordingly, their median age is quite low. In Nunavut, where a vast majority of the population is Inuit, the median age is 24, compared to 41 for Canada as a whole. Forty-two percent of the population are aged under 20, whereas 23% of all Canadians are in that age group (Heleniak, 2014: 75, 80). In the Russian Arctic, the mean age of the population (35) is younger than the national average for all of Russia (39), but among the numerically small peoples of the North, it is much younger (27.5) (Heleniak, 2014: 82).

The total population of the Arctic, as defined in this chapter, is about four million. This figure has not changed much in the past 20 years, but reductions in population of certain regions of the Russian North since the end of the Soviet period have been offset by increases in other parts of the Arctic, particularly in the North American Arctic. The total indigenous population is estimated at about half a million. Thus, their share is about 12% of the total, but this share varies widely from region to region (see Fig. 2.2).

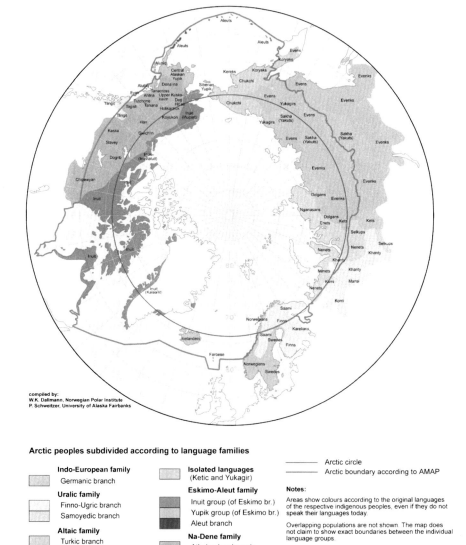

**Fig. 2.2** Arctic peoples subdivided according to language families. (Source: Dallmann and Schweitzer (2004))

The trend towards urbanization is also strongly at work in the Arctic, but indigenous peoples tend to live in the smaller settlements, from a few dozen individuals to a few thousand, whereas in bigger cities they usually are a minority. For example, more than one-third of the 53 communities in which the 47,000 Inuit of Inuit Nunangat reside, have populations under 500, and most of them can only be reached by air, or by sea during the open water season (Canadian Geographic, n.d.).

## Pillars of Identities

Through the decades of the twentieth century that were so disruptive for indigenous peoples, the ideologues and bureaucrats of social engineering and modernity, East and West, were not the only ones who promoted their assimilation and predicted the disappearance of their cultural distinctiveness. Even some of the most sympathetic ethnographers joined the chorus. In 1920, the famous Danish-Greenlandic "eskimologist" Knud Rasmussen wrote the following about the Greenlanders:

"One has no choice. The life of nature peoples in the future is entirely dependent on their development possibilities under new circumstances; the way forward may have to go over the dead body of their race. What is therefore important now, is to give nature people, in their soul and spirit, as delicate and sparing a death as possible." (Rasmussen, 1920).

Diamond Jenness, respected specialist of Canadian Inuit, stated in 1968:

> The day we Europeans allowed our traders and missionaries to settle on Eskimo shores we destroyed all possibility of shielding the aborigines from the outer world, except in the grave. ... They may speak the Eskimo language still; they may still roam the hunting-grounds of their forefathers; but most of them are Eskimos no longer. ... We have banished their old life beyond recall. (Jenness, 1968: 46)

On this basis, Jenness recommended moving all Canadian Inuit out of their territories to the southern parts of the country. The error of reasoning, which had tragic consequences, was to essentialize the indigenous "others" and deny them the capacity to maintain distinct identities under changing circumstances and influences from the dominant society. This attitude has not disappeared today, and indigenous peoples must expend considerable energy convincing authorities and mainstream society of the countries they inhabit that they simply exist, while being just as indigenous as they ever were, and that living in the modern world, and for some in urban settings, does not automatically deprive them of their heritage and distinctiveness and, most importantly, of their rights.

The status of indigenous peoples was invented by incoming Europeans. Scholars emphasized the fact that some were hunter-gatherers, meaning that they had not reached the "stage" of producers, and that others were herders (of reindeer), which implied their having reached a higher "stage", but not that of the Neolithic domestication of plants. Thus, up until quite late in the twentieth century, allusions were commonly made to them belonging to the "Stone Age," which the ancestors of the

Europeans had grown out of millennia ago. In some countries, particularly in Russia, it remains difficult for the fractions of these peoples who live in urban settings and do not participate fully in "traditional" activities to have their identity and associated rights recognized.

As Laugrand and Oosten put it, "Early ethnographers, like many modern anthropologists, were caught in an ideology that considers the incorporation of foreign elements into Western culture as cultural enrichment and, conversely, the adoption of Western culture by other societies as a loss of authenticity. From this essentialist point of view, other cultures gradually disintegrate, losing their own traditions as they succumb to modernity, whereas Western culture gradually increases in richness and cultural diversity" (Laugrand & Oosten, 2010: 372–373). The idea that one is either "traditional" or "modern", but certainly not a combination of both, has repeatedly been proven wrong; it also denies indigenous peoples the right to development without assimilation. Individuals identify themselves with different groups simultaneously, ranging from kin to village to an ethnic group, to a nation, or to common residence in the Arctic, or to being indigenous, etc. Members of an indigenous group may collectively, at any time, favor a different blend of cultural elements and values, such as those presented below, that best defines them collectively.

## *Relationship with the Natural Environment*

The relationship with their environment, expressed through the perpetuation of subsistence activities, is a central pillar of identity for most indigenous peoples. It is no less the case where snowmobiles and modern rifles have replaced dog or reindeer traction and bow and arrow, and where GPS are resorted to for navigation. Such activities are pursued even when they are economically unprofitable, for their value as central aspects of meaningful lives. No less important are domestic skills such as meat and skin preparation, sewing, and preparing country food. The sharing and common consumption of food are central cultural elements, also for those households that cannot participate in harvesting, due to physical or financial limitations, or because they live outside the home area, including in bigger cities. Among communities in which only a minority intensively practice "traditional" activities, for the others, the vicarious experience, and participation in consumption of the harvest, are important elements of ethnic belonging and cultural well-being.

## *Language*

Several dozen indigenous languages are spoken in the Arctic, most of them belonging to five main language families (see Fig. 2.1[note from author: put map 2.1 about here, and it becomes Fig 2.2]). The retention rates of these languages vary widely from group to group (Meltofte, ed. 2013: Chapter 12 and Appendix 20), as well as within groups themselves, from community to community. Among the Inuit, for example, 14% of the Alaskan Iñupiat spoke their language in 2007 (Schweitzer et al., 2014: 115, Fig. 3.1). At the other end of the spectrum, almost all Greenlanders remain fluent in their own language. Among the Canadian Inuit, at the 2016 census, the share of those able to converse in Inuktitut (the Inuit language) was 21% in Nunatsiavut, 22% among the Inuvialuit, 89% in Nunavut, and 99% in Nunavik (ITK, 2018: 21).

A number of languages are critically endangered, and some have become extinct in recent years (Meltofte, ed. 2013: 657). Few have escaped the trend towards language loss over time. For those groups who retain high fluency in their native language, this is invariably a central element of their identity. However, they cannot escape the competition from dominant languages, due to the need for most to be bilingual or multilingual in order to get by, and also because, beyond a certain level, there exists no teaching material in a given native language, and because there is no translation for many new concepts. Life in settlements, away from the land, also makes some of the specialized vocabularies linked to subsistence activities obsolete, and forgotten by the younger generations.

Numerous examples testify to the fact that even among peoples whose original language is lost or nearly so, a strong sense of indigenous identity may rest on other elements. For example, one may contrast the Kalaallit (Greenlanders) – who have a very high language retention rate across all generations, but where professional licensed hunters make only a small share of the population – with the Iñupiat of northern Alaska, among whom the share of fluent speakers is relatively low (it decreased from 23% to 14% between 1997 and 2007; Schweitzer et al., 2014: 115, Fig. 3.1), but identification with harvesting, especially of bowhead whales, and associated sharing and rituals, are important pillars of cultural affiliation.

## *Spirituality, Values and Indigenous Knowledge*

The spiritualities and world views of indigenous peoples stem from their long and intimate relationship with the environment, the animals, and the nonempirical beings that are part of their cosmology. Most indigenous peoples were submitted to Christianization efforts, but elements of animism (that is, the belief that not only humans, but also animals, plants, and other natural elements have a soul) persist. In most regions of the circumpolar North, shamanism was superseded by Christianity, often in syncretic ways (Csonka & Schweitzer, 2004: 56–58; Laugrand & Oosten, 2010). Just about everywhere, at least in some circles within indigenous societies, aboriginal spiritualities are being revitalized (Schweitzer et al., 2014: 120–121), partly in new forms adapted to contemporary settings.

Indigenous knowledge systems are being valued as precious cultural heritage by their holders. These knowledge systems not only include technologies, know-how, skills, and practices, but also language, systems of classification, resource use practices, social interactions, rituals, and spirituality. Indigenous peoples avoid the pitfalls associated with the belief in immutable traditions, and recognize that although they are rooted in past experiences, these systems and bodies of knowledge evolve with time. They are incorporated in formal and informal education of the young. In Nunavut, for example, *Inuit qaujimajatuqanngit* (roughly: "that which has long been known by Inuit")[3] has been promoted by the territorial government for years now and is included in many realms under its jurisdiction. Indigenous values are also emphasized, and often considered (at least implicitly) superior to those prevalent in the surrounding mainstream society. The strength of family ties, the sharing and the spirit of cooperation are often put forward as central values.

## *Arts, Handicrafts, Sports, and Games*

The arts are generally thriving throughout the indigenous Arctic. Dancing, singing, and music remain vibrant symbols of indigenous identity. So are costumes, and handicrafts. Many of these expressions are rooted in "tradition", but just like anywhere else, traditions evolve and new ones are created out of relatively recent elements, up until hip hop, breakdancing, and rap. Inuit carving for the world market has been encouraged in Canada since the 1950s, as a source of income when people were settled in communities. It continues to thrive today, and many of the themes remain rooted in traditional ways of life, legends, spiritualities, and values, but the forms evolve and mix with modernity. The production of literature, film, and video increases. Sports relating to traditional skills, as well as games adapted from earlier forms, are also powerful markers of belonging.

Indigenous identities today rest on particular combinations of elements. Some may be lacking, but compensated by others. Even societies that have lost such an important element of identity as language may retain a vibrant culture through emphasis on other aspects.

## Some Current Challenges

All indigenous peoples of the Arctic have experienced colonial rule, whether for shorter or longer periods, and incorporation in encroaching nation states. For most of them, the decades around the middle of the twentieth century were the most disruptive. Since then, the pace of change has not abated, and the drivers have become more numerous, including, recently, climate change, and the fully fledged irruption of global society through television and the Internet. The post-colonial times they live in continue to be influenced by the colonial past. All of these societies remain

---

[3] "Inuit beliefs, laws, principles and values along with traditional knowledge, skills and attitudes are what the Government of Nunavut and Elders refer to as Inuit Qaujimajatuqangit" (Nunavut Department of Education, 2007: 22).

underprivileged compared with the averages of the countries they live in. Among the many issues they face, which apply to most indigenous societies to varying degrees, a few are listed below in no particular order.

## Health

In terms of health, there are significant gaps between the indigenous populations and the other groups in the countries they inhabit. The poorer health status of the first inhabitants can be linked to high poverty rates, low education, food insecurity, inadequate housing and, with regard mental health, the stress related to lack of employment opportunities, low self-esteem, and lack of control over one's fate.

Infant mortality is generally higher among indigenous people than the national average. In Alaska, for instance, the infant mortality rate in 2007–2009 was 4 per thousand births for non-Natives and 11.4 for Natives. In Greenland, in 2018, the rate was 6.1, compared to 3.3 in Denmark (Grønlands Statistik, 2020a: 18; Statista, 2020).

Indigenous people live generally shorter lives than their compatriots. In 2016 in Greenland, life expectancy was 73 years for females and 70.8 years for males, while the equivalent figures for Denmark were 82.8 and 79 years, respectively (World Bank, 2020). They suffer from high rates of infectious diseases, particularly tuberculosis. As many indigenous settlements are small and isolated, and many are accessible only by air, or by water in the ice-free season, they have difficulty accessing proper health care facilities.

Death by suicide, especially among young men, has emerged as a major health issue since cultural and social disruption started accelerating, around the middle of the twentieth century. Suicide rates among indigenous residents everywhere are multiples of those of the rest of the country they live in (Young et al. 2015). They are especially high among the Inuit. Among the Canadian Inuit, in 2011–2016, the suicide rate was 72.3 deaths per 100,000 person-years at risk, nine times higher than the suicide rate among non-indigenous people (8.0). The suicide rate was three times higher among Inuit males than Inuit females (Kumar & Tjepkema, 2019). Suicide rates also supersede the national averages among the Sámi, and among the small peoples of the Russian North.

## Education and Gender Imbalance

The level of formal education achieved is generally lower among the indigenous population of the Arctic countries. The share of the population with post-secondary education is comparatively low, and it is lower in the North American Arctic and in Greenland than in the northern European countries and in Russia (Rasmussen et al., 2010: 78 sq.). In most regions, the level of education of indigenous women is higher than that of men.

Many indigenous communities, especially the smaller ones, experience gender imbalance to a higher degree than the national average. The surplus of males is generally due to the excess outmigration of females, which is linked to women seeking higher education, or jobs that better correspond to their level of education, outside their home area. Among those born in Greenland living in Greenland, for example, there are 105 men for every 100 women, whereas among those settled in Denmark, there are only 75 men for 100 women (Grønlands Statistik, 2020b; Danmarks Statistik, 2020).

## *(Un)Employment and Poverty*

Traditional activities such as hunting, herding, fishing, and trapping, even when subsidized, generally generate low income and are pursued by some mainly for the well-being they bring. In many instances, the higher income of the formally better educated female partner allows the husband to afford expensive hunting equipment. In Inuit Nunangat (the Inuit lands within Canada), in 2015, mean income varied between 23,485 CAD a year for the Inuit and 92,011 CAD for the non-Inuit, both figures diverging greatly from the Canadian national average of 34,604 CAD (ITK, 2018: 17). This pattern of non-indigenous persons, (who are generally better formally educated) occupying the better-paying jobs is a common feature of Arctic communities. Not only are incomes lower for Natives, but the prices of commodities are generally much higher in the North, due to the narrow markets and the prohibitive costs of transportation. Unemployment rates are also much higher among Natives than among the general national averages. Employment opportunities in smaller indigenous settlements are generally limited.

## *Environment, Renewable Resources, Food Security*

Apart from some bartering with neighboring groups, the resources of their environment have been the sole basis for the subsistence of indigenous peoples for thousands of years. Now, with growing populations and diversification in economic activities, local food resources do not cover all the needs. Just about everywhere, to different degrees, the environment is threatened not only by the consequences of climate change, but also by competing activities of non-renewable resources extraction, dam-building, and by sports hunters and fishers. It is a constant struggle for indigenous communities to claim and protect their rights in negotiations with their governments. Food itself, and the activities surrounding the harvest, preparation, sharing, and common consumption of food, are central symbolic elements of indigenousness. Food security is impacted by the loss of independence from imported goods and by the lack of cash to buy them in stores. Furthermore, some types of country food are also problematic as they contain unsafe amounts of contaminants.

In particular, persistent organic pollutants (POPs) and traces of heavy metals find their way into the meat of seals and other marine animals at the top of the food chain, and threaten the health of the Inuit for whom they are staples (AMAP Assessment, 2015).

## *Climate and Environmental Change*

Through their millennia of residence in the Arctic, the indigenous peoples have learned to cope with the frequent and sometimes rapid and drastic changes in weather, climate, and availability of fish and game that are typical of the Arctic environments. Therefore, one may surmise that they are well prepared to cope with the challenges of climate change. This would be a mistaken shortcut, however. As indicated above, their lives have been disrupted – exponentially so since the beginning of the twentieth century – by transformations much more pervasive than climate change has been so far, even though global warming is more strongly at work in the higher latitudes. Thus, climate change does not intrude in otherwise stable conditions; it represents an additional driver of change, combining with quite a few others (Csonka, 2020). So far, many indigenous residents of the Arctic have themselves indicated that they consider the consequences of global warming and its effects on the weather patterns to be less threatening to their culture and their way of life than the issues they already face. Many also refer to their competence at coping with changes in the environment, inherited from their ancestors. Nevertheless, all are acutely aware of these changes, which are related to global warming (e.g., Forbes & Stammler, 2009; Lavrillier & Gabyshev, 2017).

They are directly affected by the changing location and abundance of the animals they depend on for their subsistence, as well as by their accessibility. This is particularly striking in areas where hunters and fishers, and their prey, depend on the annual formation, duration, and thickness of sea ice. Not only is access to traditional food sources modified, but also travel on the ice becomes more hazardous. Terrestrial species are also impacted, as milder and more unstable winter weather is accompanied by more frequent episodes of freezing rains causing the formation of an ice crust over the snow, which the reindeer and caribou cannot break to have access to their source of food. Direct impacts of climate change include coastal erosion (due to the absence of seasonal protection from storm surges by sea ice), and the melting of permafrost, which affects infrastructure, to the point that housing, and even entire towns, must be rebuilt or relocated. The village of Shishmaref, on the Island of Sarichef in Northwest Alaska, is an oft-cited example. Climate change combines with other related developments and affects indigenous residents in indirect ways as well. The thawing of glaciers and permafrost makes some non-renewable resources more easily accessible, and their extraction often leads to conflicts with traditional land use.

This section has provided an overview of some of the challenges that are common to most indigenous peoples of the Arctic today, but these are not the only ones.

The process of urbanization goes hand in hand with a less close relationship with the natural environment. Nevertheless, the urban diaspora often remains in mutually beneficial contacts with home villages. Many regions have a lack of adequate housing, which perpetuates health issues and potential social problems. An important challenge is the struggle for their specific rights to be recognized by the government of their country, and the possibility to govern themselves at the regional level (this aspect is covered in Chap. 18).

## Perspectives

Although the basis of being "indigenous" lies far in the past, the consciousness of being indigenous clearly arose from the colonial encounter, and evolved with the unfolding of history. Today, the indigenous peoples of the circumpolar North are caught in a whirlwind of rapid and still accelerating changes. The simple facts that most of them have resisted assimilation, and that the despondency they have felt under pressure to renounce their identities is being replaced by renewed pride in their heritage, is a success story.

However, the identity that is ascribed to them, especially in official settings recognized by national authorities, may not always correspond to their own experience of their identity, or the recognition of each other within their particular group. Currently, this issue is particularly acute in the Russian Arctic, where indigenous rights are less well recognized and protected, and are currently under pressure from changing government policies (Bristkaya, 2020; Goble, 2016; Laruelle, 2019; Stammler & Ivanova, 2016). Recently, a new law was proposed that would introduce a parallel register of indigenous individuals, admitting only those who lead a traditional way of life, within the boundaries of territories officially recognized as inhabited by indigenous people. This proposed law was accepted by the State-endorsed Russian Association of the Indigenous Peoples of the North (RAIPON), but it is opposed by independent indigenous associations within Russia and has not yet been passed (IWGIA, 2020: 562–563). Thus, the future orientation of indigenousness in the Arctic will depend on how self-identification will evolve under changing conditions, and on how the identification and rights of these peoples will be recognized by the States they live in, as well as in the interplay between these two aspects (see also the "Concluding remarks" of chapter [19].

Continued and accelerating climate change will also be a multifaceted driver of social change. With the reduction in the extent of sea ice and the lengthening of the ice-free season, it is expected that commercial shipping through the Arctic ocean will increase. This may mean the expansion of large seaports throughout the region and carries the risk of catastrophic pollution in case of an accident. The industrialization of the region, and the development of resource extraction, may be accompanied by the immigration of workers from the south and further urbanization, with impacts on the local cultures. Tourism, which is partly driven by the wish to witness climate change firsthand, and to see a world that may be transformed beyond recognition within decades, creates opportunities and challenges for the local communities.

Given the vastness and the diversity of environmental conditions in the circumpolar North, the consequences of climate change are not uniform across its extent. In the short run, some regions and some situations may actually benefit. A longer growing season allows the expansion of agriculture northwards; fish, game and fauna move northwards and may become available in places where previously they were not; and exploitation of non-renewable resources may provide jobs locally, and possibly also royalties. In the medium term, however, it is clear that the overall risks exceed the benefits. Mitigation and adaptation strategies to reduce the vulnerability of indigenous communities to the challenges they face, including climate change, and increase their resilience, are central issues today (see Arctic Council, 2016).

Even though indigenous peoples have to deal with the consequences of changes that have been caused by humans outside the Arctic, they should by no means be considered just as victims. On the contrary, they are the ones to decide and implement actions at the level of their communities, and they should be included as equals in the search for solutions at global levels. Their views, based on an ethos that rejects the predominant paradigm of human domination of the natural world, will contribute to finding innovative solutions.

# References

Allemann, L. (2020). Soviet-time indigenous displacement on the Kola Peninsula: An extreme case of a common practice. In T. Koivurova, E. G. Broderstad, D. Cambou, D. Dorough, & F. Stammler (Eds.), *Routledge handbook of indigenous peoples in the Arctic* (pp. 92–105). Routledge.

AMAP. (2015). *Assessment 2015: Human health in the Arctic*. Arctic Monitoring and Assessment Programme (AMAP).

Arctic Council. (2016). Arctic Resilience Report (M. Carson & G. Peterson, Eds). Stockholm Environment Institute and Stockholm Resilience Centre, Stockholm. http://www.arctic-council.org/arr

Bogoyavlenskii, D. (2012). Dannie vserossiiskoi perepisi 2010. http://raipon.info/peoples/data-census-2010/data-census-2010.php

Bristkaya, T. (2020). They are no longer counted as indigenous people. *The Barents Observer*, October 7. Translation from Novaya Gazeta. https://thebarentsobserver.com/en/indigenous-peoples/2020/10/they-are-no-longer-counted-indigenous-people?fbclid=IwAR29D1-D8eQmO_AOhXs40FCMxdLIj6X3RGbap-QXqtIJg67eIKlHlemgfIU

Canadian Geographic. n.d. *Indigenous peoples atlas of Canada*. Inuit Nunangat. https://indigenouspeoplesatlasofcanada.ca/article/inuit-nunangat/

Csonka, Y. (2020). Indigenous peoples of the Arctic: Facing rapid change. In M. Cerny & P. Cerny (Eds.), *Art as a mirror of science* (pp. 76–77). Museum Cerny/Cerny Inuit Collection.

Csonka, Y., & Schweitzer, P. (2004). Societies and cultures: Change and persistence. In *Arctic human development report* (pp. 45–88). Stefansson Arctic Institute.

Dallmann, W., & Schweitzer, P. (2004). Arctic peoples subdivided according to language families (map). *Arctic human development report* (p. 47). Stefansson Arctic Institute. Downloadable at https://ansipra.npolar.no/english/Indexpages/Maps_Arctic%20.html

Damas, D. (2002). *Arctic migrants, Arctic villagers*. McGill-Queen's University Press.
Danmarks Statistik. (2020). *Personer født i Grønland og bosat i Danmark*. https://www.statistikbanken.dk/BEF5G
Forbes, B., & Stammler, F. (2009). Arctic climate change discourse: The contrasting politics of research agendas in the West and Russia. *Polar Research, 28*, 28–42.
Goble, P. (2016). *Numerically small indigenous peoples of the Russian north an ever bigger problem for Moscow*. UpNorth, March 30. https://upnorth.eu/numerically-small-indigenous-peoples-of-the-russian-north-an-ever-bigger-problem-for-moscow/
Government of Canada. (2015). *Inuit and the past tuberculosis epidemic*. https://www.rcaanc-cirnac.gc.ca/eng/1552073333119/1552080794636
Grønlands Statistik. (2020a). Grønland i tal 2020 https://stat.gl/publ/da/GF/2020/pdf/Gr%C3%B8nland%20i%20tal%202020.pdf
Grønlands Statistik. (2020b). Grønlands befolkning 2020. https://stat.gl/dialog/main.asp?lang=da&version=202001&sc=BE&subthemecode=t1&colcode=t
Heleniak, T. (2014). Arctic populations and migration. In *Arctic human development report II* (pp. 53–104). Nordic Council of Ministers.
ITK (Inuit Tapiriit Kanatami). (2018). Inuit statistical profile 2018. https://www.itk.ca/wp-content/uploads/2018/08/Inuit-Statistical-Profile.pdf
IWGIA. (2020). The Indigenous World 2020.
Jenness, D. (1968). *Eskimo administration* (Vol. V). Arctic Institute of North America.
Krupnik, I., & Chlenov, M. (2007). The end of "Eskimo land": Yupik relocation in Chukotka, 1958–1959. *Etudes/Inuit/Studies, 31*(1–2), 59–81.
Kumar, M., & Tjepkema, M. (2019). Suicide among first nations people, Métis and Inuit (2011–2016). https://www150.statcan.gc.ca/n1/pub/99-011-x/99-011-x2019001-eng.htm
Laruelle, M. (2019). Introduction: Indigenous peoples, urbanization processes, and interactions with extraction firms in Russia's Arctic. *Sibirica, 18*(3), 1–8.
Laugrand, F., & Oosten, J. (2010). *Inuit Shamanism and Christianity*. McGill-Queen's University Press.
Laugrand, F., Oosten, F., & Bilgen-Reinart, U. (2011). La relocalisation des Dènès sayisis et des Ahiarmiuts dans les années 1950. *Recherches amérindiennes au Québec, 41*(2–3), 99–116.
Lavrillier, A., & Gabyshev, S. (2017). *An Arctic indigenous knowledge system of landscape, climate, and human interactions: Evenki reindeer herders and hunters*. Kulturstiftung Sibirien.
Marcus, A. R. (1995). *Relocating Eden: The image and politics of Inuit exile in the Canadian Arctic*. University Press of New England.
Meltofte, H. (Ed.) (2013). *Arctic biodiversity assessment*. Conservation of Arctic Flora and Fauna. https://arcticbiodiversity.is/index.php/the-report/chapters/linguistics-diversity
Motzfeldt, J. (1999). Forord. In B. Gynther & M. Aqigssiaq (Eds.), *Kalaallit Nunaat*. Gyldendalske Boghandel.
Nunavut Department of Education. (2007). *Inuit Qaujimajatuqangit Education Framework*. https://www.gov.nu.ca/sites/default/files/files/Inuit%20Qaujimajatuqangit%20ENG.pdf
Rasmussen, K. (1920). Tanker om Grønland i Fortid og Fremtid. *Danmarksposten, 1*(1), 19.
Rasmussen, R. O., Barnardt, R., & Keskitalo, J. H. (2010). Education. In J. Larsen & P. Schweitzer (Eds.), *Arctic social indicators I*. Nordic Council of Ministers.
Robert, J. (1971). Les Ammassalimiut émigrés au Scoresbysund. *Cahiers du Centre de recherches anthropologiques*, XII, 8(1–4).
Schweitzer, P., Irlbacher Fox, S., Csonka, Y., & Kaplan, L. (2010). Cultural well-being and cultural vitality. In J. Nymand Larsen, P. Schweitzer, & G. Fondahl (Eds.), *Arctic social indicators I* (pp. 91–108). Nordic Council of Ministers.
Schweitzer, P., Sköld, P., & Ulturgasheva, O. (2014). Cultures and identities. In G. Fondahl & J. Nymand Larsen (Eds.), *Arctic human development report II* (pp. 105–150). Nordic Council of Ministers.
Stammler, F., & Ivanova, A. (2016). Resources, rights and communities: Extractive mega-projects and local people in the Russian Arctic. *Europe-Asia Studies, 68*(7), 1220–1244.

Statista. (2020). *Denmark: Infant mortality rate from 2009 to 2019.* https://www.statista.com/statistics/806792/infant-mortality-in-denmark/

Tester, F., & Kulchyski, P. (1994). *Tammarniit: Inuit relocations in the Eastern Arctic, 1939–63.* UBC Press.

UNPFII (United Nations Permanent Forum on Indigenous Issues). n.d. *Who are indigenous peoples?* (factsheet) https://www.un.org/esa/socdev/unpfii/documents/5session_factsheet1.pdf

US Census. (2010, 2013). 2010 Census CPH-T-6. American Indian and Alaska Native Tribes in the United States and Puerto Rico: 2010, Table 16, Alaska. https://www.census.gov/data/tables/time-series/dec/cph-series/cph-t/cph-t-6.html accessed 2020.12.16.

Walsøe, P. (2003). *Goodbye Thule. The compulsory relocation in 1953.* Tiderne Skifter.

World Bank. (2020). https://data.worldbank.org/country/denmark?view=chart

Young, T. K., Revich, B., & Soininen, L. (2015). Suicide in circumpolar regions: An introduction and overview. *International Journal of Circumpolar Health, 74.*

**Yvon Csonka** is a Professor Emeritus at Ilisimatusarfik, The University of Greenland, and a former president of the International Arctic Social Sciences Association (IASSA, 2004–2008). He is a member of the Swiss Committee on Polar and High Altitude Research (Swiss Academies of Arts and Sciences).

# Chapter 3
# Arctic Cities-Pioneers of Industrialization: More than "Mining Towns"

**Nadezhda Zamyatina**

## Introduction

At the end of the nineteenth century and the first half of the twentieth, a powerful wave of industrialization took place throughout the world's Arctic. The population increased several times over, dozens of new cities sprung up in different Arctic regions of the world, and previously roadless areas were connected by railways. This wave of industrial development of the Arctic was a natural – and, as it happened, final – stage of a powerful "rush" for resources, which previously gave rise to industrialization in the resource basins from the Ruhr to the Urals and Appalachia ("the Golden Age of Resource-Based Development", which was caused by the transition to the "fossil-fueled civilization" (Barbier, 2011, p. 372)). However, something changed around the 1960s. On one hand, active exploration and development of new resources continued, the Arctic shelf was involved in the development, and the importance of raw material extraction remained high for the Arctic economy (see also: Pilyasov, 2019). On the other hand, the construction of new cities with new deposits stopped, with most of the Arctic regions switching to the shift method since the 1960–1980s, and in the North of Russia after the collapse of the USSR, in the 1990s.

Of course, the change in the methods of development and the interest in new cities was a logical consequence of the development of technology and increased transport permeability of space, the development of aviation, etc., as well as of the size of new deposits. However, this raises an urgent question: is the era of Arctic industrial cities really gone? Are they the product of imperfect technologies that required a long-term mining towns for the development of deposits? Can these cities find their place in the modern conditions of shift mining, or are they destined to repeat the fate of "ghost towns"? (Fig. 3.1).

---

N. Zamyatina (✉)
Lomonosov Moscow State University, Moscow, Russia

© The Author(s), under exclusive license to Springer Nature Switzerland AG 2022
M. Finger, G. Rekvig (eds.), *Global Arctic*,
https://doi.org/10.1007/978-3-030-81253-9_3

**Fig. 3.1** *The upper picture*: The suburbs of the city of Vorkuta, which was once one the biggest Arctic city in 1980th. *The lower picture*: A part of the city of Vorkuta, the place from which Vorkura ran it's developing (houses in the left are now abandoned). (*Source*: Photos by Zamyatina)

In any case, such ideas are discussed by journalists who relish photos of abandoned housing around Vorkuta or in the villages on the Kolyma or along the Northern Sea route in Russia, where the change in technological patterns has replaced the socio-political structure, and the pioneer cities of Arctic industrialization have suffered dramatically. The population of Igarka has decreased by 75% (from 18 to 4,5 thousand people) compared to the end of the 1980s; Vorkuta and its suburbs have decreased by two-thirds; and even the population of Murmansk, the administrative and business center of the region of early industrialization of the Soviet Arctic, has decreased by a quarter.

In the 1960s, Canadian planners cited the audacity of the Swedish and Soviet Arctic exploration and the construction of cities in high latitudes.[1] Today, the number of urban residents in the Russian Arctic as a whole (2.32 million people, Rosstat is only 70.1% of what the urban population was in the same territory in 1989. The population of the Swedish Kiruna and other "pioneers of industrialization" of the Nordic Arctic is also declining, albeit not as dramatically as in Russia, as can be seen, for example, from the reports of Nordregio (Jungsberg et al., 2019). In modern conditions, if the population of Arctic cities is growing, it is only in business and university capitals, not in industrial centers (Zamyatina & Goncharov, 2020). Does this mean that the industrial cities of the Arctic are just "dinosaurs" of the era of early industrialization, living out their days?

Our hypothesis is that, paradoxically, many cities of the early years of industrialization were not just mining cities, but "post-industrial," and retained this potential to transform today into centers of knowledge-based, rather than industrial, development of the Arctic. It takes some shrinkage in population, but in many cases does not lead to the end of the city's biography.

Apparently, this was due to the specifics of how the early industrial cities of the Arctic responded to the challenges of remoteness. Remoteness is an important factor that hinders the development of the Arctic regions and especially cities (for example: Huskey & Morehouse, 1992. The Arctic cities usually fall out of the urban networks (networks could allow them to achieve agglomeration effect and/or save on specialization and exchange). In principle, for these "lonely" Arctic cities located a considerable distance from other economic centers, two types of answers are possible. The first involves strengthening the exchange of goods and services with remote centers (this is possible only with a high level of development of transportation technologies). The second involves building a relatively multicomponent, complex economy at the place. Even if early cities in the Arctic developed through mining, they often followed the second path. Many of the pioneering cities of industrialization that have survived to this day have become real bases for the development of the surrounding territory, with universities and specialized scientific organizations, with a specific cultural environment, and this is an amazing feature of cities from the early stage of Arctic industrialization.

## The Beginning of Industrialization: In Pursuit of Gold

After decades of feverish prospecting for gold in the Western United States, from Colorado to California, the wave of gold prospecting – the "gold rush" – swept to Alaska.

---

[1] Mid-Canada development corridor… a concept. [Presentation]. Available at: http://www.plancanada.com/midcanada_corridor_report.pdf

The first gold in the territory was found in Juneau in 1880 (the town was named after one of prospectors). The burst of business activity that began immediately made it possible to make this town the capital of Alaska, which Russia had sold to the United States shortly before.

Interestingly, there were earlier reports of the presence of gold, both in Alaska (controlled by the Russian American Company) and in the adjacent Canadian territory (controlled by the Hudson Bay Company). Neither company wanted to change its specialization from fur to gold, and, apparently, both were afraid of a massive influx of gold miners (Berton, 2001; Borneman, 2003; also, in Russian, Ermolaev, 2013). This gives reason to talk about the beginning of gold mining (and metal ores and other minerals in general) precisely as an important milestone both in the management system and in the drivers of economic development. It was ores that almost simultaneously replaced furs and fish on different continents as major drivers in the development of the Arctic, or its "staples", to use the term of the Canadian economist Harold Innis (Innis, 1930).

Juneau's ores were mined quite quickly (the largest Treadwell gold mine that explored Paris Lode) was on Douglas Island, where the area of Juneau's prestigious housing is now located).

Gradually, the search for new gold led its seekers deep into Alaska. Here, on relatively poor deposits, one-day towns arose: Circle in the American part of Alaska, Forty Mile in Canadian territory, and some others. Despite the relatively small volumes of gold production (although, by 1896, Alaska was already producing $800,000 of gold per year, compared to $30,000 in 1887; Borneman, 2003), the service sector developed rapidly in these towns. In 1895, the Circle was called the "Paris of the North"; it was "the largest log cabin town in the world, with 28 saloons, 8 dance halls, an opera house, a library with the complete works of Darwin, Carlyle, Irving and Macaulay, and a copy of the Encyclopaedia Britannica" (Cole, 1991: 11). All this did not prevent the city from emptying rapidly when, at the end of the summer of 1897, news came about the discovery of rich gold deposits on the Klondike stream.[2]

---

[2] The discoverer is considered to be Carmack, a hereditary gold digger who was born on the frontier (his father was a prospector during the 1949 California gold rush). The history of this discovery is confusing. Most versions agree that the prospector Robert Henderson found signs of gold on one of the streams in the Klondike River basin (a tributary of the Yukon), and when he met George Washington Carmack, he recommended that the latter look at a nearby stream, a tributary of the Klondike River (the stream was called Rabbit Creek, later named Bonanza) in the Yukon Territory of Canada. Carmack was with his family: he married a native Indian woman, she and her relatives accompanied Carmack on his way. According to some reports, the first gold was found by one of them, the brother of Carmack's wife Skookum Jim, but the discoverer's right was registered to Carmack (according to one version, he explained to his brother-in-law that no one would believe the Indian anyway; Skookum Jim and his Indian relative, Tagish Charley, however, registered "ordinary" ("non-discoverer") applications that are half the size according to Canadian law). Carmack brought news of gold to Forty Mile, where he came to register land applications (Borneman, 2003).

The famous "gold rush" began, and the following year (1898) thousands of enterprising people moved to Alaska. They either traveled by steamer along the Yukon or arrived in the town of Skagway to cross the mountains through the Chilkut (steeper) or White (which was easier to pass, but took more time to travel) passes. After passing through the mountains, potential gold prospectors loaded onto rafts and boats and floated along the lake system and further along the Yukon. An important point on their way was Whitehorse (the current capital of the Yukon Territory in Canada), which involved passing difficult rapids. By 1899, a railway passed through the White Pass from Skagway to the town of Bennet, and in 1900 it reached Whitehorse. The road was laid in difficult natural conditions; it has survived to this day, but is now used as a tourist road. The modern highway, running in parallel, is often closed to traffic today due to unfavorable weather conditions. Considering the circumstances, one marvels at the speed of construction.

Thus, Whitehorse became an important staging post on the way to the Klondike. Interestingly, we can mention that the co-owner of the hotel with the prophetic name "New Arctic" was one Friedrich Trump, who made his fortune on the frontier here. He was the grandfather of future US President Donald Trump (Holland, 2012; Obiko, 2016).

Using the services of the Trump Hotel and others like him, the gold prospectors moved along the Yukon to where the town of Dawson grew up (in the north of the Canadian territory of the Yukon), widely known from the stories of Jack London.[3]

Apart from stories such as *White Fang* and *Call of the Wild*, Jack London also wrote a curious economic article on the results of the gold rush. He showed that the main profits were gained by those who were already in Alaska by the time the Klondike was discovered, and numerous travelers through Chilkut lost rather than gained from the enterprise. However, London considered the main results of the gold rush to be the demolition of infrastructure barriers and the emergence of cities and roads in Alaska, which allowed future generations of Alaskans to extract its resources at new costs. He wrote: "In short, though many of its individuals have lost, the world will have lost nothing by the Klondike. The new Klondike, the Klondike of the future, will present remarkable contrasts with the Klondike of the past. Natural obstacles will be cleared away or surmounted, primitive methods abandoned, and hardship of toil and travel reduced to the smallest possible minimum. Exploration and transportation will be systematized. There will be no waste energy, no harum-scarum carrying on of industry. The frontiersman will yield to the laborer, the prospector to the mining engineer, the dog-driver to the engine-driver, the trader and speculator to the steady-going modern man of business; for these are the men in whose hands the destiny of the Klondike will be entrusted" (London, 1900).

---

[3] His translations were extremely popular in the USSR, judging by the interviews with the residents of the North of Russia of the older generation; they inspired a whole generation of Soviet people to conquer the "white silence" of the Soviet Arctic (according to the oral report of my colleague Elena Lyarskaya, a specialist in anthropology of the North from the European University at St. Petersburg).

This work by Jack London allowed Alaskan economist Lee Huskey, a hundred years later, to formulate the "Jack London hypothesis" about the development of frontier regions: during the boom, the local economy accumulates a critical mass that allows it to survive in the post-frontier time of depletion of the main resource (Huskey, 2017). The essence of this effect is that, during the boom phase of resource extraction and accompanying growth of the economy of the frontier city, it is possible to accumulate a critical volume of local population and local demand. When the volume of local demand surpasses the critical point (considering economies of scale), local production of certain goods and services becomes more profitable than importing them. As a result, the local economy becomes more diversified and, consequently, more stable (see also Zamyatina et al., 2020).

This was the main feature of the stage under consideration: from one region of the Arctic to another there was an explosive demolition of infrastructural barriers, growth of the new Arctic of cities near resource deposits, transport hubs, and in some cases business capitals. Anchorage became such an externality of the gold rush, although the connection with the Klondike can only be traced after a few iterations.

In August of 1901, an enterprising man named E.T. Barnett sailed across the Yukon on a steamer, hoping to open a trading station to supply gold prospectors. That year, the Yukon became shallow, and the steamer turned back, unloading Barnett and his luggage on a wild shore, where the city of Fairbanks was thus founded. To a large extent, this was facilitated by another accident; namely, that three starving gold diggers, who had been looking for gold for a long time in vain far from the already divided resources of the Klondike, went straight to the Barnett camp. Their stories gave rise to hope for the presence of gold in the area.

In 1903 Barnett announced the discovery of large deposits of gold in the area of his "town", when in fact there was no discovery yet. As it happened, gold was found and the town survived, becoming a real business center of central Alaska and named after Senator Fairbanks in order to receive some administrative support (Cole, 1991).

Unlike the Klondike, where the main economic agents were individual entrepreneurs who washed gold in cans, mines in Fairbanks were required to extract gold from lodes. This sometimes gives rise to assertions that the "real" frontier of small entrepreneurs ended in Dawson, but from Fairbanks the industrialization of the Arctic has already passed into the stage of large industrial enterprises. In addition, the Northern Commercial Company (NC, now the Alaska Commercial Company), which inherited the property of the Russian American company, essentially monopolized the supply of firewood and food to central Alaska, building an "an empire". It sold food, hardware, clothing, and operate a power-plant with a 60-foot smokestack, water system, and steam-heating system to serve the business center. By 1907, the NC was grossing $2 million a year from its various enterprises. The company emerged as one of the largest factors in the early development of Fairbanks because of its large supply of goods for sale and its policies extending credit, which allowed many miners to develop their clams (Cole, 1991: 52). Thus, the economy of American Alaska (and here there was some difference from the Yukon) was already developing as more or less large-scale capitalist.

Other infrastructure projects also required large investments. In the first decade after the founding of Fairbanks, sections of railways were built, first from Fairbanks with the Nenana Coal basin south of the city; secondly, the Copper River and Northwestern Railway from Cordova on the southern coast of Alaska; and third, from Seward on the north of the Kenai peninsula. A rough road was laid to Fairbanks.

In 1912, Alaska acquired the status of a territory, which allowed the US president, William Howard Taft, to take control of the construction of railways to a new resource region.

Taft immediately initiated railroad construction in a rapidly developing area, although project ownership issues (private or public) arose; alternative means of railway construction were also considered, taking into account the existing sections of railway tracks. The project started in 1914 when the new president, Woodrow Wilson, signed the Alaska Railroad Bill. It is interesting how the problem of choosing the path of the road was solved. The owners of the Copper River and Northwestern Railway were asking too high a price for their road ($17.7 million; Borneman, 2003: 262), and it was logical that the road would follow the western route from Seward. A settlement was founded in the Cook Bay area as a construction base. As a port, this point was inconvenient due to high tides (today, the inhabitants of the city that grew up there can see the ocean only behind a wide strip of gray mud of the tidal zone). Nevertheless, it was this point (another accident) that became the economic center of the construction of the railroad to Fairbanks, to the "heart" of the then-boom in the economic development of Alaska.

As one of the famous governors of Alaska, Walter J. Hickel, wrote: "The influence of the railroad on the development of Alaska would be hard to exaggerate. It built the city of Anchorage from scratch and fed to growth of Fairbanks as well as places in between, including Palmer, Wasilla, Talkeetna, Nenana, and Denali National Park. The railroad aided the supply of Alaska coal … Building a rail line across this remote territory took 8 years. As an investment in Alaska, it yielded extraordinary dividends, yet as a private business venture, it never would have survived. The railroad's revenues covered its operating costs in bottom line terms. The lesson to learn is that the bottom line is wrong measure to use in judging a project that opens up a frontier. Instead, these projects should be judged by the cities they build, the resources they access, the contributions they make to national security, and the civilization they create" (Hickel, 2002: 103–104).

In fact, Governor Hickle provided evidence for the Jack London Hypothesis. An interesting externality of the gold rush in Alaska was the deployment of gold mining in Kolyma (although this region does not belong to the Arctic according to the official criteria of Russia, its history and climate are so typical of the Arctic that it would be wrong not to consider it in our work). By the end of the nineteenth century, just as gold rushes were in full swing in America in the more southern states, in the Russian Empire there was rapid gold mining in the "pre-Arctic" region, on the Lena and Aldan rivers.

There were symbolic attempts to see the development of the territory of the North-East of Russia along the frontier of Alaska, and even with Alaska as a springboard. In a monograph that summarized hundreds of sources about the gold of the

North-East, mining engineer E.E. Anert wrote: "The high gold content in the deposits, cheap sea freight and the proximity of Nome with its advanced technology, confined to similar climatic and geological conditions, fully compensate for the disadvantages. Providing free mining here to all enterprising people can lead to new discoveries and the creation of a new Alaska" (quoted from: Pruss, 2017: 30).

By the mid-1920s, individual prospectors were washing small amounts of gold in streams in the Kolyma region, as it was in the interior of Alaska on the eve of the "Klondike era". However, things did not go further. Guided by preliminary information about gold in the Kolyma region (both from prospectors and geologic surveys in nearest regions), Yuri Bilibin, a graduate of the famous Mining Institute in St. Petersburg, managed to organize an expedition to this region. Despite monstrous hardships, the expedition allowed him to predict the vast gold content of this area. Bilibin possessed not only brilliant geological talent, but also rare tenacity, which allowed him to "break through" the organization of large-scale gold mining in Kolyma.

Returning from the expedition, Bilibin, as he himself later wrote, "began to actively promote the Kolyma" (Bilibin, 1964: 202). He insisted that the territory had vast gold content, but most importantly, he insisted on changing the institutional system of geological exploration. Realizing that small parties cannot achieve a powerful increase in gold production, Bilibin demanded the organization of a large-scale organization of scientifically grounded geological research: "I considered the expeditionary system of work in Kolyma to be irrational. Therefore, back in the spring of 1930, I raised the issue of organizing a permanent 'Indigirsko-Kolymskiy Geological Prospecting Bureau' with large allocations for geological exploration." The geologist drew up a "Plan for the development of exploration in the Kolyma", which involved fantastic costs, but would have also provided fantastic results: "In the very first year of its existence, I envisaged investments for exploration in the amount of 4.5 million rubles, and with their progressive increase, he considered it possible to provide reserves for placer gold mining in Kolyma in 1938 in the amount of almost four times gold mining in the Soviet Union in 1930" (Bilibin, 1964: 203). The geologist wrote that he was horrified by his forecast. (Bilibin, 1964: 202).

For the development of the gold reserves of the Kolyma, a special Dalstroy trust was created (some features of the East India Company and the like can be traced at the basis of its organizational structure). If we do not delve into the terrible ethical aspect that Dalstroy used the labor of prisoners, the trust was a completely unusual entity for the Soviet Union, both economically and performing the functions of administrative management of a huge territory, with unprecedented administrative powers and independence. Unlike the United States, the development of gold in Kolyma had another interesting difference: it was a scientifically grounded (from the point of view of the geological substantiation of reserves, of course) project for the development of the territory.

Despite all the differences, however, one can see a similarity in the development of Kolyma and Alaska; soon after the discovery of gold, a road – the Kolyma Highway – was laid to the inner regions of Kolyma. However, a railroad was also built to coal regions (as well as on young Alaska), although it was later dismantled. The earliest movements (including the Bilibin expedition itself) in this region (again, as in Alaska) were carried out along the rivers.

## Iron and Other Ores Create New City-Islands in the Arctic

At almost the same time, the "iron rush" was unfolding in Europe. The construction of railways allowed the commissioning of new powerful deposits of orc resources, which were required by the ongoing industrialization of the leading countries. The iron ores of Kiruna were so rich, and their development began so rapidly (after the construction of the railway, which broke the barrier of transport inaccessibility) that it was often compared to the Klondike. Somewhat later, the development of apatite ores on the Kola Peninsula, as well as copper-nickel ores in Norilsk, began. The development of ore deposits from the end of the nineteenth century until the middle of the twentieth century had similarities in different regions of the Arctic. Very rich ores were put into circulation, which made it effective to build new cities in these areas. It should be noted that these cities developed outside the urban network, at a great distance from other economic centers. This remote "island" position is an important feature of the new "ore" cities of the Arctic. Three main ore "city-islands" appeared.

### *Sweden: Kiruna, a Place of an "Ore Rush"*

The existence of iron ores in northern Sweden has been known since at least the seventeenth century, but their development has been blocked for centuries by the region's impassable roads. In the 1960s, the first attempt was made to build a railway from Luleå to ore-rich areas. The British Gellivara Company Ltd. received the construction rights, but went bankrupt and construction was stopped (Raw Materials Group & Robert Dewsnap, 1981).

Thus, the first Swedish Arctic railway project could almost open a number of unfulfilled railway construction projects planned in different regions of the world's Arctic in the early twentieth century. If, in Alaska, the hope for gold was able to break down the barriers of off-road, in many other regions of the world's Arctic, large-scale railway construction projects planned in the first half of the twentieth century remained unfulfilled (although they still appear on the agenda of Arctic development). This includes a project for a tunnel through the Bering Strait and a railway from Alaska to Chukotka and further to the gold-bearing regions of Siberia (this was also supposed to be a foreign concession, and it was refused for reasons of state security). Another well-known unfulfilled project was the railway that was supposed to run between Arkhangelsk and the Urals (Slavin, 1961). To some extent, its modern equivalent is the Belkomur project.

However, the fate of the Swedish project was happier, albeit difficult. To a large extent, this can be explained by the unique qualities of the local ore. The local ore is rich in iron content, and so rare that it gave its name to the type of ore ("kiruna type ore") in geological books. Its extraction is so economically effective that comparisons with gold as the engine of the development of the territory are not too far-fetched.

Between 1884 and 1888, a new concession agreement with the British "North Europe Railway Company" built a railway between Malmberget, which for a time became the main center for iron ore mining in the area, and the port at Luleå. At the same time, the Swedish capital was dissatisfied with the penetration of the British into the rich ore region and demanded that foreign firms be prevented from being allowed to "plunder our northern ore fields" (Raw Materials Group & Robert Dewsnap, 1981). In 1891, the road was nationalized (helped by the bankruptcy of the company that owned it); on March 12, 1888, the first ore train arrived at the newly established ore harbor in Luleå (It starts …).

The road finally opened up the possibility of industrial development of iron ores. For this purpose, a company called LKAB (Luossavaara-Kiirunavaara Aktiebolag) was created in 1890. LKAB insisted on extending the railway to the ice-free port of Narvik, which was carried out in 1902 (the famous and still operating Malmbanan).

The LKAB partnership (later the company became a state-owned enterprise) was the real owner of the iron mining area. It was led by the legendary geologist and founder of Kiruna, Hjalmar Lundbohm. It is important to note that the production process and development of the territory were simultaneously managed by one company: "As part of the state's permission to mine the area, LKAB had to guarantee housing and other facilities for its incoming workers. A model company town was developed, with high housing standards as well as a library, hospital, fire station, church and schools, and a tram line connecting residential areas with places of work. The town continued to develop alongside LKAB, which has been state-owned since 1907" (Sjöholm, 2020). Here we can trace some, albeit somewhat shaky, parallels with other areas of new development under the company's control, such as the Hudson's Bay Company or Dalstroy on the Kolyma basin.

Kiruna's "planning" contrasts especially with the "wild" frontier cities of Alaska, although at the early stage of development, settlements in the area of iron ore mining were quite similar to frontier camps: during the first years, workers lived in shacks built from wasted dynamite crates (for details of the early history of Malmberget, see: Svensson & Wetterberg, 2009: 86 pp).

Kiruna was built as "a model city" with a well-thought-out town plan. A model city would "counteract social deprivation through good architecture, education and good working conditions. The goal was to build an attractive community in order to recruit and retain labour in this young mining community in the Far North" (Overud, 2019). Here is a direct parallel with the first Arctic cities of the Soviet Union, although they were designed to demonstrate the victory of the Communist party over the "wildness" and "backwardness" of the Northern outskirts. First of all, this role was played by Igarka (the image of Igarka as a city that could be visited by foreign ships was given special importance; in particular, a constructivist architect named Ivan Leonidov was invited to create its master plan for it). Also Kirovsk to some extent, and later Norilsk, could be mentioned here (see below). It is no accident that today, due to the expansion of the mines, it was decided to develop Kiruna in a new location, or that efforts are being made to move key attractions, as well as to recreate the "spirit" of the city (Nilsson, 2010).

For a hundred years, Kiruna (with its surrounding territories) has been the largest iron producer in Europe. In addition to railways, the coast has been the gateway to Arctic mineral resources since about the beginning of the twentieth century. For example, the development of iron ore in Sydvaranger (Kirkenes, Norway) began in 1906. Access to the sea was obviously the main advantage of this area.

Of particular note is the development of nickel ores (Kolosjoki nickel mine) in the Petsamo region of Finland (Stadius, 2016), currently the settlement of Nikel in the Murmansk region of Russia. According to Eloranta and Nummela (2007), "Petsamo nickel in Northern Finland became a bargaining chip in the Great Powers' political game before and during the Second World War." Strategic considerations are another important factor in the industrialization of the Arctic.

## *The Murmansk Railway Opens the Way to the Development of the Kola Ores*

To the east, on Russian territory, the situation developed as follows. During the First World War, the new railroad was laid here to the shores of the Barents Sea, which are ice-free thanks to the Gulf Stream. As is almost always the case in the Arctic, the construction of the road was due to some extraordinary circumstances: if not the gold rush, then the war (during the Second World War, for example, the Alaska Highway appeared in America and railway to Vorkuta in the USSR). The speed of construction of railways in America, Sweden, and Russia was surprising especially in the context of the fact that the construction of The Northern Latitudinal Railway (Yamal region of Russia) started in 2003 is not yet completed (it seems this is due to the significantly lower importance of the railway in the modern economy compared to the period of early industrialization.)

The above-mentioned railway to the Barents Sea ended in Murmansk, which became a transshipment base on the way to Aleksandrov (where the port was founded a few years earlier). Subsequently, Murmansk became an administrative center of the region and one of the largest Arctic cities in the world, along with Arkhangelsk and Anchorage).

Among other things, Murmansk is interesting because it became the first Arctic point (at least in the USSR) where special benefits were provided for newcomers to attract settlers (From the history …), up to the provision of equipped housing. This feature can be considered a sign of the politically strategic importance of new construction, in contrast to the economic-driven gold rush, which usually did not suffer from the absence of migrants.

Having received a ready-made railway from the tsarist regime, the Soviet government that assumed the leadership of the country was not sure of the expediency of its content, and sent a scientific expedition to work out this issue. Waiting at one of the parking lots, while the firewood for the locomotive was being prepared, the scientists walked to the surrounding Khibiny mountains. The prominent geologist Alexander Fersman, who was among the members of the expedition, drew attention to the mineralogical composition of local rocks (Kamenev, 2013); subsequently he organized a number of new studies and, as a result, proved the presence of rich deposits of apatite. To develop them, the city of Khibinogorsk was soon built (after the assassination in 1934 of a popular government official S.M. Kirov, the city was renamed to Kirovsk). Today Kirovsk is the home base of Russian leading supplier of apatite, the PhosAgro company.

Copper-nickel ore was soon found not far from Kirovsk, and already in 1937 the city of Monchegorsk was founded for their developing. Complex ores of Kovdor (iron, apatite, rare mineral baddeleyite containing zirconium) were discovered a little to the west of Kirovsk; the settlement here was founded after the Second World War. The 1930s became a stage of large-scale geological research (the development followed purposefully organized geological research there, unlike Alaska but as in Kolyma). In the following decades, several more centers for the extraction of various raw materials developed on the Kola Peninsula: iron ore in Olenegorsk, copper-nickel ore in the city of Zapolyarny, laparite (raw materials for obtaining tantalum, niobium, etc.) in the settlement of Revda, etc. In the 1950s there was an attempt to develop rare earth metals in the settlement of Afrikanda village (Development …).

As a result, the Murmansk region as a whole has become one of the important industrial clusters of the country; it has even been compared to the Urals. However, the region, which was developed more than half a century ago, is now experiencing (in addition to specific Arctic restrictions) the classic difficulties of an old-industrial, depressive "mountain" region, like the non-Arctic old-industrial regions (Ural, Appalachian, and even Saar).

## *Norilsk: Nickel, Copper, and "Black Gold"*

In the history of the Norilsk industrial region, all three factors of industrialization of the Arctic have been intertwined: this is a private initiative that ensured the primary exploration of ore reserves; the extreme conditions of the civil war and politics that pushed for in-depth exploration, and finally, it is a powerful state organization working as a "machine" for the development of an entire industrial region, which, like Kiruna, administered the development of not only industry, but also the territory (adjusted for the fact that the labor was that of prisoners).

The resource base of Norilsk was no less diverse. Even at the first stage, it was interesting for reserves of both copper ores (it later turned out that copper-nickel also contained rare metals) and coal.

In many areas, the development of the Arctic was accompanied earlier by the exploration and development of coal deposits, a key energy carrier (in Russia it was called "black gold", although this nickname later passed to oil). Indeed, at the turn of the twentieth century, coal was needed for steam locomotives, steamboats, as well as for equipment used in mines.

Coal mining accompanied (and it seems, even partially contributed to) the development of the Norilsk copper ores. Copper was mined by hand in Taimyr several decades before geological expeditions confirmed the richness of local deposits. The initiative of the first professional geological research in this area belonged to Alexander Sotnikov, the son of local landowners. In 1915 he brought some examples of ores from Taimyr to Tomsk Polytechnic Institute where he studied. He asked Nikolai Urvantsev, a student of the same university but older and more experienced, to describe them. A geological report was prepared. In 1919, Sotnikov reported on the possibility of ore mining to Admiral Kolchak, who at that time (there was a civil war in Russia) took possession of a significant part of Siberia (Urvantsev and Sotnikov …). The territory governed by Kolchak was cut off from the outside world by his enemy, Soviet Russia. Under these conditions, the idea of "launching" the Northern Sea Route (NSR) was activated, which had previously been actively discussed by some entrepreneurs, especially the above-mentioned Norwegian Jonas Lead (Lead, 2009). Entering the world through the NSR could have saved Kolchak's "empire" economically – at least, that is what Lead assumed (Lead, 2009). Among other factors, ensuring long-distance shipping required coal, and the Siberian Geological Committee, established by the Kolchak government, organized a new expedition by Sotnikov and Urvantsev to Taimyr. In such conditions, one can assume that coal played no less a role than copper. Upon returning from the expedition to the North, they found the government already replaced; Sotnikov (who, shortly before being sent to the North, was fighting against Soviet power) was shot, and Urvantsev continued to work.

In the conditions of the formation of a new state, in the Far North, practically in an undeveloped area, the expeditions performed the functions of not only geological surveys proper, but a comprehensive study of the region. In his later published works, Urvantsev wrote about the need to borrow local knowledge of indigenous peoples (for example, eating frozen raw fish to prevent scurvy, as well as ways to survive in a blizzard), efforts to find effective transport routes, etc.

In 1935, construction of an industrial ore mining enterprise began. To do this, in accordance with state policy at the time, a forced labor camp, Norillag, part of the infamous GULAG, was organized. The ambitious Avraamiy Zavenyagin, who had previously worked on the construction of the famous new city of Magnitogorsk in the Urals, founded to develop rich deposits of iron ore, became the head of the project. He was distinguished by his commitment to "humanitarian principles of social policy, understanding that creating favorable conditions for recreation not only improves the mood, but also affects the productivity of employees of the main enterprise… He found opportunities to create more favorable living conditions for at least the necessary construction specialists from among the prisoners. Many of them were thus saved from death. Taking into account the human factor allowed

A. P. Zavenyagin to ensure the implementation of tasks in the most stringent time frame" (Reut, 2018: 185). Zavenyagin decided not only to organize mining, but also to build a beautiful city there. It is interesting that the first head of Kolyma project Edvard Berzin also dreamed of a beautiful city, the capital of this area, and initially intended to build it not where Magadan appeared, but on a river island (modeled on St. Petersburg or Paris). Zavenyagin's project, unlike Berzin's, was implemented. Zavenyagin used the work of architects who were among the prisoners of the camp (including outstanding specialists such as Yakov Trushins, Mikael Mazmanyan, and Gevorg Kochar-Kocharyan (see Slabukha, 2010), as well as engineers and others. He also invited the "free" young architect Witold Nepokoychitsky, who developed the General plan of the city. Even today, Norilsk is interesting from an architectural point of view: its historical part is built in the characteristic pompous "Stalinist" style ("Stalin Empire style"), symbolizing the triumph of the new political system.

During the construction of the city and in the first decades of its existence, there were many innovations in the field of organizing urban life in the Arctic. In particular, Michael Kim, a former prisoner of the Norillag and later the director of the Norilsk Research Department of the Krasnoyarsk "Promstroyproject" Institute, developed a method for constructing the foundations of multi-story buildings on the permafrost, which was subsequently copied many times in other regions. In 1939 Kim was released for his developments but remained in Norilsk as a "civilian"; in 1966 he received the most prestigious Lenin Prize in the USSR for his idea.

In terms of production, Norilsk has become the world's largest nickel producer, and absolutely unparalleled as a large metallurgical plant located in such harsh Arctic conditions. The Russian town of Monchegorsk and the Canadian towns of Thompson and Sudbury are in northern, but much milder conditions. Nickel mined today in the "real" Canadian Arctic is not melting here but exported for processing to more southern regions. However, Norilsk Nickel, which owns a plant in Norilsk, is gradually shutting down part of the Norilsk production; including the Nickel Plant, one of several plants in Norilsk, which was closed in 2016. The corresponding operations were moved to the Kola Peninsula (also an Arctic territory, but with a milder climate and better transport accessibility).

Coal was the main source of heat and power supply for a long time, until, in the second half of the twentieth century, natural gas reserves were discovered and hydroelectric power plants were built, providing such affordable electricity that in 2020 Norilsk hosted the first cryptocurrency mining farm Bitcluster Nord in northern Russia (on the territory of the closed Nickel Plant).

Another city with a similar fate is Vorkuta. Based on the discovery of coal deposits in the 1930s, a network of GULAG camps was deployed, which played the role of pioneering development of the region. The peak of construction activity fell during the Second World War: the country needed coal instead for the coal of the Donbass, which fell into the zone of fascist occupation. The settlement of Vorkuta, the center of the region's development, received the status of a city in 1943, and later became one of the largest cities in the Russian and global Arctic (together with the suburbs, its population in 1989 exceeded 200,000 people). Since the 1990s, however, due to the crisis in the coal industry, the city has lost about half of its population, and the surrounding villages are being gradually liquidated (Zamyatina et al., 2020).

Another coal region, Svalbard, had a completely different fate. Svalbard can rightfully be considered one of pioneer of the industrialization of the Arctic, which began at the very end of the nineteenth century (along with Alaska and the north of the Scandinavian Peninsula). However, there was a peculiarity here: this is due to international agreements, the availability of the island for entrepreneurs from different countries (during the period under review, Norwegian, Swedish, American, Dutch, Russian and British companies were active here), and, in general, the large role of politics in its development (Dag et al., 2011).

## Cities on the "Caravan Routes"

Despite the enormous length of Russia's northern maritime borders, even in the era of active expeditions to the North Pole, undertaken by a number of Northern countries, including Norway, Great Britain, etc. countries, the development of the Northern Sea Route in Russia was delayed for several decades.

Only the construction of Igarka in 1929 put an end to a whole series of experiments regarding the organizing of industrial supplies across the Arctic Ocean. Among these were the attempts by the Siberian merchant Michael Sidorov to organize the export of graphite from the Kureika deposit (not far from Igarka), as well as the powerful activity of the Norwegian businessman Jonas Lead in organizing regular transport links between Western Europe and Siberia through the Arctic Ocean. Among other attempts to launch the Arctic merchant navigation, Sidorov managed to carry out a delivery of graphite and some other goods on a voyage under the leadership of British captain Wiggins (the expedition experienced a number of difficulties and was not completed in one navigation). He met many obstacles from officials. In Russia, the answer of one of the tsarist officials (General Zinoviev) to Sidorov became a symbol of the position of skeptics regarding the prospects for infrastructural development of the Arctic: "Since there is constant ice in the North, and arable farming is impossible, and no other trades are possible, then, in my opinion, and my friends, it is necessary to remove the people from the North to the internal countries of the state, but you are busy on the contrary and explain about some kind of Gulfstrem, which cannot be in the North. Such ideas can only be carried out by the mad" (Zubov, 2014).

Jonas Lead, in turn, launched activities for the integrated industrial development of Siberia, where the Arctic regions were supposed to provide a transshipment base: there was a need to find a place for the transshipment of goods from river to sea ice-resistant vessels and vice versa. For this, Lead founded the Ust-Port base, but the revolution in Russia interfered with his plans.

The NSR, "reassembled" under Soviet rule, passed through a new point, Igarka, which apparently became the optimal (for its time) variant of a transshipment base. Igarka quickly became a key export point for Siberian timber to Western Europe (Zamiatina, 2020). It is curious that, to this day, the export of raw materials, mainly in the western direction, remains the main economic function of the Northern Sea Route, with only the types of raw materials changing (the export of timber has virtually stopped; from the mid-twentieth century to this day there are supplies of copper-nickel concentrate, and large-scale LNG supplies have been established in the last decade).

At the same time, the Northern Sea Route was developing as a strategic infrastructure connecting the western and eastern regions of the USSR. It is no coincidence that the head of the Northern Sea Route Main Department (Glavsevmorput) in 1939, the legendary polar explorer Ivan Papanin, said that the work of the NSR guarantees that "the times of Tsushima will not be repeated". He referred to the shameful death of the Russian military squadron, who tried to move from the Atlantic Ocean to the Pacific theater of operations in 1905 (Russo-Japanese War) in the area of Tsushima Island in the Sea of Japan (18th Congress …. 1939); the NSR would have allowed this ferry to be made without leaving territorial waters.

Several points arose along the NSR, in fact, providing navigation. Large meteorological and research bases emerged in the settlements of Amderma and Dikson. Fluorite was also mined in Amderma (however, mining was soon recognized as unprofitable). Dixon was founded under emergency conditions (as already mentioned, this is a frequent case for the Arctic). It was a temporal radio station for rescuing one of the Arctic sea expeditions, but after the completion of the rescue operation the decision was made to keep it, and the station evolved to a powerful meteorological center, a stronghold of polar aviation, a coal-bunkering point for ships, the base of the navigation service, etc. The breadth of its functions is amazing: there was even a kennel for breeding sled dogs. In recent times, both Dixon and Amderma have almost become ghost towns, practically losing their economic functions.

In the eastern part of the Arctic in 1933, the port city of Pevek and the settlement of Tiksi were founded as logistics centers.

Around the same time, another Arctic port was built in Canada for the export of raw materials – wheat rather than timber – through the sea route. Also, contrary to the objections of skeptics who doubted the possibility and economic feasibility of Arctic navigation, the port of Churchill was built on the coast of Hudson Bay to export wheat from Manitoba and other "steppe provinces" of Canada. Churchill is a vivid example of the brainchild of the increased global demand for wheat, which has led to the intensified development of the prairies of the USA, Canada, Argentina, as well as Russian Malorossia.

The fate of Churchill, like that of Igarka and many settlements on the NSR in the late twentieth and early nineteenth centuries, is unenviable. Fully dependent on global market conditions for raw materials, which are produced hundreds of kilometers from the city, port cities have experienced a difficult economic crisis, a change in the owners of enterprises, etc. The title of the section reflects their similarity to ancient cities on trade routes (like the African city of Timbuktu), which are entirely dependent on external factors. The narrow specialization in transit and transshipment services makes such cities extremely vulnerable economically. This was not always the case.

## Vegetable Gardens and Scientific Bases

In the early period of the industrial development of the Arctic, young cities were created in practically undeveloped areas. In the early years of such cities, reaching these areas often took several weeks or months (delays remained before railroads were constructed and if transportation was by sea). Unsurprisingly, such cities often suffered from food shortages, and there were urgent attempts to produce it locally. As a result, many young Arctic cities of the early specialization period became centers not only for mining, but also for the production of agricultural products; this seems like a paradox today, when the topic of Arctic food security seems to be a fresh trend.

The situation was similar in the production of building materials, scientific research, etc. The topic of local, intra-Arctic production, the use of local building materials, energy raw materials, and products is widely discussed today. However, the topic experienced a long hiatus in the second half of the twentieth century, when the powerful development of transport – primarily air transport, coupled with relatively cheap hydrocarbon fuel (it is no coincidence that Urry wrote about the onset of an era of unprecedented mobility) – enabled the massive delivery of goods to the Arctic. Economists at that time were deeply impressed by the savings potential of inter-regional specialization. It seemed that bringing food to the North, especially food that had been mass-produced in the southern regions, would create huge cost savings. The concentration of large production volumes in the south created costs in the form of reduced product quality (abandonment of local products), and also somewhat reduced food security, but in the 1960s, inter-district flows around the world won out over the principles of self-sufficiency. New deposits were increasingly often developed on a rotational basis. In the USSR, although fully fledged cities were created, no one set the task of creating "centers of cultural and industrial development of the surrounding territory" in them; cities arose that would later be called "single-industry".

Returning to the first half of the twentieth century, the most famous example of food self-sufficiency is probably the pioneer of the Northern Sea Route: the city of Igarka. In view of the scurvy epidemics in that city, its leaders almost immediately drew attention to the need to provide the population with vegetables. In the early years, the collection of pine needles (for decoction) and berries was organized, and soon an experimental vegetable growing farm was organized. The farm employed a lot of exiled peasants, the so-called "kulaks"; they were often skilled workers in agriculture, who achieved good results in their homeland and were expelled for being "too rich" to be a peasant. As a result, it was reported in 1935 that "Igarka's experiments proved that problems of growing vegetables and other agricultural plants can be successfully solved in Siberian Arctic Circle conditions. This will greatly remove the agricultural edge. Igarka's experiments are already integrated in other Northern sites … In 1934 Dudinka Ust-Port and other towns planted vegetables on their own premises with Igarka's seeds" (Ostroumova et al., 1935: 7–9). An important confirmation of the success of agriculture was the fact that, during the Second World War, this Arctic city was fully self-sufficient in potatoes.

Kirovsk became another powerful center of polar agriculture in Russia. Mainly the center of Apatite mining, it has got an agricultural experimental station, with a polar botanical garden, almost immediately; now there is a unique storage of potato tubers here.

Experimental stations of polar agriculture operated in Salekhard; in Norilsk, until the 1980s, they kept an industrial cattle farm (the efficiency of development in the Arctic of such a large-scale industrial agriculture was, of course, extremely low). In Fairbanks, the pioneer of industrial development in Alaska, agriculture also flourished (Papp & Phillips, 2007). It is no coincidence that agriculture became one of the areas of specialization of the newly opened university (in fairness, we note that the demand for food in the mining regions of Alaska also gave rise to specialized agricultural areas in the south of the region, such as Delta Junction and, somewhat later, Matanuska Valley).

An interesting feature of the early Arctic cities in the great deal of attention given to scientific research. As fossil resources were depleted, scientific research became a factor in diversifying the economy and ensuring the future sustainability of frontier cities: the above-mentioned "Jack London effect" worked not only on the growth of industrial infrastructure, but also, to a large extent, on the growth of scientific research infrastructure and competencies. It is paradoxical that powerful scientific research institutions were characteristic precisely of the earlier Arctic cities, where they served the purpose of ensuring the very possibility of building cities and enterprises in new conditions, and it was caused by the very situation of the Arctic frontier. In Norilsk, permafrost construction technologies have become the most important area of scientific research; the permafrost station was founded in the first years of Igarka's existence. The integrated scientific station Tietta (geology, agriculture, etc.) is practically the same age as the hosting city of Kirovsk. Powerful geological research was carried out in Magadan; in Vorkuta, research was also conducted in the field of Arctic construction. In Apatity (the twin city of Kirovsk), a large scientific center of the Russian Academy of Sciences was founded soon after the Second World War; it is the only center of the Russian Academy of Sciences located in a provincial town rather than in the capital of the region. Fairbanks University was founded in 1917, at the decline of the gold rush, and became one of the most important lines of economic diversification (along with servicing military bases).

Curiously, later cities, which seemed to develop in more favorable conditions, were supplied with scientific developments from more southern regions, which is understandable, but this circumstance impoverishes the spectrum of specialization of these cities. For example, there is a large research center in the field of oil production, serving the oil-producing regions of Western Siberia, in Tyumen.

The modern corporate research center LKAB Hjalmar Lundbohm Research Center is located in Luleå. In both cases the Arctic industrial cities are connected with university cities with auto bans and railways, while really remote Arctic cities tend to have own research infrastructure.

It is interesting that only cities located in the harshest conditions of all, and at the same time possessing sufficient economic potential for long-developed deposits, have grown into interesting research centers.

This is how a curious type of base cities appeared (the type is described in the example of Magadan (Zamyatina, 2020)). Their main function, in addition to housing a part of the workers in the extractive industry and the location of the headquarters of the extractive enterprises, is to provide the production process with "development services". Being deprived of the opportunity to develop manufacturing industries, these cities received a great profile of specialization in the 1930s, with a forcedly high share of post-industrial sectors. However, it was these industries that actually provided support for the viability of the Arctic regions for decades. Examples include geology that provided an increase in reserves due to the exploration of more and more new deposits (Fairbanks, Kirovsk-Apatity, Magadan, Norilsk), and research in the field of construction under Arctic conditions (Norilsk, to some extent Vorkuta, and now Fairbanks), and research in the field of polar agriculture, polar medicine, climate, etc.

Today, the established scientific base of the pioneer cities in the development of the Arctic can breathe a second wind into their economic development.

## Conclusion

During the period from the end of the nineteenth century to the outbreak of the Second World War, the economic map of the Arctic has changed significantly.

Here appeared the cities of Fairbanks and Anchorage (and a network of smaller ones) in Alaska, Whitehorse, and Dawson in the Yukon (the latter has since lost its population, but retained the right to proudly be called the City of Dawson, contrary to the Canadian criteria for urban settlement). Two parallel railways ran to Fairbanks and Whitehorse. In Sweden, an area of mass mining of iron ore was formed with a center in Kiruna and export through the Norwegian Narvik. In Russia, the railway reached the young city of Murmansk on the Barents Sea. Along the way, the city of Kirovsk, a center for the extraction of apatite raw materials, was founded, followed by a series of other cities. Industrial exploitation of the Northern Sea Route began, the main "star" of which was Igarka, an exemplary Soviet city; timber was also exported from the "revived" Naryan-Mar. Along the NSR, navigation points emerged in Amderma and Dikson, Tiksi and Pevek, as well as points for industrial fishing. In Canada, Churchill on Hudson Bay became a major grain export port. In the Far East, a large industrial region was formed in the Kolyma and Magadan regions. Finally, just before the Second World War, two more large resource regions arose: Vorkuta in Pechora Coal Bassin and Norilsk, extremely rich in ore resources, absolutely unprecedented in terms of the size and variety of city functions in the extremely harsh natural conditions of Taimyr.

All of these cities were created in the "resources rush" in the early stages of industrialization and could be recognized as "minetowns" – this would be true if they were not in the Arctic. However, in the Arctic many (but not all) of them were forced to evolve into more complex "organisms" and become drivers of the development of surround territories. The need to ensure the functioning of the city in the conditions of remoteness contributed to the development of cultural and knowledge activities, research of specific features of living in the Arctic, and/or remoteness from large industrial and knowledge centers. An analysis of the "experience" of individual cities of the early industrialization of the Arctic shows that the remoteness and harsh conditions seem to have contributed to a somewhat more diversified development of these "pioneer cities" compared to the classic mining cities in more densely populated areas. If these cities had been created in conditions of better transport links, they would probably be more "single-industry"; an example of this is the "oil" cities in Western Siberia from the late USSR. On the contrary, the potential of the pioneer cities of industrialization allows many of them to now act as drivers of local knowledge development and development of post-industrial Arctic activities.

What is particularly interesting is that similar trends are typical for the city pioneers of industrialization in completely different Arctic countries and regions: the "generational" features of these cities are often more important than regional ones.

# References

18th Congress of the All-Union Communist Party (b). March 10–21, 1939. Verbatim report (1939). OGIZ, State Publishing House of Political Literature, 740 p. (In Russian)

Dag, A., Hacquebord, L., Aalders, Y., De Haas, H., Gustafsson, U., & Kruse, F. (2011). Between markets and geo-politics: Natural resource exploitation on Spitsbergen from 1600 to the present. *Polar Record, 47*, 29–39.

Barbier, E. (2011). *Scarcity and frontiers: how economies have developed through natural resource exploitation*. Cambridge University Press.

Berton, P. (2001). *Klondike: The last great gold rush, 1896–1899*. Random House of Canada.

Bilibin, Y. A. (1964). *Selected works*. Publishing House of the Academy of Sciences of the USSR. (In Russian)

Borneman, W. (2003). *Alaska: Saga of a bold land* (1st ed.). Perennial.

Cole, T. (1991). *Crooked past: A history of a frontier gold town Fairbanks*. University of Alaska Press.

Development of the African REM deposit: white powder from rare earths. Investment portal of the Arctic zone of Russia: https://arctic-russia.ru/project/osvoenie-afrikandskogo-mestorozhdeniya-rzm-belyy-poroshok-iz-redkikh-zemel/

Eloranta, J., & Nummela, I. (2007). Finnish nickel as a strategic metal, 1920–1944. *Scandinavian Journal of History, 32*(4), 322–345. https://doi.org/10.1080/03468750701747528

Ermolaev, A. N. (2013). *Russian-American company in Siberia and the Far East (1799–1871)*. Abstract of dissertation for the degree of Doctor of Historical Sciences. (In Russian)

From the history of development of legislation on "Northern benefits". State Archives of the Murmansk Region: official website. https://www.murmanarchiv.ru/index.php/news/34-publications/327%2D%2Dl-r (In Russian)

Hickel, W. J. (2002). *Crisis in the commons: The Alaska solution*. Institute for Contemporary Studies in association with the Institute of the North.

Holland, E. (2012). How the Trumps Struck Klondike Gold. *Up Here: the voice of Canada's Far North*, September. Available at: https://www.uphere.ca/articles/how-trumps-struck-klondike-gold

Huskey, L. (2017). Alaska's economy: The first world war, frontier fragility, and Jack London. *Northern Review*, [S.l.], *44*(April), 327–346. Available at: http://journals.sfu.ca/nr/index.php/nr/article/view/639

Huskey, L., & Morehouse, T. A. (1992). Development in remote regions: What do we know? *Arctic, 45*(2), 128–137.

Innis, H. (1930). *The fur trade in Canada: An introduction to Canadian economic history*. University of Toronto Press.

It starts with the iron. LKAB: our history. Official site. https://www.lkab.com/en/about-lkab/lkab-in-brief/it-starts-with-the-iron/

Jungsberg, L., Turunen, E., Heleniak, T., Wang, S., Ramage, J., & Roto, J. (2019). *Atlas of population, society and economy in the Arctic*. https://doi.org/10.30689/WP2019:3.1403-2511

Kamenev, E.A. (2013). Alexander Evgenievich Fersman: "The most vivid impressions in my life were from the Khibiny." *Mining Magazine*, Special issue based on the materials of the Kirov Museum of History and Local Lore, 7–13. (In Russian)

Lead, J. (2009). Siberia is a strange nostalgia. Autobiography/Jonas Lead; Per. from Norv. – Moscow: Ves Mir Publishing House. 304 p. Russian translation of the edition: Lied Jonas Over deh∅ye fjelle, Jacob Dybwads Forlag, Oslo, 1946. (In Russian).

London, J. (1900). The economics of the Klondike. *The American Monthly Review of Reviews, 20*, 70–74.

Nilsson, B. (2010). Ideology, environment and forced relocation: Kiruna – A town on the move. *European Urban and Regional Studies, 17*(4), 433–442. https://doi.org/10.1177/0969776410369045

Obiko, N. (2016). Trump's family fortune originated in a Canadian gold-rush brothel. *Bloomberg*, October 26. Available at: https://www.bloomberg.com/features/2016-trump-family-fortune/

Ostroumova, V. P., Brilinskii, I. V., & Chepurnov (Eds.) (1935). *Igarka*. Igarka Municipal Council Publishing, Krasnoyarsk, Publishing House "Krasnoyarsky Rabochy". (In Russian.)

Overud, J. (2019). Memory-making in Kiruna: Representations of colonial pioneerism in the transformation of a Scandinavian mining town culture unbound. *Journal of Current Cultural Research, 11*(1), 104–123. https://doi.org/10.3384/cu.2000.1525.2019111104

Papp, J. E., & Phillips, J. A. (2007). *Like a tree to the soil: A history of farming in Alaska's Tanana Valley, 1903 to 1940*. University of Alaska Fairbanks.

Pilyasov, A. N. (2019). Arctic industry of Russia in recent decades: Industrialization, deindustrialization, industrialization 2.0. *The North and the Market: Forming the Economic, 64*(4).

Pruss, Y.V. (2017). *Geological survey of the North-East of Russia, 1931–2014*. Hunter. (In Russian)

Raw Materials Group & Robert Dewsnap. (1981). The history of the Lappland iron ore fields. *Minerals & Energy – Raw Materials Report, 1*(1), 64–69. https://doi.org/10.1080/14041048109408989

Reut, G.A. (2018). Zavenyagin's handwriting: A city-garden for a city of an industrial enterprise. In: *The Arctic 2018: International cooperation, ecology and security, innovative technologies and logistics, legal regulation: History and modernity. Proceedings of the international scientific and practical conference* May 2018, Krasnoyarsk (pp. 183–188). (In Russian)

Sjöholm, J. (2020). Moving costs: History and identity in Kiruna, Sweden. *The Architectural Review*, October 21. Available at: https://www.architectural-review.com/essays/city-portraits/moving-costs-history-and-identity-in-kiruna-sweden

Slabukha, A. (2010). Norillag architect ("group portrait" – From some statistics). Materials of the international scientific conference "MONUMENTALITÁ & MODERNITÁ. Architecture and art of Italy, Germany and Russia of the "totalitarian" period, St. Petersburg, June 30–July 2, 2010. Capital. URL: https://web.archive.org/web/20101223223714/http://kapitel-spb.ru/index.php/component/content/article/48-slabuha. (In Russian)

Slavin, S.V. (1961). Promyshlennoe i transportnoe osvoenie Severa SSSR [Industrial and transport development of the North of USSR]. Ekonomizdat, Moscow. (In Russian)

Stadius, P. (2016). Petsamo: Bringing modernity to Finland's Arctic Ocean shore 1920–1939. *Acta Borealia, 33*(2), 140–165. https://doi.org/10.1080/08003831.2016.1238177

Svensson, B., & Wetterberg, O. (Eds.). (2009). *Malmberget. Structural Change and Cultural Heritage Processes – A Case Study*. The Swedish National Heritage Board.

Urvantsev and Sotnikov: expedition 1919. Norilskfilm: Norilsk's past and present [media project]. http://norilskfilm.com/archive/statii/1919. (In Russian)

Zamiatina, N. Y. (2020). Igarka as a frontier: Lessons from the pioneer of the Northern Sea Route. *Journal of Siberian Federal University Humanitarian and Social Sciences, 13*(5), 783–799. https://doi.org/10.17516/1997-1370-0607

Zamyatina, N. Y. (2020). Northern city-base: Its special features and potential for the Arctic development. *Arctic: Ecology and Economy, 2*(38), 4–17. Doi: https://doi.org/10.25283/2223-4594-2020-2-4-17. (In Russian)

Zamyatina, N. Y., & Goncharov, R. V. (2020). Arkticheskaya urbanizaciya: fenomen i sravnitel'nyj analiz [Arctic urbanization: A phenomenon and comparative analysis]. *Vestnik Moskovskogo universiteta. Seriya 5: Geografiya [Moscow University Bulletin. 5: Geography], 4*, 69–82. URL: https://vestnik5.geogr.msu.ru/jour/article/view/718 (In Russian)

Zamyatina, N., Suter, L., Streletskiy, D., & Shiklomanov, N. (2020). Shrinking cities, growing cities: A comparative analysis of Vorkuta and Salekhard. In R. Orttung (Ed.), *Urban sustainability in the Arctic: Measuring progress in circumpolar cities* (Vol. 3, pp. 49–73). Providence [etc.], United States.

Zubov, N. (2014). *Domestic seafarers-explorers of seas and oceans*. Paulsen. (In Russian)

**Nadezhda Zamyatina** is a Associate professor from June 2021 at Lomonosov Moscow State University, Moscow, Russia. She has completed her PhD in Geography in 2001 and is a vice-director of the Institute of regional consulting (Moscow).

# Chapter 4
# Arctic as a New Strategic Region in the Soviet Union in the 1920s–1930s and Transformation of the Arctic Science

**Alexander Saburov**

## Introduction

Although global climate change, industrial pollution, and the development of indigenous peoples have been prioritized in Russia's new Arctic strategy, published in 2020,[1] the national importance of the Arctic has, for a long time, been determined by completely different factors. The region was first considered as a storehouse of natural resources, a strategic frontier in the confrontation during the Cold War, and a transport corridor of national importance (the Northern Sea Route or the Northeast Passage).

This approach to the Arctic as a specific object of management, which is highly ranked in the government priorities, was laid down in 1920s and 1930s, the first 20 years after the Bolsheviks came to power.

It was during this period that, for the first time in Russia's history, the integral governance system of the north was created. It reached extreme centralization in the mid-1930s after the creation of the Main Directorate of the Northern Sea Route (Glavsevmorput, GUSMP). During these years, using the labor of hundreds of thousands of prisoners, large-scale development of apatite-nepheline and copper-nickel ores of the Kola Peninsula,[2] Pechora coal basin, ore reserves of the Norilsk

---

[1] Ukaz prezidenta Rossiiskoi Federatsii "*O Strategii razvitiia Arkticheskoi zony Rossiiskoi Federatsii i obespecheniia natsionalnoi bezopasnosti na period do 2035 goda*". (2020, October 26). President Rossii. http://www.kremlin.ru/acts/news/64274

[2] See Belov (1959).

---

A. Saburov (✉)
Northern Arctic Federal University, Arkhangelsk, Russia
e-mail: a.saburov@narfu.ru

© The Author(s), under exclusive license to Springer Nature Switzerland AG 2022
M. Finger, G. Rekvig (eds.), *Global Arctic*,
https://doi.org/10.1007/978-3-030-81253-9_4

industrial region,[3] and gold of Kolyma[4] began. Another consequence of the Northern policy during this period was the massive construction of cities, which largely determined the modern Russian Arctic. The cities of Norilsk, Magadan, Vorkuta, Inta, Igarka, Naryan-Mar, Kirovsk, Monchegorsk and others were built in the interwar period.

During the same period, work has been underway to transform the Northern Sea Route into a regularly operating transport corridor. At the same time, the research of the Arctic in the Soviet Union was institutionalized. In 1920, two specialized organizations dedicated to northern studies were created: the Northern Scientific-Commercial Expedition (Sevekspeditsiya) and the Floating Marine Research Institute (Plavmornin). Finally, in the 1930s the North and the Arctic, with the help of state propaganda, for the first time occupied a significant place among the public.[5] The whole country was watching the news about the *SS Chelyuskin*, which involuntarily drifted in the Chukchi Sea in 1934, and then about the four Papanintsy who landed on an ice floe in 1937 to create the world's first drifting polar station, "North Pole-1". The "Arctic" interwar 1920s contained many feats and tragedies, bringing glory to some and passing into silence for many years to others after the political repressions in the Soviet Union.[6]

This chapter provides an overview of the transformation processes that took place during this period in the Soviet Arctic. These processes determined the vector of the Russian North development for many decades. It is impossible to describe and analyze in detail all of the aspects associated with the Arctic and the North reflected in hundreds of monographs and dissertations. I will give priority attention to the development of Arctic science, which developed extensively and intensively and was closely linked with other key "Arctic" areas. Research activities during this period contributed not only to the accumulation of knowledge, but also to the solution of practical tasks. Among them are foreign policy (scientific expeditions and polar stations provided presence and, as a result, sovereignty over the islands and archipelagos), economics (exploration of natural resources), development of the Northern Sea Route (hydrographic and oceanographic research, work of polar stations).

Despite relatively good knowledge of various developments in the Soviet Arctic research in the interwar period, a number of "blank spots" remain that have not been covered in the scientific literature. Almost all major generalizing monographs on the topic were published between the end of World War II and the end of the 1960s,. The fundamental works of M.I. Belov[7] are still valuable due to the thorough analysis of state policy, scientific organizations activities, and the interaction between them based on a wide range of sources, including archives. However, these publica-

---

[3] See Ertz (2004).
[4] See Zelyak (2014), Shulubina (2003).
[5] See McCannon (1998).
[6] See "Vragi naroda" za Polyarnym krugom (2007).
[7] Belov (1959), Belov (1969).

tions, like all works of the Soviet period, deny continuity between pre-revolutionary and Soviet research, and focus mostéy on the successes of Soviet science; they also poorly cover failures and political repressions.

After a relative decline in interest in this issue in the 1990s, a number of significant works devoted to the pre-war Soviet studies of the Arctic have been published since the beginning of the 2000s.[8] However, modern literature still lacks a holistic vision of the Soviet Arctic science, which would take into account new approaches and sources. In addition, several key scientific problems remain unrevealed: driving forces that determined the development of Soviet Arctic research; interaction of various actors within the framework of this process, especially among various government structures and scientific organizations.

## Terminology: "Arctic", "North", "Polar Countries", or "Zapolyarye"?

In this chapter I apply the concept of "Arctic research" to scientific works that were conducted in the waters of the seas of the Arctic Ocean, as well as on the islands and archipelagos located in the area. These include Franz Josef Land, Novaya Zemlya, Severnaya Zemlya, and Wrangel Island, which became of political and economic interest, not only scientific interest.

This understanding generally corresponds to the concept of "Arctic countries" or "Northern polar countries", which is highlighted in the first edition of the Great Soviet Encyclopedia of 1926. According to the definition, the Arctic countries "cover the Arctic Sea with the islands located there, as well as the adjacent northern outskirts of the Eurasia and America."[9] In this case, the term "country" means not a "state", but a "region".

In general, the word "Arctic" only came into widespread use at the end of the 1920s. This was a time when the Arctic Commission was established under the Council of People's Commissars (Sovnarkom, SNK), the Research Institute for the Study of the North (former Sevekspeditsiya) was transformed into the All-Union Arctic Institute, the journal *Sovetskaya Arktika* (*Soviet Arctic*) was published, and a large number of publications appeared mentioning the toponym.[10] In the 1920s, the word "Arctic" was hardly ever used and, according to the pre-revolutionary tradition, the use of the notion of "North" (for example, Northern Scientific-Commercial Expedition, Committee for Assistance to the Peoples of the Northern Outskirts, Committee of the Northern Sea Route) or "Polar" (for example, Polar Commission

---

[8] For example "Vragi naroda" za Polyarnym krugom (2007), Belyaev (2005), Dyuzhilov (2001), Koryakin (2000), Koryakin (2005), Koryakin (2007), Koryakin (2011), Krasnikova (2006), Lajus and Luedecke (2010), Lajus (2004), Saburov (2016), Vozvrashhenie imeni (2005), Zhukov (2008).

[9] Bolshaya sovetskaya enciklopediya (1926), s. 355).

[10] See Wiese (1932), Wiese (1935), Wiese (1939), Esipov (1933), Ostrovskii (1931), Ostrovskii (1933).

of the Academy of Sciences) was more common. In the mid-1930s, the toponym "Zapolyarye" also became widespread, referring mostly to the Kola Peninsula and the north of the Krasnoyarsk Krai.[11] At the same time, the well-known scientist and organizer of Arctic research, Rudolf Samoilovich, had a negative attitude towards this "innovation", believing that the term "polar countries" more correctly described this part of the world.

## Pre-revolutionary and Soviet Research of the Arctic: Continuity and Novelties

The preconditions for the rapid development of Arctic science during the Soviet period were laid before the events of 1917. The beginning of large-scale Russian exploration of the Arctic can be attributed to the first half of the eighteenth century, when the Great Northern Expedition (1733–1743) led by Vitus Bering mapped most of the north-east of Eurasia. In addition, for the first time in history, the "academic detachment" of the expedition began the natural-scientific and ethnographic study of Siberia.

For the next 150 years, there was a gradual accumulation of scientific knowledge about the region, but the active development of Arctic research in Russia began at the end of the nineteenth century. Important initiatives of this period include the establishment of a meteorological station Malye Karmakuly on Novaya Zemlya in 1882. Observations there were conducted regularly, except in 1900, 1910–1911 and 1916–1920.[12] The foundations of commercial oceanography and ichthyology in the Barents Sea were laid by the Murmansk scientific and commercial expedition (1898–1908)[13] under the leadership of Nikolai Knipovich and Leonid Breitfus. Fundamental research into the oceanography of the Kara and East Siberian Seas, as well as large-scale scientific work on the northern coast of Eurasia, were carried out by the Russian Polar Expedition of Eduard Toll (1900–1902).

To ensure the sea route to the estuaries of the Yenisei and Ob, the Arctic Ocean Hydrographic Expedition of 1894–1904 was conducted, headed by Andrey Vilkitsky, Alexander Varnek, and Fedor Drizhenko. For the same purpose, the Main Hydrographic Directorate built telegraph stations in the Yugorsky Strait, on Vaigach, at Cape Mare-Sale (Yamal Peninsula), in the port of Dikson.[14] The Arctic Ocean Hydrographic Expedition of 1910–1915, under the leadership of Boris Vilkitsky, studied the possibilities of navigation in the eastern part of the Northern Sea Route. The expedition made a voyage from Vladivostok to Arkhangelsk for the first time

---

[11] Korandei (2019).
[12] Koryakin (2000, s. 206).
[13] See Lajus (2004).
[14] Doklady na soveshhanii po izucheniyu severa Rossii (1920), s. 9–11).

in history and made the last major geographical discovery: Emperor Nicholas II Land (renamed in 1926 as Severnaya Zemlya).

A significant contribution to the study of Novaya Zemlya was made by the expeditions of Vladimir Rusanov (1909–1912) and Georgy Sedov to the North Pole (1912–1914). In 1910 Rusanov proposed measures that formed the approach for Northern Sea Route development in Soviet times: creating a polar stations network to obtain hydrometeorological data, the use of icebreakers, hydrographic support, and state organization of Arctic ocean shipping.[15] The First World War and the Revolution of 1917 did not allow these plans and ideas to be fully realized.

In the late 19th and early 20th centuries, a significant amount of scientific knowledge about the Arctic was obtained. During this period outstanding scientists and organizers of Soviet Arctic research times developed and received a "polar" experience; these included Alexey Wangenheim, Vladimir Wiese, Pavel Wittenburg, Boris Davydov, Konstantin Derjugin Nikolay Evgenov, Nikolay Zubov, Nikolay Knipovich, Alexey Lavrov, Nikolay Matusevich, Ivan Mesyatsev, Nikolay Pinegin, Rudolf Samoylovich, and Alexander Fersman. At the same time, without underestimating the importance of all the studies mentioned above, Arctic research in the Russian Empire was not carried out systematically. Such research was mostly separate, unrelated initiatives that did not provide regular exploration of the region. In a report at the Conference on the Study of the North of Russia in 1920, the outstanding oceanographer and president of the Russian Geographical Society, Yuli Shokalsky, noted that even the relatively well studied islands and archipelagos of the European part of the Arctic "Novaya Zemlya, Kolguev, Vaigach, Matveev and Dolgiy Islands were never surveyed correctly, only partially."[16]

## 1920s: Scientific Community as the Engine of the Arctic Idea

Arctic research, like the Soviet development of the Arctic in general during the interwar period, can be chronologically split into two main stages: (a) 1920–1931 and (b) 1932–1941. The year 1932 is recognized as the beginning of centralization of the Arctic development in the USSR in connection with creation of Glavsevmorput. Since that time, maintenance of the Northern Sea Route became the main goal of the Arctic science. Research activities became mostly connected with the work of the polar stations network.

The dominant role of the state in organizing of the Arctic research was not typical for the 1920s. This period was marked by a strong intensification of the scientific community activity, resulting in a large number of initiatives, to a different extent supported by the country's leadership. However, all of these initiatives were united

---

[15] Koryakin (2005, s. 204–205).
[16] Doklady na soveshhanii po izucheniyu severa Rossii (1920), s. 8).

by the fact that the Arctic region began to be considered strategically significant. The second common idea was using science to solve problems of national importance in the region.

*Firstly, the exploration of the Arctic was seen as a tool for the development of natural resources in the North, as well as for the establishment of a sea route between the European part of Russia and Siberia.*

As an example, we can cite the draft report of the Arctic Ocean Hydrographic Expedition to the Maritime Collegium dated May 8, 1918, which noted: "The economic and industrial crises and famine in European Russia are prompting the most urgent measures to use all the vital forces of the country … there is a need for widespread use of sea routes to the mouths of the large Siberian rivers and the development of fishing and hunting in the Arctic Ocean."[17] The authors of the document requested the ships for the Hydrographic Expedition from the Maritime Collegium, noting that "while sailing off the northern shores of Russia, the sailors have so far been deprived of the most elementary cultural state assistance, which consisted in the appropriate hydrographic equipment of the sea."

The submission of the Main Directorate of the General Staff to the People's Commissariat for Maritime Affairs of April 20, 1918 indicated that "the establishment of urgent steamship sailing to the estuaries of the Ob, Yenisei and Lena rivers will cause a general cultural upsurge and revitalization of the commercial and industrial life of Western Siberia and facilitate the exploitation of huge natural resources of Siberia by exporting them to European markets."[18] When considering this document, attention is drawn to the fact that, by April 1918, the Main Directorate of the General Staff had already been closed down. Thus, the submission was directed from a no-longer-existing department, which can be explained by the complex processes of the Soviet public administration system formation. Nevertheless, these examples show that the ideas of sea communication between the European part of Russia and Siberia did not disappear, but were promoted among the authorities after 1917.

The idea of using the natural resources of the north helped justify establishment of Sevekspeditsiya in 1920. The "enormous importance of northern industries as an inexhaustible source of food for the entire country" was the main point in the corresponding document of the Special Food Commission of the Northern Front.[19]

*The second task of research activities was assisting in solving foreign policy issues in the Arctic.* This question was no less relevant, since the sovereignty of the Russian Soviet Federative Socialist Republic (RSFSR; after 1922, USSR) in the north was repeatedly challenged in the 1920s. According to the results of the Treaty of Tartu between the RSFSR and Finland in 1920, large territories on the Kola Peninsula were passed to the new Finnish state. During the Civil War, Norwegian,

---

[17] Russian State Archives of the Navy (further – RGAVMF). F. P-898. Op. 1. D. 3. L. 8.
[18] RGAVMF. F. P-898. Op. 1. D. 8. L. 1.
[19] Koryakin (2007, s. 31).

Finnish and English ships in the White Sea led uncontrolled fishing and hunting marine mammals, mainly harp seals. In this regard, in 1921 Sovnarkom adopted a decree prohibiting foreign citizens to conduct fishing and hunting in the White Sea and in a 12-mile zone from the coast of the Arctic Ocean.[20] The Soviet authorities faced similar challenges in the Asian part of the Arctic. In 1921 and 1923, Canadian anthropologist and polar explorer Viljalmur Stefansson organized the colonization of Wrangel Island. In response to these actions, in 1924 an expedition on the gunboat "Red October" took out the settlers from the island and raised the Soviet flag. In 1926, a Soviet expedition with settlers headed by the famous polar explorer Georgy Ushakov was sent to avoid incidents of this kind and to ensure a permanent presence on Wrangel Island.

Document analysis shows that regular research was considered a tool for ensuring the permanent presence of the Soviet Russia in the Arctic. The explanatory note to the establishment of Plavmornin noted that "a comprehensive and systematic study of the North Sea (*Arctic Ocean*) and its islands at the present time, when the North Sea is the only outlet of the Republic to the global ocean, is especially important and urgent and has more than just scientific and economic significance, but also political consideration, specifically protection of the North."[21] The authors of the above-mentioned draft report of the Arctic Ocean Hydrographic Expedition pointed out that development of the Northern Sea Route is necessary for the development of all of Siberia and is of strategic importance for the armies "operating against China and Japan."[22]

In the early 1920s, the probability of sovereignty loss over the Novaya Zemlya, an archipelago visited by Russians before Willem Barents, was seriously considered by the Soviet leadership. Tyko Vylka, the first chairman of the Novaya Zemlya Island Council, stated that in 1920 "… no one knew will own Novaya Zemlya. Some speak for the Norwegians, some for the Bolsheviks." At the meeting of the Standing Commission on Scientific Expeditions of the Academy of Sciences, Sergei Oldenburg said that "it is necessary to take all measures to prevent the possibility of peaceful conquest of Novaya Zemlya by the Norwegians, and science should play a huge role here."[23] In these circumstances, the Norwegian geological expedition to the Novaya Zemlya under the direction of Olaf Holtedahl in 1921 was regarded as a threat to the Soviet sovereignty.

---

[20] Dekret SNK RSFSR "*Ob okhrane rybnykh i zverinykh ugodii v Severnom Ledovitom okeane i Belom more*" (1921, May 24).

[21] Organizaciya nauki v pervye gody Sovetskoj vlasti (1968), s. 277).

[22] RGAVMF. F. P-898. Op. 1. D. 8. L. 1.

[23] Wittenburg (2003, s. 68).

## The First Arctic Research Institutions

Development of Arctic research in the early years of Soviet Russia was largely associated with the establishment of a new type of organizations: research institutes specializing in studies of the Arctic and the North. Primarily, this applies to Sevekspeditsiya and Plavmornin.

### *Northern Scientific-Commercial Expedition, Institute for the Study of the North, All-union Arctic Institute*

The Northern Scientific-Commercial Expedition was created on March 4, 1920 under the Supreme Council of the National Economy (VSNKh). The project for its creation was elaborated in February 1920, while the Civil War in the north was still continuing. An interdepartmental meeting was held at the Special Food Commission of the Northern Front in Vologda. A proposal was formulated to "create an extra-departmental body in charge of all issues of scientific and commercial research of the Northern Territory. "Researchers of the north R. Samoilovich, N. Kulik, A. Zhilinsky, S. Kerzelli were the main developers of this initiative, which received V. Lenin's approval later.[24] The decision of the VSNKh Presidium stated: "Samoylovich, Kulik and Kertselli to be granted 50 000 000 rubles for the organization of the Northern Scientific-Commercial Expedition."[25]

The Sevekspeditsiya works program was approved by the Academic Council, headed by the President of the Russian Academy of Sciences Alexander Karpinsky and prominent scientists and public figures including A. Fersman, N. Knipovich, K. Deryugin, Yu. Shokalsky, and A. Gorky.[26]

On May 27, 1920, the first constituent meeting took place, where R. Samoilovich was approved as the head of the organization.[27] Samoilovich, the leader of the expedition and its successors until 1938, was a member of V. Rusanov's expeditions. Together with P. Wittenburg, in 1913 he participated in the extraction of the first Russian coal in Spitsbergen. His leadership of the Soviet expedition on the icebreaker "Krasin" in 1928, which saved the crew of the crashed airship *Italia* brought world fame for Samoilovic. The successful completion of this voyage greatly increased the prestige of the Soviet Union in the international arena. Samoilovich was awarded the newly established Order of the Red Banner of Labor. Samoilovich, according to numerous sources, proposed the plan of an expedition to Novaya Zemlya with the subsequent creation of a polar station.[28] Until the last years of his

---

[24] Organizaciya nauki v pervye gody Sovetskoj vlasti (1968), s. 313–315).
[25] Koryakin (2007, s. 31).
[26] Raboty otryadov Sevekspedicii (1922), s. 4).
[27] Belov (1959, s. 126).
[28] Koryakin (2007, s. 128).

life, he combined the leadership of a large scientific organization with research. The last expedition of R. Samoilovich was the famous high-latitude expedition of three icebreakers *Sadko*, *Sedov*, and *Malygin* in 1937–1938, which had a forced wintering in the Arctic. Samoilovich was unanimously chosen as the head of the wintering of three ships, where he brilliantly showed himself as a leader.[29]

According to the Regulations, the main task of the Sevekspeditsiya was to conduct "scientific and technical research of the natural productive forces of the Russian North for the purpose of their best practical use."[30] Activities of the Sevekspeditsiya included a wide range of works: research on fish, fossil and forest resources, reindeer husbandry, crafts, as well as hydrological, meteorological, ethnographic, and other scientific works of a general geographical nature. Applied research and participation of specialists from all major natural disciplines were the main peculiarities of the Sevekspeditsiya works.[31]

Already, the first results of the Sevekspeditsiya works were recognized as having national importance at a meeting of the VSNKh Board of the Scientific and Technical Department on December 13, 1920.[32] Sevekspeditsiya conducted large amount of research on the islands in the European part of the Arctic in 1920s. Expeditions to Novaya Zemlya under the leadership of R. Samoilovich eliminated many "blank spots" on the archipelago map, and extensive geological material was collected.[33] Samoilovich pointed out in the mid-1920s that the study of Novaya Zemlya "gives us not only interesting scientific results, but also more and more economically secures this remote outskirts for the USSR."[34]

Activities of the Northern Scientific-Commercial Expedition were not limited to the Arctic islands and archipelagos. In 1921, 23 scientific groups with a total of 287 people worked in the European north of the country.[35] The Khibiny mountain-geological group, led by A. Fersman, discovered the unique Monchegorsk copper-nickel deposit and the Khibinsky apatite deposit on the Kola Peninsula. Large oil fields of the Timan-Pechora oil and gas province were discovered in the late 1920s near Ukhta. At the same time, high-calorific coal deposits of the Pechora coal basin were discovered in the Vorkuta region. Due to the significant expansion of its operations in 1925, Sevekspeditsiya was transformed into and Institute for the Study of the North.[36]

---

[29] Samoilovich Rudolf Lazarevich (01(13).09.1881–04.03.1939). (n.d.). Arkticheskaia Toponimika. Retrieved December 31, 2020, from http://www.gpavet.narod.ru/Names3/samoilovich.htm

[30] Central State Archive of Scientific and Technical Documentation of St. Petersburg (further – TSGANTD SPb). F. 369. Op. 1-1. D. 1. L. 1.

[31] Raboty otryadov Sevekspedicii (1922), s. 3–4).

[32] Dyuzhilov (2001), s.31).

[33] Koryakin (2007, s.88).

[34] Kanevskij (1982, s. 37).

[35] Dyuzhilov (2001), s.31).

[36] TSGANTD SPb. F. 369. Op. 1-1. D. 8. L. 1.

## Floating Marine Research Institute, State Oceanographic Institute

The idea of creating the Floating Marine Research Institute appeared in 1920 within a group of Moscow biologists headed by I. Mesyatsev, who had significant influence in the All-Russian Communist Party (Bolsheviks). He directly addressed this proposal to the Council of People's Commissars.[37] Once the idea was approved in Sovnarkom, he addressed the People's Commissariat of Education (Narkompros) and, on March 10, 1921, a corresponding decree was signed. The purpose of Plavmornin's establishment was "a comprehensive and systematic study of the Northern Seas, their islands, coasts, which are currently of national importance." According to the decree, the area of operation of the new institution was determined by the "Arctic Ocean with its seas and estuaries, islands and adjacent coasts of the RSFSR in Europe and Asia."[38] The explanatory note to the decree emphasized: "The scientific sector of the Narkompros recognizes the organization of the Floating Marine Research Institute as an urgent task of paramount state importance."[39]

Oceanographer A. Rossolimo became the head of the Committee of the Floating Marine Research Institute, the governing body of Plavmornin. The leadership also included I. Mesyatsev, who headed the Institute's expeditions for many years, and other world-renowned oceanologists in L. Zenkevich, S. Zernov, and N. Zubov.

In 1929, Plavmornin was merged with the Murmansk Biological Station and transformed into the State Oceanographic Institute,[40] and in 1934 into the Polar Research Institute of Marine Fisheries and Oceanography.

In the period from 1921 to 1932 Plavmornin's work covered the entire Barents Sea, studying depths, soils, temperature, salinity, and data on marine biology, particularly on the distribution and behavior of fish, which was especially important for the development of the fishery industry. The Floating Marine Research Institute held expeditions to Novaya Zemlya in 1921, 1923, 1925, and 1926, mainly conducting topographic surveys, hydrological, and hydrobiological works, collecting botanical and biological samples.

## Arctic Ocean Hydrographic Expedition, Northern Hydrographic Expedition

Hydrographic works along the Northern Sea Route were provided by the structures of the Main Hydrographic Directorate of the Naval Forces. In 1920, the Arctic Ocean Hydrographic Expedition of the Main Hydrographic Department was

---

[37] Lajus (2004, s. 163).
[38] Organizaciya nauki v pervye gody Sovetskoj vlasti (1968), s. 276–277.
[39] Vasnetsov (1974).
[40] Organizaciya sovetskoj nauki (1974, s. 311).

launched. It was regarded as a continuation of a successful same-name pre-revolutionary expedition led by B. Vilkitsky. The decision to hold it was made by the Council of People's Commissars back in 1918, but it did not take place due to the outbreak of the Civil War. In 1920–1921 the expedition carried out hydrographic works, including off the coast and in the straits of Novaya Zemlya. In 1922, due to financial difficulties, funding for the Main Hydrographic Directorate was reduced, and as a result, activities of the Arctic Ocean Hydrographic Expedition were terminated.

Hydrographic research in the Soviet Arctic continued a year later. Working under a different name, the Northern Hydrographic Expedition[41] since 1923 conducted hydrographic surveys off the coast and islands of the Arctic ocean annually until 1932.

Construction of the Matochkin Shar polar station at Novaya Zemlya in 1923 was the expedition's most significant achievement. It was caused by the need to provide information on sailing conditions (meteorological reports, ice conditions) on the way from the European part of the USSR to the mouth of the Ob and Yenisei. In addition, the installation of the station facilitated the work of hunters on Novaya Zemlya. Since 1883 there had already been a polar station in the Malye Karmakuly on the archipelago, which was built for the First International Polar Year (IPY). However, it was located inside the southern island of Novaya Zemlya and could not provide valuable information for navigation.

At the meeting of the Presidium of the State Planning Committee (Gosplan) on November 16, 1922, the creation of a polar station on Novaya Zemlya was recognized as an important way of securing the archipelago for the Soviet Russia.[42] N. Matusevich, who headed the Northern Hydrographic Expedition, noted in 1923 that "the construction of radio stations, drawing up of new maps, the setting of identification marks and the constant navigation of Russian ships off the shores of Novaya Zemlya for several years would be proof of Russia's economic relationship to Novaya Zemlya, and will serve … the fact of recognition of its sovereignty over Novaya Zemlya by foreigners".[43]

As a result of 5 years of work of the Northern Hydrographic Expedition, navigational maps of the entire coast of the Barents Sea, the western coast of the southern island of Novaya Zemlya and the Matochkin Shar Strait were created. In 1930, N. Evgenov created the first pilot chart of the Kara Sea and Novaya Zemlya.[44] All important navigational places were equipped with beacons and distinctive signs.[45] In 1931, the Hydrographic Department organized the construction of a polar station at the northern tip of Novaya Zemlya – Cape Zhelaniya.

---

[41] RGAVMF. F. P-548. Op. 1. D. 22. L. 75.

[42] Kratkij otchet o deyatelnosti otdelnogo severnogo gidrograficheskogo otryada (1924), s. 3).

[43] Belov (1959, s. 117).

[44] Belov (1959, s. 121).

[45] Pinegin (1935, s. 49).

Some sources note that research on the Novaya Zemlya in 1920s was conducted mainly for the purpose of scientific reconnaissance, without practical application, and the shores were surveyed insufficiently for navigation.[46] In particular, R. Samoilovich pointed out the inaccuracy of the maps of the Main Hydrographic Directorate during his expeditions to Novaya Zemlya.[47] Some works have also noted that geological research was fragmentary.[48] Nevertheless, all these works laid the foundation for large-scale expeditions of the 1930s and collected extensive material that formed the basis for the first fundamental publications, prepared over the next decade.

## Cooperation, Competition, and Coordination Challenges in the Arctic Science

Although the main part of the Arctic research in 1920s was organized and carried out by the three above-mentioned institutions, other scientific institutions took part in the expeditions. Mostly they attached staff to expeditions of Sevekspeditsiya, the Northern Hydrographic Expedition, or Matochkin Shar polar station winterings.

Given the large number of scientific institutions that worked in the Arctic, the question of effective coordination inevitably arose. According to Otto Schmidt, until 1928–1929, their activities "partly had the character of fragmentation."[49] Already at the level of statutory documents, some contradictions can be found: several organizations at once would take a leading role in the studies of the north. According to the Sevekspeditsiya Regulation, it was entrusted with the leadership and coordination of research activities, which was carried out by other bodies and institutions: "all Sevekspeditsiya orders concerning the regulation and organization of scientific and applied research in the Russian North are mandatory and are subject to unswerving execution, both for private and government public organizations.".[50] The decree on the creation of the Floating Marine Research Institute stated that it was to conduct "a comprehensive and systematic study of the Northern Seas, their islands, coasts, which are currently of national importance". Coordination functions were also indicated in the statutory documents of the Academy of Sciences Polar Commission.[51]

There was also no single government body responsible for polar research issues. Sevekspeditsiya was a subordinate organization of the Supreme Council of the National Economy. The Main Hydrographic Directorate was subordinate to

---

[46] Wiese (1935, s. 70–71).

[47] Novaya Zemlya (1929), s. 35).

[48] Geologicheskie issledovaniya Novoj Zemli i Vajgacha (1936, s. 37).

[49] Za osvoenie Arktiki. (1935), s. 6).

[50] Dyuzhilov (2001), s.27).

[51] Krasnikova (2006, s. 54).

the Navy, Floating Marine Research Institute – to the People's Commissariat of Education.

Moreover, relationships between scientific institutions were in some cases competitive. In the early 1920s there was a struggle between Sevekspeditsiya and Plavmornin for the Murmansk biological station,[52] vessels *Delphin* and *Persei*.[53]

At the same time, one cannot fail to note close interaction between various scientific organizations. At least seven Arctic expeditions between 1920 and 1931 were carried out with the participation of several research institutions. Work of the Matochkin Shar polar observatory on Novaya Zemlya was organized jointly by the Main Hydrographic Directorate, the Main Physical Observatory, and various structures of the Academy of Sciences.[54]

Despite the individual episodes of tough competition, we can also refer to the consolidation of the scientific community. This was also reflected in the fact that reputable scientists were involved in the work of various organizations. A. Karpinsky, president of the Academy of Sciences, was the chairman of Sevekspeditsiya and a member of the Academy of Sciences Polar Commission. R. Samoilovich, the head of Sevekspeditsiya, was also a member of the Polar Commission for more than 15 years. Y. Shokalsky, N. Knipovich, A. Fersman, N. Matusevich, and P. Wittenburg took part in the work of several institutions in a similar way.

One of the largest and most significant collaboration cases was preparation of the Second International Polar Year (IPY-2). In 1929, under the Academy of Sciences Polar Commission, the Committee for the Preparation of the Second IPY was formed. It included representatives of the Academy of Sciences (Fersman, Tolmachev, and Wittenburg), the Institute for the Study of the North (Samoilovich), and the Main Geophysical Observatory (N. Rose), and N. Knipovich.[55]

An important aspect of cooperation was mutual assistance with providing scientific equipment among institutions. Scientific organizations often jointly published results of research activities. Mutual assistance was also provided in relation to transportation and fuel.[56]

## Late 1920s: Sovnarkom Arctic Commission and the First Attempts to Centralize Arctic Science

The first attempts of centralized planning of Arctic research in the Soviet Union date back to the late 1920s. In the resolution of the Council of People's Commissars dated July 31, 1928, "On strengthening research work in the Arctic possessions of

---

[52] Lajus (2004, s. 164).
[53] Lajus (2004, s. 174).
[54] Kratkij otchet (1924), s. 4).
[55] Lajus and Luedecke (2010), p. 150–151).
[56] Saburov (2016, c. 46).

the USSR" it was ordered "to form a commission under the Council of People's Commissars of the USSR for the organizational and financial study of a five-year research plan in the Arctic possessions of the USSR...".[57] The created Arctic Commission was entrusted with developing a plan to set up polar stations on Franz Josef Land, Novaya Zemlya, and Severnaya Zemlya. Its tasks also included the construction of mooring masts for Arctic research expeditions on airships.

Construction of the first polar stations on Franz Josef Land and Severnaya Zemlya was connected with the fact that these territories remained no man's land, even after the decree of the Central Executive Committee Presidium "On declaring the lands and islands in the Arctic Ocean as territory of the USSR" in 1926.[58] Indeed, a Norwegian expedition was organized in 1929 with the aim of private occupation of Victoria Island and Franz Josef Land.[59] However, it was not able to reach the coast due to difficult ice conditions and the absence of an icebreaker. Information that the Norwegian ship was going to approach the polar archipelago prompted the SNK to allocate 207,000 rubles for the setting up of a geophysical observatory and a radio station in the archipelago in 1929.[60] Main organization works were carried out by the Institute for the Study of the North, headed by Samoilovich and his deputy V. Wiese. The expedition leader was Otto Schmidt, the future head of Glavsevmorput – his first Arctic experience. The expedition built the polar geophysical observatory on Franz Josef Land in Tikhaya Bay. It was the northernmost polar station at that time, which carried out a wide range of scientific works. Although Norway did not officially recognize Soviet sovereignty over Franz Josef Land until 1955, no foreign ship entered the waters around the archipelago without permission from the USSR since 1930, except the Second World War period.[61]

Creation of the Arctic Commission was one of the turning points in the organization of Soviet Arctic research. Firstly, it was the first attempt to centralize not only research activities, but also the development of the Arctic as a whole. Since this moment, the Arctic has received attention at the highest state level. Secondly, Arctic Commission accumulated numerous proposals of new approaches and methods of work, including active use of icebreakers and aviation. The Commission considered issues related to the introduction of new instruments for polar observations. In particular, it recommended the use of the radiosonde of Professor P. Molchanov for studying the upper atmosphere. Schmidt wrote in his memoirs: "The commission of S. Kamenev (*head of Commission*), where scientists, sailors, and pilots participated, played a huge role. For the first time, the Arctic work was organized in a systematic

---

[57] Postanovlenie SNK SSSR, "Ob usilenii nauchno-issledovatelskoj raboty v arkticheskih vladeniyah Soyuza SSR" (1928, July 31).

[58] This decree declared the territory of the USSR all open and undiscovered lands and islands in the Arctic Ocean, located between 32 and 168 degrees east longitude. See Postanovlenie Prezidiuma TSIK SSSR, "Ob obyavlenii territoriej Soyuza SSR zemel i ostrovov, raspolozhennyx v Severnom Ledovitom okeane" (1926, April 15).

[59] Drivenes (2004, s. 212).

[60] Zhukov (2008, s. 249).

[61] Belyaev (2005, s. 129).

manner; for the first time research on land, at sea and in the air was combined."[62] The Commission contributed to the formation of permanent sources of funding for the Arctic science. In January 1929, at the suggestion of the Commission, Sovnarkom adopted a resolution on the deduction of 1.5–2.25 percent of income from fisheries, trade, and transport operations beyond the Arctic Circle for research financing.[63]

The Arctic Commission started centralization of the Soviet Arctic research. The chairman of the Commission, S. Kamenev, wrote: "It seems to me that the moment is ripe when some of the scattered institutions, whose task is to study Arctic issues, should now be united. In particular, it would undoubtedly be useful to unite the Institute of the North with Plavmornin, as well as with scientific organizations in Murmansk (biological station) and Arkhangelsk. In my opinion, the work of these organizations should be strictly coordinated and unified, this will only introduce a certain systematic and useful work …" Kamenev saw Schmidt at the head of the joint institute.[64] Despite the VSNKh's plans to close down the Institute for the Study of the North, Samoilovich managed to convince V. Kuibyshev, VSNKh chairman, Kamenev, and SNK chief administrator N. Gorbunov to create a new central structure responsible for the Arctic research on the basis of his institute.[65] As a result, in 1930, the All-Union Arctic Institute (henceforth, Arctic Institute, VAI) was created under the USSR Central Executive Committee as "the central organizing and leading research institution for a comprehensive study of the polar countries of the USSR."[66]

Arctic Commission did not fully manage to solve its main task, which was to realize its ambitious research plan in the Soviet Arctic realm. The draft plan was prepared by a group of scientists led by Fersman. The main activities within its framework were the creation of a polar stations network, the construction of an ice-class vessel, and the construction or purchase of an airship.[67] However, the review of the plan dragged on for 5 months, and it was not implemented over the 2 years of the Arctic Commission existence, apart from setting up polar stations.

The issue of Arctic research coordination also was not finally resolved. The Second Conference on the Productive Forces Studies in the North in 1931 still noted the presence of parallelism in the research held by institutions working in the North.[68] The Transport Section of the Severnyi Krai planning commission noted that, "there is no link between Ubeko-Sever and the Hydrographic Expedition, on the one hand, and the Institute for the Study of the North, on the other" when performing hydrographic work in the north.[69]

---

[62] Bulatov (1989, c. 45–46).
[63] Belov (1959, s. 348).
[64] Zhukov (2008, s. 260–261).
[65] Zhukov (2008, s. 276–277).
[66] TSGANTD SPb. F. 369. Op. 1-1. D. 48. L. 1.
[67] Zhukov (2008, s. 261–262).
[68] Gorbatskii (1931, s. 98).
[69] RGAVMF. P-739. Op. 1. D. 129. L. 17.

The last meeting of the Arctic Commission took place on June 13, 1930. It was finally closed down only after establishment of the Main Directorate of the Northern Sea Route in 1932. Although its existence gave a powerful impetus to the Soviet Arctic research, it did not become a permanent body coordinating development of the Arctic in USSR. In my opinion, the main reason for termination of the commission's activities was connected with the fulfillment of original plans to secure the Arctic archipelagos for the USSR through construction of polar stations. The absence of other tasks included in the Arctic Commission responsibility led to its gradual extinction.

By order of the Gosplan of June 3, 1932, No. 177, a Council for the Study and Development of the North was created under the Gosplan Presidium to "unify and plan all work on the exploration and study of the North, conducted by research and economic organizations and drawing up a plan for research and survey in the North for the next five years." Also, the council was supposed to summarize the results of research of past years and conduct an analysis of the natural-productive forces of the Northern regions and national districts with an outline of their economic profile in the second five-year period.[70] However, its activity was actually terminated less than a year after the creation of Glavsevmorput.

## Centralization of the Soviet Arctic Science in 1932–1941

The 1930s were characterized by centralization and growing state participation in the development and study of the Arctic. The creation of the Main Directorate of the Northern Sea Route in 1932 became the high point of this process.

The main reason for such attention to the Arctic was the potential military threat that arose after the occupation of north-eastern China by Japan in 1931. The North-East Passage began to be viewed as a strategic transport route, considering the remoteness of the Far East from the European part of the USSR and the tragic experience of the Russian-Japanese war of 1904–1905. The new place of the Arctic region in defensive plans is also confirmed by the creation in 1933 of the Northern Military Flotilla (since 1922 there were no large navy units in the North).[71] Military infrastructure, particularly airfields and naval bases, was developed from 1933 onwards, when the construction of the Polyarnoye naval base started in the Yekaterininskaya harbor.[72] In 1936, warships of the Baltic Fleet made the first passage to the Far East along the Northern Sea Route with ice escort from the GUSMP.[73]

The need to increase the country's defense capability in the region and the general centralization processes of the 1930s greatly influenced the Arctic research

---

[70] TSGANTD SPb. F. 369. Op. 1-1. D. 80. L. 94.
[71] Kozlov and Shlomin (1983).
[72] Kozlov and Shlomin (1983).
[73] Kozlov and Shlomin (1983).

organization. The Main Directorate of the Northern Sea Route was established under Sovnarkom by a decree dated December 17, 1932. The new super-organization was tasked with "finally laying the Northern Sea Route from the White Sea to the Bering Strait, equipping this route, keeping it in good condition and ensuring the safety of navigation along this route."[74] To solve this problem, all existing meteorological stations and radio stations located on the coast and islands of the Arctic Ocean were transferred to the GUSMP. In an expanded decree of December 20, 1932, the Glavsevmorput was entrusted with the task "to form coal bases in the necessary places and to produce elementary port equipment …, to form aircraft bases in the necessary places for constant observation of ice and assistance in escorting ships …, to develop the type of icebreakers required for the Northern Sea Route …, develop a plan for aerial photography of the coast of the Arctic Ocean and correcting maps."[75] During 1932–1933, Glavsevmorput concentrated almost all functions and resources for the Arctic research: it was given powers of the Hydrographic Directorate, the Arctic Institute, all polar stations, a large icebreaker fleet and other ships, and aircraft for polar flights.

From the mid-1930s, GUSMP was expanded to a large industrial and trade organization that united all the processes of the Arctic development. According to Yu. Zhukov, the excessive concentration of functions, primarily economic, in the hands of one organization became one of the main reasons for the 1937 failure. Almost the entire Soviet polar fleet was stuck in the Arctic Ocean ice in the winter of 1937.[76] In 1938, the downsizing of the GUSMP began; as a result, only the initial functions of supporting the Northern Sea Route remained under its responsibilities.

The head of Glavsevmorput Otto Schmidt (09/18/1891–09/07/1956) the organization from the moment of its creation until the end of the 1930s. Until 1929, Schmidt was not associated with the north, and the "Arctic" stage of his career began unexpectedly for him. Schmidt recalled this moment: "While watching the film about last year's Pamir expedition in March N. Gorbunov told me about the expedition to Franz Josef Land and suggested me as its leader… In May I agreed, received the appointment of the Sovnarkom, and in June I came to Leningrad, the Institute for the Study of the North, where agreed on the main issues with R. Samoilovich and V. Wiese."[77]

In 1930–1932 Schmidt was director of the All-Union Arctic Institute, and in 1932 he was appointed head of the GUSMP. Samoilovich became an Arctic Institute director again with Wiese as his deputy. In 1932 and in 1933, Schmidt led the famous expeditions on the ships *A. Sibiryakov* and *Chelyuskin*. In 1937 Schmidt led an air expedition to organize the North Pole-1 drifting station (awarded the title of Hero of the Soviet Union) and, in 1938, a rescue operation to remove station

---

[74] Postanovlenie SNK SSSR N 1873 "Ob organizacii pri Sovete narodnyh komissarov Soyuza SSR Glavnogo upravleniya severnogo morskogo puti" (1932, December 17).

[75] Novikov (1956, s. 116).

[76] Zhukov (2008, s. 348).

[77] Koryakin (2011, s. 74–75).

personnel from the ice floe. Unlike many polar explorers, he escaped purges after the failure of the Arctic navigation in 1937–1938, but was removed from his position. From 1939 to 1942 he served as vice-president of the USSR Academy of Sciences.

During the 1930s the role of the highest bodies of state power and the Communist Party in decision-making regarding Arctic research grew significantly. This is evidenced, firstly, by GUSMP's position in the state hierarchy: its status corresponded to the level of the People's Commissariat, and it was directly subordinate to the SNK chairman. Secondly, in 1934, political departments were created in GUSMP, and the Main Political Directorate of the GUSMP had the status of the Communist Party Central Committee department. Functions of the political departments included selection, training, and placement of personnel; organization of ideological and educational work aimed at improving the political and labor activity of polar explorers; strengthening labor discipline; and political and general education.[78] All personnel affairs in the GUSPM system were resolved only after approval with political bodies.[79]

The setting of a new task in the Arctic marked new approaches to research planning. The leadership of Glavsevmorput formulated an orientation towards the applied research. On March 31, 1933, the GUSMP Board decided: "In connection with the industrial development of the Arctic and transformation of the Northern Sea Route into a normal waterway, the Arctic Institute needs to restructure its future work radically, by making emphasis on applied character. ... Each expedition should have very specific practical tasks, clear target settings and upon completion… give certain practical conclusions that contribute to the further development of the polar spaces of the USSR."[80]

The 1930s became the time of the largest expansion of the polar stations network, which ensured not only the presence of the USSR in remote territories, but were also rescue points in the Arctic and became a springboard for studying high latitudes. The active construction of new polar stations in the Soviet sector of the Arctic began in 1932 as part of the Second International Polar Year (1932–1933). Until 1917, 10 polar stations operated in the Russian sector of the Arctic, and from 1932 to 1934, 28 new stations were constructed.[81]

Centralization of the Arctic development and creation of the Glavsevmorput determined serious changes in the system of scientific institutions. The All-Union Arctic Institute was included in the GUSMP. Hydrographic works in the Arctic passed from the Navy Hydrographic Directorate to the Hydrographic Directorate of Glavsevmorput, created in 1933. From January 1, 1935, the Arctic Department of the Central Hydrometeorological Service Directorate became part of the Glavsevmorput Polar Directorate as the "Department of Weather and Ice

---

[78] Bulatov (1989, s. 58).
[79] Bulatov (1989, s. 61).
[80] TSGANTD SPb. F. 369. Op. 1-1. D. 83. L. 52.
[81] Schmidt (1934, s. 6).

Information."⁸² The general management of all polar stations was transferred to the GUSMP Department of Polar Stations, while the methodological and scientific leadership remained with the Arctic Institute. A number of institutions were abolished. In 1933, the Arctic Commission was eliminated. In 1936 the same happened with the Academy of Sciences Polar Commission. Thus, almost all Arctic scientific activities were carried out within the Main Directorate of the Northern Sea Route. The exception was research in the field of marine biology carried out by the State Oceanographic Institute (from 1933, the Polar Research Institute of Marine Fisheries and Oceanography).

In order to better coordinate Arctic research, all scientific work in the Soviet sector of the Arctic, including the work of polar stations, was united by the Arctic Institute in Leningrad. According to the regulation of June 9, 1933, it became "the organizing and leading research institution for a comprehensive study of the polar territories of the USSR."⁸³ In the 1930s there was a significant increase in the number of the institute staff. In the early 1920s, the permanent staff of the Arctic Institute was not much more than 10 people. In 1933 there were 251 employees at the Institute, including the Arkhangelsk, Far Eastern, and Yakutsk branches.⁸⁴ The Council of the Arctic Institute consisted of representatives of various institutions and government bodies.⁸⁵

## Centralization and Efficiency?

State participation in research planning, centralization of Arctic science, the concentration of resources, and the setting of a common goal ensured success in the development of the region in the 1930s. Observational data from polar stations were used not only for solving local research problems, but also for a systematic study of the Soviet sector of the Arctic. The hydrological and ice regime of the Arctic seas was studied. Materials of hydrological observations were published in the "Catalog of basic hydrometeorological data", "Yearbook of the Arctic Ocean tides" "Ice yearbooks", and "Atlas of ice cover". The materials of meteorological observations were published in the "Meteorological observations yearbooks".

The methodology for seasonal and monthly forecasts for the Arctic developed by G. Wangenheim became an important practical result. The features of the Arctic climate were clarified. In particular, the myth of the large amount of precipitation in the region was refuted. The process of the central part of the Arctic warming was explained.⁸⁶ The Arctic Institute's regular monitoring afforded grounds for conclu-

---

[82] Dzerdeevskij (1935, s. 32).
[83] TSGANTD SPb. F. 369. Op. 1-1. D. 78. L. 1.
[84] TSGANTD SPb. F. 369. Op. 1-1. D. 68. L. 137.
[85] Samoilovich (1931), s. 3).
[86] Belov (1969, s. 368).

sion on the fast climate warming of the Arctic, which began around 1920. The area of the Arctic sea ice was reduced approximately two times from 1920 to 1940.[87] Since 1936 Arctic Institute began to systematically publish "Materials on the climatology of the USSR polar regions". Accumulated materials of magnetic field observations in the Arctic made it possible to publish magnetic maps in 1944.

Despite these successes, the potential of science in the region was not fully used. Firstly, there was no general research plan for GUSMP,[88] and the work plans of the Arctic Institute were loosely linked to the activities of other Glavsevmorput directorates. For example, a detailed work plan was developed for the institute for the period 1933–1937. It included more than 200 pages, with a detailed indication of work, resource requirements, and personnel. However, it almost did not consider issues of interaction with other scientific and economic organizations working in the Arctic.[89]

Areas of responsibility were not clearly delimited within Glavsevmorput. Despite the fact that the Arctic Institute was declared "the organizing and leading research institution," the GUSMP Polar Stations Directorate, Mining and Geological and Hydrographic Directorates were not subordinate to the Arctic Institute in terms of research and were not obliged to align their scientific work. In 1936, Nikolai Urvantsev fairly noted the need to develop a promising multi-year plan for the Arctic geological exploration.[90]

General planning of the Arctic research and coordination of the various institutions work began to improve in the late 1930s. By this time, the general plan of geological work in the North had been drawn up,[91] as had the polar stations network development plan.[92] Transformations took place in the structure of the Arctic Institute. For example, the preparation of ice forecasts was transferred to the institute. In 1938 the Arctic Institute developed a plan for studying the Northern Sea Route for the period 1938–1942.[93]

## Dark Side of the Party Intervention and Arctic Explorers' Repressions

Interference of the Communist party in the organization and the personnel policy of the Glavsevmorput also became a factor that negatively influenced Arctic research in the second half of the 1930s.

---

[87] Bujnitskiy (1945).
[88] Soveshhanie hozyajstvennyh rabotnikov (1936), s. 210–211).
[89] TSGANTD SPb. F. 369. Op. 1-1. D. 80.
[90] Soveshhanie hozyajstvennyh rabotnikov (1936), s. 181).
[91] Urvancev (1937), s. 21–24).
[92] Mikhailov (1937, s. 25–28).
[93] TSGANTD SPb. F. 369. Op. 1-1. D. 238. L. 1.

The first vivid manifestation of this trend was the meeting of Glavsevmorput workers in January 1936. On one hand, criticism of the Arctic Institute was partly fair: there remained inconsistency in research activities in the Arctic and there was neither general planning nor effective coordination of science (see above). At the same time, numerous negative comments from the political leadership of Glavsevmorput were voiced without convincing arguments. For example, the head of the Tobolsk Political Department, A.P. Mikhailov, stated: "… The Arctic Institute is far from deep, serious approach expeditions preparations. Hence the handicraft, hence the pleasant walks of the Arctic Institute workers in the north". The instructor of the GUSMP political department, M.M. Mikula, noted that "it was possible to register a large number of facts when the parties were sent to places where they were not supposed to go. And those places where they had to go were left without expeditions." The head of the Leningrad political department also drew attention to the "disgraceful attitude towards funds and materials released by the government".[94]

Probably the fairest assessment was provided by N. Knipovich who spoke at the session of the Arctic institute Academic Council in January 1937, stating: "It is impossible that in such a short time the institute with such complicated tasks would suddenly become free of all shortcomings and could immediately foresee everything." He noted that the large-scale activities of the Arctic Institute, actively supported by the state, had been going on for only 3 years since the establishment of GUSMP.[95]

Gradually, criticism of the Arctic Institute grew on the pages of the Glavsevmorput magazine "Sovetskaya Arktika" (Soviet Arctic). Along with practical remarks and suggestions for improving organizational and research work, many politicized assessments appeared in the magazine: "The entire management system of the polar meteorological service was organized in such a way that irresponsibility and the opportunity for sabotage found the best opportunities there,"[96] "Enemies of the people, operating in the Northern Sea Route system, made their way into the radio network."[97] In the May 1938 issue of *Sovetskaya Arktika*, an article entitled "On the All-Union Arctic Institute" was published. The author of the article argued: "The main reason for the lagging behind of the institute is "activity" of its director Samoilovich, who littered the apparatus of the institute with people who were clearly questionable in business and political terms, who fenced himself off from the public with a wall of inaccessibility and unquestionable authority"; and "One of the negative aspects in the work of the institute director Samoilovich is his admiration for foreign countries."[98] In the article entitled "On the study of the Northern Seas" published in May 1939, its author B. Ivanov wrote: "Enemies of the people who made their way to the Arctic Institute in the past tried in every possible way to

---

[94] Soveshhanie hozyajstvennyh rabotnikov (1936), s. 207).
[95] TSGANTD SPb. Ф. 369. Оп. 1–1. Д. 200. Л. 73.
[96] Komov (1938, s. 13).
[97] Uluchshim rabotu polyarnoj radioseti (1938, s.15).
[98] Sysoev et al. (1938, s. 31).

hide the most valuable historical materials."[99] In 1939, the journal published an article on the hydrological research of the Northern Sea Route: "The subversive work of the former enemy leadership of the Arctic Institute inflicted a lot of harm on the polar stations. Enemies of the people tried to disrupt the work of the polar stations, to belittle their role in the development of the Northern Sea Route."[100]

Unfortunately, famous scientists and ordinary polar explorers, and Arctic science as a whole, were seriously affected by the political repressions of the 1930s. In 1931, P. Wittenburg, secretary of the Academy of Sciences Polar Commission and a renowned polar geologist, was convicted in the "case of the Academy of Sciences" and sentenced to death, later commuted to 10 years imprisonment.[101] His experience in geological research was in demand on Vaigach island and in Amderma, where Wittenburg discovered a fluorite deposit. In 1935 he was released early. However, despite his professionalism and authority in the scientific world, upon returning to Leningrad, Wittenburg failed to gain a foothold in the Arctic Institute. In his daily notes, he wrote: "How bitter it is for me that the Arctic Institute has turned its back on me, and I am not in the family of Russian Soviet polar explorers."[102]

The fate of A. Wangenheim, creator and first head of the United Hydrometeorological Service of the USSR and organizer of the Soviet Union's participation in the IPY-2, was tragic. Wangenheim was at the forefront of polar stations network, the main tool for the Arctic development in the 1930s, but in 1934 he was arrested on charges of a counter-revolutionary sabotage organization in the hydrometeorological service and sent to the Solovki special camp. It is difficult to read Wangenheim's Solovki letters to his wife: optimism and faith coexists with a bitter sense of injustice. In a letter dated June 5, 1934, he wrote: "And the expedition of O. (Schmidt) and N. (Zubov) and others are only part of the whole polar year, to which I gave so much effort, time and thoughts. They receive orders, honor, and I cannot achieve only one trifle – to be listened to."[103] Despite numerous requests to his wife to find out the fate of the statements to Stalin, Kaganovich, Kalinin, to meet with A. Gorky, G. Ushakov, and especially Schmidt, Wangenheim remained in the camp. On October 9, 1937, by a resolution of the NKVD Troika in the Leningrad oblast, he was sentenced to death. Less than a month later, he was shot in the notorious Sandormokh location in Karelia.

The years 1937 and 1938 became especially difficult, aggravated by the failure of navigation on the Northern Sea Route. The authorities clearly identified the cause of the incident: "USSR Sovnarkom recognizes the work of the Northern Sea Route for 1937 ... as not satisfactory. Reason for such serious fails of Glavsevmorput during navigation in 1937 ... are: bad organization in the Northern Sea Route, complacency and conceit, as well as completely unsatisfactory recruitment, which created

---

[99] Ivanov (1939, s. 48–49).
[100] Lappo (1939, s. 35).
[101] "Vragi naroda" za Polyarnym krugom (2007), s. 16).
[102] Wittenburg (2003, s. 189).
[103] Vozvrashhenie imeni (2005), s. 33).

a favorable environment for the criminal anti-Soviet activity of disruptors in a number of GUSMP bodies."[104] As a result, 12 employees of the Hydrographic Directorate of Glavsevmorput were sentenced to imprisonment for terms of 5 to 8 years.[105] In 1937, S. Bergavinov, head of the GUSMP Political Directorate, was arrested and executed (according to other sources, committed suicide) on charges of a counter-revolutionary conspiracy.[106] In 1939, the director of the Arctic Institute, Samoilovich, was sentenced to death on charges of treason, sabotage and participation in a counter-revolutionary organization. His student M. Ermolaev was sentenced to 10 years for sabotage, anti-Soviet agitation and participation in a counter-revolutionary organization. From 1938 to 1945, a well-known geologist, discoverer of unique deposits in the Norilsk region, deputy director of the Arctic Institute N. Urvantsev, was imprisoned.

## Conclusion

Arctic research in the RSFSR-USSR in 1920–1940 underwent significant developments. This transformation process can be characterized by a number of trends.

Firstly, scientific activity in the region was mainly aimed at solving domestic and foreign policy tasks. The prospects for annexation of previously no-man's territories in the Arctic (Franz Josef Land and Novaya Zemlya) and use of the natural resources became the major factor that determined regular scientific activities. Successes were achieved largely due to the fact that scientific institutions, through expeditionary activities and the creation of polar stations, ensured a permanent presence in the region. In addition, scientific activities contributed to the economic development of the territories. Hydrographic works, which were especially widespread in the Arctic sector of the USSR, led to the creation of maps and pilots for the Northern Sea route. Meteorological observations in the Arctic ensured more accurate weather and ice forecasts, mainly for shipping.

The second trend was the creation of scientific organizations specializing in Arctic research (Sevekspeditsiya, Plavmornin). The Northern Hydrographic Expedition was a separate unit within the naval department, specializing in carrying out hydrographic works, as well as servicing a number of polar stations in the 1920s.

Another feature of the Arctic research organization in Soviet Russia was the consolidation of a significant part of the scientific community. This trend was especially noticeable in the 1920s, when there was no coherent state policy in the region, and the centralization of the science management did not take place. Interactions among scientists, joint promotion of ideas in government bodies, and joint research

---

[104] Koryakin (2011, s. 333–334).

[105] "Vragi naroda" za Polyarnym krugom (2007), s. 16.

[106] "Vragi naroda" za Polyarnym krugom (2007), s. 28.

activities increased the efficiency of research, helping to obtain additional funding for expeditions and polar stations.

The leading role of the state became a characteristic feature of the Soviet Arctic science in the 1930s. This was preceded by attempts of authorities to unite a large number of institutions engaged in the Arctic research in the late 1920s. Such attempts included the creation of the Arctic Commission under the Council of People's Commissars in 1928, as well as the transformation of the Institute for the Study of the North into the All-Union Arctic Institute in 1930.

Maximum centralization of Arctic research ensured its consistency with state policy in the Arctic, combined scientific, economic and defense tasks. The state approach concentrated enormous resources on Arctic science, which was reflected in the growth of the number of polar stations, the staff of scientific institutions, and the provision of vehicles, primarily icebreakers and aircraft. Fundamental studies of the natural system of the region were carried out and a methodology for weather and ice forecasts in the Arctic was developed and introduced. The profession of a polar explorer received high social status and a system of training and selection of personnel along the party line was formed. Special educational programs were implemented on the basis of the Arctic Institute, the directorates of the GUSMP, the Hydrographic Institute under the Glavsevmorput.

The enormous influence of the authorities and the extreme centralization of the Arctic development had some negative aspects. The creation of a large bureaucratic system often led to inefficient spending of funds and problems with maintenance and supply to polar stations. The centralization of research could not completely eliminate the challenges of coordination, in particular, the All-Union Arctic Institute, the Department of Polar Stations of the GUSMP, and the Hydrographic Department of the GUSMP. The activities of the Arctic Institute in the second half of the pre-war decade were often unfairly criticized by party officials, who, not being professional researchers, were probably trying to get political dividends through this criticism. Polar explorers did not escape the political repressions of the 1930s, which caused serious damage to Russian Arctic science and became a great human tragedy.

An analysis of the Soviet scientific development of the Arctic allows us to formulate recommendations or, at least, to outline the key provisions that must be taken into account when organizing scientific activities in high latitudes today.

Historical experience shows the effectiveness of a national and international strategies for Arctic research, formulated in accordance with the tasks of developing the region and maintaining its sustainable development. Elements of such a strategy existed within the framework of the Glavsevmorput in the 1930s and proved to be effective, although a separate strategic planning document did not exist.

At the same time, it is important to find a reasonable balance between the scope of the assigned tasks and the creative freedom of scientists, as well as between the volume of fundamental and applied research. State support for the initiatives of Russian scientists in the 1920s laid the foundations for the widespread development of the Arctic in the 1930s. In turn, the interference of party bodies in scientific activity during this period often had negative consequences.

The need to coordinate the activities of scientific institutions at the research planning stage is obvious. This issue is especially relevant in the context of a large number of organizations conducting research in the region. Coordination contributes to the formation of integrated interdisciplinary and multidisciplinary approaches and avoids duplication of scientific work, which was typical for Soviet studies in the 1920s and 1930s. At the same time, there is probably no need to create a separate centralized structure in the system of authorities, like the Main Directorate of the Northern Sea Route. It seems expedient to organize, through the efforts of scientific institutions, authorities, economic organizations and other interested parties, a platform for exchanging information, comparing plans and developing common approaches. This experience has successfully proven itself within the framework of the Arctic Commission under the Council of People's Commissars, as well as the Council of the Sevekspeditsiya – Institute for the Study of the North – All-Union Arctic Institute, which included representatives of various institutions working in the Arctic.

# References

*"Vragi naroda" za Polyarnym krugom (sbornik statej).* (2007). Moskva: IIET im. S.I. Vavilova RAN.
Belov, M. I. (1959). *Sovetskoe arkticheskoe moreplavanie 1917–1932 gg.* Morskoi transport.
Belov, M. I. (1969). *Nauchnoe i hozyajstvennoe osvoenie Sovetskogo Severa. 1933–1945 gg.* Gidrometeorologicheskoe izdatelstvo.
Belyaev, D.P. (2005). *Gosudarstvennaya politika Rossii v oblasti izucheniya i osvoeniya arhipelagov akvatorii Barentseva morya vo vtoroj polovine XIX – pervoj treti XX vekov* [Dissertasiya na soiskanie stepeni kandidata istoricheskih nauk]. Murmansk.
*Bolshaya sovetskaya enciklopediya. T. 3: Anrio – Atoksil.* (1926). Moskva: Sovetskaya enciklopediya.
Bujnitskiy, V. H. (1945). *Osnovnye itogi i perspektivy nauchno-issledovatelskih rabot Arkticheskogo instituta.* Izd-vo Glavsevmorputi.
Bulatov, V. N. (1989). *KPSS – organizator osvoeniya Arktiki i Severnogo morskogo puti (1917–1980).* MGU.
*Doklady na soveshhanii po izucheniyu severa Rossii, sozvannom Rossijskoj Akademiej Nauk v Petrograde s 16 po 24 maya 1920 g.* (1920). Petrograd: Desyataya gosudarstvennaya tipographiya v Glavnom Admiralteistve.
Drivenes, E. (2004). *Ishavsimperialisme.* Gyldendal.
Dyuzhilov, S.A. (2001). *Razvitie nauchnyh issledovanij na Kolskom Severe, 1920–1941* [Dissertasiya na soiskanie stepeni kandidata istoricheskih nauk]. Petrozavodsk.
Dzerdeevskij, B. L. (1935). Sluzhba pogody i ee rabota v Arktike. *Sovetskaya Arktika, 1*, 31–35.
Ertz, S. (2004). Stroitelstvo Norilskogo gorno-metallurgicheskogo kombinata (1935–1938 gg.): stanovlenie krupnogo obekta ekonomicheskoj sistemy GULAGa. *Ekonomicheskaya istoriya: Ezhegodnik, 2003*, 140–176.
Esipov, V. K. (1933). *Ostrova Sovetskoj Arktiki. Novaya Zemlya – Vajgach – Kolguev – Zemlya Francza–Iosifa.* Sevkraigiz.
*Geologicheskie issledovaniya Novoj Zemli i Vajgacha.* (1936). Leningrad: Izd-vo Gl. upr. Sev. morsk. puti.
Gorbatskii, G. (1931). Vtoraya konferenciya po izucheniyu proizvoditelnyh sil Severnogo kraya. *Byulleten Arkticheskogo Instituta, 6*, 91–98.
Ivanov, B. (1939). Ob izuchenii severnyh morej. *Sovetskaya Arktika, 5*, 48–50.

Kanevskij, Z.M. (1982). Vsya zhizn – ekspediciya. M., 1982.
Komov, N. N. (1938). Meteo dlya meteo? *Sovetskaya Arktika, 3*, 11–13.
Korandei, F.K. (2019, February 1). Kak izobreli Zapolyarye. GoArctic. https://goarctic.ru/regions/kak-izobreli-zapolyare/
Koryakin, V.S. (2000). *Istoriya izucheniya prirodnoj sistemy Novoj Zemli: Do serediny XX veka* [Dissertasiya na soiskanie stepeni doktora geograficheskih nauk]. Moskva.
Koryakin, V. S. (2005). *Rusanov*. Molodaya gvardiya.
Koryakin, V. S. (2007). *Rudolf Lazarevich Samoilovich, 1881–1939*. Nauka.
Koryakin, V.S. (2011). *Otto Schmidt*. Moskva, 2011.
Kozlov, I. A., & Shlomin, V. S. (1983). *Krasnoznamennyj Severnyj flot*. Voenizdat.
Krasnikova, O.A. (2006). *Akademiya nauk i issledovaniya v Arktike: nauchno-organizacionnaya deyatelnost Polyarnoj komissii v 1914–1936 gg.* [Dissertasiya na soiskanie stepeni kandidata istoricheskih nauk]. Moskva.
*Kratkij otchet o deyatelnosti otdelnogo severnogo gidrograficheskogo otryada za kampaniyu 1923 goda. Postrojka radiostancii na Novoj Zemle*. (1924). Leningrad: Red.-izd. otd. Morskogo vedomstva.
Lajus, J., & Luedecke, C. (2010). The second international polar year 1932–1933. In S. Barr (Ed.), *The history of the international polar years (IPYs)* (pp. 135–173). Springer.
Lajus, J.A. (2004). *Razvitie rybohozyajstvennyh issledovanij Barentseva morya: vzaimootnosheniya nauki i promysla, 1898–1934 gg.* [Dissertasiya na soiskanie stepeni kandidata istoricheskih nauk]. Moskva.
Lappo, S. (1939). Gidrologicheskie issledovaniya na severnom morskom puti. *Sovetskaya Arktika, 7*, 31–38.
McCannon, J. (1998). *Red Arctic: Polar exploration and the myth of the north in the Soviet Union, 1932–1939*. Oxford University Press.
Mikhailov, A. P. (1937). Set polyarnyh stancij v tretej pyatiletke. *Sovetskaya Arktika, 10*, 25–28.
*Novaya Zemlya. Ekspediciya 1921–1927 gg. pod nachalstvom R.L. Samoilovicha*. (1929). Moskva: Izd. nauch.-tehn. upr. VSNH.
Novikov, V. D. (1956). *Iz istorii osvoeniya Sovetskoj Arktiki*. Gospolitizdat.
*Organizaciya nauki v pervye gody Sovetskoj vlasti (1917–1925)*. (1968). Leningrad: Nauka. Leningr. otd-nie.
*Organizaciya sovetskoj nauki v 1926–1932 gg.: sbornik dokumentov*. (1974). Leningrad: Nauka. Leningr. otd-nie.
Ostrovskii, B. G. (1931). *Sovetskaya Arktika*. Leningr. obl. izd-vo.
Ostrovskii, B. G. (1933). *Forposty sovetskoj nauki v Arktike: polyarnye stancii SSSR*. Severnoe izd-vo.
Pinegin, N. V. (1935). *Novaya Zemlya*. Severnoe kraevoe izdatelstvo.
*Raboty otryadov Sevekspedicii v 1921 g.: predvaritelnyj otchet*. (1922). Peterburg: Znanie-sila.
Saburov, A.A. (2016). Organizaciya sovetskih arkticheskih issledovanij v 1920-e gody: planirovanie i koordinaciya nauchnoj deyatelnosti. *Vestnik Severnogo (Arkticheskogo) ederalnogo universiteta, seriya «Gumanitarnye i socialnye nauki», 3*. 41–48.
Samoilovich, R.L. Vsesoyuznyj Arkticheskij institut. (1931). *Byulleten Arkticheskogo Instituta, 1–2*. 1–4.
Schmidt, O. Y. (1934). *Issledovanie Arktiki v Sovetskom soyuze*. Nauchnoe izdatelstvo instituta Bolshogo sovetskogo atlasa mira pri TSIKe SSSR.
Shulubina, S.A. (2003). *Sistema Sevvostlaga. 1932–1957 gg.* [Dissertasiya na soiskanie stepeni kandidata istoricheskih nauk]. Tomsk.
*Soveshhanie hozyajstvennyh rabotnikov sistemy Glavsevmorputi pri SNK SSSR 13–15 yanvarya 1936 g.* (1936). Leningrad: Izd-vo Glavsevmorputi.
Sysoev, G., Shiryaev, I., & Nazarov, V. (1938). O Vsesoyuznom Arkticheskom institute. *Sovetskaya Arktika, 5*, 25–31.
Uluchshim rabotu polyarnoj radioseti. (1938). *Sovetskaya Arktika, 3*. 15–16.

Urvancev, N. N. (1937). Geologoissledovatelskie raboty na Krajnem Severe v tretem pyatiletii. *Sovetskaya Arktika, 10,* 21–24.

Vasnetsov, V. (1974). *Pod zvezdnym flagom «Perseya». Vospominaniya.* Gidrometeoizdat.

*Vozvrashhenie imeni. Aleksej Feodosevich Wangenheim.* (2005). Moskva: Tablitsy Mendeleeva.

Wiese, V. Y. (1932). *Mezhdunarodnyj polyarnyj god.* Izd-vo Vses. arktich. in-ta pri TSIK SSSR.

Wiese, V. Y. (1935). *Istoriya issledovaniya Sovetskoj Arktiki. Karskoe i Barentsevo more.* Sevkraigid.

Wiese, V. Y. (1939). *Morya Sovetskoj Arktiki: Ocherki po istorii issledovaniya.* Izdatelstvo Glavsevmorputi.

Wittenburg, E. P. (2003). *Pavel Vladimirovich Wittenburg: geolog, polyarnik, uznik GULAGa (vospominaniya docheri).* Izdatelstvo Sankt-Peterbergskogo Istituta istorii RAN "Nestor-Istoriya".

*Za osvoenie Arktiki.* (1935). Leningrad: Izd-vo Glavsevmorputi.

Zelyak, V.G. (2014). *Gornopromyshlennyj kompleks Severo-Vostoka Rossii: stanovlenie i razvitie (konets 1920-x – nachalo 1990-h gg.)* [Dissertasiya na soiskanie stepeni doktora istoricheskih nauk]. Tomsk.

Zhukov, Y. N. (2008). *Stalin: arkticheskij schit.* VAGRIUS.

**Alexander Saburov** is a Director of the Institute for strategic development of the Arctic at Northern Arctic Federal University (NArFU).

# Chapter 5
# Norway in the Cold War – A Contemporary Case for Today

**Lars Saunes**

## Introduction

The prospects for conflict between two nuclear powers, the United States and the Soviet Union, during the Cold War demanded that the Kingdom of Norway quickly and carefully adapt its foreign policy and military strategy. Norway, neighbor to the Soviet Union and an Arctic coastal state, found itself at the geographical center of this struggle for military supremacy. Norway did adapt, balancing domestic political interests with the concerns of the great powers. Despite joining NATO, Norway's foreign policy of building bridges was both necessary and effective at maintaining a minimum level of cooperation between the United States and Soviet Union. The role and integration of the Norwegian military in the 1980s, especially joint and combined planning, exercises, and operations, are still relevant today, as the United States, Russia, and other great powers seek to secure their national interests in the Arctic region.

To explain why the Norwegian government, chose this strategy during the Cold War, and some of the political challenges it faced, scholars have pointed to how Norway as a young nation pursued neutrality and a bridging policy in its relations with great powers. Available sources show that Norway's policy and behavior changed in three stages, and that the Norwegian leadership struggled with implementation at each stage. First, the Norwegian leadership's initial policy choice after the end of the Second World War was to continue as a neutral, forge stronger relations with the west and bridge between Western and Eastern great powers. However, by 1948, Norway's pursuit of foreign policies that would not alienate either side or increase tension between them collapsed with the Marshall Plan. Second, the pressure from the Soviet Union on the status of the Spitsbergen treaty forced the

---

L. Saunes (✉)
U.S. Naval War College, Newport, RI, USA
e-mail: lars.saunes.no@usnwc.edu

© The Author(s), under exclusive license to Springer Nature Switzerland AG 2022
M. Finger, G. Rekvig (eds.), *Global Arctic*,
https://doi.org/10.1007/978-3-030-81253-9_5

Norwegians to seek security by joining the Western powers. Thirdly, recognizing the importance of the geopolitical locations in a conflict between the Western powers and the Soviet Union, it was important to encourage allied forces to be present in Norway to deter from Soviet military pressure. This chapter identifies the central dilemma confronting Norwegian leadership from 1947–1991 and explores its effects on decision-making and implementation related to its military strategy and capabilities in the North Atlantic and Arctic region. The analysis is based on extensive sources, including my own experience during and after the Cold War including personal archives of senior Norwegian officials, foreign Ambassadors, and declassified NATO documents.

I argue that Norwegian leadership succeeded in adapting to the great power strategies during the Cold War and, with the return of global great power competition, there is a need to revisit the previous policies developed for the Cold War. I develop these arguments in three sections. First, I revisit existing explanations for Norway's policy choices after the Second World War. Secondly, I describe how Norway struggled to cooperate with the United States and other Allies after the Second World War and had to adapt to security challenges from the Soviet Union. I examine Norway's cooperation with the US and Soviet Union from 1947–1991, focusing on changes in Norway's foreign policy regarding its cooperation with both powers. Third, I explore what scholars and policymakers can learn from these findings for understanding foreign policy dilemmas in the context of today's great power competition in the Arctic region.

## *Existing Explanations, and a New One*

Arctic geopolitics played an important role between United States and the Soviet Union during the Cold War. As a small Arctic coastal state, Norway was forced to adapt to the strategies and ideological differences of two great powers. With the Cold War ending in 1990 without a regional conventional military conflict, Norway showed how critical small states can be in navigating competition and preventing conflict between major powers. At the forefront of this dynamic relationship was nuclear deterrence. Specifically, this chapter discusses how Norway's foreign policy adapted to US and Soviet national strategies and the role of Norway's military in securing the country's national interests. Many historians have studied the cold war and documents from the period are declassified and available. This chapter will examine how Norway's foreign policy adapted during the Cold War, it build on historical cases, and aim to extract what is applicable for sustainable and secure development of the High North as we again return to a strategic period of great power competition.

Norway was one of the founding nations of the North Atlantic Treaty Organization (NATO) on May 12, 1949. Why did Norway join a military alliance instead of continuing its neutrality or build stronger ties to its Scandinavian neighbors? Joining an Atlantic military alliance was not the natural choice for Norway and the decision must be put in a historic context, but also in an ideological and political context. When studying the Cold war, it is tempting to limit the case to a unique regional case and neglect the global context of boxes of war. The global competition between the United States and the Soviet Union lead to proxy wars in Asia, but a conflict in Europe was avoided because of the risk of nuclear war.

Norway in the nineteenth century was a maritime nation, an Arctic nation, and had global maritime interest. By the end of the century, the Norwegian global maritime industry and Atlantic economic partners created a diversion with Sweden. The dissolution of the Norwegian and Swedish union in 1905 increased national awareness and unity and Norway established an independent diplomatic foreign policy to support its national interest. As a small coastal state, the armed forces were focused on securing the path for independence and protection of its sovereignty territory as it emerged on the Northern European map. Norway had to choose a maritime strategy to secure its future[1] and the geostrategic position in a European theater of military conflict was not well understood. As a small power with global interests, Norway could join a bilateral or collective alliance, establish functional ties with a powerful nation like Britain, or choose to remain neutral, benefitting from the general protection from international powers. At the outbreak of the First World War, Norway was supporting both sides with critical resources and pursuing a policy to remained neutral in the conflict. A rather modern Navy, built up to secure its independence from Sweden only 10 years earlier, protected Norwegian neutrality, but its merchant fleet faced challenges in the North Sea theater of operations. The British naval blockade on German trade forced the Norwegian to support the Entente and Norway came out of the war referred to as the "the Neutral Ally".[2] Former studies of Norway's foreign relations during the war indicates two main factors to explain why Norway could remain neutral in the First World War but not in the Second World War. One factor was the appearance of remaining neutral when forced to support one of the belligerent parties. The German counter to the blockade and the unrestricted submarine warfare caused a heavy toll on Norwegian merchant ships and, by the end of the war in 1918, more than half of the Norwegian merchant fleet had been sunk, a substantial loss despite the country's declared neutrality. By comparison, the Norwegian Navy remained neutral, was never challenged, and mostly patrolled local waters for the duration of the war. Secondly the understanding of Norway's exposed geostrategic position related to securing access, sea lanes of communication (SLOC) to and from the North Sea. The reason why Norway was able to remain neutral for the duration of the First World War because Germany was not prepared to seize and exploit the Norwegian coast to challenge the British naval blockade and, subsequently, Britain did not see a need to challenge Norwegian sovereignty as long as Norway supported the entente. In the inter-war period, these two factors created a dilemma in Norwegian foreign and security policy and the

---

[1] Hobson, R., & Kristiansen, T. (Eds.). (2004). Navies in Northern Waters (1st ed.). Routledge. https://doi.org/10.4324/9780203005866

[2] Riste Olav. 1963, "Norway's Relations with Belligerent Powers in the First World War." PhD thesis, University of Oxford.

ambition of remaining neutral in the conflict between the Axis powers and the allies failed. In the early stages of the Second World War, Germany invaded Norway to gain access to resources and take advantage of its geostrategic position to minimize the consequences of a British blockade, as well as to attack the Atlantic and Arctic sea lines of communication between the allies. After the Second World War the policy of neutrality was again the preferred option for the Norwegian labor government. Adapting to the current situation, Norway was exploring a "bridging" policy with the communist Soviet Union and the Western powers.

The geopolitical position of Norway and its northern territory, including Svalbard, had the potential to make it key terrain in the emerging divide between the Soviet Union and the Western powers, Britain, and the United States. In this situation, the aim of the Norwegian diplomats was to achieve cooperation between the great powers and give concessions to the Soviet point of view.[3]

## Decision to Join North Atlantic Treaty Organization (NATO) 1946–1949

There are many reasons why Norway moved from a war alliance to bridge building neutrality, to a peacetime military alliance. One major issue was the dispute between Soviet and Norway regarding the Spitsbergen treaty. The Spitsbergen treaty (Svalbard) was signed February 9, 1920, during the Versailles negotiations after the First World war. The Soviet Union was not part of the treaty and protested as they had not been consulted in the process. An agreement was made in 1924 between Norway and Russia, but the Soviets was blocked to accede the treaty by the United States, as they did not recognize the Soviet Government. The former no-man's-land gained much attention from the Soviets after the Second World War for its resources and geostrategic position for the Soviet Northern fleet. The Soviet Navy has limited access to the global oceans from their naval bases in the Baltic and the Black sea. The role of the Northern Fleet located at the Russian Kola peninsula in the Barents Sea became more important as it provided ice-free access to the world oceans through the Norwegian sea. After the Soviet liberation of Northern Norway in 1944, the Soviets proposed a fundamental change in the international status of the Svalbard archipelago.[4] The Soviet Minister of Foreign Affairs, Vyacheslav Molotov, proposed to share the responsibilities of Svalbard jointly by Norway and the Soviet as a condominium and suggested that Norway should abandon the treaty from 1920. The Norwegian government played its cards carefully and diplomatically proposed to meet many of the Soviets' demands regarding

---

[3] Arne Ordings diary, June 27, 1946, manuscript division, Norwegian National library, Oslo Helge Pharo, together again, Anglo-Norwegian relations and the early Cold War, *Scandinavian Journal of History*, 37, 2, May 2012, pp. 261–277.
[4] Sven G. Holtsmark A Soviet Grab for the High North,IFS, Forsvarsstudier 7/1993.

establishing a joint defense at Svalbard. By 1947, Soviet pressure on the Svalbard issue was discussed in the international media and the British and the Americans expressed no concerns against a Norwegian Soviet arrangement. However, in February 1947 the Norwegian "Storting" rejected the idea of a Norwegian-Soviet common defense of Svalbard but opened up for a dialogue about a revision of the 1920 treaty. At this stage, the Norwegians were skeptical of the British and American will to protect them and continued a policy of "bridge building" between the three great powers. As a relatively small country with limited military resources, Norway had to adapt to a policy of mitigating the risk regarding its relations with the Soviets. One could argue that a small state like Norway is punching above its weight when dealing with a powerful neighbor, but the lack of support from the British and Americans at this stage forced the Norwegians to adapt to a policy of finding a solution with the Soviets. The concern was real and growing after the Communist coup in Prague in 1948, which convinced the Norwegian government to take a clear stand regarding its room for maneuver. It left its path of neutrality and joined NATO. Norway had to look to the West for protection and the lessons from the Second World War clearly demonstrated that, in the case of another war in Europe, Norway would not be able to remain outside a conflict or be able to defend itself. The growing ideological divide and the US strategy of containment put Norway in a vulnerable position without the military resources necessary to maintain national security.

## Containment and the Emerging Bipolar World (1950–1959)

The geopolitical divide and NATO's military strategic plans focused on securing a European central front from an attack by the Soviet Union. The Norwegian armed forces, with its ties to Britain and post-war challenges of force reduction, were preparing for a strategic military assault. The primary areas for operation were in the south and reflected that Central Europe was the primary theater of military operation, and the military balance between Soviet Union and NATO forces was seen to be in favor of the Soviet Union. The US national strategy NSC-68[5] evaluated that the Soviet military at this time was capable of overrunning Western Europe, with the possible exception of the Scandinavian peninsula. The Soviets were capable of launching air attacks against the British Isles and air and sea attacks against the lines of communications of the Western Powers in the Atlantic. The Norwegian government interpreted NATO's overall strategy of "collective" balanced forces within the alliance as meaning that a "national" balanced force was not required, and the defense of the littoral coast of Norway should be

---

[5] Naval War College Review, Vol. XXVII (May–June, 1975), pp. 51–108. Also in U.S. Department of State, Foreign Relations of the United States: 1950, Volume I.

entrusted to the Western powers and not to Norway.[6] The Norwegian understanding of a NATO collective balanced forces was originally rejected by allies. The division of responsibility between the national territorial Commanders and the allied Commanders in war time clearly stated that defending coastal areas were the responsibility of the nation. However, following strong pressure from the Norwegian government, the NATO commander-in-chief of the Allied Forces in Northern Europe became responsible for the territorial waters of Norway, pending a territorial boundary between the Supreme Allied Commander Atlantic (SACLANT) and the Supreme Allied Commander Europe (SACEUR). Norwegian politicians realized that the initial defense of their country from an invasion was their national responsibility. NATO allied forces, and in particular the US carrier groups, were not available on short notice and the British government was not willing to commit forces from their strike fleet for the initial defense of Norway. Norway had accepted NATO's strategy of massive nuclear retaliation and, in the event of war, Norwegian bases would play an important role. NATO strategic concept (MC14/2) of May 1957 was based on first use of nuclear weapons and an air offensive of retaliation as the primary military response against an attack on NATO territory. In this period, the Army and the Airforce was prioritized in the initial defense of Norway and the Navy was neglected. A reassessment of the geostrategic risk to Norway in 1957, identified a need to enhance the naval defense as part of a national defense from a seaborn invasion.[7] It identified Northern Norway as the primary area for defense and the need to establish a standing balanced defense force defending Norway from the sea in the Arctic. At the same time, the government realized that the establishment of a strong Norwegian anti-invasion defense depended on allied military support and indirect economic support to build up military bases and infrastructure. This led to a National security concept that still is framing the military defense today. It had a predictable reassurance policy towards the Soviet Union, regarding self-imposed restrictions combined with deterrence through a strong military presence to demonstrate NATO resolve and cohesion. The primary concept for avoiding war was to establish a balanced joint denial defense to hold back an invasion until allied forces could reinforce Norway. At the same time, Norway was signaling a defensive posture in peace to avoid a military provocation regarding the Soviets. Prior to joining NATO, the Norwegian government declared that Norway would not open bases on its territory for the armed forces of foreign nations as long as Norway was not attacked or exposed to threats of attack.[8] However, in February of 1951, Minister of Defense Jens Christian Hauge clarified that there were some exceptions to that statement and that Norway could: (1) open bases for allied forces at the time of armed attack on the North Atlantic area or at a time Norwegian

---

[6] Hobson, R., & Kristiansen, T. (Eds.). (2004). Navies in Northern Waters (1st ed.). Routledge. https://doi.org/10.4324/9780203005866, pp. 233.

[7] Forsvarsdepartementet St.prp.nr.23. (1957) Om hovedretningen for Forsvaret i aarene fremover (in Norwegian)

[8] LODGAARD and GLEDITSCH 1977, 209–219.

authorities felt exposed to threats of attack and invite allied armed forces to Norway; (2) in the prescribed constitutional forms, conclude conditional agreements with allied countries with such a situation in mind; (3) construct military facilities that may be necessary to receive and support the allied forces which are necessary to assist in the defense of the country; and (4) participate in common allied exercises or receive brief visits by allied air or naval forces in peace-time.[9] The second self-imposed arms restriction was announced by Prime Minister Gerhardsen in a NATO meeting in December 1957, which concluded that Norway in peace would not permit nuclear weapons arsenals or intermediate-range ballistic missiles on Norwegian territory. This position created a capability gap in the Norwegian military force structure and increased the need to build a larger conventional force with necessary readiness. These policies of not provoking the Soviet Union, and adapting to its proximity, aimed to reduce the risk from horizontal escalation and acknowledged the geostrategic challenges of defending NATO Northern Flank in the High North. The main intention of military exercises in Northern Norway was to stop Soviet armed forces from making a strategic assault on Norway. Allied forces were restricted in movement when operating or exercising in Norway. Allied Air and Naval forces operating from Norway were not permitted to cross the 24 longitude east and would stay 500 km from the Soviet border. This gave the appearance of not provoking the Soviets, but these minor restrictions were based on military exercises in peace and did not apply in crisis or war.

## Militarization and Security Dilemma in the High North (1960–1979)

For the NATO alliance, the threat of military hostilities on NATO's northern flank was very real. In the aftermath of the Cuban Missile Crisis in 1962, Fleet Admiral Gorshkov, the commander-in-chief of the Soviet Navy, was able to convince President Nikita Khrushchev that it was necessary to possess a fighting force that could operate in nuclear and subnuclear context at sea.[10] The Soviet Navy introduced a new concept called the balanced fleet.[11] This is different from NATO's definition and the "Gorshkov doctrine" defined the force structure in terms of missions that it could accomplish in war and peace. The fleet should be balanced to perform nuclear and non-nuclear battle missions, peace- and war-time missions, balanced in types of vessels in service and a balance of forces within the military. As a dominant

---

[9] ibid. (LODGAARD and GLEDITSCH 1977, 209–219).

[10] Hudson, G.E. (1976). Soviet Naval Doctrine and Soviet Politics, 1953–1975. *World Politics*, 29(1), 102, Cambridge University Press, Stable URL: https://www.jstor.org/stable/2010048, Accesses 19-02-2021.

[11] Ibid, pp. 104.

land power, the Soviet Union realized it needed naval capabilities to provide strategic and conventional responses in case of a war with NATO. The Soviet naval policy was to challenge the United States' "Command of the Sea" by denying US and NATO absolute control of the SLOC.

NATO responded to the emerging threat with a new maritime strategy that fueled the security dilemma in the High North. NATO identified a possible Soviet military campaign to include a military campaign against Norway, Sweden, and Denmark. The Soviet nuclear submarines entering the Atlantic shipping lanes through the Greenland–Iceland–United Kingdom (GUIK) gap posed a real threat to the transatlantic routes between Europe and America. The build-up of the Soviet Navy, demonstrating its ability to deny NATO freedom of operations, pushed NATO to review its maritime strategy. Securing the Northern flank including access to the Baltic approaches became a more important task for Allied Forces Northern Europe (AFNORTH). The territorial waters of Norway remained within the AFNORTH command, and the synchronization with the bordering Supreme Allied Commander Atlantic (SACLANT) located in Norfolk Virginia United States became a challenge. The Norwegian forces were under the command of their service chiefs and there was a need to synchronize the national defense with NATO contingency plans, both in Europe and in the Atlantic theaters of operations. This created a dilemma for Norwegian defense and the establishment of the dual-hatted national joint headquarters Defence Command North Norway and Defence Command South Norway in 1971 ensured a more holistic plan for defense of the country. These commanders were responsible for planning all national operations, but at the same time were principal subordinates to CINCNORTH in the European NATO command. When operating submarines and maritime patrol aircraft (MPA) outside Norwegian territorial waters, the national commanders were functional commanders in the SACLANT chain of command.

Since the early 1970s, the northern flank of NATO and Norway became a focal point in NATO's military strategic planning. The NATO contingency planning was synchronized with a national conventional anti-invasion concept and the total defense concept of Norway. By "dual-hatting" the national command, Norway had national flag officers in NATO's chain of command, which ensured Norwegian interests were present in the decision in war and in the preparations for war. NATO's nuclear "flexible response"[12] doctrine recognized that the United states and the Soviet Union both had the capability to destroy each other and their allies in a nuclear confrontation. This implied that it was necessary to demonstrate NATO's conventional military capabilities through exercises and operations to deter from hostilities or try to end a military confrontation prior to escalation to nuclear war.

For Norway, this was an existential battle of survival and the military concept was based on mobilizing the Army in Troms and holding the line as long as necessary to get Allied reinforcement to push back the Soviets and subsequently for

---

[12] Sokolsky, Seapower in the Nuclear Age, pp. 92–93.

NATO to restore national territorial integrity. Air power was the main focus in this operation, with contribution from the Norwegian and several allied air forces. Defensive counter air operations were vital to protect ground forces, deny Soviet amphibious attacks and secure the Norwegian ability to receive allied reinforcement. In 1975 the Norwegian government decided to acquire 72 F-16s in addition to maritime patrol aircrafts. The Navy had a similar support role with a layered sea denial concept in the Norwegian littorals. By the latter part of the 1960s, the new fleet plan was completed and the Navy now composed of modern ships, integrated with allied navies and had a high degree of specialized coastal and arctic training. They became a competent littoral fighting force that, when integrated with a mobilized coastal artillery, was able to close the most important axis from an amphibious attack. The Navy included squadrons of coastal patrol ships equipped with torpedoes or 76 mm guns. The squadrons were later developed into a combination of missile and torpedo boats. The 15 Kobben class coastal submarines, with surface and submarine torpedoes, were primed to close the sea lanes of communication along the coast to prevent an amphibious landing and support special forces operations behind the enemy lines. The combination of mines, coastal artillery, and coastal tailored forces was a formidable sea denial force, challenging Soviet maritime ambition of a successful invasion of Northern Norway.

The most important national and NATO task was logistic support to secure the reinforcement of Norway. Follow-up forces from NATO needed to secure the SLOC to Norway and along the inner leads along the coast to northern Norway. This was the primary task of the Oslo class frigates and mine clearance forces reinforced by allied naval and air forces. Securing choke points along the coast, ports of departure and arrival, became essential military tasks for achieving the mission. The increased capabilities of the Soviet navy to challenge and, to a certain degree deny, NATO reinforcement through the GUIK gap was a growing concern. The Soviet Navy was probably capable of closing down the ports and the SLOC along the Norwegian coast by using mines and their submarines, which was a credible challenge to the trans-Atlantic SLOC. There was always a risk of escalation if a US carrier group deployed into the Norwegian sea and a consequence of US President Jimmy Carter's "swing" strategy was that fewer US forces was available for NATO exercises in Europe. The main purpose of this policy was to reduce the risk of unnecessary escalation that could have led to a nuclear response. Norway realized that the US policy of keeping NATO naval forces below the GUIK gap limited deterrence and NATOs capabilities to protect Norway from Soviet occupation and pushed hard to get the United states to commit Navy and Marine forces to its defense. This led to pre-positioning of US Marine corps equipment in Norway and exercises to support a major US reinforcement. In the "Cold Winter" exercise of 1977, a biannual exercise in Northern Norway, the US marines learned the lessons of winter training and were shocked to find that they were not prepared or equipped for Arctic winter conditions.

Another important process created conditions for a more pragmatic policy regarding the Soviet Union. The United Nation Convention on Law of the Sea (UNCLOS) was developed by consensus among more than 160 participating nations in the 1970s and was signed on December 10, 1982. For Norway, this was an opportunity to develop and frame relationships with neighboring coastal states on good governance at sea. In December 1976, Norway established the 200 nautical mile exclusive economic zone (EEZ) and initiated negotiations with the Soviets to delimit its borders in the high North. This was a challenging situation, as the Norwegian EEZ was rich in biological resources and the discovery of oil and gas reserves required negotiation with neighboring states on borders and potential sharing of resources. In the North, the Cold War frame made such negotiations extremely difficult.[13] The Soviets claimed a "sector line" and Norway claimed a "median line", creating a disputed area in the Barents Sea.

For the Soviet Northern Fleet, any concession to a NATO country might have strategic consequences and the Barents Sea was rich on fishery resources and potential gas and oil reserves. The Norwegian-Soviet Grey Zone Agreement of 1978 did not solve the delimitation, but instead established a practical regulation for facilitating a bilateral fishery agreement. This demonstrated how Norway and the Soviets were able to find pragmatic solutions even under the umbrella of the Cold War. In 1975, the joint Russian–Norwegian Fisheries Commission was established, and the two countries agreed on regulating the most important fish stocks in the Barents Sea. The agreement built on a long-standing research and the commission managed to establish principles to solve problems and agree on a common approach to fisheries management. As part of this pragmatic people-to-people cooperation, Norway established the Norwegian Coast Guard in 1977, separate but integrated in the Royal Norwegian Navy, with the main responsibility of enhancing maritime governance in the established EEZ. Establishment of the Coast Guard was key to avoiding a potential conflict escalation with the Soviet Union. If state sovereignty or sovereign rights are threatened by another state, it is easy to link this to state security. The primary function of the defense and foreign service war and the defense sector's primary responsibility is to secure the state sovereign rights by use of force when necessary. As a small nation, Norway had to adapt and create strategies to manage its "asymmetric" military relations with the Soviet Union. The security of the state was built on deterrence provided by NATO, supported by diplomatic initiatives between Oslo and Moscow. The combination of coast guard constabulary functions and navies are especially useful for a small coastal state as tools to manage an asymmetric military situation and prevent that legal disputes turns into an armed conflict. The synergy between a coast guard and the navy builds on overlapping but different mission sets. The coast guard helps to maintain law and order at sea by safeguarding the coastal state's rights and obligations in peace and during security episodes and crisis. It builds on the international legal framework and national legislation. Managing resources and solving disputes are based on nations respecting the coastal states rights and responsibilities. By establishing a police

---

[13] The Grey Zone Agreement of 1978, Fishery Concerns, Security Challenges and Territorial Interests, by Kristoffer Staburn, FNI Report 13/2009, https://www.files.ethz.ch/isn/112916/FNI-R1309.pdf, accessed February 22, 2021.

authority in the Coast Guard, the decision to start legal prosecution remained in the chain of justice, reducing the potential for escalation into an armed conflict. The Norwegian Navy's role is to conduct integrated operations with Norwegian Armed Forces and Allies in peace, crisis, and war. The risk of escalating a dispute into armed conflict is higher when using military forces to enforce law in parallel with defending national sovereign rights. Bringing Soviet fishing vessels to court, focusing on the private legal subject and at the same time increasing cooperation regarding managing resources, kept tension relatively low, allowing diplomatic solutions to prevail.

## US and NATO Naval Strategies and Consequences for Norway in the High North (1980–1990)

Entering the 1980s was a dangerous time for NATO in the Cold War. The rapidly growing Soviet fleet was increasing its global presence, demonstrating that it was capable of challenging American Command of the sea. Soviets had a growing knowledge of the capabilities of US and NATO forces and was increasingly confident in its own ability to counter NATO maritime forces at sea. The Soviets saw the US "swing strategy" as a shift in the bipolar balance of forces. By the end of 1979, 70 American hostages had been seized at the US Embassy in Iran, Soviet military forces had invaded Afghanistan, and there were military incidents with forces from Libya and Syria regarding US naval forces operating in the Mediterranean.

For Norway and the NATO alliance, the Soviet Union's ability to push the maritime defensive zone out to and beyond the Greenland Iceland UK (GIUC) gap, potentially involving a ground invasion of Norway, was a dangerous situation. By securing forward positions in Norway, the Soviet Union could deploy land-based air cover to support naval operations in the Atlantic, which could be decisive in the battle of the Atlantic. The naval support to secure the allied reinforcement of Norway was critical and, while the war could not be won at sea, it definitely could be lost at sea. The shift in US naval policy was demonstrated in Exercise Ocean Venture 81[14] and it represented a turning point in the Cold War. The US Secretary of the Navy John Lehman was leading the centerpiece of President Ronald Reagan's campaign of naval rearmament and maritime superiority. The main aim of the "forward strategy" of deploying aircraft carrier groups near the Soviet Union was to challenge the Soviet growing influence by sea.

The main objective of the new US maritime doctrine, incorporated with the NATO's concept of maritime operations, was to contain the Soviet Forces through operating forward close to the Soviet naval bases, by maintaining the initiative and command at sea. The doctrine was based on NATO maintaining collective

---

[14] Lehman, J. (2018). *Winning the Cold War at Sea*, W.W. Norton and Company: New York, NY.

deterrence and if deterrence failed, it was designed to operate forward to bring the battle as close to the Soviet Union as possible in order to protect the territory of the alliance. The concept of operations was subdivided into geographical theaters of operations: the Mediterranean SLOC, the eastern Mediterranean, the Atlantic SLOC, and the Norwegian Sea. NATO developed contingency plans to support the operation and a military force were allocated with in place national and supporting allied forces. An exercise concept was developed as part of the contingency plans to train, integrate, and evaluate war-time plans, as well as serve as deterrence as it demonstrated the alliance's resolve.

The Norwegian Sea campaign aimed to prevent the Soviet fleet from deploying into the Norwegian sea and securing the trans-Atlantic SLOC, and also aimed to prevent an amphibious assault and invasion in Northern Norway. The campaign provided naval support to allied air and ground operations in Norway. The concept of maritime operations was drawn from the US national military strategy of "operating forward", as well as NATO's maritime concept. The concept was based on deploying a US carrier battle group with allied air and ground support to the coastal waters of Northern Norway. Sheltered by the Norwegian fjords, NATO's objectives were to repel a Soviet amphibious assault and support Norway against a land invasion through Finland and Sweden. The primary aim of NATO was to prevent the Soviets from using Norwegian facilities and contain or destroy the Northern Fleet. However, the risk was high and could easily have led to a horizontal escalation and a preemptive nuclear respond from the Soviet Union. The US Carrier groups of the Norwegian coast may look like a diplomatic tool to demonstrate NATO deterrence as a modern gun boat diplomacy in peace. For NATO, the Carrier group represented flexibility, mobility, and endurance and played a role in nuclear deterrence. For the Soviets and the Warsaw Pact countries, the forward demonstration of US NATO forces into their local waters may have been perceived as preparation to launch nuclear strikes, as they represented a substantial threat to the Warsaw Pact. The Soviets knew they were vulnerable to long-range cruise missiles and assumed that NATO would be forced to use nuclear weapons as a first response in war. The US carrier group, with its cruise missile arsenal delivered from submarines, aircrafts, and ships approaching within the striking range of the Northern Fleet, was perceived as an extreme military challenge and the Soviets could have selected to strike first in order to protect its nuclear strategic capabilities. In military operations, accepting high risk when expecting high reward can be acceptable and the three-phase US maritime strategy was a success as it contributed to end the Cold War. In the first phase, transition to war, forces were forward deployed with nuclear powered attack submarines in positions deep inside the Soviet bastion in the Barents Sea and into the Arctic. Multinational battlegroups would form into carrier groups and amphibious task groups,[15] securing SLOC in the Atlantic and supporting allied reinforcement along the Norwegian coast. The US Marines brigades would deploy to Norway and use their prepositioned equipment for onward

---

[15] Børresen, p. 4.

deployment, together with Dutch and British marines. The sealift of multiservice reinforcement would commence, and the British Royal Navy would send its nuclear-powered submarines and anti-submarine task groups into the eastern Atlantic to protect the SLOCs. The allied Army reinforcements were important to the maritime strategy and included protection of the British mobile force and the Canadian Air-Sea-Transportable (CAST) Brigade Groupe to Norway.

The next phase was to seize the initiative as far forward as possible, carrying the fight to the enemy. The air campaign would seek to suppress enemy air defenses and carrier-based air and cruise missiles would establish air defense forward – "an outer air battle" – and cause maximum attrition on enemy land-based air capabilities. Available land-based tactical aircrafts would complement these efforts in the Norwegian sea and provide air protection for deployed land forces in Northern Norway. Surveillance, intelligence, and raiding operations against command, control, and communication sites by allied special forces would be valuable supplements. The last phase, carrying the fight to the enemy, was based on heavy strikes on the Northern flank, culminating in attacks on Soviet territory combined with securing vital choke points with amphibious operations. The aim was to push the Soviets towards war termination and recover territory lost to Soviet attacks.[16] The theory of victory was to achieve war termination on favorable terms using conventional weapons and avoid a nuclear war. This required setting up a dominating conventional force to convince the Soviet Union that it would not benefit from continuing aggression and would choose to retreat, while giving it no incentives to escalate to nuclear war.[17]

There was a considerable military presence in Norway in the 1980s: on average, around 10,000 personnel in training and more than 15,000 in exercises each year. The impression of reassurance regarding the Soviets was fading and allies continued to put pressure on the Soviets from Norway. The need for allied logistic support and a forward base structure in Norway reached its limit and there was constant pressure on the Norwegian Government from the Allied nations to establish peacetime support. The Striking Fleet Atlantic (STRIKEFLT) operations in Norwegian waters required a logistic tail that required both national host nation support and military resources as they operated along the coast of Norway. The STRIKEFLT consisted of a carrier group with an air wing, anti-submarine forces (ASW), amphibious and marine, strike forces, and was capable of establishing sea control and air superiority as a basis for force projection ashore. The forward-deployed carrier group demonstrated NATO deterrence and the US's determination to defend Norway, but it also represented the forward defense of the US as it brought the fight to the enemy.

The new US strategy challenged the Norwegian adaptation to carefully balance policies of deterrence and reassurance towards the Soviet Union and integration

---

[16] Hattendorf and Swartz, eds., *U.S. Naval Strategy in the 1980s*, pp. 85–86.
[17] Ibid, p. 92.

with NATO.[18] First, the US strategy challenged the Norwegian base policy of reassurance and the limitation on how many allied soldiers, aircrafts or ships could be present in Norway at any given time. Secondly, there was a risk of "horizontal escalation". The Norwegian government feared that if a conflict between the United States and the Soviet Union should occur in another region, the US would attack Soviet strategic resources at the Kola peninsula. This put pressure on the Scandinavian countries, with a potential for nuclear war. In 1986, Norway's defense minister, Johan J. Holst, warned against a permanent presence by the US 6th Fleet. At the same time, he was engaged in securing regular allied presence in the high north to reduce Soviet dominance and induce restraint on both sides. This again demonstrated the long-term adaptation regarding Norwegian security policies.

## *The Cold War: A Contemporary Case for Today*

The Cold War ended with the collapse of the Soviet Union in 1990 and the Arctic became a region of low tension, but high attention to the emerging climate changes. From the end of the Cold War until the Russian annexation of Crimea in 2014, there existed an Arctic exceptionalism regarding security confrontation in the High North. Norway again developed its "bridging policy" regarding Russia and the United States and Arctic dialogue, and cooperation was achieved mainly through organizations like the Arctic Council founded by the Ottawa Declaration in 1996. Other frameworks with avenues for Arctic cooperation were people-to-people relations, like "Pomor",[19] which created cooperation within areas like education, trade, and even military relations. Maritime governance and Arctic stewardship were at the forefront of Navy and Coast Guard cooperation and the Arctic Coast Guard forum was established in 2015 to foster safe, secure, and environmentally driven maritime activities in the Arctic region.[20] From 2008, annual exercises between the Norwegian Navy and the Russian Northern Fleet created conditions for dialogue and transparency regarding Arctic security development, but this activity was stopped due to the Western sanctions on military cooperation with Russia from 2014.

Even with the recent shift in the geopolitical security environment, Arctic coastal states continue to promote an assessment of low tension on the top of the world and seek to isolate the Arctic from the current geopolitical challenges. The world is changing, and the Arctic is becoming increasingly prominent in the defense and security policies of major powers, both inside and outside the Arctic

---

[18] Børresen, p. 5.

[19] A Russian expression meaning "People living by the coast"; refers to historic trade cooperation between the people of Northwest Russia and people living along the coast of Northern Norway.

[20] https://www.arcticcoastguardforum.com/about-acgf, accessed March 28, 2021.

region. Arctic states are increasingly stressing the need to defend their maritime approaches, exercise their sovereign rights, and protect their Arctic waters. Norway again faces the challenge of adapting to great power competition in the High North between the United States and Russia. China's growing interests in the Arctic from their Polar Silk Road initiative creates additional long-term challenges. The United States has identified the growing rivalry with China and Russia[21] as a global power competition that creates strategic challenges from an assertive China and a destabilizing Russia. Will the experience from the Cold War be applicable for adaption to this new strategic competition in the High North and what role will Norway play in this Arctic great power rivalry? From a military point of view, the strategic cooperation with the United States in the frame of NATO demonstrates a military deterrence that facilitates a political dialogue on equal terms with Russia. Secondly, will the predictable behavior, transparency, and reassurance regarding Russia during the Cold War have the same effect in today's global security environment? Thirdly, how can Norway play a role in bridging the rivalry between the great powers in the Arctic as it opens up for more communication between Asia and Europe and a potential military confrontation in the Arctic between the United States and Russia?

For Norway, the extensive Russian renewal of military capabilities and build-up of Arctic military bases from the Barents Sea to the Bering strait is a growing concern. These bases, with air defense and missile systems, complemented with icebreakers and coastal infrastructure, are defensive systems that aim to deny any conventional military challenge from NATO in the Arctic. The strategic interests for the Russians are to protect their nuclear strike bastion in the Arctic, to secure free access to the world's oceans and maintain a strategic balance with the United States or other powers like China. The key military objective is to protect the Russian secondary nuclear strike capabilities, which were recently demonstrated by their new generation of strategic submarines. The Norwegian Minister of Defense, Bakke Jensen, summed this situation up as follows: *"The key task of the Russian capabilities on the Kola Peninsula is global deterrence, making horizontal escalation a lasting concern for Norway. There is no indication that Russia will slow down its engagement in the Arctic – rather, the opposite."*[22]

Norway is again put on the spot as a stepping stone to monitor the new Northern Fleet military district and its deployments from the Barents Sea. The NATO exercise "Trident Juncture" in 2018 demonstrated NATO's defense of Norwegian territory in a potential conventional conflict similar to the maritime campaign from the Cold War. Defending Norway on NATOS's northern flank is a maritime campaign and 55,000 aviators, sailors and army soldiers provided a convincing deterrent towards Russia. Russia, as the geostrategic competitor, perceived this demonstration as

---

[21] https://www.whitehouse.gov/wp-content/uploads/2021/03/NSC-1v2.pdf, accessed March 29, 2021.

[22] Defensenews, Norway's defense minister: We must ensure strategic stability in the high North, https://www.defensenews.com/outlook/2021/01/11/norways-defense-minister-we-must-ensure-strategic-stability-in-the-high-north/, accessed Feb 2, 2021.

aimed against Russia, confirming that NATO and the United States is a strategic threat against Russia. In this new geostrategic competition, Norway again experienced how difficult it is to balance "the Article 5" deterrence mission with reassurance towards Russia. The increased military demonstrations are currently fueling a security dilemma and, following the 2018 NATO exercise, Russia has responded with trans-Arctic nuclear war games as well as large naval exercises in the Norwegian Sea. Military signaling, combining the forces from the Northern and the Baltic fleet in demonstrating their bastion defense and conducting weapon drills of the Norwegian Coast, is a clear political message, not only towards Norway but to any power challenging Russian Arctic strategy. NATO exercises and allied training in Norway represent a credible deterrence, raising the threshold for use of military aggression by Russia in a potential conflict. Even though the military capabilities are increasing the tension, it still remains relatively low, allowing diplomacy and political dialogue to prevail.

In the bipolar system of competition between NATO and the Warsaw pact predictable security, behavior was measured in a dialogue between two parts. During the Cold War, restrictions on allied forces operating in Norway were based on what was perceived as a military threat from weapon and sensor systems against the Soviet Union in the 1950s. Today, US forces and its NATO alliance have global reach in all service domains with cyber, space, air, surface, and subsurface capabilities. In the strategic competition, the competition or conflict in peace is more a matter of global influence than creating a military kinetic effect. The backbone of this effect is to maintain a global nuclear deterrence and a credible conventional military response. To Russia, the transparent predictable behavior from Norway could be confirming the existing threat assessment that the United State is an enemy and Norway is a road for the United States to attack its strategic capabilities in a global conflict. Russia is launching an information campaign against Norway, presenting Norway as a useful tool for US aggression against Russia, using the former reassurance regime of self-imposed restriction from the Cold War as an important part of the campaign. For Norway, the long-term challenge will be to maintain a level of allied presence as security challenges to the United states and other allies occurs in other regions of the world. The United States' pivot to the Pacific and changes in the defense strategy may have consequences for future operations in Norway. Currently, the US Airforce, Navy, and Army have issued new Arctic strategies that include cooperation and shared training with Norway in the Arctic. However, with limited military capabilities, unnecessary restriction in mobility and reach will limit the efficiency of military operations.

The increasing effect of climate change is more visual in the Arctic Ocean and will subsequently increase access to resources, but also connect 70% of the world's population. The opening of the Arctic connects the security architecture of the Pacific region with the Atlantic European region. The rise of China and its Polar Silk Road is a manifestation of these changes and it is necessary to recognize a circumpolar security challenge. Norway has previously demonstrated that dialogue and the bridging of conflicting views is possible in the Arctic. Arctic coastal states are aware of the Arctic challenges, regarding climate change, risk to Arctic governance and sustainable development, as well as the emerging Arctic security dilemma. The cooperation on Arctic stewardship is working well and includes non-Arctic states in a rule-based system that builds on international law. The increasing arctic security dilemma requires a new circumpolar security dialogue that includes both Arctic and non-Arctic states. As the importance of the Arctic grows, mechanisms to manage tension and enhance cooperation in the security realm might contribute to regional peace and stability. Norway could play a bridging role in establishing a new security architecture. With increased maritime activity in the Arctic Ocean, including surface, subsurface, and aviation activity, coupled with the lack of a forum to discuss issues related to hard security, may lead to unintended escalation. Considering the consequences of climate change from 2050 to the end of the century the trans polar bridge will be a geostrategic different space than today. Non Arctic states in Europe and the Asian-Pacific region will have strong national interests in the Arctic not only for access to resources and communication, but also due to the consequences of global climate change on their regions. A long term Arctic security strategy should recognize the bridge between the Pacific and the European security architecture and maritime trade and develop forums for mitigating tension, risks that include not only the Arctic as such but include future stakeholders from a circumpolar perspective including Asian Pacific as well as the European region. There is an opportunity to create an Arctic Ocean Maritime Symposium to enhance transparency and dialogue. The forum would be designed to tackle issues related to maritime security and defense. This forum would be vital due to the limitations on addressing these topics in other leading Arctic forums. It builds on similar arrangements in the Pacific and the Indian Ocean and promotes security, transparency, and cooperation among international naval leaders to meet common Arctic security challenges. The forum would exclude no one but should limit management rights to the eight Arctic states. It would be an open and inclusive platform for discussion of globally relevant maritime security and defense issues related to the Arctic region. Its mission would be to generate a flow of information between naval professionals, leading to common understanding and cooperative solutions to promote maritime security partnerships, preserve maritime access, and provide ready naval forces to respond to crises and contingencies.

## Bibliography

*Forsvarsdepartementet t prp nr 23 (1957) Om hovedretningslinjer for Forsvaret i årene framover* a.
*Soviet Naval Doctrine 1953–1975* b.
*United States (2018) National Defense Strategy Summary* c.
*United States Interim National Security Strategic Guidance 1* d.
Brooks, L. F. (1986). Naval power and national security: The case for the maritime strategy. *International Security, 11*(2), 58–88. https://doi.org/10.2307/2538958. https://www.jstor.org/stable/2538958.

Børresen, J. (2011). Alliance naval strategies and Norway in the final years of the Cold War. *Naval War College Review, 64*(2), Article 7. https://digital-commons.usnwc.edu/nwc-review/vol64/iss2/7.

Espenes, Ø., & Naastad, N. E. (2000). The Royal Norwegian Air Force: A multipurpose tool during the Cold War. *Air Power History, 47*(1), 40–51. http://www.jstor.org/stable/26288548.

Hobson, R., & Kristiansen, T. (2004). *Navies in northern waters*. London: Routledge. http://usnwc.idm.oclc.org/login?url=http://search.ebscohost.com/login.aspx?direct=true&db=nlebk&AN=116291&site=ehost-live.

Holtsmark, S.G. (1993). A Soviet Grab for the High North? *Forsvarsstudier*, 7. Oslo: Inst. for Forsvarsstudier.

Lehman, J. F. (2018). *Oceans ventured: Winning the Cold War at sea*. New York: W.W. Norton & Company.

Lindgren, W. Y., & Græger, N. (2017). The challenges and dynamics of alliance policies: Norway, NATO and the High North. In M. Wesley (Ed.), *Global allies* (pp. 91–114). ANU Press. http://www.jstor.org/stable/j.ctt1sq5twz.10.

Lodgaard, S., & Gleditsch, N. P. (1977). Norway – The not so reluctant ally. *Cooperation and Conflict, 12*(4), 209–219. http://www.jstor.org/stable/45083209.

Pharo, H. (2012). Together again: Anglo-Norwegian relations and the Early Cold War. *Scandinavian Journal of History, 37*(2), 261–277. https://doi.org/10.1080/03468755.2012.667968. http://usnwc.idm.oclc.org/login?url=http://search.ebscohost.com/login.aspx?direct=true&db=30h&AN=75370418&site=ehost-live.

Stabrun, K. (2009). *The Grey Zone Agreement of 1978 fishery concerns, security challenges and territorial interests* (FNI Report 13/2009). https://www.fni.no/getfile.php/132053-1469870055/Filer/Publikasjoner/FNI-R1309.pdf

Zimmerman, M. (2018). High North and High Stakes; the Svalbard Archipelago could be the epicenter of rising tension in the Arctic. *Prism, 74*, 106–123. https://www.jstor.org/stable/26542710.

**Lars Saunes** is a professor and distinguished international fellow at the U.S. Naval War college, Newport Rhode Island, where he is the co-lead for Newport Arctic Scholars initiative. He retired as Rear Admiral in 2017 and is a former Chief of the Royal Norwegian Navy and Commandant of the Coast Guard.

# Chapter 6
# The Post-Cold War Arctic

**Lassi Heininen**

## Introduction

At the end of the Cold War in the 1980s, when the Arctic was transferred into a "military theatre" with systems of nuclear weapons and those of early-warning radars, the region faced the threat of a global nuclear war and a "nuclear winter" as a global cooling of the planet's climate. Following from this, the peace movement with its campaigns and peace rallies (e.g., Greenpeace, 1989) was increasingly vocal. At the same time, increasing pollution in the Arctic, in particular radioactive wastes in northern and Arctic waters, was becoming an additional trigger for a significant transformation "from confrontation to cooperation" (Heininen, 1992).

Among traditional images, narratives and visions of the 1980s' Arctic was defined a homeland (for Indigenous peoples) vis-à-vis a peripheral place on the world's margins, a land for discovery, a storehouse of resource vis-à-vis a nature reserve (e.g. AHDR, 2004; Steinberg et al., 2015), and a military theater of the Cold War vis-à-vis a "Zone of Peace".

Now, in the 2020s, the Arctic region faces growing risks and threats of climate change and environmental degradation due to global warming, pollution and collapse of biodiversity as a wicked problem, both regionally and worldwide. Among images, narratives and visions are the Arctic as an environmental linchpin for, and a "climate archive" of, the planet, an "early-warning system" or a canary in a coalmine for climate research, an exceptional area in international politics, and a/the "global" Arctic.

Due to the 2020s' political rhetoric and great power rivalry, issues dealing with the Arctic have often involved misinformation and disinformation. Examples of such issues are the Arctic as a "political space" either emphasizing "high

L. Heininen (✉)
University of Lapland, Rovaniemi, Finland
e-mail: lassi.heininen@ulapland.fi

geopolitical stability and constructive cooperation" (stated by ministerial declarations), or a region with emerging conflicts (stated by media, some experts). Therefore, it is politically important for policy-shapers/makers and scientifically valuable for (Arctic) literature to have deeper and multi-dimensional analysis on relevant phenomena, which would provide a platform with which to evaluate the current situation and compare it to that of the Cold War period.

Political sciences, and international relations in particular, have sometimes been criticized – deservedly or otherwise – for not taking history into consideration, or even for rejecting it. Indeed, a few publications from the early twenty-first century reveal that there is a lack of a sense of history, as authors are not sufficiently familiar with literature covering the Cold War period (on issues such as environmental awakening, the global deterrence of nuclear weapons, or the Airman's perspective). Therefore, the near past is briefly discussed here.

This article aims to draw up a holistic picture on a state of the post-Cold War Arctic, Arctic geopolitics in particular, and the significant geopolitical changes behind and followed from the shift from the Cold War to the 2020s. According to my studies, it is even possible to interpret that the post-Cold War era has come to its end (in the Arctic). The method of analysis is a comparative study on transformation; on one hand, the changing dynamics of main themes and trends of Arctic geopolitics and governance, and on the other, how the end of the Cold War's "Peace" vision and its initiatives are been realized and modified. The motivation behind the analysis is to focus on stability-building and cooperation, as it takes bravery to not be skeptical but to believe in confidence – the most important means in (international) politics and diplomacy –, looking specifically at the importance of common interests. Here I lean on a few studies and analysis on Arctic policies, as well as, retrospectively, my studies on Arctic geopolitics and governance in the last 30 years.

## Visions for, and Triggers Ensuing Significant Geopolitical Changes in, the Arctic

The "Zone of Peace", launched by President Mikhail Gorbachev in his speech of October 1987 in Murmansk on the Kola Peninsula, was a new and contradictory vision. It was rare and a real contrast to the Cold War image of the Arctic as a military theater. Unsurprisingly, it received a mixed reception, ranging from positive responses to the environmental and scientific aspects to generally cool ones to the security aspects, by Western countries (Archer, 1989), although 35 years later it could be considered to be successful. This article could also be interpreted as a story from the vision of a "Zone of Peace" of the 1980s to the "Arctic Paradox" of the 2020s.

The Murmansk speech consisted of six initiatives (Pravda, 1987). Two aspects pertain to arms control: first, to establish a nuclear-free-zone in Northern Europe; and second, to stimulate confidence-building measures in the Arctic seas by "banning naval activities in mutually agreed-upon zones of international straits and in intensive shopping lanes in general". The other four aspects pertained to civilian cooperation: first, to cooperate peacefully "in developing the resources of the North and Arctic" for example, "agreement on a single energy program for the north of Europe"; second, to cooperate on scientific research in the Arctic "of immense importance for the whole mankind" for example, "setting up a joint Arctic scientific council"; third, to cooperate with the Nordic countries in environmental protection, as "the urgency of this is obvious"; and fourth, that the Soviet Union "could open the North sea route to foreign ships by the services of our ice-breakers".

As kind of preconditions behind the Murmansk Speech's vision and its initiatives were the new era of *Glasnost* and *Perestroika* in the Soviet Union by President Gorbachev and his open-minded, new kind of thinking, which included new proposals for arms control/disarmament and confidence-building. This policy included a new habit of sending young observers abroad to hear and learn new ideas and giving them the courage to rethink, and based on these reports to make policy initiatives like those by the Murmansk Speech. On the other hand, there was the problem of increasing concern about pollution, particularly radioactive waste, in the Arctic by northern residents and local societies (e.g., AMAP, 1997; Heininen, 1990). This led to an environmental awakening among Arctic Indigenous peoples, international non-governmental organizations, sub-national governments and social movements, as well as (forerunner) scientists and scholars.

All of this, supported by Gorbachev's overall vision, consisted of the reasons and preconditions for the first significant geopolitical changes in the Arctic – it became a trigger for a paradigm shift in international politics. Following this, surprisingly soon, the Arctic states (Canada, Finland, Iceland, Kingdom of Denmark, Norway, Russian Federation, Sweden, USA) started the process of international cooperation on environmental protection in 1989, based on a Finnish initiative, which is correspondingly based on the Murmansk speech. The states signed the Arctic Environmental Protection Strategy (AEPS) in their first intergovernmental ministerial meeting in 1991. Thus, the existing international Arctic cooperation was originally formulated at the Arctic Environmental Protection Strategy and adopted by the eight Arctic states. Importantly and interestingly, they were open-minded about beginning transboundary cooperation for environmental protection after being pushed by the environmental awakening of non-state actors, as well as clever about initiating parallel region-building with the ultimate aim of decreasing military tension and increasing political stability.

As a result, the Arctic states managed not only to start international (functional) cooperation for environmental protection, but also to institutionalize the functional cooperation by creating the Arctic Council (AC) for "*sustainable development in the Arctic and the protection of the Arctic environment*" (Ottawa Declaration, 1996), and to affirm the importance of maintaining the achieved political stability. This was a real paradigm shift in the Cold War state politics, particularly given that, since then, the Arctic has been in a state of high geopolitical stability (e.g., Heininen, 2018), highlighted by the Ministerial declarations/statements (e.g. Reykjavik,

2021; Rovaniemi, 2019; Fairbanks, 2017) as the first preamble, and explicitly mentioned by national policies.

Behind all of the significant change, as well as the environmental awakening and movement, was growing concern about increased long-range and regional transboundary pollution – radioactive wastes, DDT, POPs, Arctic haze – in the Arctic and Arctic seas (Heininen, 2020). Air pollution, including "anthropogenic pollutants originating in the heavily industrialized, mid-latitude regions of Eurasia which are transported in the Arctic region by the wind currents that prevail during the winter" (Soroos, 1990), was defined as a problem by many residents and local societies. On the global agenda of environmental problems and challenges, concern was growing about atmospheric (long-range) pollution "as scientific evidence mounted on the scope and consequences of acid rain,… and a trend toward global warming". In addition to international regimes to address acid rain and ozone depletion, "negotiations began on a regime to limit climate change", particularly the Arctic haze (including soot), as a distinctively Arctic phenomenon in the 1990s (Soroos, 1990). Interestingly, scientifically and partly secretly, the impacts of global warming have been known for at least 70 years or so, as "climate change became a U.S. national security concern even before the Cold War became hot" (Doel, 2009, 17).

This means that in the Arctic region there were, on one hand, new non-military security risks and challenges threating people and societies of the High North, and on the other hand, as a result of this, new active actors became involved in influencing the situation and their destiny, primarily due to new security threats and challenges. Because actors are never alone, –and instead have cooperative, competitive or conflicting interests, the "new" Arctic actors (even though some of them were old and indigenous) had their own multifunctional interests.

It is valid to interpret that despite the initiatives of Gorbachev and other states, the first significant geopolitical change of the Arctic region, towards a phase of cooperation and political stability, was not only (or even mostly) due to the two great powers and the Arctic states' growing concerns about the environment. Such change was originally demanded and promoted, and later assisted, by non-state actors – Indigenous peoples' organizations (including the ICC and the Saami Council), NGOs, northern civil societies and scientific community – due to growing environmental concerns. Further, it was based on functional cooperation on a field of low-politics, largely according to Mitrany's theory on Functionalism (Heininen, 1999). An introduction to the history of contemporary Arctic cooperation would make it easier to understand the current situation, the structure of the Arctic Council, and the strong involvement of Arctic Indigenous peoples' organization in Arctic governance and the Council's work.

Parallel to changes in Arctic geopolitics are those in the phase of Arctic security nexus from the hegemony competition of the Cold War (between the Soviet Union and the USA) to a transition period, including a few arms control and disarmament aspects, and further to the post-Cold War security nexus. The current state of stability to balance military presence and functional cooperation in globalization, and accept and adopt new (non-military) security premises, indicates a significant change in Arctic security nexus towards a more comprehensive security discourse, although it is not yet a paradigm shift (see Heininen, 2020).

If the first wave of transformation was within the circumpolar north, as a distinctive region (e.g., AHDR, 2004), and about the Arctic states (e.g., Armstrong et al., 1978), and inspired and pushed by environmental awakening, the second significant geopolitical change of the Arctic is global in nature and deals largely with global warming.

Rapidly advancing climate change in the Arctic region has largely been discussed after the launch of the Arctic Climate Impact Assessment Report in 2004 (ACIA, 2004). Having a global scale as the main feature, the second transformation has finally made it clear that the Arctic is an important part of the planet, as well as the global dynamics of multifunctional changes. Since the early twenty-first century, beyond and despite the cooperation between the Arctic states and the Arctic Council, the Arctic has been globalized and the "global Arctic" is a new, multi-dimensional geopolitical context (Heininen & Finger, 2017) and a "global embedded space" (Keil & Knecht, 2017).

The zenithal map projection (as opposed to the traditional Mercator map) reveals that the globalized Arctic has a deeper global approach, consisting of twofold dynamics: global impacts within the region, and worldwide implications by the globalized Arctic (The GlobalArctic Project, 2014). However, this is not a totally new observation, as the region is been facing global impacts since the seventeenth century from whaling, sailing, exploration, trade, as well as influences of pan-Arcticness by Indigenous peoples (e.g., Southcott, 2010). Also, scientific projects on "Global Change" and pan-Arctic cooperation, including studies on sea ice, paleoenvironmental record, response of ecosystems, and analysis and prediction of changes by IASC, started in the 1990s. These may have been due to the end of Cold War period, or due to the thaw between the Soviet Union and the USA, which was greatly influenced by the Murmansk speech (e.g., Irwin, 1993).

As part of global environmental issues, and following the "Arctic Paradox", new ethical questions have emerged in the future development of the Arctic when it comes to the exploitation of oil and gas, particularly offshore oil drilling. A major risk of oil spills in cold, icy waters, and consequently oil-spill response, as well as attempts by Greenpeace activists to board the *Prirazlomnaya* oil rig (which resulted in them being captured), are probably familiar to many experts and activists, as is the strategic nature of Arctic energy security. Interestingly, it is also a US interest to strengthen its Arctic policy through US–Russian maritime cooperation. The creation of knowledge-based energy sectors also plays an important role in cooperation among Iceland, the Faroe Islands, and Greenland.

Due to their many impacts and consequences, the grand environmental challenges are transferring into wicked problems and a/the *"Arctic paradox"* (e.g., Palosaari, 2019). Indeed, together with pollution and loss of biodiversity, the impacts of global warming (in the Arctic) have been considered as a wicked problem. The basic knowledge on the global environment and climate reveals that, instead of military threats, this combination might become a trigger for environmental catastrophe in the region and worldwide. Or, an ultimate reason, the sort of hey straw, for a paradigm shift, as there is already a search for a shift (Heininen and Exner-Pirot, 2020).

As an important part of the first change, there has also been a significant change of actors, particularly broadening the spectrum of relevant Arctic actors, or "stakeholders", from the two superpowers' dominance of the Cold War period into a rich variety of local, regional, and international actors influencing Arctic governance, geopolitics, and security. This has meant a larger number of actors acting in, and influencing different circles of internationalization with their multi-functional (cooperative, competitive, and conflicting) interests, as well as (re)mapping and defining their country related to the Arctic and even self-identifying themselves as "Arctic stakeholders". The changes in actors and their (changing) interests are interrelated with the two geopolitical changes. The first change brought the concept of a "multiple-use region" (Young, 1992), and that of "a region of multifunctional and conflicting interests" (Heininen, 1999), while the second brought the slogan of "global Arctic" including the remapping of non-Arctic states (Heininen et al., 2019).

All in all, the two changes are significant and closely interconnected with each other, consisting of a process. As an outcome, the Arctic has several layers of activities and circles of internationalization, ranging from local and regional to international and global. The region is becoming increasingly integrated into world affairs, while it is simultaneously experiencing the growing impacts of climate change and pollution. Following the geopolitical changes and the broadening spectrum of actors, it is possible to identify and formulate main themes and trends of Arctic geopolitics and governance in the early twenty-first century.

## Characteristics of Post-Cold War Arctic Geopolitics: Main Trends and Dynamics

According to the Arctic Human Development Report (2004; Heininen, 2004), there were three main trends of Arctic IR and geopolitics at the beginning of the twenty-first century. The first was increasing circumpolar cooperation by indigenous peoples' organizations, sub-national governments, and international non-governmental organizations, encouraged and enhanced by their success in pushing the Arctic states' governments to start cooperation for environmental protection. The second was the start of "region-building" and thus developing the Arctic as a distinctive region discourse with unified states as major actors and intergovernmental organizations (such as the Arctic Council) at major platforms (for functional cooperation in field of low-politics). The third trend was a new kind of relations between the Arctic and the rest of world, emphasized by the growing importance of international scientific Arctic research and an increasing role of non-Arctic states and their interests in the region.

If the two first trends were obvious in the early twenty-first century, the last one was still emerging. Fifteen years later, there is a slight change dealing with and between these trends, and the emerging trend has started to dominate as the Arctic has been globalized. Correspondingly, the two others still exist, although they are diminishing, for various reasons.

Arctic Indigenous peoples' organizations are not always happy about having limited resources and not always being recognized. These organizations are skeptical towards non-Arctic states' involvement, and do not recognize them as "Arctic stakeholders". They also criticize Arctic states for having (neo)colonial policies towards them and in their Arctic regions. At the same time, these peoples and organizations highlight the importance of, and lean on, international treaties and international organizations, such as UNs and AC, and in general international (human right) law (e.g., Inuit Arctic Policy, 2009; Sami Arctic Strategy, 2019). All in all, an Indigenous approach is very relevant and unique in the global context, and is becoming even more important in the Arctic.

Correspondingly, "region-building", after been thoroughly discussed in Arctic literature, applied in practice and flourished for a while, as well as supported by regional actors – less so by Indigenous peoples – has been little discussed. The trend is still kept among the means and tools of states to increase political stability, although the original mission of decreasing military tension and increasing political stability (in the post-Cold War Arctic) is considered to have been accomplished. The Arctic states are seemingly less enthusiastic for supporting regional cooperation by and between sub-national governments. Interestingly, it was already in 1978 that Armstrong et al. (1978) discussed and (re)mapped the Arctic region, and discussed and analyzed its future in world affairs, not explicitly as "distinctive region" but as a region in general.

Finally, as a new research approach (initiated by the *GlobalArctic* Project in 2014), the "Global Arctic" has become a new discourse with a deeper global approach, a valuable contribution to Arctic research, as well as a slogan that has been repeated by several experts and policy-makers. Also, while Arctic Indigenous peoples used to lean on international treaties on their rights, the "Global Arctic" has become an image and perception for non-Arctic observer states to (re)self-identify themselves towards the Arctic (as Arctic stakeholders).

Based on the IIASA's comprehensive study and analysis on Arctic policies and policy documents – of the Arctic states, Indigenous peoples' organizations, the Arctic Council and its observer states (56 policy documents adopted in 2010–2019) – and their priorities, there are four new and emerging overall trends of Arctic governance and geopolitics in the 2020s (Heininen et al., 2019, 249–253). First, as if the wicked problem, in particular rapid global warming, would not be enough, there is an ambivalence or paradox of Arctic development whenever a balance is sought between environmental protection and climate change mitigation and new economic activities, largely due to "political inability". The second trend is state domination supported by geopolitical stability and sovereignty vis-à-vis globalization based on international treaties, UNCLOS and maritime law, and UN declarations regarding indigenous rights and self-determination. The third is a focus on science, as to lean on scientific research and international cooperation in science, for problem-solving due to the pressure of the rapidly advanced climate change and the above-mentioned paradox. The fourth and final trend is the new interrelationship between the Arctic and Space (such as digital security, meteorology, WMO) due to climate change, globalization, and the global economy.

According to these state policies, climate change as a threat multiplier is a uniting and merging factor, as well as the main reason for focus on science, as one of the overall trends of Arctic governance and geopolitics. There is also a clear tendency to increase economic activities, made more possible by the impacts of global warming in the Arctic region, although environmental protection should be, and officially is, the main priority. This conclusion is supported by those who have warned about "Arctic hype"; that is, optimistic hopes of economic opportunities in the region. Thus, despite the focus on science, explicitly discussed synergy between scientific and indigenous knowledge to tackle climate change, and (long) lists of priorities/goals/objectives, states' policies on the Arctic reflect hesitation, when facing and trying to solve grand environmental challenges and the wicked problem (combination of rapid climate change, pollution and declining biodiversity).

Finally, while it is interesting to approach to the Arctic and Arctic issues from the point of view of (Outer) Space, or that of bathymetric of the Arctic Ocean, it is not totally novel, as they both have been discussed (in literature) for decades, such as vertical perspectives along altitudes instead of latitudes. The Airman's perspective (e.g., Henrikson, 1975), as part of classical geopolitics, was due to the development of aircraft technology and that of missile technology, was revolutionary, and applied to map drawing like the above-mentioned zenithal map projection from Space onto a plane surface at the North Pole. Correspondingly, the bathymetric of the Arctic Ocean measured by (nuclear) submarines, based on the advanced submarine technology, has been known for decades, although it was highly confidential information in the superpowers' Cold War competition, and the first bathymetric maps of the Arctic Ocean were published in the 1980s.

There are also relevant interrelations between these trends and several Arctic narratives, such as ambivalence vis-à-vis race for resources, state domination vis-à-vis geopolitical stability and state-controlled development, focus on science vis-à-vis climate ethics, and the Arctic and Space vis-à-vis "Global Arctic".

All in all, the trends reveal the existing "political inability" to act, although maintaining the achieved high geopolitical stability based on constructive cooperation would support the tackling of climate change and pollution in the Arctic region, if only that would be made as the first priority. This kind of "best practice" would be mutually and globally beneficial, and could be a foundation for "political ability" to make a paradigm shift in mind-set in world politics, as a precondition for problem solving globally. A bigger picture drawn by these trends is more complicated, as it is less about the "cooperation – conflict" dilemma, or geostrategic implications of economic opportunity due to rapid warming, than it is about political ability to decide real priorities and implement them by making hard decisions.

## Reinterpreting Geopolitics and Security by Broadening the Scope

If the Arctic was heavily militarized during the Cold War, the post-Cold War Arctic has been heavily politicized due to rich resources and their mass-scale exploitation, pollution and climate change as a growing concern, mutually beneficial common interests – Arctic politics is with dynamics and lively debates. Consequently, there are tendencies towards growing interest and involvement in Arctic governance and skepticism about worldwide involvement from outside the region, as well as slowness in progress at Arctic cooperation due to reflections of turbulences and growing tension in world politics. This is according to the features of critical geopolitics, that geopolitics is not only about physical geography, natural resources and power politics of states, but also about other actors and their interests, immaterial issues, knowledge (as power), and the politicization of a physical space (e.g. Heininen, 2018).

Despite analysis that Arctic stability, as a common interest of the major actors, is resilient, and the Arctic seems to be an "ideal stage to cooperate on environmental protection", there are cautious evaluations, false alarms of political rhetoric, or predictions that Arctic politics is in, or approaching, a crossroads concerning the current state of cooperation. There are competition and tensions between different perceptions on Arctic security, such as comprehensive security vs. traditional narrow one, as well as that the Arctic is neither exceptional nor insulated from power politics, but a part of the world politics and the global tensions between major powers (for example, NATO and Russia/the US and China) (e.g., Hoogenson & Hodgson, 2019). This may be the case if Arctic stability, and factors behind it, are interpreted in a traditional realist way; that is, that due to "anarchy", world politics is becoming worse and moving towards (emerging) conflicts, and consequently high geopolitical stability and the "zone of peace" vision could be (soon) lost.

A politically relevant and academically interesting debate and speculation about the future of Arctic stability and cooperation started after the Crimea annexation by Russia in 2014. The "new Cold War" rhetoric actually started several years earlier after the Russian expedition to the bottom of the North Pole, which was interpreted by Canada and Denmark first of all as a geopolitical maneuver. The "new Cold War"/"great power rivalry" rhetoric is (only) one of the discourses and perceptions on a state of the post-Cold War Arctic; the "high stability/"zone of peace" rhetoric is another. It does not, however, take into consideration the severe (grand) environmental challenges and wicked problems, nor constructive multi-functional cooperation on issues, such as environmental protection and science, as common interests.

Interestingly, but not surprisingly, several studies and books have examined and discussed the main sections of Arctic politics from the point of view of the five littoral states of the Arctic Ocean (Canada, Kingdom of Denmark, Norway, Russia, the USA) after the related Ilulissat Meeting of the Arctic littoral states in 2008 (see, Ilulissat Declaration 2008), less than a year after the Russian expedition. This was first time that the Arctic Five (A5) had been constructed as an exclusive club; since then, the A5 vs. the Arctic states plus Indigenous peoples (A8+) has been discussed and speculated on. Due to the meeting, there are said to be tensions between the Arctic states, as Finland, Iceland and Sweden are excluded from the group of coastal states (Iceland located in another ocean, and Finland and Sweden as landlocked countries concerning the Arctic Ocean). However, this claim, and the argument that due to, and since, the meeting the Arctic states have just repeated the mantra of cooperation, has no real evidence, as only Iceland protested the procedure of the Ilulissat meeting. In fact, the cooperation started when the Arctic Environmental Protection Strategy was negotiated and signed, and strengthened by the establishment of the Arctic Council.

Based on national policies of the Arctic states, the dividing line on military-security, including state sovereignty and defense, goes between the five littoral states (including Russia as a non-NATO member state) and the three others (Finland and Sweden as non-aligned member states, and Iceland as a NATO-member state). Behind is that the littoral states highlight sovereignty and defense, unlike the other three, which highlight international cooperation and comprehensive security. Thus, it is about geography and geopolitics – that is, whether or not to be a littoral state of the Arctic Ocean – not necessary military-security policy and membership of a bloc. The same was seen before the Kiruna Ministerial of Arctic Council (in 2013), when the dividing line about accepting or not accepting new observer states did not go between the A5 and the "left-overs", but between the Nordic states and the rest.

Finally, the Arctic Council is still the most important platform, where Arctic states are equal and six Arctic Indigenous peoples' organizations are involved in (A8+) and, most importantly, the cooperation in its auspices continues. In this context, US Secretary of State Mike Pompeo's "great power rivalry" speech in May 2019 in Rovaniemi was a small episode, despite the speculations, which neither changed the basics of Arctic politics nor the Arctic Council's basic foundation for cooperation, as it was strongly stated by the seven Arctic states and the permanent participants in the Rovaniemi Ministerial and its joint Statement (2019).

When it comes to the biggest Arctic and Arctic Ocean littoral State, the Russian Federation is viewed as an active, but also revisionist, Arctic military power (e.g., Konyshev and Sergunin, 2014). At the same time, it is surprising how little the Russian economic interests in Arctic region are taken into consideration and discussed. To state that Russia, as another major nuclear weapon power with heavy military presence and activities in the Arctic, is mostly seeking to remilitarize the region is an understatement and does not provide the whole picture of the economic importance of the Russian Arctic Zone for the entire federation. This is largely why Russia is investing heavily in its Arctic Zone infrastructure such as its ice-breaker fleet, and consequently, Russia's policies on Arctic shipping and the Northern Sea Route are advanced and further developed.

Therefore, Russia is "ahead in the race to control the Arctic" as *Newsweek* magazine put it (March 17, 2017), which explains Russia's strong military presence there to defend state sovereignty and economic interests in the Arctic region, as Canada, Kingdom of Denmark, and Norway also do. Interestingly, Russia's Arctic policies and their implementation include military security (hard) and search and rescue (soft) aspects, as well as aspects of legal status and administration. For example, based on the lists of "development tasks" of the recently approved state policy (President of the Russian Federation, 2020) safety and security are high on the agenda, which can be interpreted as referring to the comprehensive security concept in addition to traditional security.

Furthermore, the Arctic is sometimes depicted as a "security region" by some scholars and policy-makers, and some others are interpreting and repeating the Arctic as a zone of peace. A few experts have been struggling with the "security region" concept, as a "security region" and a "zone of cooperation" are not the same thing. The Arctic could be interpreted not as a typical security region, but a "cooperative region" including stability and confidence-building. Often behind is the basic weakness that "security" is neither discussed nor defined, when the basic questions – What kind of security and security premises are there? Who are security actors and whose security is in question? – should be asked and security (re)defined accordingly.

When it comes to traditional security/military-security in the Arctic and the military capabilities of the USA and Russia in the region, it is important to consider the nuclear weapons systems deployed there and their capability for a revenge strike. The weapons and structures, as well as their capability, are still there, as is their original purpose and nature, as a direct continuity to the Cold War military presence of the two superpowers. For example, a possibility that the USA could permanently deploy a nuclear submarine fleet and sea-based ballistic missile defense systems is not new, as the US nuclear submarines deployed and mobile in the Arctic are capable of projecting power into and operating within the Arctic Ocean region. Correspondingly, a Russian air base on the Franz Josef Lands would not be the only place to reach the northernmost parts of North America and the US Thule Air Base, as this could be done by a multi-warhead ballistic missile from a strategic nuclear submarine patrolling in the Arctic Ocean.

Here, the Arctic, as well as Arctic geopolitics/security, is viewed through the lenses of military security, or the analysis is stuck with "great power rivalry" rhetoric. Furthermore, when dealing with security and geopolitical factors, it is no longer possible to take as given that the security trajectory of the Arctic region means only, or mostly, military-security, or that it is all about state security. Neither it is valid nor acceptable to neglect the role of external non-military threats and growing concern on the environment and human security. This perspective is too narrow, as long-range pollution, climate change, and collapse of biodiversity – in general, environmental degradation – have, 30 years after the Murmansk Speech, become the focus on Arctic cooperation and (geo)politics. Further, the Arctic Council has recognized environmental degradation as a security dimension and taken environmental and human (comprehensive) security into consideration as an important factor.

It is not necessarily about the "either conflict or cooperation" dilemma, which is interpreted as contradictory, as it could be about dualism. The "Arctic Paradox" is a good term with which to describe the dilemma, so it deserves to be recognized. Also, as global warming continues causing environmental awakening, functional cooperation on environmental protection and "ambivalence" of Arctic development, new non-military threat pictures and challenges should be taken into consideration and defined as security aspects, as well as indicators/variables of a "security region". Then there will be other criteria for division lines.

From the point of view and context of environmental degradation (combined by pollution, climate change and mass-scale utilization of resources), the Arctic could be interpreted – and would make sense to interpret – as a distinctive region sharing similar conditions and challenges. This does not mean "security community" *per se* as a traditional (state-based) military-political complex, but more a cooperative region based on functional cooperation on certain fields (such as environmental protection, science) and the idea of "New Regionalism" (Hettne, 1994). Interestingly, the current international Arctic cooperation, which started due to similar problems (pollution, nuclear safety), included "region-building" as indicated by the Arctic Council, and managed to create high geopolitical stability. This state has been maintained despite growing tension in world politics (for example Crimea 2014 or the Syrian War), largely due to shared common interests, and is a reason why policy-makers and experts often state that direct conflict *over* the Arctic in itself is "very unlikely".

All in all, in the Cold War period the Arctic region reached a high geostrategic position globally due to the deployed nuclear weapon systems. In the post-Cold War era, the systems are still there, but more important is the politicization of the region due to non-military aspects related to the environment and climate change, and new (non-states) security actors. Based on the manmade transformation from confrontation to cooperation, the Arctic is with high geopolitical stability and multilateral cooperation, as well as valid concepts of comprehensive security.

## Common Interests of the Arctic States and Special Features of the Arctic Geopolitics

The transformation from confrontation to cooperation has been maintained as the Arctic states use to affirm their commitments to "*maintain peace, stability and constructive cooperation in the Arctic*", as repeated by the declarations of the biennial ministerials of the Arctic Council (such as Fairbanks Declaration of 2017, Rovaniemi Joint Ministerial Statement of 2019, Reykjavik Declaration of 2021). This state of Arctic geopolitics, which is supported by Arctic Indigenous peoples' organizations (as permanent participants of the Arctic Council) and several non-Arctic states (as Arctic Council observer states), including China, Japan, and Singapore in Asia, and France, Germany, and the UK in Europe, is also interpreted to mean "Arctic exceptionalism", although this interpretation is also debated (e.g., Hoogenson & Hodgson, 2019).

One might expect that instead of discussing and speculating about terms or concepts of phenomena, and noting the rhetoric of foreign ministers, the phenomena – here 'high geopolitical stability' and 'constructive cooperation' – could been discussed and analyzed. In particular, functional cooperation (on the fields of low politics), based on Mitrany's (1966) Functionalism, is defined as increasing confidence that will decrease military tension and increase political stability, between former rivals and enemies. This is exactly what is going on in the post-Cold War Arctic, and interestingly, after 30 years, it is possible to evaluate that the theory has functioned well and been efficient. Thus, one could also argue that it is more important to analyze what is behind geopolitical stability and constructive cooperation than to speculate and discuss whether the Arctic of the early twenty-first century is a security region or exceptional. As well as, how this state has been maintained despite tensions, turbulence, and uncertainties in international politics.

The current state of high geopolitical stability and constructive cooperation is neither accidental nor entirely a matter of chance, but a manmade change that has been thought throughly, as well as intentional, and slowly implemented. As I have analyzed (Heininen, 2018), this state is based on common interests of the Arctic states, first of all "to decrease military tension and increase political stability", but also due to certain prerequisites for high Arctic stability. Among the common interests which the Arctic states share and Indigenous peoples' organizations support, six stand out. The first and most fundamental is to decrease the military tension of the Cold War period and increase political stability of the Arctic region as the ultimate aim after the end of the Cold War. The second is to agree on Arctic environmental protection and start transboundary (functional) cooperation on environmental protection due to growing (long-range) pollution in the Arctic (as discussed earlier). The third is to have modern region-building with states as major actors and non-states as supporters, such as Barents Euro-Arctic Region and AC. The fourth is to support the parallel, though de facto already started, circumpolar cooperation by Indigenous peoples and sub-national governments. The fifth is to strengthen international scientific cooperation, or focus on science, in the Arctic. The sixth and final common interest is to stimulate and enhance long-term business relations and trade, as well as (new) economic activities, in the Arctic.

Also, according to a recent analysis on national strategies and policies (Heininen et al., 2019), the *de facto* priorities of the Arctic states include governance, economy, international cooperation, as well as environmental cooperation (including pollution and climate change) – all of which indicate common interests, less so security.

This would not be enough for high geopolitical stability, but together with certain prerequisites and special features of Arctic geopolitics it is. Among the prerequisites for high stability are the following. First, the original nature of the Cold-War military presence, in particular the nuclear-weapon systems (of the Soviet Union and the USA), for global nuclear deterrence. Second, neither armed conflicts nor serious disputes of state sovereignty in the Arctic between the Arctic states since the end of the Second World War. Third, much opposite the region is with a high degree of legal certainty. Fourth, devolution and other soft ways to govern, particularly the Nordic model of governance (as five of the eight Arctic states are Nordic countries). Finally, flexible agenda setting in Arctic cooperation, as the Arctic states from the beginning separated issue areas and concentrated policy-shaping on fields of low politics (military-security matters, fisheries excluded on Arctic Council agenda), and remain flexible in keeping them so.

Based on this analysis it becomes increasingly clear that behind the "peaceful" and "cooperative" features of the region are common interests between the Arctic states, such as environmental protection, scientific cooperation, economic interests, and region-building. Here, the first significant geopolitical change in the Arctic, due to long-range pollution, was a test and paradigm shift from confrontation to cooperation. It led to the rise of new actors (Indigenous peoples, sub-national governments), and brought in the high geopolitical stability as an ultimate aim, and also revealed new geopolitical trends, such as region-building and new relations between the Arctic and the rest of the world. The common interests are good reasons to maintain the high geopolitical stability, not least due to the pressure of globalization (such as climate change and growing (economic and other) interests of non-Arctic states towards the Arctic), particularly as the state of stability has shown that it can be mutually beneficial. Behind is conventional wisdom, also acknowledged by several International Relations theories, including new realism, that stability is good for rulers, governance and business, as well as people, particularly after (significant) transformations, such as the end of the Cold War period.

It is possible to argue that the common interests and special features of Arctic geopolitics have made the Arctic a special case, even an exceptional one in international politics and IR. Indeed, in turbulent times and uncommon instabilities, this kind of state of "Arctic consensus" goes beyond "Realpolitik" and the hegemonic "Great Power" politics. In so doing, the Arctic states, supported by Indigenous peoples and northernmost regions, have reshaped and reconstructed their geopolitical reality by considering environmental protection and focus on science as means of achieving the common aim of stability and cooperation as "soft law" governance. This could be interpreted as "exceptional" or a model of peace, or at least as an example of stability-building.

In addition, world politics has a few other exceptions, such as the International Space Station (ISS). On November 16, 2020, NASA, in cooperation with the SpaceX Company, launched the *Crew Dragon* spaceship to the ISS. This was the first time in nine years that the USA was able to launch a rocket into orbit instead of leaning on the Russian Sojuz as a carrier rocket. Despite disagreements and growing tension, the ISS has been jointly used by Russia and the US to host astronauts and cosmonauts, as well as space pilots from other countries.

Overall, Arctic geopolitics, as well as the ISS, represents a common interest of these two competitive states, which could be interpreted to be exceptional in international politics, but shows the mutual benefit of common interests. However, the reconstruction of Arctic geopolitical reality has not, yet, changed security premises by security-political elites nor caused a paradigm shift of security policy by the Arctic states. Nor is this state resilient, when traditional images of confrontation and conflict are still there as a (hidden) ghost of the Cold War.

## Gorbachev's Vision Vis-à-Vis Current State/Today's Vision

Among the visions, images, and perceptions of the Arctic, the "Zone of Peace" was a new one in the 1980s, and it would be interesting to analyze how the 1987 Murmansk initiatives have been realized since then. In brief, they have mostly been successful, as half of the initiatives have been completed, but the vision has not yet been accomplished: the aspects of civilian cooperation are in progress, while the two concerning military security have seen less progress.

Progress has neither been made with "nuclear-free-zone in Northern Europe", when the initiative is seldom mentioned, nor with legally binding measures "banning naval activities in mutually agreed-upon zones of international straits". However, UNCLOS, as the constitution of the seas, is both respected and implemented. The Arctic Military Environmental Cooperation (AMEC) started (in 1996) between Norway, Russia and the USA, soft-law confidence-building measures have been implemented, and there are also a few platforms for discussing Arctic military security matters (e.g., Luszczuk, 2016). All this supports the two initiatives, as well as the Arctic high stability.

By contrast, the four aspects pertaining to civilian cooperation have been successful, as environmental challenges were driving cooperation, the eight Arctic states signed the AEPS, based on which the Arctic Council was established with environmental protection as the main pillar. Even though Finland initiated the AEPS in 1989 and started the Rovaniemi process, as the main foundation for the multilateral Arctic cooperation and the AC, the Finnish initiative owes a debt of gratitude for the Murmansk Speech.

"The urgency of" cooperating in environmental protection with the Nordic countries and other Arctic states soon became "obvious", and long-range and regional pollution in the Arctic was the trigger for (functional) international Arctic cooperation. Environmental protection is now one of the main tasks and highlights of the Arctic states, and one of the (most important) criteria for observer states. All actors that would like to become involved in Arctic cooperation, and hope to be recognized as "Arctic stakeholders", are expected to support environmental protection as well as sustainable development, the other pillar of AC. Another closely related and successful initiative is scientific research in the Arctic, as "of immense importance for the whole mankind". Preparations for "a joint Arctic scientific council" started same time and were set already in 1990, when the IASC was established. There is also peaceful cooperation "in developing resources", including energy ones, of the Arctic. Even more so with the Northern Sea Route, which is with growing international traffic, as Russia has opened the route "to foreign ships by the services" of its ice-breakers.

The impacts of common interests and special features of Arctic geopolitics, which are related to, and partly based on, the vision and initiatives, have made the Arctic a unique case, if not an exception, in world politics and IR. Behind is that in the world of turbulences, growing tensions and uncertainties, as well as rapid climate change and pandemics, it is worth studying the value of mutually beneficial common interests, recognizing non-state actors and their security premises, and acknowledging immaterial things as indicators of the Arctic's growing role in world politics.

Following from this, it is possible to brainstorm an idealized vision of how today's – post-Cold War – Arctic continues the tradition. For example, the Arctic as a (unique) model of peace based on the commitment to "*maintaining peace, stability and constructive cooperation in the Arctic*", characterized by circumpolar cooperation, empowerment of the indigenous peoples, and strengthened sub-national governments. Or, as a (unique) forerunner of environmental protection in response to the wicked problem (pollution, loss of biodiversity, climate change), and an international platform for innovations and scientific research cooperation. Or, as a (unique) opportunity for resilience based on expertise, sustainable business, and immaterial things.

All in all, the Murmansk Speech was a manifestation of new thinking and impacts of environmental awakening in the entire North, as well as the trigger for intergovernmental cooperation of the Arctic states, pushed by Arctic Indigenous peoples who became involved due to their own initiative, courage, and capacity. The lesson learned is that immaterial things – knowledge and human capital, innovations in political and legal arrangements, paradiplomacy, and stability and confidence – as well as values, are important and valuable, particularly as many of them are indicators of the implications and drivers of the globalized Arctic affecting the rest of the globe.

## Conclusion

Despite threatening pictures and false alarms in political rhetoric, the basic knowledge about the current state of Arctic geopolitics reveals high geopolitical stability and constructive cooperation. Being a result of significant transformation from military tension of the Cold War to comprehensive security of the post-Cold War Arctic, it is mutually beneficial and based on common interests of the Arctic states, Indigenous peoples, and the northernmost regions. Here, the Arctic Council has acted as the main international cooperative platform for environmental protection, science and expertise, and confidence-building, and in particular for policy-shaping. Established in 1996 to provide a means of promoting cooperation among the Arctic states, with the involvement of Indigenous peoples and other Arctic residents, on issues of environmental protection and sustainable development, the Council has exceeded all expectations. Instead of having a strong mandate in policy-making, the Council has been inclusive, influential, and responsible, and thus idealistic for policy-shaping (see, The Arctic Yearbook 2016 devoted to the Arctic Council's 20 years of regional cooperation and policy-shaping).

The strategy of reconstruction of the geopolitical reality by going beyond the hegemonic game of the Arctic states, and being supported by Indigenous peoples and northernmost regions, has been successful. Beyond the Arctic states and the

Arctic Council cooperation, the 2020s have seen the "Global Arctic" as a new, multi-dimensional geopolitical context consisting of global impacts within the region, and worldwide implications by the globalized Arctic.

The basic knowledge on the global environment and climate reveals that, instead of military threat and having an emerging conflict unlikely, the combination of climate change, pollution, and loss of biodiversity is been considered as a wicked problem. This is accelerated by the ambivalence or paradox "whenever a balance is sought between environmental protection and climate change mitigation vis-à-vis an increase of (new) economic activities for Arctic development … largely due to 'political inability'", as a new overall trend of Arctic governance. At the same time, there is a focus on science, as another overall trend.

However, the Arctic Paradox is not inevitable, as much depends on the criteria by which the Arctic states make their decisions: hopefully to maintain the high geopolitical stability and constructive cooperation and to tackle global warming by starting real mitigation, as well as to continue the reconstruction of geopolitical reality. So far, the soft law regime built around the Arctic Council and international Arctic cooperation has done an excellent job. However, the hard decisions related to tackling climate change, pollution, and loss of biodiversity would require more legislative power and legally binding agreements to restrict or ban a few activities – first of all, offshore oil drilling – as well as a political will of all relevant stakeholders.

Considering the new trends, particularly the Arctic paradox tied with state dominance, the current state of the Arctic raises two concluding questions. First, how resilient is "Arctic stability/exceptionalism" in light of power politics (such as the rise of China and Asia in the Arctic), growing global economic interests for, and activities in, the Arctic, and, above all, rapid environmental changes in the Arctic? Second, how would the Arctic states reach and enhance "political ability" to overcome "Arctic ambivalence" and "political inability", as well as involve Arctic Indigenous peoples and sub-national actors as full partners in climate change mitigation and Arctic governance.

# References

ACIA. (2004). *Impacts of a warming Arctic: Arctic climate impact assessment* (ACIA Overview Report. Summary). Cambridge University Press.
AHDR. (2004). *Arctic human development report 2004.* Akureyri & Reykjavik, Iceland: Stefansson Arctic Institute.
AMAP. (1997). *Arctic pollution issues: A state of the Arctic environment report.* Arctic Monitoring and Assessment Programme (AMAP): Oslo. https://www.amap.no/documents/doc/arctic-pollution-issues-a-state-of-the-arctic-environment-report/67.
Archer, C. (1989). Western responses to the Murmansk initiative. *Defence Studies Centre, Centre Piece, 14.*
Armstrong, T., Rogers, G., & Rowley, G. (1978). *The circumpolar North. A political and economic geography of the Arctic and Sub-Arctic.* Cambridge: Methuen & Ltd. London.
Doel, R. E. (2009). What's the place of the physical environmental sciences in environmental history? *Revue d'histoire modern et contemporaine [English language version], 56*(4), 137–164.

Greenpeace. (1989). *Nuclear Free Seas*. (mimeo).

Heininen, L. (Ed.). (1990). *Arctic environmental problems* (Occasional Papers No. 41). Tampere: Tampere Peace Research Institute.

Heininen, L. (1992). The conflict of interests between the environment and military strategy in Northern waters and the Arctic. In J. Käkönen (Ed.), *Perspectives on environmental conflict and international politics* (pp. 55–71). London & New York: Pinter Publishers.

Heininen, L. (1999). Euroopan pohjoinen 1990-luvulla. Moniulotteisten ja ristiriitaisten intressien alue. In *Acta Universitatis Lapponiensis 21 – Arctic Centre reports 30*. Rovaniemi: Lapin yliopisto.

Heininen, L. (2004). *Circumpolar international relations and geopolitics* (AHDR (Arctic Human Development Report) 2004) (pp. 207–225). Akureyri: Stefansson Arctic Institute.

Heininen, L. (2018). "Arctic geopolitics from classical to critical approach – Importance of immaterial factors." *Geography, Environment, Sustainability,* 11(1), 171–186. PEEX Special Issue. Ed. by S. Chalov. https://doi.org/10.24057/2071-9388-2018-11-1-1771-186

Heininen, L. (2020). Before climate change, 'nuclear safety' was there – A retrospective study and lessons-learned of changing security premises in the Arctic. In L. Heininen & H. Exner-Pirot (Eds.), *Climate change and Arctic security. Searching for a paradigm shift* (pp. 107–129). Cham: Palgrave Macmillan/Palgrave Pivot.

Heininen, L., & Exner-Pirot, H. (Eds.). (2020). *Climate change and Arctic security. Searching for a paradigm shift*. Cham, Switzerland: Palgrave Macmillan, Palgrave Pivot.

Heininen, L., & Finger, M. (2017). The 'Global Arctic' as a new geopolitical context and method. *Journal of Borderlands Studies*. https://doi.org/10.1080/08865655.2017.1315605.

Heininen, L., Everett, K., Padrtova, B., & Reissell, A. (2019). *Arctic policies and strategies – Analysis, synthesis, and trends*. Helsinki: International Institute for Applied Systems Analysis (IIASA) & Ministry for Foreign Affairs of Finland. https://doi.org/10.22022/AFI/11-2019.16175. http://pure.iiasa.ac.at/id/eprint/16175/.

Henrikson, A. (1975). The map as an "idea": The role of cartographic imagery during the second world war. *The American Cartographer,* 2(1), 19–53.

Hettne, B. (1994). The new regionalism: Implications for development and peace. In *Hettne B. – A. Inotoi: The new regionalism* (pp. 1–49). Forssa: WIDER.

Hoogenson, G., & Hodgson, K. K. (2019). 'Arctic exceptionalism' or 'comprehensive security'? Understanding security in the Arctic. In L. Heininen, H. Exner-Pirot, & Barnes (Eds.), *The Arctic yearbook 2019 – "Redefining Arctic security"* (pp. 218–230). Iceland: Thematic Network on Geopolitics and Security of University of the Arctic and Arctic Portal. On-line – http://www.arcticyearbook.com.

ICC. (2010). Inuit Arctic Policy. Inuit Circumpolar Council (ICC). https://iccalaska.org/wp-icc/wp-content/uploads/2016/01/Inuit-Arctic-Policy-June02_FINAL.pdf

Ilulissat Declaration. (2008). Arctic Ocean Conference, Ilulissat, Greenland, 27–29 May 2008.

Irwin, A. (1993). *Arctic benefits of cold war thaw*. The Time Higher.

Keil, K., & Knecht, S. (Eds.). (2017). *Governing Arctic change. Global perspectives*. London: Palgrave Macmillan.

Konyshev, V., & Sergunin, A. (2014). Is Russia a revisionist military power in the Arctic? *Defense & Security Analysis*. https://doi.org/10.1080/14751798.2014.948276.

Luzczuk, M. (2016). Military cooperation and enhanced Arctic security in the context of climate change and growing global interest in the Arctic. In L. Heininen (Ed.), *Future security of the global Arctic: State policy, economic security and climate* (p. 3554). Basingstoke: Palgrave Macmillan.

Mitrany, D. (1966). *A working peace system*. Chicago.

*Newsweek*, March 17, 2017.

Ottawa Declaration. (1996). On the Establishment of the Arctic Council, September 19, 1996, Ottawa, Canada. (mimeo).

Palosaari, T. (2019). The Arctic paradox (and how to solve it.) oil, gas and climate ethics in the Arctic. In M. Finger & L. Heininen (Eds.), *The global Arctic handbook* (pp. 141–152). Cham: Springer. Online: https://www.springer.com/gb/book/9783319919942.

Pravda. (1987). Speech by Mikhail Gorbachev at a formal meeting in Murmansk on October 1, 1987. No. 138, October 5, 1987.

President of the Russian Federation. (2020). Basics of the State policy of the Russian Federation in the Arctic for the Period till 2035. Approved by the Decree of the President (No. 164), March 5, 2020. (unofficial translation) (mimeo).

Soroos, M. S. (1990). *Arctic Haze: A case study in regime formation*. Prepared for the Arctic Cooperation Project directed by G. Osherenko and O. Young, Dartmouth College. August. (mimeo).

Southcott, C. (2010). History of globalization in the circumpolar world. In L. Heininen & C. Southcott (Eds.), *Globalization and the circumpolar north* (pp. 23–55). Fairbanks: University of Alaska Press.

Steinberg, P. E., Tasch, J., Gerhardt, H., Keul, A., & Nyman, A. E. (2015). *Contesting the Arctic. Politics and imaginaries in the circumpolar North*. London: I.B. Tauris Publishers.

The Fairbanks Declaration. (2017). *On the occasion of the tenth ministerial meeting of the Arctic Council*. https://oaarchive.arctic-council.org/handle/11374/1910

The GlobalArctic Project. (2014). (mimeo).

The Reykjavik Declaration. (2021). *On the occasion of the twelfth ministerial meeting of the Arctic Council*. https://oaarchive.arctic-council.org/bitstream/handle/11374/2600/2021%20 Reykjavik%20Declaration%2020-5-2021.pdf?sequence=1&isAllowed=y

The Rovaniemi Joint Ministerial Statement. (2019). Arctic Council. https://arctic-council.org/ images/PDF_attachments/Rovaniemi_Joint_Ministerial_Statement_2019_Signed.pdf

The Sámi Arctic Strategy. (2019). The Sámi Arctic strategy/Sami Arktalas Aigumusat/Samisk Strategi for Arktiske saker. Securing enduring influence for the Sami people in the Arctic through partnerships, education and advocacy. Saami Council, September 2019. http://www.saamicouncil.net/fileadmin/user_upload/Documents/Eara_dokumeanttat/FINAL_Saami-Arctic-Strategy_with_attachment.pdf

Young, O. (1992). *Arctic politics. Conflict and cooperation in the circumpolar North*. University Hanover, NH: Press of New England.

**Lassi Heininen** is Professor Emeritus from the University of Lapland, Finland and a Professor at the Northern Arctic Federal University, Arkhangelsk, Russia. He is the editor of the Arctic Yearbook and Chair of the Arctic Circle Mission Council on the GlobalArctic.

# Chapter 7
# The Arctic as the Last Frontier: Tourism

**Edward H. Huijbens**

## Introduction

> *"The Great North is the new exotic."* (Mian, 2018, quoted by Foresight, 2018)

Almost every policy document for regional development in the Arctic territories has identified tourism development as an opportunity. This platitude is neatly captured in the opening words of a recent report made for the Nordic Council of Ministers by a group of Arctic tourism and geography scholars.

> For more than two decades, Arctic destinations have experienced ever-growing tourism figures and an increasing global interest in the North and its attractions. This has contributed to the establishment of alternate livelihoods and new hope, at least in those places and regions that have recently suffered from deindustrialization and out-migration. Indeed, at some locations tourism development has become so dominant that it has been perceived as problematic and a phenomenon to be managed properly, while in other places of the Arctic there is still an aspiration to facilitate further growth. (Müller et al., 2020, p. 4; see also: Jóhannesson, 2016).

This quote shows the sporadic nature of tourism growth in the region. At the same time as tourism is being evoked as an alternative livelihood and one of new hope, it is mainly represented as the lowest-hanging fruit for economic diversification without any substantial engagement with its implicit potential for cultural exchange, impacts on community, and social values or natural ecosystems. If these challenges are managed properly, tourism can be a force for good on all accounts. In order to attain this, however, tourism needs to be conceived as part and parcel of myriad processes integrating the Arctic in an increasingly globalized world. As tourism hinges on access, and thus transport infrastructure, these need to be considered as key drivers of tourism and carefully negotiated as they open gateways into

E. H. Huijbens (✉)
Wageningen University, Wageningen, The Netherlands
e-mail: edward.huijbens@wur.nl

societies, nature areas and communities that need to be prepared and involved. The future of Arctic mobility presumes that as more of the Arctic opens up (due to factors such as melting sea ice and transport infrastructure investment), growing numbers of people (tourists, but also short-term and seasonal workers, etc.) will continue to arrive in the region. The key question to be addressed in this chapter is how and under which circumstances tourism can be a force for good.

Although the Arctic's exotic appeal, and that of the North more generally, has been recognized for many years (see, e.g., Ísleifsson & Chartier, 2011), tourism endeavors across the region are currently transforming it. On the most general level, this transformation involves capitalizing on the region's natural assets, peoples, landscapes, and remoteness. This chapter will bring these four aspects together under the terms of the last one – remoteness and access thereto – framing the Arctic as the last frontier to be overcome. As mentioned, all tourism development is dictated by access to a place. Coupled with attractiveness, access can turn a place into a destination and the more accessible it is, the greater the possibility for popularity and destination development. Recognizing the role of access pays heed to the fact that what constitutes a place is not a simple point on a map, or a place to 'go to', but a complex amalgam of situational factors and relations, some of which have global stretch and duration. Transport and mobilities infrastructure is typical of these relations and the ways in which these converge upon and make for a place and are actively maintained and perpetuated in locally specific manners (Massey, 2005). Tourism is just one of many frames for these converging relations that make for a place. As such, tourism can be one reason for transport infrastructure development and, thereby, the making of a place into a tourism destination. However, the identification of and possible development of tourism as an unplanned consequence of transport infrastructure development for other reasons is usually not the case, which limits the positive impact of tourism development.

The Arctic *per se* cannot be conceived of as a place, or a particular destination, unless some sweeping generalizations are deployed. Tourism flows to the Arctic constitute a vastly disparate region with several nodes of transport and infrastructure, some of which function as gateways to the surrounding hinterland of Arctic allures, whilst others become destinations in themselves. Despite this diversity within the Arctic, the region can be characterized, as in the opening quote, as 'the exotic'. In addition to its focus on access, this chapter will unravel these characteristics and explore which particular constituent parts hold merit when it comes to Arctic tourism development and how these can be framed for sustainable Arctic tourism development.

The remainder of this chapter unfolds in five sections. First, some facts and figures underlining the sporadic nature of tourism development in the Arctic will be presented. Second, I look at the role of transport, accessibility, and mobilities infrastructure in place or being developed which facilitates tourism development. Thirdly, I detail the resource factors that determine the development of tourism. Fourthly, the discussion of the chapter will move to the tourism resources of the region framed under the idea of "Arctic allures" and tie that with the research of the International Polar Tourism Research Network (IPTRN). Lastly, and somewhat building on the emergent research agenda of polar tourism, I address some of the challenges to sustainability of tourism in the Arctic from the perspectives of governance and economic development.

## Facts and Figures on Arctic Tourism Development

Reliable and regionally comparable statistics about Arctic tourism development are hard to come by and even more difficult to compile (Huijbens & Lamers, 2017, p. 2). This section details the available tentative numbers of inbound tourists and problematizes the compilation of these in the context of international tourism development. Having a handle on how many people are coming, who they are, and where they are coming from is crucial to underpin sound tourism development.

The common understanding of tourism in Arctic policy documents, as in so many national tourism strategy documents, is to see it a source for foreign revenue, employment, and economic benefits. All the benefits of tourism accrue through those visiting and the activities they undertake and services they use. Thus, these benefits arrive at the hands of so-called 'inbound tourists,' people coming from somewhere else to stay for leisure and recreation in a location they do not consider home. The tourism industry moves over 1.4 billion tourists across international boundaries and this figure is set to grow 4 percent year on year in the foreseeable future (UNWTO, 2018), barring major events like pandemics, war, international terrorism, economic recessions, or natural disasters. Even such events have proven to put only a short-lived dent in tourism growth (Gil-Alana & Huijbens, 2018). The United Nations World Tourism Organisation (UNWTO) statistics rely on data delivered by national tourism authorities in sovereign countries. However, the Arctic is an amalgam of sovereign territories, nations, and peoples. This diversity is also reflected in the intermittent availability of data on inbound tourism numbers and other indicators of tourism growth and development across the region. Fay and Karlsdottir (2011) provided an overview of useful Arctic tourism indicators (Table 7.1).

Beyond listing most of the key indicators of tourism development, Table 7.1 shows that the administrative regions of the Arctic range from Alaska via Canada and Greenland through Iceland, Norway, Lapland (northern Sweden and northern Finland) and Russia. Table 7.1 places primacy on the number of inbound tourists; that is, how many non-locals visit a region represented. How to count them is challenging, however, especially as there are multiple entry points and multiple means of entering throughout the Arctic and its individual administrative units. Moreover, not all visitors are tourists in the traditional sense of simply coming for leisure and recreation. Seasonal migrant workers (such as those in the tourism industry), scholars and students, people visiting friends and relatives, and those having second homes in the Arctic periphery, all constitute inbound people who could potentially be counted as tourists, but are generally overlooked as such, even though many of them are the most stable and loyal visitors to a given location.

**Table 7.1** Arctic tourism indicators

| Tourism indicators | Arctic tourism indicators | | | | | | |
|---|---|---|---|---|---|---|---|
| | Alaska | Canada | Greenland | Iceland | Lapland | Norway | Russia |
| *Visitors* | | | | | | | |
| Total visitors by year | ✓ | ✓ | | | | | ✓ |
| Accommodation nights by year | | | ✓ | ✓ | ✓ | ✓ | |
| Total visitors by month by year | | ✓ | ✓ | ✓ | | | |
| Visitors by mode (e.g. Air, cruise ship) by year | ✓ | | | ✓ | ✓ | | |
| Visitors by origin by year: domestic | | ✓ | ✓ | | | ✓ | |
| Visitors by origin by year: foreign | | ✓ | ✓ | ✓ | | ✓ | ✓ |
| Visitors by origin by year: Scandinavian | | | ✓ | ✓ | | ✓ | |
| Cruise ship passenger numbers by port by year | ✓ | | | ✓ | | ✓ | |
| Visitor expenditures by year | | | | | | ✓ | |
| *Domestic and foreign combined* | | | | | | | |
| Domestic | | | | | | | |
| Foreign | | | | ✓ | | | |
| Accommodation nights | | | ✓ | ✓ | ✓ | ✓ | |
| *Employment* | | | | | | | |
| Visitor related employment by month by year | ✓ | ✓ | | | | | |
| Visitor related peak July employment by year | ✓ | ✓ | | | | | |
| *Income* | | | | | | | |
| Visitor related average annual/monthly earnings | ✓ | ✓ | | | | | |

Source: Fay & Karlsdottir, 2011, p. 65

To get a sense of who is who when it comes to inbound tourism, it is important to gauge the country of origin and purpose of travel. This can only be done by surveying at the entry points upon arrival or departure, an exercise that is both costly and human-resource-intensive. In Iceland, a clever solution has been in use due to beneficial situational factors, namely to have all departing passengers state their nationality on the way to the security check at Keflavík International Airport. Touch screens with the flags of the most common visiting nationalities are located at two manned posts, so the country of origin and number of people of each nationality is well documented. However, this is possible as almost all in- and outbound tourism from Iceland goes through this one air terminal. A proxy to physically counting and surveying all inbound tourism can be accommodation statistics; that is, how many nights a person of non-local origin spends at an accommodation provider.

**Table 7.2** Inbound tourism in the Arctic regions, 2015

| Estimates of tourist numbers to a variety of Arctic Regions now by using the most current numbers available | |
|---|---|
| Country/region/province | Tourist numbers (Estimates) |
| USA (Alaska) | 2,066,800 |
| Canada | |
|    Yukon | 255,000 |
|    Northwest territories | 93,910 |
|    Nunavut | 16,750 |
|    Nunavik (Northern Quebec) | 1,000 |
|    Nunatsiavut (Northern Labrador) | 19,840 |
| Greenland | 80,862 |
| Iceland | 1,289,100 |
| Svalbard (Norway) | 118,614 |
| Norway (Nord Norge - northernmost 3 counties | 1,045,538 |
| Sweden (Norrbotten county) | 2,152,000 |
| Finland (Finnish Lapland) | 2,523,897 |
| Russia | 500,000 |

Source: Maher, 2017, p. 216

That requires the accommodation provider to report such information, which depends on several factors, such as the size and capacity of the accommodation provider, organized gathering of data, and the perceived severity of taxation in the region. However, in order to understand the purpose of travel, the only option is to ask those people considered to be tourists.

To get an idea of the sizes and figures usually promoted when it comes to Arctic tourism, Maher (2017) summarized official inbound tourism in the Arctic for 2015 (Table 7.2). Therein, the regions of the Arctic are somewhat more finely defined than in Table 7.1.

The general trend in the Arctic, as elsewhere when it comes to tourism, is that through the democratization of travel, transport technology developments, and increasing accessibility, some parts of the Arctic are witnessing increasing inbound tourism. As Table 7.2 shows, around ten million people visited regions of the Arctic in 2015. This is still a small figure in the global scheme of things, where a city like Amsterdam (with 800,000 inhabitants) receives around 19 million visitors a year. To compare globally, in 2015 more than 25 million people took boat cruises worldwide.

Surveys of visitors to the Arctic do not exist in any comparable fashion, nor is there a detailed breakdown of their origin, although it is safe to say that these are mostly people of European and North American origin, like most tourists worldwide, although China is emerging as a growing source of inbound tourism, both in the Arctic region and elsewhere (Jørgensen & Ren, 2015; Jørgensen et al., 2018; Jørgensen & Bertelsen, 2020; Natasha & Ren, 2020). So, as can be seen, the Arctic, with its four million inhabitants, has tourism, but it is regionally differentiated.

Moreover, and not reflected in Table 7.2, its growth is sporadic. In some places it is growing very rapidly whilst in others it is in decline and in others it is relatively stable. Moreover, Maher's figures only tally inbound tourism totals for those staying on land in one particular year. Another common form of visitation to the Arctic, which is growing in popularity and visibility in public debate, is cruising.

Going to the poles is a niche within cruising that has grown slowly but surely ever since the Swedish-American trail-blazer Lars-Erik Lindblad started offering tours to the Arctic in the early 1960s and Antarctica in 1966. Detailing the history of Lindblad Expeditions, the company's website states that these trips have always been:

> All with the focus of creating experiences that foster an understanding and appreciation of the most remote and pristine places on the globe. These legendary pioneering adventures led him to be defined as the father of 'eco-tourism'. (Lindblad Expeditions, 2019)

Eco-tourism's paternity notwithstanding, Lindblad was indeed the first to lead a 43-day voyage by a tourist ship through the Northwest Passage in 1984, using an ice-strengthened vessel. To date, cruise ships of growing sizes have penetrated ever deeper into the Arctic. In 2016, with much fanfare, history was made when the *Crystal Serenity* passed through the Northwest Passage with more than 1000 people on board. The trip took 32 days and was repeated in the summer of 2017, but thereafter cancelled. Larger cruise ships of this kind are not easy to accommodate in Arctic communities and safety issues abound in navigating Arctic waters; search-and-rescue capacity is limited and usually located far away. Consequently, the cruising industry is developing in the Arctic on the back of so-called expedition-type vessels, such as the *Fram*, *Hanseatic*, *Expedition*, *Bremen*, *Sea Explorer*, *Delphin*, *MS Hamburg*, *National Geographic Explorer*, *Le Soleil*, *Le Boreal*, *Sea Spirit*, *Silver Explorer*, *Polar Pioneer,* and *Albatros*. These vessels have a range of 50 to 200 passengers and are usually less than 10,000 GTs. A new generation of these expeditions vessels is in development and, in March 2016, at the annual SeaTrade Cruise Global convention in Fort Lauderdale, Florida, Crystal Cruises introduced the first-ever expedition mega-yacht to sail with a PC6 Polar Class designation. The *Crystal Endeavour* has over 100 suites for passengers, is equipped with helicopters and Zodiacs, and plies routes in the Arctic and Antarctic, along with some more equatorial destinations in the spring and autumn seasons as it repositions between the poles. It is difficult to gauge the total number of inbound tourists to the Arctic region that these vessels carry. In a given year, the number of ships venturing north in the short Arctic summer would need to be counted and its passenger capacity then multiplied by the number of trips that season. To give some indication, the number of cruise passengers coming to the port of Akureyri in northern Iceland in the summer season of 2019 were 152,692, in addition to 62,657 crew. By far the largest majority of these were German, British, and American (see www.port.is).

The quotation from Lindblad Expeditions' homepage also captures the key demand factors for Arctic tourism: pioneering to the remote and pristine exotic. However, wrapped up in the image of the cruise ship is one of the key factors determining the development of tourism anywhere prior to any branding, marketing, or image creation. This is access. It is said that for tourism to develop you need three As: accommodation, attraction, and access. Cruise ships get you there and thereby provide access. They are themselves the accommodation, which is otherwise missing in many parts of the Arctic. Regardless, aviation remains the key means of providing access to most parts of the Arctic.

## Arctic Transport Infrastructure

This section focuses on the remoteness of the Arctic and how access is being developed to ever more locations becoming destinations in the region, underlining how challenges to overcome remoteness engender questions of sustainability in the region.

In the report for the Nordic Council of Ministers cited in the introduction, the key point of departure and emphasis of the report was on the oft-neglected role and prominence of urban centers in the Arctic when it comes to tourism. The authors cite their amazement at their neglect in general tourism discussions in the Arctic as '… these centers offer not only the critical transport and hospitality infrastructure required to accommodate tourists on a large scale …' (Müller et al., 2020, p. 6). They are also tourism hotspots in themselves, such as Reykjavík in Iceland, Umeå in northern Sweden, Rovaniemi in northern Finland, and Tromsø in Norway. As such, the urban areas are destinations themselves and nodes of transport functioning as gateways to the surrounding hinterlands, which is much evident in the cruising industry.

One of the main authors of that report, Müller, had earlier (2011, p. 132) provided an overview of the challenges if tourism was to be used as a tool for regional development in the Arctic. These can be summed in the following six bullet points:

- Geographical remoteness from demand markets
- Weak internal economic linkages, dependency on imports
    - Poor information flows
- Lack of political and economic control over decision making
    - Lack of innovation
    - State intervention
- Decreasing and aging populations
    - Lack of human capital
- Poor infrastructure
- High aesthetic value due to underdevelopment

The Arctic is generally peripheral and geographically remote from the main source markets of international tourism. The region mostly depends on imports for many of the supplies to cater to visitors, and information and innovation does not flow easily through the sporadic communication networks of the Arctic. Being remote and often under the auspices of some central government in a capital city far further south, many Arctic communities lack political and economic control over their decision making. This last point is pertinent when it comes to transport infrastructure. Providing access for tourism requires infrastructure investment in, for instance, airports or harbor facilities. The communities themselves cannot finance such investment and state intervention becomes the norm.

The Arctic communities are generally small and far apart, often lacking road connections and direct aviation links between places. For instance, circumnavigating the Arctic by scheduled flights is impossible and will always lead one to zig-zag across the region, with transits in each of the regional or national capitals. Challenges of creating access and infrastructure development often result in high levels of outside investment and influence. This means that those steering the course of tourism development in the Arctic come from outside the region, from the large hubs or capitals further south. An example here in the industry context could be Greenland (Bjørst & Ren, 2015; Ren, 2016; Ren & Chimirri, 2017). In Greenland, no roads connect communities on the world's largest island and aviation links converge on the capital Nuuk or the Kangerlussuaq International Airport, which was originally built as an American military base. Whilst only sporadic helicopter links connect communities within, Greenland aims to develop its airport in Nuuk, but the costs are prohibitive to a national economy of 60,000 people, necessitating outside investment and thereby intervention. In Iceland, all domestic flights go via the capital city airport (originally built as a British military air field during World War II), whilst the international airport near Keflavík is also an old post-war American military base. Talk about having domestic and international flights in the same airport – as is the case for the Scandinavian capitals and greatly facilitates links to their respective hinterlands – is again hampered by prohibitive costs for a nation of 345,000 people.

All investment in larger infrastructure and transport links in the Arctic has come from elsewhere, nationally or in some cases internationally. So, by its nature, Arctic tourism inevitably becomes enmeshed within the global geopolitical "rush" to the Arctic. For instance, it is clear that the Arctic is gradually becoming increasingly embroiled in the geopolitics surrounding the region's energy resources and minerals (Norum, 2016; Ren, 2016). Competing Arctic policies manifest global superpowers ambitions, as Brady (2017) detailed in the case of China and Steinberg et al. (2018) documented more generally. The potential for conflict and crisis seems ripe and the US is surely preparing (Boring, 2014). While tourism features in recent Arctic policies from both the US and China, in both cases it is mentioned in passing reference as the beneficiary of infrastructure investment and access development. Indeed, the Arctic is sliding into an era featuring jurisdictional conflicts, increasingly severe clashes over the extraction of natural resources, and the emergence of a new 'great game' among the global powers; tourism plays a role therein through proposed access development (Huijbens & Alessio, 2015). As such, tourism development in the Arctic is not simply about creating access. What has to be considered is who creates it, for whom, why, and how, and this should frame debate around tourism in the Arctic region and its challenges to sustainability.

## Tourism Resources of the Arctic

This section details the resources beyond transport infrastructure development that need to be harnessed and aligned for the successful sustainable development of tourism in the Arctic.

At the opening of this chapter, the Italian journalist Marzio Mian was cited, claiming the great North is the new exotic. This exoticness is wrapped around notions of pristine wilderness at the edge of the world. The polar regions are framed as representing the only remaining great wildernesses on the planet, augmented by their remoteness and anonymity (Stewart et al., 2005; Hall & Saarinen, 2010). In addition to these, the Arctic's particular biogeographic characteristics, extreme climatic conditions, widespread perceptions of being a relatively inhospitable environment (for humans), and high levels of marine biodiversity and productivity, have been shown to be readily identified and widely recognized attractions in the region. Tourists are attracted by the pristine character of the Arctic, its sparsely or non-populated wilderness, and unique historical and cultural assets. The already cited report for the Nordic Council of Ministers talks about the "Arctification" of northern tourism in global industry and policy discourse (Müller et al., 2020, p. 5). These 'Arctified' exotic images form the staple for marketing and branding and thereby represent key resources for the promotion of tourism.

What the notion of 'Arctification' makes clear is that tourism is a people's industry, capitalizing on place-bound resources through framing these and packaging by locals or global tour operators. As such, it is about entering the minds of potential visitors, getting them to travel to the Arctic to experience attractions in-situ. As a people's industry, the valuing of Arctic tourism resources is reliant upon both locals' and visitors' ever-changing perceptions, tastes, and frames of reference. Being elusive and subjective in this manner, I refer to these Arctic resources as Arctic allures. The key challenge of tourism is to balance the constituent parts of this allure, their resonance with the local populations, and allowing for people there to experience it.

However, tourism is people's industry from the supply side as well. As the tourism assets of the Arctic are place-bound, anyone who lives in the Arctic and is able and willing to create tourism products based on the Arctic's allure is permitted to do so. Within the Arctic, however, populations are in decline, not least outside the urban hubs, and the remaining population is aging and transitioning from a resource-based economy, such as fisheries or modern versions of hunting and gathering, towards the dominant and globalized service-based economy. This results in a distinct lack of human capital to cater to tourism due to persistent outmigration from the peripheries and a skills gap with remaining populations, making tourism development in the Arctic even more dependent on outside influence and control. An example here in the industry context could be Greenland, where most tourism services are owned and operate by Danes or Icelanders and infrastructure development needs approval and subsidy from Copenhagen. Being such a people's industry, both from the perspectives of supply and demand, understanding the perceptions, expectation, and hopes invested in tourism is key to underpinning its potential benefits.

## Arctic Allures

This section focuses on the elusiveness of the Arctic tourism resources as they are constituted as attractions in the minds of the visiting tourist, locals, and the tourism entrepreneurs. In the case of the Arctic (and the Antarctic, for that matter), purity and pristine wilderness form the basis for their allure and it can indeed be problematic or even directly harmful to exploit these resources by visiting. Therefore, this section also outlines the research agendas emerging on polar tourism.

All places that become destinations go through what Butler (1980) described as the Tourist Area Life Cycle (TALC), whereby the initial "discovery" of a place by "explorers" paves the way for industry development and later consolidation of the place as a tourist destination. As visitor numbers swell, destinations grow and mature, ultimately going into decline or managing to repackage or redevelop themselves for a new cycle to start or build on the previous cycle. It is safe to say that the Arctic as a tourism region is in its "discovery" phase, with an ever-growing number of people seeking out this final frontier of exploration on the planet. Some of the trips even draw directly on the tropes of "the Heroic era" of polar exploration in the late nineteenth and early twentieth centuries. This is neatly captured in the promotion for travel with the newly polar-class *Crystal Endeavour*:

> Journeying to Earth's most remote polar regions and faraway destinations, these are true expert-led expedition voyages, combining rugged adventure and Zodiac landings with rare discovery, extraordinary wildlife and stunning landscapes few have ever seen. https://www.crystalcruises.com/experience/expedition

Tourists from across the world are attracted by the pristine character, the sparsely or non-populated wilderness areas, and the unique historical and cultural assets of the Arctic. This is often at odds with global conservation ambitions and the Arctic's key qualities when it comes to tourism; namely, that they remain relatively untouched by human activity and, for the most part, can be regarded as wilderness areas. Additionally, the Arctic is among the regions in the world where climate change is having the greatest impact, turning it into a focal point of the global environmental debate. The ramifications of climate change on the delicate ecosystem balances of the Arctic regions and the idea of them as the final frontiers of the planet elicits a type of "last chance tourism" (Lemelin et al., 2012), which is the desire for tourists to witness vanishing landscapes or seascapes and disappearing species. Going to see nature in its 'natural' state and 'authentic' indigenous communities before both get tainted or spoiled by the modern world of Western consumer capitalism is becoming a pressing last chance in the minds of many. Moreover, seeing polar bears before they become extinct, or visiting melting glaciers to experience the profundity of global climate change is also a potent driving force for Arctic tourism exploitation. The complex social relationships loosely inscribed under the umbrella term of the 'tourism encounter' with nature and natural environments (Gibson, 2010; Abram & Lien, 2011; Pearce et al., 2017; Cheer et al., 2019) takes on a particular guise in the Arctic. Visitor expectations that revolve around seeing what is pure nature or some ur/authentic wilderness stand at odds with both the ways in which these expectations have been historically framed and socially constructed through, for example, the legacy of Arctic exploration and the fact that several measures and investments need to be made to allow the presence of people right there.

Moreover, international tourism can be considered the greatest cultural confluence of our times and a world-making force (Hermann et al., 2019). The cross-cultural exchanges that take place when visitors, often from far away, encounter a particular place or community have generally been theorized and analyzed from the tourism industry perspective in terms of entrepreneurship, innovation, product development, or, more broadly, in terms of its economic impact or regional development (Viken & Granås, 2016). If wrested from the rather simplistic framing of visiting 'authentic' indigenous communities, tourism development and visitation can build on understanding local community expectations, their conduct of engaging with visitors, and understanding visitor expectations and conduct. As such, tourism has the potential to enhance community pride and the valuing of traditions (Higgins-Desbiolles, 2006), something that is especially pertinent in the context of the many indigenous communities and nations in the Arctic, living within the Arctic states. In this sense, community empowerment through tourism in the face of legacies of colonial exploitation holds a particular appeal for the tourism encounter in the Arctic.

## *An Emerging Research Agenda*

In order to bring all the above considerations together, the International Polar Tourism Research Network (IPTRN) has developed since 2006, focusing on tourism and its intersection with environmental, cultural, and economic issues in Polar Regions. The network has held six community-embedded workshop conferences: Montreal/Kangiqsujuaq, Canada (2008), Abisko, Sweden (2010), Nain, Canada (2012), Christchurch/Akaroa, New Zealand (2014), Raufarhöfn, North-East Iceland (2016), and, most recently, in the Yukon, Canada in June of 2018. The seventh was planned for November 2020 in Ushuaia, southern Argentina, but was postponed till 2022 due to the COVID-19 crisis. A publication is available from each of the six events, but Stewart et al. (2017) emphasized that the development of the polar tourism scholarly network is highly fragmented to date, with the exception of a dense principal core group of researchers around the IPTRN. These researchers thematize the focus of polar tourism scholarship in eight general categories: development, management, experience, global change, governance, impact, community, and reviews. Exemplified in these themes of research are ways to tackle issues of sustainability, but these need to be further consolidated and will be addressed below. Table 7.3 reproduces Stewart et al.'s (2017, p. 77) key questions pertaining to five consolidated themes of research into polar tourism.

The five proposed themes for research into polar tourism, shown in Table 7.3, summarize the key issues that will dictate the terms of tourism development in the Arctic. Questions about who is coming and why, who is catering to them and how, along with questions of the governance of Arctic tourism and its relation to other

**Table 7.3** Advancing new thematic areas for polar tourism research

| Theme | Key questions |
|---|---|
| 1. Changing tourism demand | How have polar tourism visitors changed?<br>How have tourism products changed and evolved? |
| 2. Understanding new polar actors | How does nationalism influence polar tourism development?<br>What are the values of new polar actors and how might these influence future tourism development? |
| 3 Governance and regulation | What is effective and efficient governance for tourism in the polar regions?<br>How will regulatory mechanisms evolve to keep pace with rapid developments in polar tourism? |
| 4. Global change | What is the relationship between polar tourism and climate change?<br>What role could tourism play in observing environmental changes in the polar regions?<br>What is the need, provision, and use of environmental information necessary for safe travel within the polar regions? |
| 5. Influence of new technology | What is the influence of new technology on tourist experience and behavior in the polar regions?<br>How might the management of polar tourism be enhanced through novel, remote monitoring, and observing solutions? |

Source: Stewart et al.'s (2017, p. 77)

globalized flows and relations, are key focus areas to understand the development of the sector and whether it can live up to promises in terms of culture and nature. The path chosen in the answers to each of these will dictate the degree to which tourism will develop in a manner we can consider sustainable and good.

## Challenges of Sustainability

This last section will summarize the challenges to sustainability of tourism in the Arctic that have been highlighted throughout the chapter from the perspectives of governance and economic development.

In a *New York Times* editorial dated 8 August 2017, the main focus was a tenacious question in the Arctic realm: "What is the best way to make certain that the newly exposed potential of the Arctic region is exploited peacefully and without further damage to the environment?" Tourism is indeed a newly exposed potential of Arctic exploitation, but, as an industry, tourism sets itself apart from those that mine for minerals or harvest wildlife. As a people's industry reliant on communication, image building and people's perceptions and expectations, tourism development can be seen both as a contributor to and a victim of its own development, with potential implications for natural resource use, ecosystems, and peripheral communities. The Nordic Council of Minister's report emphasized this capricious nature of tourism. Moreover;

> ... increasing Arctification and internationalization may be quite volatile in the North. An extensive focus on tourism for international export markets, at the expense of other forms of tourism development, makes the sector extremely vulnerable to externally caused "boom and bust" cycles. (Müller et al., 2020: p. 15)

Beyond vulnerability to external factors, Snyder (2007, p. 5) summed up the sustainability challenge of tourism to the Arctic as:

- Conserving environmental quality
- Preserving cultural and social values by means of participatory decision-making
- Creating sustainable economies
- Ensuring positive visitor behavior, safety, and enjoyment

From a global perspective, challenges pertaining to sustainable development have been framed as the Sustainable Development Goals (SDGs) launched in 2015. These built on the measured successes and evaluation of the Millennium Development Goals set in 2000. Following the launch of the SDGs, the UNWTO set about focusing on its tourism context. During the International Year of Sustainable Tourism for Development, the UNWTO identified three goals particular for the tourism industry. These are goal 8 (on decent work and economic growth), goal 12 (about responsible production and consumption), and goal 14 (focused on the sustainability of life under water) (see Saarinen, 2020; Bramwell et al., 2017).

However, the challenge remains that tourism in the Arctic is focused on growth, as it is globally, and many tourism scholars find it hard to reconcile this dominant growth paradigm with notions of sustainability (see, e.g., Bianchi & de Man, 2020). As Boluk et al. (2019, p. 859) pointed out:

> ... sustainable tourism was the industry's way to rationalise the consumption of the environment, commodifying it for the tourists' gaze and enforcing its preservation as an exclusive amenity for advantaged tourists.

Bianchi and de Man (2020) argued that, overall, the goals pertinent to tourism fail to address the actual links between tourism and poverty, environmental degradation, exploitation of resources and inequalities; all inherently aspects created and sustained within the current capitalistic neoliberal 'growth' economy.

This growth paradigm certainly plays out in the Arctic context. Tourism is praised for creating jobs, diversifying the hitherto resource dependent economy and providing income to local communities. However, a question that is often not asked is, "what kind of employment it creates, and for whom" (Saarinen, 2020, p. 5). When the ownership and management decisions for tourism development, access creation, and service provision are in the hands of proprietors and political decision makers in capital cities further south or different countries, potential equitable job creation is removed from locals and economic leakages abound. The paying jobs that locals have a chance to obtain often remain low-skilled service tasks and/or 'front-line' hospitality work. Tourism development dictated by terms set elsewhere not only taps into local communities as mere labor resources, but also the natural resources of a site and its infrastructures to an extent that locals are potentially deprived. In the

Arctic context, this could be, for instance, the buying up of land that is valued for other purposes or means by the locals, to build a hotel or other services. Tourism development can tap into existing service resources such as food provision, electricity, and water beyond their built capacity, hiking prices with increasing demand by large-scale tour operators to the detriment of the quality of life for locals. As such, the 'growth' of the tourism sector as proposed in SDG goal 8, does not always work positively when looking at the reality of local communities in a tourism destination. As Bianchi and de Man (2020) pointed out, it is often the locals who pay the price for environmental degradation and resource exploitation that can be the inherent consequence of the 'growth' and continued development of the tourism industry. The sustainability challenge for the Arctic then revolves around what tourism development in the region could look like whilst paying heed to decent work. The challenge would be how to pace the growth of the industry to avoid exceeding local community capacity and to foster responsible production and consumption. Moreover, goal 14, focused on the sustainability of life under water, is particularly relevant when it comes to large tracts of the Arctic region, as most communities are coastal and rely on the sea for sustenance. The challenge is how to balance the demand for wildlife watching and experiencing with the fact that this wildlife is hunted and traditionally consumed by many local communities. The controversies and challenges thereof have been, for instance, explored in the context of whaling and whale watching. In a recent book chapter entitled 'feasting on friends', the fact that puffins and whales are also eaten in Iceland by the very people who go to marvel at them in their habitat leads to the following conclusion:

> Developing responsibilities to our common earthly future and attuning to more than human rhythms of life are arguably afforded to those going whale and puffin watching. However, for this to happen, these affordances and encounters with marine and avifauna must be informed by storytelling recognizing this potential. The bodily consumption of this marine and avifauna arguably holds less potential for this attunement, yet informs us of our earthly cohabitation with non-humans from a more historical perspective. Both have particular values in developing ethics of hospitality, but it is imperative that these values be recognized if tourism's encounter with wildlife in their habitat or on a plate is to be hospitable. The former affordances connect us to our common future on this Earth, while the latter prompts us to thinking how we are one with this planet and all that lives on it. (Huijbens & Einarsson, 2018, p. 23)

The above quote pertains to the fact that beyond the SDGs and local articulations of sustainability challenges for Arctic tourism, the Arctic is one of the world's regions where climate change is shown to be the most profound and most readily observed, turning them into an exemplar of the global environmental debate. Indeed, the Arctic is often touted as the 'canary in the coalmine' when it comes to climate change. At the same time, a growing number of tourist visits in the Arctic regions by long-haul travelers increases the use of high-emission air transportation, which is at odds with the climate change mitigation goals (Scott et al., 2012). To counter these somewhat, it is imperative that newly accessible areas receive visitors in a way that can help preserve communities' natural and cultural resources and facilitate some sense of historical continuity and contribute to their future viability within a rapidly globalizing world. Thus, balancing the allure of the Arctic with the rush to exploit it relates to our current planetary state of environmental emergency and it is imperative for Arctic tourism to be developed in tenable ways in this light (Gren & Huijbens, 2016).

The Arctic and the Antarctic are the last holistic frontiers of capital accumulation and consumption on the planet. From this perspective, an analysis of the driving forces of tourism offers us a great way to peak into the cultural framing of the polar regions for commercial purposes. Moreover, from the perspective of climate change and how the polar regions are exoticized through last chance tourism, tourism's problematic nature emerges. That is, what is fueling travel to the Arctic, literally and metaphorically, is the very stuff that is bringing about its demise. In sum, growing tourism numbers bring a host of environmental challenges, including long-recognized wildlife disturbance (Erize, 1987), invasive species and pathogens (Frenot et al., 2005), and fossil fuel emissions (Eijgelaar et al., 2010), such as in relation to wildlife disturbance (Dawson et al., 2010). Geiling (2016) pointed to the inherent value of the Arctic landscapes and animal life:

> It's really a matter of balancing the value of that experience and the educational opportunity of that experience with the inherent value of nature and species that are not simply there for our use and our entertainment. To try and balance those two is difficult. (Geiling, 2016, quoting Judith Stark)

Thus, framing the Arctic as a venue for commercial exploitation through tourism inherently ties the Arctic to our environmental consciousness and challenging ethical questions about our current co-habitation with planet Earth and its other inhabitants.

## Concluding Remarks

This chapter has demonstrated that tourism development across the Arctic realm remains sporadic, largely unplanned, and not premised on an engagement with the myriad sustainability challenges wrapped up in its development. Facts and figures are difficult to compile and paint a sketchy picture of the industry across the region. The key challenge to tourism development is the remoteness of potential destinations in the Arctic realm and the fundamental difficulty of providing access to these and overcoming other aspects of peripherality, as defined by Müller (2011, p. 132). Therefore, this chapter has focused on the challenges raised when aiming for the successful development of the tourism industry in the Arctic from the perspective of transport and mobilities and how this links with a particular geography of the Arctic. At the same time, creating access poses major challenges to tourism development in terms of who controls that key. Invested therein too boot are the sustainability challenges surrounding the fact that all access to the remote Arctic is long-haul and fossil-fuel based, contributing directly to the climate change transforming the world that people are so eager to see before it disappears. But there are also other concerns. Through the commodification of indigenous cultures, to make them consumable and visitable, a loss of identity and cultural sustainability becomes a concern. And with ownership of tourism enterprises generally in the hands of outsiders, the risk is great that the economic and employment effects of tourism in the Arctic remain bound to menial and low-paid service jobs with limited prospects of income or innovation for those involved. In this sense, the three pillars of sustainability remain a formidable challenge to address in the Arctic tourism context, not least the environmental one.

At the same time, the promises of tourism are great. The development of tourism products can enhance community pride and the economic viability of many communities in the Arctic through diversifying their income basis. Tourism as a tool for cross-cultural interaction and encounters also holds great promise in terms of bringing traditional and indigenous means of world-making and understanding into communication with the rest of the world, possibly reorienting our values and relations to land to the benefit of the environment. However, these promises fundamentally rest on the ways in which tourism is perceived and how it is implemented and by whom. Making tourism an equitable tool for regional development in the Arctic remains a challenge, with few best practices to share so far. The starting point would be to make tourism development an explicit part of infrastructure development and, in doing so, ensure the interests of the communities opened up to the world through this development have a clear say in how it is meant to unfold.

# References

Abram, S., & Lien, M. E. (2011). Performing nature at World's ends. *Ethnos: Journal of Anthropology, 76*(1), 3–18.
Bianchi, R. V., & de Man, F. (2020). Tourism, inclusive growth and decent work: A political economy critique. *Journal of Sustainable Tourism, 29.* https://doi.org/10.1080/09669582.2020.1730862
Bjørst, L. R., & Ren, C. (2015). Steaming up or staying cool? Tourism development and Greenlandic futures in the light of climate change. *Arctic Anthropology, 52*(1), 91–101.
Boluk, K. A., Cavaliere, C. T., & Higgins-Desbiolles, F. (2019). A critical framework for interrogating the United Nations sustainable development goals 2030 agenda in tourism. *Journal of Sustainable Tourism, 27*(7), 847–864.
Boring, K. T. (2014). *Operational Arctic: The potential for crisis or conflict in the Arctic region and application of operational art*. School of Advanced Military Studies.
Brady, A.-M. (2017). *China as a polar great power*. Cambridge University Press.
Bramwell, B., Higham, J., Lane, B., & Miller, G. (2017). Twenty-five years of sustainable tourism and the journal of sustainable tourism: Looking back and moving forward. *Journal of Sustainable Tourism, 25*(1), 1–9.
Butler, R. W. (1980). The concept of a tourist area cycle of evolution: Implication for management of resources. *Canadian Geographer, 24*, 5–12.
Cheer, J. M., Milano, C., & Novelli, M. (2019). Tourism and community resilience in the Anthropocene: Accentuating temporal overtourism. *Journal of Sustainable Tourism, 27*(4), 554–572.

Dawson, J., Stewart, E. J., Lemelin, H., & Scott, D. (2010). The carbon cost of polar bear viewing tourism in Churchill, Canada. *Journal of Sustainable Tourism, 18*(3), 319–336.

Eijgelaar, E., Thaper, C., & Peeters, P. (2010). Antarctic cruise tourism: The paradoxes of ambassadorship, "last chance tourism" and greenhouse gas emissions. *Journal of Sustainable Tourism, 18*(3), 337–354.

Erize, F. J. (1987). The impact of tourism on the Antarctic environment. *Environment International, 13*(1), 133–136.

Fay, G., & Karlsdottir, A. (2011). Social indicators for Arctic tourism: Observing trends and assessing data. *Polar Geography, 34*(1–2), 63–86.

Foresight. (2018). *Arctic tourism: How the Great North is becoming the new exotic*. From: https://www.climateforesight.eu/future-earth/arctic-tourism-how-the-great-north-is-becoming-the-new-exotic/. Accessed 14 Apr 2020.

Frenot, Y., Chown, S. L., Whinam, J., Selkirk, P. M., Convey, P., Skotnicki, M., & Bergstrom, D. M. (2005). Biological invasions in the Antarctic: Extent, impacts and implications. *Biological Reviews, 80*(1), 45–72.

Geiling, N. (2016). *Visiting melting glaciers can be Profound. But is it morally wrong?* Smithsonian.com From: https://www.smithsonianmag.com/science-nature/visiting-melting-glaciers-can-be-profound-it-morally-wrong-180960514/. Accessed 20 Feb 2018.

Gibson, C. (2010). Geographies of tourism: (un)ethical encounters. *Progress in Human Geography, 34*(4), 521–527.

Gil-Alana, L. A., & Huijbens, E. H. (2018). Tourism in Iceland: Persistence, seasonality and long memory behaviour. *Annals of Tourism Research, 68*, 20–29.

Gren, M. G., & Huijbens, E. H. (Eds.). (2016). *Tourism and the Anthropocene*. Routledge.

Hall, C. M., & Saarinen, J. (2010). *Tourism and change in polar regions: Climate, environments and experiences*. Routledge.

Hermann, I., Weeden, C., & Peters, K. M. B. (2019). Connecting the dots: Ethics, global citizenship and tourism. *Hospitality and Society, 9*(1), 3–8.

Higgins-Desbiolles, F. (2006). More than an "industry": The forgotten power of tourism as a social force. *Tourism Management, 27*(6), 1192–1208.

Huijbens, E. H., & Alessio, D. (2015). Arctic 'concessions' and icebreaker diplomacy? Chinese tourism development in Iceland. *Current Issues in Tourism, 18*(5), 433–449.

Huijbens, E., & Einarsson, N. (2018). Feasting on friends. Whales, puffins and tourism in Iceland. In C. Kline (Ed.), *Animals as food: Ethical implications for tourism* (pp. 10–27). Routledge.

Huijbens, E. H., & Lamers, M. (2017). Sustainable tourism and natural resource conservation in the Polar Regions: An editorial. *Resources, 6*(45). https://doi.org/10.3390/resources6030045

Ísleifsson, S., & Chartier, D. (Eds.). (2011). *Iceland and images of the North*. Presses de l'Université du Québec and Reykjavík Academy.

Jóhannesson, G. T. (2016). A fish called tourism: Emergent realities of tourism policy in Iceland. In R. van der Duim, C. Ren, & G. T. Jóhannesson (Eds.), *Tourism encounters and controversies: Ontological politics of tourism development* (pp. 181–200). Ashgate.

Jørgensen, M. T., & Bertelsen, R. G. (2020). Chinese tourism in the Nordic Arctic – Opportunities beyond the economic. *Scandinavian Journal of Hospitality and Tourism, 20*(2), 166–177.

Jørgensen, M. T., & Ren, C. B. (2015). Getting 'China ready': Challenging static and practice based configurations of the Chinese tourist. *Asia Europe Institute – Insights, 1*(1), 19–35.

Jørgensen, M. T., Law, R., & King, B. E. (2018). Beyond the stereotypes: Opportunities in China inbound tourism for second-tier European destinations. *International Journal of Tourism Research, 20*(4), 488–497.

Lemelin, H., Dawson, J., & Stewart, E. J. (2012). *Last chance tourism. Adapting tourism opportunities in a changing world*. Routledge.

Lindblad Expeditions. (2019). *Our history. The lindblad legacy of respectful cruise tourism*. From: https://world.expeditions.com/about-us/lindblad-history/. Accessed 15 Jan 2019.

Maher, P. T. (2017). Tourism futures in the Arctic. In K. Latola & H. Savela (Eds.), *The interconnected Arctic — UArctic congress 2016* (pp. 213–220). Springer Polar Sciences.

Massey, D. (2005). *For space*. Routledge.
Mian, M. G. (2018). *Arctic. The battle for the Great White North*. Neri Pozza.
Müller, D. K. (2011). Tourism development in Europe's last wilderness. In D. K. Müller & A. A. Grenier (Eds.), *Polar tourism. A tool for regional development* (pp. 129–153). Presses de l'Université du Québec.
Müller, D. K., Carson, D. A., de la Barre, S., Granås, B., Jóhannesson, G. T., Øyen, G., Rantala, O., Saarinen, J., Salmela, T., Tervo-Kankare, K., & Welling, J. (2020). *Arctic tourism in times of change: Dimensions of urban tourism*. Nordic Council of Ministers.
Natasha, M., & Ren, C. (2020). Adventurous Arctic encounters? Exploring Chinese adventure tourism. *Scandinavian Journal of Hospitality and Tourism, 20*(2), 126–143.
Norum, R. (2016). Barentsburg and beyond: Coal, science, tourism and the geopolitical imaginaries of Svalbard's "New North". In G. Huggan & L. Jensen (Eds.), *Postcolonial perspectives on the European high north: Unscrambling the Arctic* (pp. 26–52). Palgrave Pivot.
Pearce, J., Strickland-Munro, J., & Moore, S. A. (2017). What fosters awe-inspiring experiences in nature-based tourism destinations? *Journal of Sustainable Tourism, 25*(3), 362–378.
Ren, C. (2016). Cool or hot Greenland? Exhibiting and enacting sustainable Arctic futures. *Journal of Cleaner Production, 111*(Part B), 442–450.
Ren, C., & Chimirri, D. (2017). *Turismeudvikling i Grønland: Afdækning og inspiration (tourism development in Greenland – Identification and inspiration)*. Aalborg University.
Saarinen, J. (Ed.). (2020). *Tourism and sustainable development goals. Research on sustainable tourism geographies*. Routledge.
Scott, D., Hall, C. M., & Gossling, S. (2012). *Tourism and climate change. Impacts, adaptation and mitigation*. Routledge.
Snyder, J. (2007). *Tourism in the polar regions: The sustainability challenge*. UNEP/Earthprint.
Steinberg, P. E., Tasch, J., & Gerhardt, H. (2018). *Contesting the Arctic politics and imaginaries in the circumpolar north*. I.B. Tauris.
Stewart, E. J., Draper, D., & Johnston, M. E. (2005). A review of tourism research in the polar regions. *Arctic, 58*, 383–394.
Stewart, E. J., Liggett, D., & Dawson, J. (2017). The evolution of polar tourism scholarship: Research themes, networks and agendas. *Polar Geography, 40*(1), 59–84.
UNWTO. (2018). *UNWTO tourism highlights 2018 edition*. UNWTO.
Viken, A., & Granås, B. (2016). *Tourism destination development: Turns and tactics*. Routledge.

**Edward H. Huijbens** is an Icelandic geographer and chair of Wageningen University's research group in cultural geography (GEO) since early 2019. Before this Edward was directing the Icelandic Tourism Research Centre, 2006-2015 and department head at the University of Akureyri 2017-2019. Edward works on spatial theory, issues of regional development, landscape perceptions, the role of transport in tourism and polar tourism.

# Chapter 8
# A Primary Node of the Global Economy: China and the Arctic

**Liisa Kauppila**

## Introduction

Over the last decade, China's entrance into the Arctic has become a hot issue among local communities, academics, and practitioners of politics and businesses alike. On one hand, China's domestic environmental problems, practices in Africa, and growing tensions with the United States are raising concerns on the nature and consequences of China's Arctic engagement. On the other hand, the world's second-largest economy's interest in the region is seen as an opportunity to carry out capital-intensive projects of national interest, improve socio-economic conditions of less developed areas, and invigorate such industries as polar tourism. Whether viewed more as a risk or an opportunity, it seems clear that, along with climate change, non-Arctic China's increasing interest in the High North – and its national plan to build a Polar Silk Road (冰上丝绸之路) as part of its Belt and Road Initiative/一带一路 (BRI) – is perhaps the most transformative feature shaping the future of the Global Arctic.

The aim of this chapter is to shed light on what it is that drives China to participate in Arctic affairs by setting the country's engagement in the High North in its

---

The author wishes to thank Outi Luova, Rasmus Gjedssø Bertelsen, Juha Vuori, Yue Wang, Egill Thor Nielsson, and the editors for the useful comments that have enabled her to improve the chapter.

L. Kauppila (✉)
University of Turku, Turku, Finland
e-mail: lllkau@utu.fi

© The Author(s), under exclusive license to Springer Nature Switzerland AG 2022
M. Finger, G. Rekvig (eds.), *Global Arctic*,
https://doi.org/10.1007/978-3-030-81253-9_8

wider global scheme – what I call *primary node vision*.[1] Manifesting a spatial imagination parallel to that of geoeconomics, China's primary node vision seeks to (re) position the country at the center of global flows in order to secure the continuation of economic growth at home. Echoing the empirical fact that China has a "distinct predilection" to organize its global engagement on a "regional basis" (Kavalski, 2009, p. 4) – that is, vis-á-vis Africa or the Arctic, for example, rather than single nation-states – the execution of the primary node vision can be cut into analytical pieces by employing the idea of *China's functional economic regions*: distinct 'spaces of flows' (Castells, 1989), dynamic systems that connect China with major corners of the world. Within this conceptual framework that challenges the dominant Euro-centric reading of space, China's engagement with the High North can be studied by tracing the flows that constitute the China–Arctic functional economic region: a complex system that provides China with liquefied natural gas (LNG), knowledge of and technology for shipping in polar waters, and satellite data for China's satellite navigation system.

Despite the China-specific nature of the conceptual framework employed in this chapter, the empirical phenomenon of functional economic regions itself is by no means unique to China's rise. Instead, the general idea of countries as nodes of large-scale functional economic regions – master clusters of flows – extending beyond their borders is both a universal and historical phenomenon that globalization has accelerated. Moreover, similar domestic economic systems of flows can equally be found within the borders of each country. As for this chapter, it is crucial to recognize the *coexistence and overlap* of these state-level and domestic functional economic regions *with nodes inside and outside Arctic localities* since their interplay is essentially what constitutes the Global Arctic. Situating China's vision and region-building efforts within this larger flow-centric Arctic political economy (cf. Aaltola et al., 2014) brings into play the aspirations of other nodes and thus underlines that, although it is collaboration that spawns functional economic regions, competition is an inevitable undercurrent of global life.

The remainder of this chapter is structured as follows. Section 2 begins by introducing the flow-centric conceptual framework through which China's global economic outreach (that is, its primary node vision) and its region-based operationalization, are studied. Section 3 provides an overview of the political goals of China's primary node vision; which sets the scene for Sect. 4, an empirical part that discusses China–Arctic flows of (1) natural resources, (2) seaborne goods, and (3) technology, knowledge, and data. As both China's Arctic engagement and the primary node vision itself are as much about future potential than about actual developments, this section also discusses possible and probable developments that

---

[1] The idea of China becoming a 'primary node' of a multi-*nodal* world was coined by Womack (2014); however, Womack's use of the concept differs from the interpretation presented here. For Womack, reaching the status of a primary node is not an aim of an intentional vision, but a characteristic of the future world order: "one in which concerns of conflicts of interests drive interactions, but no state or group of states is capable of benefiting from unilaterally enforcing its will against the rest" (Womack, 2014, p. 265).

might shape China–Arctic flows in the short, medium, and long term. Finally, the conclusion elaborates on the future prospects and implications of the emerging China–Arctic functional economic region.

## China's Primary Node Vision and Functional Economic Regions

In many ways, China has been a true winner of globalization. Had its economic reforms and opening up (改革开放) not coincided with the accelerated compression of time and space between societies (Harvey, 1990), the country's economy would not have expanded at an average rate of 10% throughout the 1980s, 1990s, and 2000s, and yet well above 6% each year throughout the 2010s. In practice, becoming connected to global flows of trade, natural resources, and technological innovations has enabled China to reach the status of an upper middle-income economy, lift over a billion people out of absolute poverty, and raise the general standard of living of all Chinese people (Naughton, 2018, pp. 1–7). Although China-bound investments and exports are gradually giving way to domestic consumption as the driver of growth, access to both material and immaterial overseas resources is just as, if not more, important than ever. Indeed, compared with the high tide of growth, China is now much more dependent on overseas energy resources and minerals, partly because the country is seeking to reduce its dependency on coal. Furthermore, China's grand task to transform itself from the 'world's factory' to an innovative economy cannot succeed without collaboration with global trailblazers of science and technology.

China is currently seeking connectivity and access to global flows through a region-based strategy that organizes the country's global economic engagement vis-á-vis major *macro regions*; or, to use a flow-centric term, *macro localities:*[2] units consisting of groups of nation-states or parts of them (Söderbaum, 2016, p. 109). In other words, China seeks to position itself at the center of global flows by building relations with clusters of states, which it organizes in units of multiple scales quite creatively, ranging from entire *continents* (such as Africa) to what are known in mainstream international relations (IR) theories as *sub-regions*, parts of continents (for example, Central and Eastern Europe). This tendency to favor macro localities as key units of China's encounters on a global stage is visible in the way China's BRI, the actual political initiative, is both advertised and carried out. For example, most Chinese BRI maps do not have national borders, but regions, corridors, and cities (Godehardt, 2016, p. 18).

Previous literature has often studied China's encounters with macro localities in general, and its Arctic engagement in particular, as a form of cross-regional diplo-

---

[2] To avoid confusion with *functional economic region*, the term *locality* is used here to refer to 'traditional' – what geographers call *formal* – regions (cf. Aaltola et al., 2014).

macy, an approach that treats China and, for example, the Arctic as two separate units whose actors (mainly governments) encounter with each other (e.g., Lanteigne et al., 2020; Kopra & Kauppila, 2018; Bennett, 2014 for a different approach). The analytical framework used here challenges this view by bringing forth the idea of China's functional economic regions, an alternative perspective that is both flow-centric and China-centric.

Unlike mainstream IR theories, the approach of China's functional economic regions does not view regions as blocks of terrestrial space belonging to the sovereignty of clusters of nation-states (cf. Buzan, 2012, p. 22). Instead, it draws from geographers' functional-relational reading of regions, which challenges the idea of established, 'pre-existing', and somewhat static regional spaces (Agnew, 2013, p. 12). Indeed, China's functional economic regions are *functional* due to their operating logic: they take shape around specific activities, such as shipping of goods. In other words, these regions *function as units* and, in a similar vein, cease to exist once the functions that hold them together stop. Given their dynamic nature, these regions have no recognized boundaries, but are best described, in the words of Castells (1989), as *spaces of flows*. While geographers often distinguish issue-specific functional regions (for example, a city and its commuting zones), China's functional economic regions should instead be viewed as *master clusters of issue-specific systems*; that is, sets of flows forming larger and more complex units manifesting a relational dynamic, but maintaining a degree of territoriality.

These regions are *China's* because China is the node/center of these master clusters and they exist *solely* to connect the territory of China with the rest of the world through flows of different types. In other words, all the movement along the linkages of these regions is "to do with China". In this sense, these regions are subjective constellations (see Agnew, 2013, p. 8), regions *for* China, or for the Chinese economy in particular. Indeed, these regions are *economic* because components and drivers of growth move along their links to (and from) China and the needs of Chinese economy are what keep them together. At the risk of oversimplification, it can be argued that from China's perspective these regions are global extensions of China's national economy, whereas from a global perspective these zones can perhaps be seen as spheres of China's economic influence. Finally, and to crack the theoretical core of the concept, they are *regions* because they can be seen as *distinct spatial units*, distinguishable from the rest of the world (cf. Buzan, 2012, p. 22). In other words, they are regions because the world can be cut into meaningful (analytical) pieces by organizing space in these distinct systems (cf. Agnew, 2013, pp. 7–8).

Against this backdrop, a broader argument can be made. In general, China's global economic engagement is guided by what can be called a primary node vision: an idea of a desirable future world in which China is positioned as a central place, a node to which major macro localities are connected through functional economic regions. Each of these regions is different in its make-up since different types of flows dominate China's exchanges with each macro locality. Although virtually all macro localities possess natural resources that are of interest to China, there are great differences in China's ability and willingness to benefit from them. In a similar vein, some corners of the world are more attractive for their level of technological knowledge and skills than for their extractive industries. Furthermore, these regions themselves also change over time, sometimes very quickly, to best serve the needs of the Chinese economy *at any given time*. Therefore, and as is the case with all functional regions, a traditional map – still image – does not do justice for these regions; instead, a real-time video would capture their ethos.

Finally, the flows making up China's functional economic regions are by no means unidirectional, although China-bound flows are the focus here. Flows of capital constitute by far the most important outbound dimension of these regions, as Chinese economic actors invest in industries that the government encourages (for example, by issuing regulations) and thus views as suitable regarding the goals of its primary node vision. However, money is not all that China contributes to these systems; goods, labor, and flows of tourism also move from China to the world within the scope of these regions. Furthermore, the movement that makes up these regions is not entirely in China's control; unwanted and potentially dangerous flows are inevitably created as their by-products. Especially the flows of knowledge constitute a potential risk factor for realizing the goals of the primary node vision since they are the hardest to control for unwanted content.

## Political Aims of the Primary Node Vision

As China's primary node vision seeks to secure continuation of economic growth at home, it is relevant to ask what ends this growth ultimately serves. The vision has two coexisting, somewhat overlapping but yet different, political goals. First, the continuation of growth bolsters the legitimacy of the Chinese Communist Party (CCP); specifically, its right to rule the country. Ever since the economic reforms began in 1978, growth has constituted the most important source of the CCP's performance-based legitimacy, "passive acceptance that the party will deliver economic growth for the people on their behalf" (Breslin, 2013, p. 44). However, it is not the *growth itself* that makes the CCP legitimate, but the *material wellbeing* that has followed growth: elimination of absolute poverty, employment, rising disposable incomes, and general improvements in overall socio-economic conditions. Nevertheless, it is the party itself that largely *sets and changes* the standards against which both growth and material wellbeing are measured. For example, the current lower rates of growth are the "new *normal*" only because the CCP has adjusted its targets to make them reachable. In a similar vein, the party's acceptance of environmental protests, pollution reduction campaigns (for example, the *War on Pollution*) and issuing of such plans as *Healthy China 2030* have made the relationship between growth and pollution widely known, thus constructing a clean environment as a perceived and aspired source of material wellbeing. In practice, this has come to mean that the fight against pollution is a matter of regime survival – a question of life-and-death – for the CCP.

Second, growth also serves another purpose that brings political ethics to the discussion: it helps the Chinese state to act out its *fundamental responsibility* to provide material and social conditions for good life, in a manner that promotes unity and stability, the foundational values of the Chinese state (Kopra & Kauppila, 2018, n.d.; cf. Jackson, 2000, pp. 170–172). If the first goal was a realist one, giving only instrumental value to human wellbeing, this second aim is an idealist one, lifting humans and values to the ultimate core. For this aim, growth is equally instrumental for advancing material wellbeing, but here material wellbeing is a prerequisite for keeping the society united and stable, a task that corresponds with the historically constructed and evolved core priorities of the Chinese people (Kopra & Kauppila, 2018, n.d.). In sum, growth is positively linked to the very attributes that, as Fairbank (1983) argued, Chinese people view as constitutive of their community and being as Chinese – not in the narrow Western-style nationalist sense, but in a culturalist sense (p. 461).

Thus, given this two-fold task, China's primary node vision can be located between realist and idealist realms. This is hardly surprising since both the needs of the state and one's population are always present in the conduct of economic foreign policy and should not be viewed as mutually exclusive aims (cf. Jackson, 2000, pp. 170–172). However, pinpointing the idealist motive is particularly important, since recognizing the population as the moral referent object of states' global conduct somewhat strips off the authoritarian stigma of China's global economic engagement. From this perspective, China's global outreach is similar to Western democracies in that they both simply seek to foster the 'needs and creeds' of their citizens and, like all great powers throughout modern history, China goes overseas to get what it needs to fulfil this task (cf. Economy & Levi, 2014). Hence, paying attention to the idealist side of the coin has the potential to promote genuine dialogue between different political systems.

Finally, making the case that China's vision is primarily motivated by domestic concerns is not the same as arguing that power struggles would not have *any* role in Chinese thinking. It is clear that the ability to exert influence across the world through functional economic regions is largely beneficial for the CCP, not least because it further bolsters the party's performance-based legitimacy, one of the components of which is prowess on the international stage (cf. Shullman, 2019). However, this does not mean that China's functional economic regions could be reduced to mere instruments of Luttwakian geoeconomics. Despite the dynamic of the current US–China rivalry, it would be a misinterpretation to argue that the overall rationale behind China's primary node vision is to use 'methods of commerce' to 'outdo' others on the global scene (Luttwak, 1990, p. 19).

## China–Arctic Functional Economic Region

The world's northernmost corner, the Arctic, is one of the most recent additions to which China is applying its macro locality-based strategy. Often viewed as the land and waters above the Arctic Circle (approximately 66° north), the Arctic has begun to attract the interest of many states. Due to climate change, the impacts of which escalate in the High North, as well as the effects of globalization and development of technology, the Arctic is gradually becoming a scene of ambitious economic activities. Ever since the 2000s, China has increased its presence in the High North to meet the goals of its primary node vision. In addition to possessing vast reserves of natural resources and offering a relatively stable and surprise-free operating environment (Stepien et al., 2020, pp. 109–110), virtually all Arctic economies are characterized by their advanced science and technology sectors. Furthermore, the Arctic's location vis-á-vis the North Pole makes it lucrative for obtaining satellite data, making short-cuts by ships and researching the impacts of climate change. This subchapter analyses Chinese efforts to utilize Arctic opportunities by tracing China–Arctic flows of (1) natural resources, (2) seaborne goods, and (3) technology, knowledge, and data.

## *Natural Resources*

Securing a sufficient supply of natural resources is the key to China's economic development and a prerequisite for material wellbeing. All in all, China's total demand for energy and minerals is the largest in the world and, despite the country's predilection for self-sufficiency, domestic production alone cannot meet the needs of its gradually wealthier population of 1.4 billion people. The country has been a net importer of energy since the 1990s, and in 2018 imports accounted as much as 72% of China's total consumption of oil and 43% of natural gas (BP, 2019). The Chinese economy is equally dependent on overseas minerals (iron ore, copper, uranium etc.), with imports amounting to an estimated 40% of the country's total demand (AZO Mining, 2019).

The natural resources of the global North are gradually gaining a more significant role in China's primary node vision – although the global South still dominates in this aspect. In part, the CCP wants to prevent such image losses that the country's engagement with Africa has created. Being a latecomer in the global resource game, China has had to rely on the deposits neglected by Western energy giants – a strategy that has forced Chinese companies to operate in conditions that favor questionable practices (Ziegler, 2011, pp. 197–198). A more obvious reason, however, is the simple fact that new opportunities are now emerging across the global North – most notably in the Arctic, which is estimated to hold 13% of the world's undiscovered oil and 30% of natural gas supplies (US Geological Survey, 2008).

Interestingly, the very challenges that have prevented many Western countries from benefiting of Arctic natural resources are allowing China to strengthen this dimension of its functional economic region. First, Western investors with top-notch know-how have been relatively uninterested in the Arctic since sanctions were set on Russia and the price of petroleum sank in 2014. This has created a window of opportunity for China to benefit from Russia's Arctic visions. Second, unlike their counterparts in democratic countries, Chinese actors do not face domestic opposition from environmental groups. Although this is not to argue that the special risks of Arctic operations would not affect Chinese companies' perceptions of the High North as a business environment (Interview with a Chinese energy company representative, Beijing, March 2016), it is somewhat more straight-forward for China to proceed with Arctic resource extraction.

In terms of natural resources, flows of LNG dominate the make-up of the China–Arctic functional economic region for the time being. This is hardly surprising given the Chinese government's plans to increase the share of natural gas to 10% by the end of 2020 and to 15% by 2030 (State Council of the PRC, 2016a). This move aims to reduce China's high dependence (58% in 2018) (BP, 2019) on environmentally harmful coal and ultimately tackle the risk that pollution poses to the CCP's legitimacy and the wellbeing of Chinese citizens. In 2018, as much as 60% of the volume of China's natural gas imports comprised of LNG (O'Sullivan, 2019), the least polluting fossil fuel. LNG has much smaller volume than the gaseous state of the substance, which makes it an ideal fuel for ships. Furthermore, it can be shipped from places where pipelines cannot be built either due to the long distance or political risks (U.S. Energy Information Administration, 2020). However, shipping of LNG over long distances is expensive and the short-cuts that the opening Arctic sea routes offer can contribute to significant savings. Furthermore, given that LNG must be kept below $-161\ °C$ to maintain its liquid form, the Arctic climate also makes the transportation process automatically more energy efficient.

Currently, flows of LNG are mainly created in the context of the Yamal LNG project, the first Arctic LNG mega scheme that started its operations in 2017. Run by the Russian independent gas-producing company Novatek (50.1% ownership), Yamal LNG conducts all stages of LNG production from extraction and liquefaction to transportation. The project's Chinese shareholders are the state-owned China National Petroleum Corporation (CNPC) (20% ownership) and the Silk Road Fund (9.9%), an investment vehicle of the Chinese government. In addition to Russian and Chinese partners, the French publicly owned Total S.A. has an ownership share of 20% (Novatek, 2020a). With its estimated output totaling around four billion barrels of oil equivalent, the Yamal LNG project is not only a pioneering scheme, but it also is producing significant flows of LNG. Moreover, it has so far, exceeded its annual capacity: in 2019, the project's output was as high as 18.4 million tons of LNG (Yamal LNG, 2020). Although the exact details of China's share of the project's output are not disclosed, the country is its largest recipient market and China-bound shipments have been regular since July 2018 (Novatek, 2018).

In the future, flows of LNG from the Russian Arctic to China will increase further. CNPC and another state-owned Chinese company, China National Offshore Oil Corporation (CNOOC), are also 10% shareholders each in another LNG project operated by Novatek: Arctic LNG 2. The project started in 2018 with early-stage site preparations and the construction of infrastructure, but it is not expected to start operations until the end of 2022. Arctic LNG 2 is even more ambitious than Yamal LNG in its production capacity, with its expected total output being around seven billion barrels of oil equivalent. On an annual basis, this accounts for 19.8 million metric tons (Novatek, 2020b).

Both Yamal LNG and Arctic LNG 2 utilize the Northeast Passage (NEP), particularly its Russian Northern Sea Route (NSR) and the connected Asia-Pacific transport corridor, for shipping LNG to China during the summer season. The NEP is a waterway that connects China to Europe through waters off the Russian north coast, offering a 30% shorter distance than the Suez Canal route. Under very favorable conditions, shipments can be made twice as fast as through the Suez route, even in 15 days. Although the traditional summer season on the NEP has lasted from July to November, in 2020 a successful sailing took off as early as mid-May, transporting a load from Yamal to China within just 25 days – faster than would be possible through any available alternative route (Novatek, 2020c). Although this was only possible with the assistance provided by a nuclear-powered ice-breaker, this test sail indicates that, in the future, the China–Arctic functional economic region could consist of year-round flows of LNG.

Given the current downturn in Sino-US relations, the China–Arctic functional economic region is not expected to expand to the East (from China's geographic position) through flows of LNG any time soon. Although a memorandum of understanding (MoU) on China-Alaska LNG collaboration was signed in 2017, negotiations between Sinopec, Chinese financial institutions and Alaska Gasline Development Corporation are on hold (as of April 2021). If realized under the initial conditions in the future, three-quarters of the project's LNG output would flow to China, equaling the country's share of the scheme's funding (Chang, 2018). This case illustrates the difficulties that China will encounter in establishing its functional economic regions and seeking to act out its primary node vision globally.

In addition to LNG, flows of minerals are likely to contribute to the make-up of the China–Arctic region in the foreseeable future. For now, Greenland's rare earth elements (REE) are most attractive to Chinese companies. Notably, several projects are in planning stages, a MoU has been signed, and rights have been obtained to extract resources in different parts of the island (Greenland Minerals Ltd., 2018; NS Energy, 2020). As for meeting the goals of China's primary node vision, REEs are needed in the technology sector and electronics industry. They are essential, for example, in developing green technology to support China's "War on Pollution" and as components of batteries, color TVs, and smartphones, many of which are made in China. Uranium in particular is used in nuclear power plants to produce what the Chinese view as green energy.

Although the actual Arctic part of Canada (above 66 °N) has seemed less attractive to Chinese mining companies, there was a significant surge in Chinese interest in Canadian resources during the period of 2010–2016 (Lajeunesse & Lackenbauer, 2016). However, all negotiations were stalled for reasons that are worth discussing here. Based on media reports (e.g., Beeby, 2016; see also Stepien et al., 2020, pp. 112–113), it seems that Canadian legislation, strict impact mitigation requirements, lengthy decision-making processes, and the importance of contributing to community relations were making the projects less attractive. Furthermore, in one planned project, the idea of bringing Chinese labor to Canadian mines met strong resistance from locals. Finally, the Canadian government does not allow foreign entities to act as a majority shareholder in the mining of certain elements, most importantly of uranium, which restricts the scope of resources that China could access – all other issues notwithstanding. Chinese companies are likely to encounter similar problems in the future in, for example, Finland's Lapland, the country's Arctic part that is naturally rich in copper and nickel.

Chinese investments in Arctic natural resources and related infrastructure may also advance a phenomenon that can perhaps be called as 'Finlandization with Chinese characteristics' (cf., e.g., Kivimäki, 2015). Traditionally, Finlandization has referred to a process in which a smaller country acts in a manner that supports the foreign policy goals of its powerful *neighbor*, as an attempt to manage and safeguard the bilateral economic relationship. In a globalized flow-centric world, however, the term neighbor has somewhat lost its meaning as an indicator of a particularly high degree of economic dependence. Furthermore, given that *economic aspects of security* are at the heart of China's primary node vision, not only – or even necessarily – those possessing a shared border but also many other small countries across the world may constitute crucial links that should "toe China's line". Arguably, especially Nordic countries' increasing economic interdependence with China does constitute a setting in which governments may begin to shy away from advancing liberal democratic values in their foreign policy with China. However, there has not been much clear evidence of such a tendency so far; on the contrary, both Norway and Sweden have had diplomatic quarrels with China over human rights issues – despite the fact that expressing such views has clearly affected Norway's salmon and Sweden's tourism industries (e.g., Milne, 2013; Elmer, 2019).

## Seaborne Goods

Ensuring a smooth flow of seaborne goods is crucial not only for China's foreign trade, but also for the Chinese-backed construction of infrastructure across the world, in localities with which China builds its functional economic regions. Especially raw materials and semi-finished goods, mostly transported in bulk by sea, are crucial for the Chinese economy, which continues to gain around 27% of its gross domestic product (GDP) from manufacturing (World Bank, 2020), despite the country's attempts to undergo a structural transformation and shake off a reputation of being the world's factory.

China has traditionally utilized routes running through the Pacific, the Indian Ocean and the Atlantic, but terrorism and piracy are creating a strong impetus for Chinese companies to look for alternative routes (e.g., Hong, 2012). Furthermore, the global shipping industry has been struggling since the Global Financial Crisis reduced cargo flows; this trend was further strengthened by the sudden reduction of volumes caused by the COVID-19 outbreak. Although state subsidies have buffered the hit, Chinese companies are under pressure to cut the costs of their operations. One method of obtaining savings is to reduce the need for fuel by choosing shorter routes, such as the NEP. Reducing the distance travelled may also be a viable strategy to meet the increasingly strict global environmental standards set for shipping, as shorter distance may mean less fuel and less pollution (Kauppila & Kiiski, 2020, pp. 466, 481). This logic favors Arctic sea lanes, even if it seems irrational to propose Arctic shipping as a "green" alternative.

The Arctic shipping routes were officially attached to the BRI on two occasions; first in June 2017 (State Council of the PRC, 2017) and again in January 2018 when the country's first Arctic policy paper came out. While the former document only picked out the NEP, or the seaway "leading up to Europe via the Arctic Ocean" (State Council of the PRC, 2017), the actual Arctic strategy covered all three Arctic passages under the term Polar Silk Road (State Council of the PRC, 2018). In addition to the most promising sea lane NEP, the other two passages are the Northwest Passage (NWP), linking China with North America through Canadian and American waters, and the Transpolar Passage (TPP), which cuts straight across the Central Arctic Ocean near the North Pole. Although these shipping routes are, for the time being, only either emerging or prospective options, the decision to elevate the Arctic waterways with the status of the Polar Silk Road means that Arctic shipping is viewed as a viable part of the primary node vision. This positive attitude towards Arctic seaways was further expressed by including polar shipping in the list of development aims of China's transport sector (State Council of the PRC, 2019) and the development of the Polar Silk Road in the *14th Five-Year Plan* (State Council of the PRC, 2021).

Currently, flows of China–Arctic seaborne goods are still infrequent and few. In fact, it can be argued that China's Arctic shipping activity has hardly reached the stage where flows are genuinely created; instead, Chinese companies are *testing* Arctic shipping to learn how to transport goods in a feasible, safe, and sustainable manner. China's first test sailing on the NEP took place in 2012, as its research icebreaker *Xuelong* made a round-trip from Shanghai to Iceland. Although Xuelong is not a commercial ship, lessons learned about the route's conditions during the pioneering voyage have been significant for the successful completion of the following business-oriented test sailings (Kauppila & Kiiski, 2020, p. 475). The first of these voyages took place the following year, in 2013, when China Ocean Shipping Group Company's (COSCO) *Yongsheng* sailed from Busan to Rotterdam via the NEP. In October 2015, China's Arctic shipping took a new turn, as COSCO published its plans to open a frequent shipping service on the NSR (SCMP, 2015). Although this plan is yet to be realized, the company has continued its test sailings actively and, by October 2018, COSCO had conducted as many as 26 trials on the NSR. Specific details cannot be accessed, but the fact that COSCO used general cargo and heavy lift vessels on these voyages gives grounds to suggest that, in addition to break bulk (goods loaded individually in containers, bags, barrels, etc.), loads were related to construction materials for the Yamal LNG project (Kauppila & Kiiski, 2020, p. 475; cf. Humpert, 2018). Furthermore, in 2019, COSCO's general cargo vessels transited the NSR (crossed the area without calling into ports) seven times (NSR Information Office, 2020).

Only two Chinese test sailings have been conducted on the other two Arctic passages, mainly due to the extremely harsh operating conditions that prevail along these routes. In 2012, *Xuelong*'s first test sailing through the NEP was combined with a voyage that utilized the routes making up the TPP (Arctic Portal, 2012). As the TPP runs through international waters, its opening for frequent shipping in the long term – perhaps only in the 2050s – can potentially shape the China–Arctic region most dramatically. Via TPP, seaborne goods could be shipped between China, Europe, and North America through the shortest Arctic passage without having to rely on the jurisdiction and regulations of any Arctic state. As for the NWP, Xuelong conducted China's first test sailing on the route in 2017. Although the sailing took place under a research permit issued by the Canadian authorities, the Chinese state media emphasized the value of the experience for the country's future cargo shipping activities. This created turbulence in Canada, where the right to navigate the NWP is a sensitive issue due to long-lasting sovereignty disputes with the US and where environmentalists are actively voicing their concerns over shipping in Arctic waters (Fife & Chase, 2017). Despite this incident, Chinese commercial interest in the route was further confirmed the following year by a NWP seminar held by COSCO (COSCO, 2018). However, these developments may be less indicative of a real plan to open a frequent route on the NWP than of a Chinese shipping company seeking to create a media buzz that attracts potential customers for Arctic shipping in general (Kauppila & Kiiski, 2020, pp. 475–476).

From the perspective of Arctic economies, the Chinese vision to utilize the Arctic shipping routes may mean investments and, consequently, new risks. An illustrative case to discuss is the prospective Chinese-funded *Arctic Corridor*, which would link a potential NEP hub, the Arctic harbor town of Kirkenes in northern Norway, with Finland's northern "capital" Rovaniemi through a new railway – and then extend all the way to continental Europe via the Helsinki-to-Tallinn ("Talsinki") tunnel (Arctic Corridor, n.d.; Yle, 2019). While the Chinese capital would enable the realization of this mega plan and possibly alter the socio-economic development prospects of the involved northern towns and areas, such investments would also make the so-called *debt trap* a possible long-term scenario. A debt trap refers to a situation in which a powerful lender has significant leverage over an indebted smaller country (Chellaney, 2017) – an outcome that is linked to China's overseas lending, since the country is not bound by the rules of the OECD, which are often seen to protect the borrowing nations (e.g., Massa, 2011). Furthermore, the recent 99-year lease of the Sri Lankan port of Hambantota as an attempt to settle unpaid loans has brought such worst-case scenarios into global attention (e.g. Ondaatjie & Sirimanne, 2019). Although the countries directly involved in the Arctic Corridor (Finland, Norway, and Estonia) are less likely to end up in Sri Lanka's position given their status as developed economies, the intensified debate over such risks has already curbed enthusiasm about attracting Chinese investments in the Nordics (cf. Mattlin, 2020). Furthermore, the Chinese-funded railway would also run across the reindeer herding areas of the Sámi people and affect their culture in myriad ways (e.g., Dickie, 2019), making the future of the already vulnerable population even more uncertain. Finally, any economic project of this scale, as well as the increasing flows of cargo, would inevitably have an impact on the fragile Arctic nature – whether the plan involves Chinese actors or not.

## *Technology, Knowledge, and Data*

Flows of technology, knowledge, and data are in a crucial role in facilitating China's transformation from the world's factory to an innovative economy that produces high-tech value products. This shift is necessary not only to maintain growth, but also to tackle the risk that polluting heavy industries pose to the legitimacy of the CCP and the material wellbeing of the Chinese people. The goal of making the Chinese economy more knowledge-intensive was made very visible in the country's *Made in China 2025* strategy (State Council of the PRC, 2015), *13th Five-Year Plan* (State Council of the PRC, 2016a), and again in the *14th Five-Year Plan* (State Council of the PRC, 2021). Most recently, partly because of the uncertainties that the Sino-US trade war has created, the highest leadership-led discussion has shifted from China becoming an innovative economy to as far as steering the Chinese economy towards high-tech independence (Zheng, 2020).

As mentioned above, the Arctic economies are known for their science and technology prowess. Furthermore, both the special global challenges created by climate change and the exceptional operating conditions of the High North attract global science and technology leaders to participate in Arctic economic projects. For this reason, the Chinese view the Arctic context as a fruitful environment for collaboration that facilitates technology, skill, and knowledge transfers.

For now, China–Arctic flows of technology stem mainly from the above-mentioned energy projects in northern Siberia, as well as shipbuilding collaboration. The Yamal LNG project has attracted both Chinese and global interest due to its pioneering nature. As a project that includes both LNG production and transportation in Arctic conditions, the Yamal LNG has allowed the Chinese side to learn about the special challenges and chosen solutions, which have been developed based on the knowledge of Novatek and other project partners. The actual LNG plant itself is built on permafrost, ground that is constantly frozen, in an area with winter temperatures as low as −50 °C. Operating in such conditions requires special engineering solutions, for example regarding pile foundations (Total, 2020). As for shipping LNG, Chinese companies lacked the experience to pass through (specific areas of) Arctic waters before entering the project – skills that Russian, northern European, and Japanese companies were seen to possess (interview with a representative of a Chinese energy company, Beijing, March 2016). The joint venture of China Shipping Ltd. (part of COSCO since 2015) and Japanese Mitsui, in particular, allowed the Chinese side to learn how to ship LNG safely with the multipurpose ice-breaker tankers that were specifically designed for the project (interview with a representative of a Japanese shipping company, Tokyo, October 2016).

Novatek's second LNG project in the Russian Arctic differs from the Yamal LNG project most strikingly in that the Arctic LNG 2 is a combined offshore-onshore project. It utilizes onshore resources but constructs an over-the-water liquefaction plant that is grounded by gravity-based structures typical of offshore oil platforms (Novatek, 2020b), thus requiring a different kind of technological expertise. The ambitious project has attracted a notable range of leading international companies despite the Western sanctions imposed on Russia. Along with the main shareholders Novatek (60%), CNPC (10%), CNOOC (10%), Total (10%), and a Japanese consortium of Mitsui and Japan Oil, Gas and Metals National Corporation (Jogmec) (10%), German Siemens provides advanced equipment for the project (Novatek, 2019). In many ways, it is fair to argue that just like the Yamal LNG project did for Yamal Peninsula, Arctic LNG 2 has turned Gydan Peninsula into a hot spot of high-class engineering that Chinese actors want to be involved in.

As for shipbuilding, the Chinese are now actively studying the construction of a specialized fleet for Arctic ventures. China's first domestically built ice-breaker, *Xuelong II*, was constructed in collaboration with Aker Arctic, a Finnish engineering company that designs, develops, and tests ice-going vessels. Aker Arctic won the basic design for the ship in 2012 and supported the construction process at Jiangnan Shipyard in Shanghai by providing expertise in structural and technical design and ice performance prediction from late 2016 onwards. Furthermore, the company participated in Xuelong II's sea trials in May–June 2019 to ensure that the design and performance objectives were met. During the process, Aker Arctic collaborated closely not only with the Polar Research Institute of China (PRIC), who ordered the ship, but also with China Marine and Ocean Engineering Design and Research Institute, the Chinese design establishment (Niini, 2019). In this way, the Sino-Finnish collaboration created not only flows of technology but also related knowledge.

The collaboration with the Finnish ice technology leader has given impetus to further domestic construction of different polar vessels, which also was explicitly encouraged in the National Development and Reform Commission's (2019) instructions for industrial structuring. The country's state-owned shipbuilding companies are currently planning on building vessels for transporting LNG and cargo; in other words, for business. For example, Hudong-Zhonghua Shipyard presented a model of its ice-going LNG carrier in January 2020, a first of its kind in China. The design process involved collaboration with both Aker Arctic, which contributed by providing ice basin test facilities, and the Russian Krylov Research Centre Institute, which participated in optimizing the design (Eiterjord, 2020). Furthermore, in June 2018 the China National Nuclear Corporation, a state-owned enterprise that oversees China's nuclear projects, opened a bid for shipbuilders to build a nuclear-powered icebreaker and provide technological support for the government (Chen, 2018). Given the nature of the bid, as well as the fact that construction of nuclear icebreakers is illegal in many countries, it is not surprising that a domestic actor, Shanghai Jiaotong University, won the bid (Eiterjord, 2019).

In addition to these joint ventures and collaboration arrangements, science activities and academic exchanges contribute to the China–Arctic functional economic region for flows of knowledge and data. As for science, Arctic expeditions and science and satellite stations are crucial venues for facilitating Chinese attempts to understand the effects of climate change, which poses a significant risk to Chinese economic development and people through extreme weather events and changes in patterns of food production. China's Arctic (and Antarctic) expeditions, known as

CHINARE, are run by the PRIC, an institution operating under the Ministry of Natural Resources (自然资源部). More than 10 voyages in different Arctic areas have been made between Xuelong's first sailing (1999) and the latest expedition that was conducted with Xuelong II in July–September 2020. As discussed in the previous sections, despite their scientific nature, CHINARE missions have served the needs of the Chinese Arctic engagement rather broadly, thus corresponding with the mandate of the PRIC to conduct scientific, technological and strategic research and monitor the environment in polar regions (PRIC, 2011). Regardless of such national importance of CHINARE, the missions have been rather international by nature since they have engaged a number of overseas scientists.

Currently, there are three Chinese state-owned or -controlled Arctic science and/or satellite stations: Yellow River Station (黄河站) in Ny-Ålesund in Norway's Svalbard, China Remote Sensing Satellite North Polar Ground Station (CNPGS) near Kiruna in Sweden and China Icelandic Arctic Observatory (CIAO) near Akureyri, northern Iceland. The science stations in Svalbard (established in 2003) and Iceland (2018) are both run by the PRIC, although CIAO is a collaboration project with the Icelandic Centre for Research (Rannís) as the local partner. The two establishments have a similar focus: to explore Northern lights, glaciers, ice, oceans, and the atmosphere; in other words, to conduct research that can enhance Chinese understanding of the effects of climate change and the universe in general.

China's first fully owned overseas land satellite station, CNPGS, was set up by the Institute of Remote Sensing and Digital Earth, an arm of China's national science think tank Chinese Academy of Sciences (CAS) in 2016. CNPGS seeks to improve China's capability to receive global remote sensing data, which is made possible by the station's location near the North Pole: satellite data can be downloaded better because the polar orbiting satellites (machines passing above the two poles) fly over the station (Chen, 2016). Generally, satellite data is needed for providing weather forecasts and researching climate change, but China also wants to develop its BeiDou Navigation Satellite System (BDS), an equivalent of the American Global Positioning System (GPS). In the early 2020s, the system is supposed to reach a stage of providing its users geolocation and time information globally (State Council of the PRC, 2016b). This would mean that navigation systems in vehicles, maps in mobile phones, robots and emergency services could rely on the Chinese system – instead of the American version. This way, CNPGS is not only contributing to flows of data and knowledge, but also technology.

All three Chinese Arctic outposts, as well as those that are only in the planning stages, have raised concerns over their dual-use potential, which refers to the use of these stations for military purposes; that is, monitoring polar regions and sharing satellite data with the armed forces. These concerns, and the *China threat* discourse in general, seem to be gaining more ground as China's primary node status becomes consolidated and the rivalry with the US is intensified. Unsurprisingly, the most vocal criticism has come from the US (see, e.g., Pompeo, 2019), but concerns have also been expressed in the Nordic countries (e.g., Mattlin, 2020).

Some conferences and joint research settings should also be mentioned as key arenas for creating China–Arctic flows of knowledge. Out of all Arctic conferences, China has perhaps been most visible in Iceland's Arctic Circle Assembly, the most important Arctic discussion forum. Iceland also had a decisive role in establishing China-Nordic Arctic Research Centre (CNARC), a 'bilateral' platform for academic collaboration that has been operational since 2013. CNARC organizes annual symposiums and an exchange program, which is coordinated from CNARC's headquarters located at the PRIC. In addition to facilitating knowledge production and sharing, as well as cultural exchanges, CNARC is notable in that it has a practice of organizing a business roundtable as part of the annual symposium (Níelsson, 2019, p. 59). Similar bilateral focus is also evident in the plan to establish a Sino-Russian joint laboratory to ensure Arctic-related knowledge sharing, technology transfer, and, notably, talent cultivation (Yan, 2019), which is a strengthening trend in Chinese economic thinking.

Finally, given the importance of the primary node vision for its legitimacy, it is clear that the CCP faces a strong pressure to ensure that public opinion across its network of functional economic regions remains favorable towards its goals. For this reason, it cannot be ruled out that the above-mentioned contexts in which Sino-Arctic flows of knowledge are created would not constitute venues for the use of *sharp power*, authoritarian governments' attempts to shape the public opinion across the world in a manner that is normal in their own countries, but what in Western democracies is deemed as questionable (Walker & Ludwig, 2017). In practice, Arctic academics may face pressure to censor their research and public presentations in the media, for example, to secure the chance to enter China in the future (cf. Stepien et al., 2020, p. 118). In Australia, there has also been speculation whether some Chinese university students keep track and report back home on anti-China activities and stances within academia (Hamilton, 2018). So far, there are no reports of such instances in the Arctic context.

## Conclusion

This chapter has illustrated various ways in which China is shaping the future of the Global Arctic. By adopting a flow- and China-centric approach to Arctic regional affairs, it has proposed an alternative perspective that – unlike the dominant discourse – does not view China as an "external" actor in the Arctic. This standpoint underlines the empirical fact that China already is an important Arctic stakeholder that is in the High North to stay. Furthermore, the approach encourages observation of global life from the perspective of Chinese spatial imagination, thus challenging the dominant Euro-centric reading of space. At the same time, however, the idea of the China–Arctic functional economic region is not completely disconnected from the conventional reading of the Arctic region: many formal regions have, in fact, started off as functional units. Therefore, it makes sense to ponder whether, in the medium-to-long term, belonging to the Arctic (formal) region will continue to be based on possessing Arctic geography, or whether the shared attribute that makes the Arctic a distinct unit will be something that "lets China in".

China–Arctic exchanges, varying from energy projects to science activities, create flows that contribute to the continuation of growth, structural transformation of the Chinese economy and, ultimately, material wellbeing of the Chinese people. The Arctic is attractive as a scene of technologically advanced business projects that allow technology transfer and learning to take place – a crucial goal for a country that seeks to become an innovative economy. The Arctic also possesses ample reserves of natural resources, which are increasingly produced into such refined forms of energy as LNG – the most sustainable fossil fuel with a clear role in the "greenification" of the Chinese economy. Furthermore, as the effects of global warming culminate in the Arctic, the locality is a "hot spot" of climate-change-related science projects that create flows of knowledge and data, which are essential in tackling the risks that extreme weather events cause to China's economy and population.

As discussed above, China is not going anywhere from the Arctic. On the contrary, it is likely that the scope of the China–Arctic functional economic region will expand in the future, even if nothing else but the current projects in Russia proceed in a business-as-usual manner. Furthermore, there is considerable interest in attracting Chinese investments, especially in Greenland, Finland, and Iceland – the three Arctic countries that have, so far, had unproblematic ties with China. Chinese companies have already signed MoUs and contracts that enable them to gain a foothold in Greenland's mining industry. At the same time, it is important to underline that the future development of the China–Arctic region is also shaped by many factors that China must simply accept. Clearly, the most obvious uncertainty is climate change itself. Although it is an unfortunate fact that humankind has somewhat lost the fight against climate change, it may advance slower than anticipated. For this reason, the feasibility of the Polar Silk Road is subject to great uncertainties and, especially, the TPP might not become navigable even in the long term.

Several issues may also hinder Chinese engagement with Arctic communities. These include resistance from local people and activists, strict legislation of Arctic countries, long processing times of permissions, and fears of the dual-use potential of Chinese stations. Also, diplomatic frictions can quickly alter the direction of promising developments, as the case of Alaska LNG illustrates. This chapter has also raised Finlandization with Chinese characteristics, as well as debt-trap diplomacy and sharp power, two methods that may be used to *intentionally* "Finlandize" countries, as risks that especially small Arctic countries may potentially face. While there is little or no clear evidence of such practices in the Arctic context, it is noteworthy that the global debate revolving around such authoritarian forms of influencing has already shaped the Nordic perceptions of China. Moreover, such mega-plans as the Arctic Corridor and the Helsinki-to-Tallinn tunnel have highlighted the negative consequences that Chinese-backed infrastructure may have for the fragile Arctic nature and the indigenous population and traditional lifestyles.

Finally, although the purpose of the China–Arctic functional economic region is to serve the continuation of growth within China, its existence and scope is also, in the first place, completely dependent on the performance of the Chinese economy.

Simply put, China's Arctic engagement will only increase if its companies, financial institutions, and scholars continue to have funding. Virtually, this means that a prolonged, deep recession in China would not only impact the material wellbeing of Chinese citizens, but also threaten the livelihoods of those Arctic communities that begin to rely on Chinese investments. Naturally, this also works the other way around: good performance of Chinese economy may save livelihoods, create jobs, and provide income in more underdeveloped parts of the Arctic.

## References

Aaltola, M., Käpylä, J., Mikkola, H., & Behr, T. (2014). *Towards the geopolitics of flows. Implications for Finland*. Finnish Institute of International Affairs. https://www.fiia.fi/wp-content/uploads/2017/01/fiia_report_40_web.pdf

Agnew, J. A. (2013). Arguing with regions. *Regional Studies, 47*(1), 6–17.

Arctic Corridor. (n.d.). *Growth through Arctic resources. A rising cross-border economic area*. Region of Northern Lapland. https://arcticcorridor.fi

Arctic Portal. (2012, August 23). *Xuelong to sail through future central route*. https://arcticportal.org/ap-library/news/827-xuelong-to-sail-through-future-central-route

AZO Mining. (2019, October 16). *China: Mining, minerals and fuel resources*. https://www.azomining.com/Article.aspx?ArticleID=53

Beeby, D. (2016, June 7). *Chinese mining companies feel misled by Canada, report says*. CBC News. https://www.cbc.ca/news/politics/china-mining-ambassador-investors-infrastructure-1.3619228

Bennett, M. M. (2014). North by Northeast: Toward an Asian-Arctic region. *Eurasian Geography and Economics, 55*(1), 71–93.

BP. (2019). *BP statistical review – 2019: China's energy market in 2018*. https://www.bp.com/content/dam/bp/business-sites/en/global/corporate/pdfs/energy-economics/statistical-review/bp-stats-review-2019-china-insights.pdf

Breslin, S. (2013). *China and the global political economy*. Palgrave Macmillan.

Buzan, B. (2012). How regions were made, and the legacies for world politics. In T. V. Paul (Ed.), *International relations theory and regional transformation* (pp. 22–46). Cambridge University Press.

Castells, M. (1989). *The informational city: Information technology, economic restructuring, and the urban regional process*. Blackwell.

Chang, J. (2018, January 30). Alaska natural gas project is promising and a win-win. *China Daily*. http://Www.Chinadaily.Com.Cn/A/201801/30/WS5a7091c5a3106e7dcc137ae9.Html

Chellaney, B. (2017, January 23). China's debt-trap diplomacy. *Projectsyndicate*. https://www.project-syndicate.org/commentary/china-one-belt-one-road-loans-debt-by-brahma-chellaney-2017-01

Chen, N. (2016, December 16). *China's first overseas land satellite receiving station put into operation*. http://www.english.cas.cn/newsroom/archive/news_archive/nu2016/201612/t20161215_172471.shtml

Chen, Y. (2018, June 27). 我国首艘核动力破冰船揭开面纱——将为海上浮动核电站动力支持铺平道路 [*China's first nuclear-powered icebreaker unveiled – paving the way for powering offshore floating nuclear power plants*]. Science and Technology Daily/Xinhuanet/科技日报. http://www.xinhuanet.com/politics/2018-06/27/c_1123041028.htm

COSCO. (2018). *COSCO SHIPPING SPE. Organized seminar for navigation along Arctic Northwest Passage*. http://www.coscol.com.cn/En/News/detail.aspx?id=11720

Dickie, G. (2019, June 5). A proposed railway in the Arctic has investors excited and – And indigenous groups terrified. *Pacific Standard*. https://psmag.com/environment/kirkenes-proposed-railway-from-europe-to-asia-investors-excited-indigenous-groups-terrified

Economy, E., & Levi, M. (2014). *By all means necessary: How China's resource quest is changing the world.* Oxford University Press.

Eiterjord, T. A. (2019, September 5). *Checking in on China's nuclear icebreaker.* The Diplomat. https://thediplomat.com/2019/09/checking-in-on-chinas-nuclear-icebreaker/

Eiterjord, T. A. (2020, January 14). *China's shipbuilders seek new inroads in arctic shipping.* The Diplomat. https://thediplomat.com/2020/01/chinas-shipbuilders-seek-new-inroads-in-arctic-shipping/

Elmer, K. (2019, August 13). *Chinese tourist numbers dip, but trade with Sweden continues to grow despite tensions.* SCMP. https://www.scmp.com/news/china/diplomacy/article/3022602/chinese-tourist-numbers-dip-trade-sweden-continues-grow

Fairbank, J. K. (1983). *The United States and China* (4th ed.). Harvard University Press.

Fife, R., & Chase, S. (2017). *China used research mission to test trade route through Canada's Northwest Passage.* The Globe and Mail. https://www.theglobeandmail.com/news/politics/china-used-research-mission-to-test-trade-route-through-canadas-northwest-passage/article36223673/

Godehardt, N. (2016). *No end of history. A Chinese alternative concept of international order.* German Institute for International and Security Affairs. https://www.swp-berlin.org/fileadmin/contents/products/research_papers/2016RP02_gdh.pdf

Greenland Minerals Ltd. (2018, August 21). *Greenland minerals sets growth plan with Chinese shareholder.* https://ggg.gl/investors/news-release/20180821-businessnews/?l=da_GL

Hamilton, C. (2018). *Silent invasion: China's influence in Australia.* Hardie Grant Books.

Harvey, D. (1990). *The condition of postmodernity: An enquiry into the origins of cultural change.* Blackwell.

Hong, N. (2012). The melting Arctic and its impact on China's maritime transport. *Research in Transportation Economics, 35*(1), 50–57.

Humpert, M. (2018, September 3). *Record traffic on Northern Sea route as COSCO completes five transits.* High North News. https://www.highnorthnews.com/en/record-traffic-northern-sea-route-cosco-completes-five-transits

Interview with a representative of a Chinese energy company (personal views), Beijing, March 2016.

Interview with a representative of a Japanese shipping company, Tokyo, October 2016.

Jackson, R. (2000). *The global covenant: Human conduct in a world of states.* Oxford University Press.

Kauppila, L., & Kiiski, T. (2020). The red dragon in global waters: The making of the Polar Silk Road. In E. Pongrácz, V. Pavlov, & N. Hänninen (Eds.), *Arctic marine sustainability. Arctic maritime businesses and the resilience of the marine environment* (pp. 465–485). Springer.

Kavalski, E. (Ed.). (2009). *China and the global politics of regionalisation.* Ashgate.

Kivimäki, T. (2015). Finlandization and the peaceful development of China. *The Chinese Journal of International Politics, 8*(2), 139–166.

Kopra, S., & Kauppila, L. (2018, April). *China, fundamental responsibility and Arctic regionalisation* [Paper presentation]. The International Studies Association (ISA) Annual Convention. San Francisco, The United States.

Kopra, S., & Kauppila, L. (n.d.). *Human-centric pluralism and the fundamental responsibility of the state: A Post-Westphalian approach to the English School theory.* An unpublished manuscript under review.

Lajeunesse, A., & Lackenbauer, P. W. (2016). Chinese mining interests and the Arctic. In D. A. Berry, N. Bowle, & H. Jones (Eds.), *Governing the North American Arctic* (pp. 74–99). Palgrave Macmillan.

Lanteigne, M., Koivurova, T., & Nojonen, M. (2020). China's rise in a changing world. In T. Koivurova & S. Kopra (Eds.), *Chinese policy and presence in the Arctic* (pp. 5–24). Brill.

Luttwak, E. (1990). From geopolitics to geo-economics: Logic of conflict, grammar of commerce. *National Interest, 20,* 17–24.

Massa, I. (2011). *Export finance activities by the Chinese government*. European Parliament Directorate-General for External Policies of the Union, Directorate B Policy Department Briefing Paper. http://www.europarl.europa.eu/RegData/etudes/note/join/2011/433862/EXPO-INTA_NT(2011)433862_EN.pdf

Mattlin, M. (2020). Kanariefågeln som tystnade. Finlands *gestalt shift* om kinesiska investeringar [The Canary that fell silent. Finland's *gestalt shift* on Chinese investments]. *Internasjonal Politikk, 78*(1), 54–67.

Milne, R. (2013, August 15). Norway sees Liu Xiaobo's Nobel Prize hurt salmon exports to China. *Financial Times*. https://www.ft.com/content/ab456776-05b0-11e3-8ed5-00144feab7de

National Development and Reform Commission. (2019). 发展改革委修订发布《产业结构调整指导目录(2019年本)》 [*NDRC issued the revised 2019 edition of Industrial structure adjustment guidance catalogue*]. http://www.tzxm.gov.cn/flfg/deptRegulations/201911/t20191106_12490.html

Naughton, B. (2018). *The Chinese economy. Adaptation and growth*. The MIT Press.

Níelsson, E. T. (2019). China Nordic Arctic research center. In A. B. Forsby (Ed.), *Nordic-China cooperation. Challenges and opportunities* (pp. 59–63). NIAS Press.

Niini, M. (2019). *Xue Long 2 enters into service*. [Brochure]. https://akerarctic.fi/app/uploads/2019/10/Xue-Long-2-Trials-1.pdf

Novatek. (2018, July 19). *Novatek shipped first LNG cargos to China* [Press release]. http://www.novatek.ru/en/press/releases/archive/index.php?id_4=2528&afrom_4=01.01.2018&ato_4=31.12.2018&from_4=4

Novatek. (2019, February 1). *Arctic LNG 2 and siemens sign equipment supply contract* [Press release]. http://www.novatek.ru/en/press/releases/index.php?id_4=2975

Novatek. (2020a). *Yamal LNG*. http://www.novatek.ru/en/business/yamal-lng/

Novatek. (2020b). *Arctic LNG 2*. http://www.novatek.ru/en/business/arctic-lng/

Novatek. (2020c, June 1). *LNG tanker "Christophe de Margerie" completes unique voyage along the Northern Sea Route* [Press release]. http://www.novatek.ru/en/press/releases/index.php?id_4=3949&from_4=2&mode_20=506&quarter_3=2

NS Energy. (2020). *Citronen lead-zinc project*. https://www.nsenergybusiness.com/projects/citronen-lead-zinc-project/

NSR Information Office. (2020). *NSR Shipping traffic – Transits in 2019*. https://arctic-lio.com/nsr-shipping-traffic-transits-in-2019/

O'Sullivan, S. (2019). *China: Growing import volumes of LNG highlight China's rising energy import dependency*. Oxford Institute for Energy Studies. https://www.oxfordenergy.org/wpcms/wp-content/uploads/2019/06/China-growing-import-volumes-of-LNG-highlight-China's-rising-energy-import-dependency.pdf

Ondaatjie, A., & Sirimanne, A. (2019, November 28). Sri Lanka wants to undo deal to lease port to China for 99 years. *Bloomberg*. https://www.bloomberg.com/news/articles/2019-11-28/sri-lanka-seeks-to-undo-1-1-billion-deal-to-lease-port-to-china

Pompeo, M. (2019, May 6). *Looking North: Sharpening America's Arctic focus* [Speech]. U.S. Department of State. https://www.state.gov/looking-north-sharpening-americas-arctic-focus/

PRIC. (2011). *Who we are*. https://www.pric.org.cn/EN/detail/sub.aspx?c=29

SCMP. (2015, October 25). *Chinese shipping firm COSCO Plans to launch services to Europe through Arctic Northeast passage, saving days in travel time*. https://www.scmp.com/news/china/economy/article/1872806/chinese-shipping-firm-plans-launch-services-through-arctic

Shullman, D. O. (2019, January 22). *Protect the party: China's growing influence in the developing world*. Brookings Institution. https://www.brookings.edu/articles/protect-the-party-chinas-growing-influence-in-the-developing-world/

Söderbaum, F. (2016). *Rethinking regionalism*. Palgrave.

State Council of the PRC. (2015). 国务院关于印发《中国制造2025》的通知 [*Announcement on the publication of Made in China 2025 report*]. http://www.gov.cn/zhengce/content/2015-05/19/content_9784.htm

State Council of the PRC. (2016a). *The 13th five-year plan for economic and social development of the People's Republic of China 2016–2020*. http://en.ndrc.gov.cn/newsrelease/201612/p020161207645765233498.pdf

State Council of the PRC. (2016b). *China's BeiDou navigation satellite system.* [White paper]. http://en.beidou.gov.cn/SYSTEMS/WhitePaper/201806/P020180608507822432019.pdf

State Council of the PRC. (2017). *Vision for maritime cooperation under the belt and road initiative.* http://english.gov.cn/archive/publications/2017/06/20/content_281475691873460.htm

State Council of the PRC. (2018). *Full text: China's Arctic policy* [White paper]. http://english.www.gov.cn/archive/white_paper/2018/01/26/content_281476026660336.htm

State Council of the PRC. (2019). 中共中央 国务院引发《交通强国建设纲要》[*The Central Committee of the CCP and the State Council issued Outline for building a strong transport country*]. http://www.gov.cn/zhengce/2019-09/19/content_5431432.htm

State Council of the PRC. (2021). 中华人民共和国国民经济和社会发展第十四个五年规划和2035年远景目标纲要 [*The 14th five-year plan for the national economic and social development of the People's Republic of China and the outline of the long-term goals for 2035*]. http://www.gov.cn/xinwen/2021-03/13/content_5592681.htm

Stepien, A., Kauppila, L., Kopra, S., Käpylä, J., Lanteigne, M., Mikkola, H., & Nojonen, M. (2020). China's economic presence in the Arctic: Realities, expectations and concerns. In T. Koivurova & S. Kopra (Eds.), *Chinese policy and presence in the Arctic* (pp. 90–136). Brill.

Total. (2020). *Liquefied natural gas: The Yamal LNG foundations on permafrost.* https://www.ep.total.com/en/areas/liquefied-natural-gas/our-yamal-lng-project-russia/foundations-permafrost-engineering

U.S. Energy Information Administration. (2020). *Natural gas explained. Liquefied natural gas.* https://www.eia.gov/energyexplained/natural-gas/liquefied-natural-gas.php

U.S. Geological Survey. (2008). *Circum-Arctic resource appraisal: Estimates of undiscovered oil and gas North of the Arctic Circle* [Fact sheet]. U.S. Department of the Interior. http://pubs.usgs.gov/fs/2008/3049/

Walker, C., & Ludwig, J. (2017). *Sharp power. Rising authoritarian influence.* National Endowment for Democracy. https://www.ned.org/wp-content/uploads/2017/12/Sharp-Power-Rising-Authoritarian-Influence-Full-Report.pdf

Womack, B. (2014). China's future in a multinodal world order. *Pacific Affairs, 87*(2), 265–284.

World Bank. (2020). *World development indicators: Structure of output* [Data set]. http://wdi.worldbank.org/table/4.2

Yamal LNG. (2020). *Yamal LNG reaches thirty million tons milestone.* http://yamallng.ru/en/press/news/38151/

Yan. (Ed.). (2019, June 20). *China, Russia to establish polar research laboratory.* Xinhuanet. http://www.xinhuanet.com/english/2019-06/20/c_138159589.htm

Yle. (2019, March 8). *Jos Helsinki–Tallinna-tunneli toteutuu, Suomesta tulee osa Kiinan arktista silkkitietä – Peter Vesterbacka toivoo rahoitusta myös länsimaista* [If the Helsinki-Tallin tunnel becomes reality, Finland will be part of the Polar Silk Road – Peter Vesterbacka hopes to get funding also from the West]. *Yle.* https://yle.fi/uutiset/3-10679141

Zheng, W. (2020, October 13). Chinese President Xi Jinping urges push towards hi-tech independence. *SCMP.* https://www.scmp.com/news/china/politics/article/3105317/chinese-president-xi-jinping-urges-push-towards-hi-tech?utm_medium=email&utm_source=mailchimp&utm_campaign=enlz-scmp_china&utm_content=20201013&tpcc=enlz-scmp_china&MCUID=338276fecd&MCCampaignID=659c73d141&MCAccountID=3775521f5f542047246d9c827&tc=5

Ziegler, C. E. (2011). China's energy relations with the global south: Potential for great power realignment. In C. Liu Currier & M. Dorraj (Eds.), *China's energy relations with the developing world* (pp. 195–212). Continuum.

**Liisa Kauppila** is a PhD Candidate at the Centre for East Asian Studies, University of Turku, Finland (UTU) and, since May 2021, a senior researcher at the Department of Philosophy, Contemporary History and Political Science, UTU.

# Part II
# Geography, Environment, and Climate

# Chapter 9
# Climate Effects of Other Pollutants – Short-Lived Climate Forcers and the Arctic

**Kaarle Kupiainen, Mark Flanner, and Sabine Eckhardt**

## Introduction

Globally, $CO_2$ and other long-lived greenhouse gas emissions are key components that affect climate, but some of the more shorter-lived air pollutants also either warm or cool the climate on timescales depending on the species. Therefore, emission reduction policies from a climate perspective need to take into account the net effect of multiple pollutants (UNEP/WMO, 2011; Stohl et al., 2015). The pollutants considered to have most climate relevance are termed short-lived climate pollutants (SLCPs) or short-lived climate forcers (SLCFs), depending on the context. For example, the Intergovernmental Panel on Climate Change's (IPCC) special report Global Warming of 1.5 C (IPCC, 2019) defined SLCFs to refer to both cooling and warming species that include methane ($CH_4$), ozone ($O_3$) and aerosols (including, black carbon, BC, organic carbon, OC, and sulfate) or their precursors, as well as some halogenated species. SLCPs refer only to the warming SLCFs (IPCC, 2019).

Due to the relatively short atmospheric lifetime of SLCFs, which ranges from about a week for aerosols like black carbon to a decade for methane, the mitigation efforts would lead to relatively fast responses in impacts. This is why policies focusing on SLCPs could serve as supplements to greenhouse gas reductions (UNEP/WMO, 2011; Shindell et al., 2012, 2017; Rogelj et al., 2014; Stohl et al., 2015).

K. Kupiainen (✉)
Finnish Ministry of the Environment, Helsinki, Finland
e-mail: Kaarle.Kupiainen@ym.fi

M. Flanner
University of Michigan, Ann Arbor, MI, USA
e-mail: flanner@umich.edu

S. Eckhardt
Norwegian Institute for Air research, Kjeller, Norway
e-mail: sec@nilu.no

Modeling studies by UNEP/WMO (2011) and Stohl et al. (2015) have demonstrated that the climate response of SLCF mitigation is strongest in the Arctic region. The Arctic region is of particular interest since, in the past 50 years, the Arctic has been warming more than twice as rapidly as the world as a whole and has experienced significant changes in ice and snow covers as well as permafrost (AMAP, 2017).

The Arctic Council's AMAP (Arctic Monitoring and Assessment Programme) working group has been providing scientific studies about the status and role of SLCFs for the Arctic since 2008 (AMAP, 2008a, b). A policy call for robust scientific information was also included in the "Arctic Council Framework for Action on Black Carbon and Methane Emissions Reductions" document, adopted in 2015 by the Arctic Council (Arctic Council, 2015). The framework calls for "continuing monitoring, research and other scientific efforts, with the inclusion of traditional and local knowledge, to improve the understanding of black carbon and methane emissions, emission inventories, Arctic climate and public health effects, and policy options" and supports "a four-year cycle of periodic scientific reporting, including the assessment of status and trends of short-lived climate pollutants such as black carbon and methane with a focus on the impacts of anthropogenic emissions on Arctic climate and public health. This should include estimates of associated costs of mitigation, as well as enhancing our state of knowledge regarding natural sources."

The AMAP assessments are being issued in the context of the Arctic Council, which is why the regional focus is particularly on the Arctic region and the eight Arctic Council member countries (Canada, Denmark, Finland, Iceland, Norway, Poland, Russia, Sweden, and the United States) as well as the 13 observer countries (China, France, Germany, India, Italy, Japan, the Netherlands, Poland, Republic of Korea, Singapore, Spain, Switzerland, and the United Kingdom). While acknowledging that the Arctic area has many definitions, the AMAP SLCF assessments have studied the climate impact analyses particularly for the area north of 60 deg N.

This chapter compiles the main messages and analyses from AMAP's scientific assessments on SLCFs. The first assessment, published in 2011 (AMAP, 2011), focused particularly on assessing the impact of black carbon on Arctic climate. The second assessment, published in 2015 (AMAP, 2015a, b), took a more integrated view on short-lived climate pollutants and estimated the role of the warming species and their aggressive mitigation in Arctic climate. The assessment published in 2021 is the third of its kind and revisits and updates the analyses of climate effects of multiple SLCFs, but also studies their role as air pollutants, informing particularly on the health effects.

This chapter starts by presenting the sources and emissions of SLCFs affecting the Arctic and analyzing whether the emission trends in line with the observed concentrations. Next, a review of the processes explaining the climate impacts of SLCFs is presented, followed by a compilation of impact estimates based on particularly the results from the AMAP 2015 and 2021 assessments. The impact sections also revisit the health impacts discussed in the AMAP 2021 work. Towards the end of the chapter, the Arctic relevant policy landscape of mitigating emissions and impacts of

SLCFs is discussed; finally, the main findings and messages are presented in the Conclusions section.

## Where Do They Come from? Emissions and Sources of SLCFs

Almost all SLCF species have both anthropogenic and natural origins, though with different proportions and seasonality. In this chapter, we discuss a variety of SLCFs, particularly the emissions of black carbon, sulfur species, methane and the formation of ozone, which have been estimated to induce a relatively strong climate impact.

Aerosol emissions, such as black carbon and organic carbon, are emitted from anthropogenic sources like burning of fossil and bio fuels in inefficient combustion conditions, but also have significant natural sources like forest and grassland fires. Sulfur species are also emitted from combustion, but only if the fuel includes sulfur. Methane, in turn, is formed in anthropogenic activities in waste management, fossil fuel extraction and agriculture, but also in natural biota like wetlands and lakes. Tropospheric ozone is formed in the atmosphere via chemical reactions of precursor species, including nitrogen oxides, volatile organic compounds, as well as methane or mixed down from the stratosphere.

Anthropogenic emissions and mitigation potential of SLCFs in the Arctic and the Arctic Council countries.

According to the 2021 AMAP assessment (AMAP, 2021), the Arctic Council member and observer countries currently comprise approximately half of global anthropogenic emissions of black carbon, sulfur dioxide, and methane. The Arctic Council member countries account for 8, 13, and 20% of black carbon, sulfur dioxide, and methane, respectively, and the Arctic Council observer countries account for 40, 40, and 30%, respectively. Surface transport is the most important source of black carbon emissions in the Arctic Council member countries, followed by residential combustion for heating, flaring in the oil and gas extraction, agricultural waste burning, and industrial combustion. In the Arctic Council observer countries, residential combustion is followed by surface transport, combustion in industry, as well as burning of household and agricultural wastes. Sulfur dioxide emissions are dominated by combustion in industry and energy production in both Arctic Council member and observer countries. Anthropogenic methane is emitted in the Arctic Council countries mostly from fossil fuel production, storage and distribution; that is, intended venting and unintended leakage during extraction and transportation of oil and gas, as well as release of ventilation air methane during coal mining. Other significant sources of methane are anaerobic decomposition of organic materials in waste and wastewater management, as well as livestock metabolism and manure management (AMAP, 2021).

The AMAP assessments have included comprehensive analyses of anthropogenic emissions of SLCFs, including historical data as well future projections. The analyses about future projections have focused on studying the potential for emission reductions of SLCFs, via introducing a full set of subsets of best available technologies to reduce emissions (BAT). Compiling an emission dataset of multiple SLCFs has been part of the assessment and the dataset has been used as input in impact analyses with atmospheric models.

According to AMAP (2015a, b, 2021), black carbon and sulfur dioxide emissions have been in decline in the last two decades among the Arctic Council member states, whereas methane emissions have increased slightly; these trends are expected to continue until 2050 with current legislation to reduce emissions. In the observer countries, the black carbon and sulfur dioxide emission trends were relatively stable until 2010, after which reductions started to occur. Anthropogenic methane emissions have increased since 2000, also in observer countries. Also in the case of the observer countries, these trends are expected to continue until 2050 with current legislation. Since the AMAP 2015 assessment, some changes in air pollution legislation and adoption of new policies have been introduced, particularly in Asia, including more stringent vehicle standards and policies to reduce coal and fuelwood use for heating and cooking, which have been included in the analyses of AMAP 2021.

An integral part of both the AMAP 2015 and the 2021 assessments has been evaluating the possibilities for further emission reductions beyond the current legislation, which has formed the baseline. The AMAP 2015 assessment studied a mitigation scenario that focused on and prioritized measures to reduce the warming SLCFs, particularly black carbon and methane (AMAP, 2015a, b). The results indicated that, by 2030–2050, black carbon emission reductions of up to 80% could be achievable in the Arctic Council countries, which would bring about reductions of other air pollutants, such as organic carbon, carbon monoxide, and volatile organic compounds (VOCs). However, for sulfur dioxide and nitrogen oxides, such a mitigation scenario would only introduce relatively minor reductions. According to AMAP (2015a, b), further reductions in anthropogenic methane emissions of up to two-thirds could be achieved by implementing the best available existing technologies.

The AMAP 2021 assessment discussed the emission reduction potential, focusing particularly on the Arctic Council and with a broadened scope to include the observer countries, and addressed a larger set of SLCFs, including mitigation of the sulfur species. The results demonstrated significant potential for further emission reductions of black carbon through the introduction of the best available technologies. In the Arctic Council member countries, the adoption of those emission mitigation technologies would halve the emissions by 2050, compared with the baseline. In the Arctic Council observer countries, black carbon emissions would be reduced more than two-thirds by targeting the same sectors as in the member countries. Implementing best available technologies to reduce sulfur dioxide could remove more than 60% of sulfur dioxide emissions in the Arctic Council member and 40% in observer countries, mostly due to measures in the energy and industrial sectors, and to a lesser degree in the residential sector. In the Arctic Council member countries, the full technical mitigation potential of anthropogenic methane was evaluated at about 50% below baseline in 2050, mostly related to measures targeting oil and gas extraction, storage, and distribution, as well as waste and wastewater management. In the observer countries, the introduction of the best available technologies would almost halve the methane emissions by 2050 (AMAP, 2021).

SLCF sources also exist within the Arctic area. Shipping, and oil and gas activities in the high Arctic emit significant amounts of SLCFs. Additionally, there are local pollution sources in Arctic communities and industrial sites. The AMAP assessments (2015a, 2021) identified boreal forest fires, outdated equipment used for energy production (diesel generators and old boilers), burning of waste, or land transport and shipping, as well as industrial operations as relevant local sources that contribute to atmospheric concentrations. Community-level emission sources may not be visible in the national emission inventories and the contributing emission sources may not be well identified and quantified.

The AMAP 2015 and 2021 assessment used emission datasets provided by the International Institute for Applied Systems Analysis (IIASA) GAINS-model (Amann et al., 2011). These datasets were chosen because they provide information on multiple key SLCFs, have global coverage on a national or subnational level with a spatial resolution of 0.5 degrees, and include historical data since 1990. The datasets also included future scenarios with alternative mitigation options until 2050. Such a broad scope has enabled comprehensive and integrated impact analyses and IIASA's datasets have served as the key source for the discussions about anthropogenic emissions in this chapter.

Official national emission data submitted to key international conventions, such as the UN ECE Convention on Long-range Transboundary Air Pollution (LRTAP Convention) and the United Nations Framework Convention on Climate Change (UNFCCC), also provide an important source of information for evaluations on key sectors emitting SLCFs and policy options to reduce emissions. Currently, all Arctic Council member countries and several observer countries have provided national emissions to the Arctic Council (Arctic Council, 2017 and Arctic Council, 2019). AMAP (2021) compared the IIASA emission inventory with official national submissions and other independent inventories, concluding that there is variation in the inclusion and handling of important emission sectors, as well as activity and emission parameters between the inventories. There is missing or incomplete information or limited understanding of calculation parameters, including activities and emission factors. However, the datasets generally indicate similar key sectors and emission trends to support the planning of further mitigation policies.

Natural sources of SLCFs.

Natural sources make up a significant share of several SLCFs or their precursors. The AMAP assessments (2011, 2015a, b, 2021) have focused on discussing natural releases of methane from terrestrial, freshwater, and marine sources, as well as black carbon emissions from biomass burning. The assessments have also included volcanic ash, mineral dust, sea salt, forest-derived volatile organic compounds (VOCs), and sulfur releases from sea areas.

AMAP (2021) reviewed estimates of current global natural methane releases and concluded that they have a similar order of magnitude as anthropogenic emissions and that 13% of the natural releases are estimated to be emitted in the Arctic area, North of 60 degrees N. Large reservoirs of carbon are potentially released as methane remains in the Arctic tundra regions, the wetlands, permafrost, freshwater, as well as carbon frozen in gas hydrates in the sea beds (Christensen et al., 2019). The magnitude of possible future releases of these reservoirs remains uncertain and depends on the future hydrological conditions of the Arctic with connections to Arctic warming and, in case of marine releases, also to future decreased sea-ice coverage (Christensen et al., 2019).

Biomass burning is also a major source of SLCF emissions. It is often difficult to distinguish clearly whether a forest or grassland fire is of natural origin or has been ignited due to human activities; nevertheless, wildfires remain a significant burden of air pollutant reaching the Arctic (Stohl et al., 2007). According to AMAP (2021), on a global level, about a quarter of total black carbon and two-thirds of total organic carbon emissions originated from fires. Based on satellite detections of biomass burning events in the Northern hemisphere, Boreal, and Arctic areas, most fire activity and emissions occur between 50° and 60° N, relatively few between 70° and 80° N, and none above 80° N (AMAP, 2021). For the Arctic, biomass burning is the most important source for summer concentrations of BC. Climate change, lightning, fuel conditions, and human activities are the main drivers for high latitude fires. In the last years, new sources have been detected, which are caused by the drying of Arctic soils and are likely to increase in frequency in the future (McCarty et al., 2021). For example, Greenland peat fires in 2017 released 23.5 t of BC and 731 t of OC at a latitude of 68°N (Evangeliou et al., 2019).

## Are the Emission Trends Visible in Concentrations of SLCFs in the Arctic?

Direct transport of air masses into the Arctic is limited by the polar dome. The concept of a "polar dome" refers to a dome-like stable synoptic structure around the pole with relatively little mixing of air masses between the inside and outside of it. Temperature gradients between the source region and the Arctic lead to the lifting of air masses along the levels of potential temperature forming the dome. This means that high latitude emissions can be transported quickly and at low levels into the Arctic, especially when it is cold in source regions. Aerosol concentrations in the Arctic follow a distinct seasonal cycle, with maxima in late winter and spring (Barrie, 1986; Eckhardt et al., 2015). Less transport and more intense scavenging lead to a summer minimum.

At several locations, sulfate and black carbon concentrations are conserved in ice cores. Black carbon concentrations in ice core samples from the Arctic are relatively low and have great temporal variability. The concentrations were mainly driven by biomass burning emissions before the Industrial Revolution, which led to the highest concentrations over a few decades. Now the levels of black carbon are at the same level as in preindustrial times or a bit higher. Sulfate in ice cores shows peaks in connection with volcanic *eruptions* and a maximum around 1970/1980, when

# 9  Climate Effects of Other Pollutants – Short-Lived Climate Forcers and the Arctic

sulfur enriched fuel was introduced. The peak lasted several decades and leveled out with new regulations at the end of the twentieth century.

AMAP (2021) studied the trends in atmospheric concentrations of black carbon, sulfate, and tropospheric ozone between 1990 and 2015 in the Arctic, as seen by models and in observations. The ambient sulfate decreased from 1990 to 1995 and thereafter remained at a constant level through to 2015. There were seasonal differences in the trend and comparing the difference between the first and last 5-year periods indicated that wintertime (JFMA) concentrations decreased by 17–57%, depending on the location. Smaller decreases were estimated for the summertime. Arctic black carbon concentrations also decreased over the period from 1990 to 2015, particularly in the winter Arctic haze season, where models estimate a 22–53% decrease, depending on the location. The trends are in line with the estimated reductions in sulfate and black carbon emissions. Tropospheric ozone concentrations have increased in the winter by approximately 1% at some Arctic locations, whereas no trend has been observed for the other seasons.

Methane emissions mix through the atmosphere on time scales shorter than its atmospheric lifetime, which is why its trends in concentrations are approximately the same everywhere on Earth (AMAP, 2015a, b). Global methane concentrations had a stable period around 2000–2007, but generally the concentrations have risen from around 1700 ppb in the 1980s to almost 1900 ppb measured today (reference).

AMAP (2021) outlined a circum-Arctic view of observations of black carbon concentrations in snow, which demonstrates that the highest concentrations are measured in North-Western Siberia and Scandinavia, whereas the lowest are observed in Central Greenland. Measured concentrations from Svalbard and northern North America are in between those extremes. The atmospheric concentrations on Greenland have a highest monthly average of 20 ng/m$^3$, which is one-tenth of measurements at Tiksi in the Russian Arctic. These results relate relatively well with the location of source areas, with high emission intensity and favored transport pathways.

Ozone concentrations decrease in the European sector, but increase in the Canadian sector. The mechanism behind this is not fully understood yet.

## What Does the "Climate" Stand for in Short-Lived Climate Forcers?

SLCFs include atmospheric constituents that have a short lifetime in the atmosphere and have the potential to affect Earth's radiative energy budget, either by directly absorbing or scattering solar or infrared radiation, or indirectly, for example, through the modification of cloud properties. SLCFs include gaseous compounds like ozone and nitrogen oxides, along with aerosols such as black carbon, sulfate, and mineral dust. The lifetime for a pollutant to qualify as "short-lived" is somewhat ambiguous; most SLCFs have lifetimes of weeks to months, and all SLCFs have lifetimes shorter than that of carbon dioxide. An additional feature of the SLCFs is that their atmospheric concentrations depend primarily on emissions rates, rather than on accumulated emissions, as is the case with carbon dioxide. Methane is a greenhouse gas, but can also be considered being part of SLCFs. It has an atmospheric lifetime of about a decade. An argument for classifying methane as a SLCF is that it both influences and is influenced by chemical reactions with other SLCF species, notably ozone.

SLCFs affect climate through several mechanisms. Gas-phase SLCFs, like methane, absorb infrared radiation and thus act as greenhouse gases. Methane is also a weak absorber of solar radiation, which supplements its global warming potential on top of its greenhouse effect (Etminan et al., 2016). In the upper layers of the atmosphere, particularly in the stratosphere, ozone absorbs ultraviolet light from the sun, leading to warming of the stratosphere. However, this absorption has less impact on surface temperatures due to limited thermal mixing between stratosphere and the lowest layer troposphere. For the purposes of quantifying the climate impacts associated with emissions, it is necessary to consider how precursor emissions of ozone and methane affect their atmospheric concentrations. Agents that react chemically and with sunlight to affect tropospheric ozone concentrations include nitrogen oxides (NOx), volatile organic compounds (VOCs), and carbon monoxide (CO). These agents also influence methane, although its atmospheric concentration is controlled more strongly by the direct emissions of methane.

Aerosols primarily – but not exclusively – influence climate through their direct and indirect interaction with solar radiation. This interaction generally results in climate warming when it increases the amount of absorbed solar energy by Earth and produces climate cooling when it results in a decrease of planetary absorbed solar energy. Black carbon is an effective absorber of sunlight and therefore warms the climate system. Sulfate aerosols, on the other hand, primarily scatter sunlight while absorbing very little, thus cooling climate by reflecting sunlight back to space that otherwise would have been absorbed in the Earth system. Organic carbon absorbs an intermediate fraction of interacting sunlight, causing cooling in some regions, and warming in others. Environments with high surface albedo, like the snow- and ice-covered Arctic, are more likely to experience a warming impact from organic carbon (e.g., Myhre et al., 2013).

Depending on the material of origin, mineral dust and volcanic ash particles can also warm or cool the climate. A key determinant of how strongly dust aerosols absorb sunlight is their content of iron-containing minerals. Furthermore, dust and volcanic ash particles are often quite large, enabling them to also absorb infrared radiation of longer wavelengths and thereby warm the troposphere through a greenhouse effect (e.g., Miller & Tegen, 1998; Flanner et al., 2014). However, larger particles tend to settle out of the atmosphere more quickly, reducing their lifetime and total radiative impact. The average atmospheric lifetimes of black carbon, organic carbon, and sulfate are in the order of days to weeks, whereas lifetimes of larger dust and ash particles are considerably shorter.

Aerosols also influence climate through indirect mechanisms. Light-absorbing particles like black carbon deposited to snow and ice surfaces can reduce albedo (in other words darken the surface) and eventually increase melt. This process is especially important for Arctic climate impacts (e.g., AMAP, 2015a). Aerosols can also alter cloud properties, including their droplet sizes, lifetimes, thicknesses, and vertical distribution, thereby impacting the radiative influence of clouds on climate. Globally, the indirect effect of aerosols on clouds is of a comparable magnitude to the direct radiative effect of the aerosols themselves, although indirect aerosol impacts are arguably less well known and thus the largest source of uncertainty in assessing anthropogenic impacts on climate (e.g., Myhre et al., 2013). The radiative impacts of Arctic clouds are unique because of the strong seasonality of insolation and surface albedo in the Arctic, frequent occurrence of mixed-phase clouds, and high stability of the Arctic atmosphere. These features lead to unique aerosol-cloud impacts in the Arctic and greater uncertainty in the sign and magnitude of the impacts than elsewhere.

Another important consideration is that SLCFs residing outside the Arctic affect Arctic climate by influencing the poleward heat flux into the Arctic. The meridional heat flux into the Arctic, transported mostly through the atmosphere and to a lesser extent through the oceans, constitutes a larger share of the surface energy budget in the Arctic than in other environments, especially during winter, and thus perturbations to this energy source can produce substantial temperature variations. Previous studies have found that SLCF emissions from large-emitting regions like East Asia have a greater impact on Arctic climate through their radiative impacts exerted outside the Arctic than through direct Arctic radiative impacts exerted by the relatively small share of the emissions that are transported into the Arctic (AMAP, 2015a; Sand et al., 2016). This contrasts with high-latitude emissions, which are more likely to deposit on Arctic snow and ice surfaces and to directly alter the energy budget of the Arctic troposphere.

The Arctic climate and environment continue to change rapidly, which also means the SLCFs' impacts will manifest differently in the future. Diminishing sea ice, increasing humidity, melting permafrost, expansion of shrubs, and increasing prevalence of liquid-phase clouds are all examples of changing conditions that can alter the lifetime and radiative forcing of SLCFs within the Arctic. The fact that these changes are occurring throughout the Arctic environment highlights the importance of multi-disciplinary and collaborative efforts to study Arctic impacts from SLCFs.

## SLCF in the Arctic Context – Climate and Health Impacts

In 2015, AMAP published comprehensive scientific reports about the impacts of black carbon and tropospheric ozone (AMAP, 2015a) and methane (AMAP, 2015b) on atmospheric chemistry and climate in the Arctic. The 2015 assessments acknowledged the dominant role of anthropogenic carbon dioxide emissions in driving Arctic climate change, but estimated that, by mitigating short-lived climate forcers (SLCFs) globally, it could be possible to avoid approximately 0.5 degrees of the projected 2 degrees of Arctic warming by 2050, in a scenario where the emission reduction measures are targeted specifically to address the warming species, black carbon, and methane. $CO_2$ followed a trajectory similar to the RCP6.0 climate scenario (see, for example, IPCC, 2014). The climate effect is not only related to the SLCFs in the Arctic area. Air masses heated by SLCFs outside the Arctic, particularly in the mid-latitudes, can be transported and mixed into the area and thus contribute to the warming. However, as AMAP (2011, 2015a) and Sand et al. (2016) have noted, the emission reductions of black carbon in the more northern areas have the largest temperature response to the Arctic climate, per unit of emissions reduced. The Nordic countries (Denmark, Finland, Iceland, Norway, and Sweden) and Russia have a larger impact than the other Arctic countries: the United States and Canada (AMAP, 2015a; Sand et al., 2016).

The AMAP 2021 assessment focused particularly on the Arctic Council countries and included the observer countries as a separate region. In 2021, AMAP revisited the impacts of SLCFs to the Arctic climate, but focused on analyzing a mitigation scenario with strong mitigation of all SLCFs, including the sulfur species. The reason for studying such a scenario was that many of the SLCFs also adversely influence air quality and human health and policies to reduce emissions of SLCFs have also been justified by air quality concerns. Thus, AMAP's 2021 assessment placed more emphasis on the role of SLCFs than the prior assessment. The negative health effects are connected with all components of particulate matter, especially smaller-sized particles, and precursors of ozone. Recognition that knowledge of both climate and air quality impacts is important for policy considerations led to the expansion of the scope of the AMAP 2021 assessment, to include air quality impacts, relative to the 2015 assessment (2015a, b).

The climate effect of SLCF reductions should be studied in connection with emissions of carbon dioxide. The AMAP 2021 assessment used for all SLCF scenarios a $CO_2$ trajectory similar to that of the SSP2–4.5 in which the Arctic warming rate associated with $CO_2$ is estimated to be 0.29 °C/decade in the period 1990–2015 and 0.35 °C/decade in 2015–2030. For the period 1990–2015, the reduced emissions of sulfur warmed the Arctic with a rate approximately equal to that of carbon dioxide. Contributions of changes in emissions of black carbon and methane to net Arctic warming were minor.

In the baseline scenario, implementing current emission mitigation legislation for SLCFs for the period 2015–2050, the associated emission reductions of sulfur species continue to warm the Arctic at a rate of 0.13 °C/decade. The emission changes of black carbon and methane in the baseline scenario only have a minor effect on the warming rate of the Arctic atmosphere.

The mitigation scenario, implementing the best available technologies to reduce all SLCFs, including sulfur species, introduces significant emission reductions for black carbon, methane and sulfur. The sulfur reductions introduce further warming of 0.1 °C/decade compared with the baseline. However, the cooling impacts connected with the emission reductions of black carbon and methane are similar in magnitude, which leads to a climate-neutral net effect. Overall, in the studied scenarios, the Arctic is likely to continue to warm from 2015 to 2050, mainly due to a combination of strong warming impacts of $CO_2$ and diminishing sulfur emissions on Arctic climate. Related to the impact of natural emission changes to atmospheric methane concentrations by 2100 Christensen et al. (2019) concluded that the feedback in the Arctic remains minor compared to the effect of global anthropogenic emission cuts that could occur in a mitigation scenario of the type used by AMAP (2021).

AMAP 2021 concluded that there is strong evidence that maximum feasible reductions in SLCF emissions produce measurable improvements in air quality in all Arctic Council member and observer countries compared to current legislation.

The AMAP 2021 assessment indicates that $PM_{2.5}$ mortality globally could drop by 24% in 2030 and 4.5% in 2050 compared with 2015 under current legislation. For the Arctic Council nations in particular, current air pollution legislation reduces the $PM_{2.5}$-attributable mortality by 59% and 57% in 2030 and 2050, respectively, compared to 2015. However, the global ozone mortality could increase, particularly in Asia, due to steady ozone concentrations coupled with growing exposed populations. If best available technologies were implemented, global PM2.5 mortality could be reduced by up to a quarter by 2030 and 2050 compared with the baseline. Significant further benefits would be observed in the Arctic Council nations as well as observer nations. The best available technologies would also reduce the ozone mortality compared with current legislation.

AMAP 2021 has demonstrated that the estimated air quality improvements by future emission reductions of SLCFs are very beneficial for health. However, AMAP (2021) pointed out that they may occur simultaneously with changes in other risk factors that also affect population health, including climate change and other non-environmental risk factors. Actions that improve public health overall would result in healthier and more resilient populations that are more prepared to overcome the negative consequences of anthropogenic climate change and other environmental risk factors (AMAP, 2021).

Utilizing the same scenarios as the AMAP 2021 assessment, the OECD conducted a quantitative assessment of the biophysical and economic benefits of air pollution policies in Arctic Council countries (OECD, 2021). The report concludes that improvements in air quality can be achieved without affecting economic growth, because the policy costs relative to the investments in best available technologies are fully balanced by the economic benefits resulting from lower air pollution impacts on health and the environment (OECD, 2021). The air pollution benefits lead to higher labor productivity, lower health expenditures, and higher agricultural productivity. To supplement the analyses with accounting for the economic values of reduced mortality and pain and suffering from illness, the welfare improvements that follow air pollution policies highlight the positive economic consequences of air pollution policies.

Having summarized some of the main findings from the AMAP assessments, particularly from 2015 and 2021 it is important to note that the estimates still suffer from substantial uncertainties and more work is needed in many areas of the SLCF science to provide a better scientific basis for policy decisions. Further developments are needed in Earth System modeling and monitoring capabilities in the

Arctic, emission inventories, including those of anthropogenic and natural emissions at present and associated with future mitigation options.

## Policy Landscape of SLCF Mitigation – An Arctic Perspective

The AMAP assessments (2011, 2015a, b, 2021) have indicated that, with targeted choices of existing mitigation measures of SLCF-rich sources, it could be possible to mitigate the projected global and Arctic climate impacts significantly in the coming few decades, provided that they could be fully implemented globally. However, achieving such global reductions can be politically demanding since no mechanisms or policy processes are currently in place.

It is possible to list several fora that address SLCFs. The Gothenburg Protocol of the Convention on Long-Range Transboundary Air Pollution (Gothenburg Protocol, 1999) encourages countries to target the black-carbon-rich sources when planning their emission reductions of fine particulate matter ($PM_{2.5}$). The Climate and Clean Air Coalition, the World Bank's Global Gas Flaring Reduction Partnership, and the Global Methane Initiative are examples of partnerships that foster action to mitigate SLCFs based on voluntary membership. The International Maritime Organization (IMO) is studying and discussing the possibility of a binding mechanism to reduce black carbon from Arctic shipping, but the work is still on-going. A common feature of the current commitments is that they are voluntary in nature (Khan, 2016, Khan & Kulovesi, 2018). The Arctic Council has been one of the frontrunners in formulating policies to mitigate SLCFs.

### SLCF Policy Efforts by the Arctic Council

The work on SLCF policies within the Arctic Council started with scientific evaluation of the role of SLCFs (AMAP 2008a; b). However, relatively early on, in 2009, a policy track was created by establishing Task Forces that were mandated to study the opportunities to mitigate SLCFs and to eventually negotiate a framework that particularly targeted black carbon and methane in the Arctic. Eventually, in 2015 the Arctic Council ministerial meeting adopted a framework document "Enhanced Black Carbon and Methane Emissions Reductions, An Arctic Council Framework for Action" (Arctic Council, 2015) in which the countries committed "to take enhanced, ambitious, national and collective action to accelerate the decline in black carbon emissions and significantly reduce our overall methane emissions". The Arctic council member countries also called upon the Arctic Council observer states to join them in the actions, given the global nature of the challenge. All member countries and many observer states have participated in the process. The framework document established a vision for future work on the issue and provided the organization for the work, with specific requirements for the countries to report on their actions.

In 2017, the Arctic Council adopted an aspirational collective goal to reduce black carbon emissions collectively by at least 25–33% below 2013 levels by 2025 (Arctic Council, 2017). This goal has provided a relevant setting for further scientific analyses on emission reduction strategies. Eventually, global efforts will be needed to achieve the full climate and health benefits laid out in the AMAP assessments (2015a, b, 2021). This is why, in the framework, the Arctic Council member countries call upon the observers to join in the actions.

## Can the Arctic Countries Achieve the Vision and the Black Carbon Goal?

AMAP (2021) included an estimate of emission development against the Arctic Council's black carbon goal and concluded that, in the baseline case that reflects the emission development with already decided legislation, a reduction of 19 to 60 Gg remains to be achieved with additional mitigation measures by 2025 to reach that goal. This represents approximately 4–12% of current emissions. The failure to implement the post-2015 legislation would move the black carbon emission trajectory from meeting the 2025 goal and lead to approximately 15% higher emission levels of black carbon towards the end of the period compared with the baseline. However, there is room for significant additional emission reductions, as indicated by the scenario analyzing a situation when all best available technologies are adopted. The increasing emissions of methane in the baseline do not comply with the Arctic Council's vision "…to significantly reduce our overall methane emissions".

## Conclusions

The Arctic Council's working group AMAP (Arctic Monitoring and Assessment Programme) has been issuing periodic scientific studies about the status and role of SLCFs in the Arctic. This chapter has compiled the main results from the assessment reports issued by 2021, with a particular focus on those published in 2015 (AMAP 2015a, b, 2021).

SLCF species have both anthropogenic and natural sources. An outlook of the selected key SLCF species demonstrates that black carbon and sulfur dioxide emissions have been declining in the Arctic Council member countries, but have remained relatively stable in the Arctic Council observer countries during the last two decades. Methane emissions have increased. Those trends are also visible in the atmospheric measurements and ice core records. With current air pollution legislation, the black carbon and sulfur dioxide emissions are expected to decline by 2050 in both Arctic Council member and observer countries. The increase in methane emissions is expected to continue.

Analysis of the baseline development included in the AMAP 2021 assessment indicates that the Arctic Council could almost reach the black carbon goal with current legislation to reduce emissions of SLCFs. Furthermore, the increasing methane emission trend should be redirected to match the Arctic Council vision "to significantly reduce emissions".

The analyses indicate that there is room for significant further emission reductions by implementing the best available mitigation technologies of SLCFs. These technologies are numerous and differ by pollutant and source, but they generally improve combustion efficiency, remove the pollutants from flue gas or exhaust, or reduce leakages and increase recovery and usage. However, in order to achieve significant reductions in Arctic climate warming and protect human health, these technologies should be introduced globally, which provides a challenge for policy efforts. Despite the global nature of the SLCF-induced challenges, regional bodies like the Arctic Council can show leadership in their own area and serve as catalysts to initiate broader regional interest to the SLCF issue also via the participation of the observer countries.

By way of conclusion, one could say that mitigation policies focusing on reducing SLCFs could serve as supplements to greenhouse gas reductions and, due to the relatively short atmospheric lifetime of SLCFs, the mitigation efforts would lead to relatively fast responses in impacts. Emission reductions of the warming SLCF species could together mitigate up to several tenths of degrees of the projected warming. However, the policies need to be balanced, taking into account the sometimes counteracting impacts of the cooling and warming SLCF species. The emission reductions would also bring about significant reductions in concentrations of air pollutants and consequent benefits to human health. The health benefits would be strongest within and near the source areas with high population densities, but Arctic emission reduction would also have an impact on the Arctic population.

# References

Amann, M., Bertok, I., Borken-Kleefeld, J., Cofala, J., Heyes, C., Höglund-Isaksson, L., Klimont, Z., Nguyen, B., Posch, M., Rafaj, P., Sandler, R., Schöpp, W., Wagner, F., & Winiwarter, W. (2011). Cost-effective control of air quality and greenhouse gases in Europe: Modeling and policy applications. *Environmental Modelling & Software, 26*, 1489–1501.

AMAP. (2008a). AMAP/Quinn et al., 2008, *The impact of short-lived pollutants on arctic climate* (AMAP technical report No. 1) (23 p). Arctic Monitoring and Assessment Programme (AMAP). https://www.amap.no/documents/doc/the-impact-of-short-lived-pollutants-on-arctic-climate/15

AMAP. (2008b). AMAP/Bluestein et al., 2008, *Sources and mitigation opportunities to reduce emissions of short-term arctic climate forcers* (AMAP technical report No. 2) (8 p). Arctic Monitoring and Assessment Programme (AMAP). https://www.amap.no/documents/doc/sources-and-mitigation-opportunities-to-reduce-emissions-of-short-term-arctic-climate-forcers/13

AMAP. (2011). In P. K. Quinn, A. Stohl, A. Arneth, T. Berntsen, J. F. Burkhart, J. Christensen, M. Flanner, K. Kupiainen, H. Lihavainen, M. Shepherd, V. Shevchenko, H. Skov, & V. Vestreng (Eds.), *The impact of black carbon on Arctic climate*. Arctic Monitoring and Assessment Programme (AMAP). 72 pp. ISBN 978-82-7971-069-1.

AMAP. (2017). Snow, Water, Ice and Permafrost in the Arctic (SWIPA), Arctic Monitoring and Assessment Programme (AMAP), Oslo, Norway. xiv + 269 pp. ISBN 978-82-7971-101-8.

AMAP Assessment. (2015a). *Black carbon and ozone as Arctic climate forcers*. Arctic Monitoring and Assessment Programme (AMAP). vii + 116 pp. ISBN 978-82-7971-092-9.

AMAP Assessment. (2015b). *Methane as an Arctic climate forcer*. Arctic Monitoring and Assessment Programme (AMAP). vii + 139 pp. ISBN 978-82-7971-091-2.

AMAP Assessment. (2021). *Impacts of short-lived climate forcers on Arctic climate, air quality, and human health*. Arctic Monitoring and Assessment Programme (AMAP). In press.

Arctic Council. (2015). *Annex 4. Iqaluit 2015 Sao report to ministers. Enhanced black carbon and methane emissions reductions*. An Arctic Council Framework for Action. Available at: https://oaarchive.arctic-council.org/handle/11374/610. Last access: 18 Dec 2020.

Arctic Council. (2017). *Expert group on black carbon and methane; summary of progress and recommendations*. Available at: https://oaarchive.arctic-council.org/handle/11374/1936. Last access: 18 Dec 2020.

Arctic Council. (2019). *Expert group on black carbon and methane; summary of progress and recommendations 2019*, available at: https://oaarchive.arctic-council.org/handle/11374/2411. Last access: 18 Dec 2020.

Barrie, L. A. (1986). Arctic air-pollution – An overview of current knowledge. *Atmospheric Environment, 20*(4), 643–663.

Christensen, T. R., Arora, V. K., Gauss, M., Höglund-Isaksson, L., & Parmentier, F.-J. W. (2019). Tracing the climate signal: Mitigation of anthropogenic methane emissions can outweigh a large Arctic natural emission increase. *Scientific Reports, 9*, 1146. https://doi.org/10.1038/s41598-018-37719-9

Eckhardt, S., Quennehen, B., Olivié, D. J. L., Berntsen, T. K., Cherian, R., Christensen, J. H., Collins, W., Crepinsek, S., Daskalakis, N., Flanner, M., Herber, A., Heyes, C., Hodnebrog, Ø., Huang, L., Kanakidou, M., Klimont, Z., Langner, J., Law, K. S., Lund, M. T., Mahmood, R., Massling, A., Myriokefalitakis, S., Nielsen, I. E., Nøjgaard, J. K., Quaas, J., Quinn, P. K., Raut, J. C., Rumbold, S. T., Schulz, M., Sharma, S., Skeie, R. B., Skov, H., Uttal, T., von Salzen, K., & Stohl, A. (2015). Current model capabilities for simulating black carbon and sulfate concentrations in the Arctic atmosphere: A multi-model evaluation using a comprehensive measurement data set. *Atmospheric Chemistry and Physics, 15*(16), 9413–9433. https://doi.org/10.5194/acp-15-9413-2015

Etminan, M., Myhre, G., Highwood, E. J., & Shine, K. P. (2016). Radiative forcing of carbon dioxide, methane, and nitrous oxide: A significant revision of the methane radiative forcing. *Geophysical Research Letters, 43*, 12,614–12,623. https://doi.org/10.1002/2016GL071930

Evangeliou, N., Kylling, A., Eckhardt, S., Myroniuk, V., Stebel, K., Paugam, R., Zibtsev, S., & Stohl, A. (2019). Open fires in Greenland in summer 2017: Transport, deposition and radiative effects of BC, OC and BrC emissions. *Atmospheric Chemistry and Physics, 19*, 1393–1411. https://doi.org/10.5194/acp-19-1393-2019

Flanner, M. G., Gardner, A. S., Eckhardt, S., Stohl, A., & Perket, J. (2014). Aerosol radiative forcing from the 2010 Eyjafjallajökull volcanic eruptions. *Journal of Geophysical Research: Atmospheres, 119*, 9481–9491. https://doi.org/10.1002/2014JD021977

Gothenburg Protocol. (1999). Protocol to the 1979 CLRTAP to abate acidification, eutrophication and ground-level ozone (adopted 30 November 1999, entered into force 17 May 2005) 2319 UNTS 81.

IPCC. (2014). In Core Writing Team, R. K. Pachauri, & L. A. Meyer (Eds.), *Climate change 2014: Synthesis report. Contribution of working groups I, II and III to the fifth assessment report of the intergovernmental panel on climate change* (151 pp). IPCC.

IPCC (2019). Annex I: Glossary. In V. Masson-Delmotte, P. Zhai, H.-O. Pörtner, D. Roberts, J. Skea, P. R. Shukla, A. Pirani, W. Moufouma-Okia, C. Péan, R. Pidcock, S. Connors, J. B. R. Matthews, Y. Chen, X. Zhou, M. I. Gomis, E. Lonnoy, T. Maycock, M. Tignor, & T. Waterfield (Eds.), *Global warming of 1.5 C. an IPCC special report on the impacts of global warming of 1.5 C above pre-industrial levels and related global greenhouse gas emission pathways, in the context of strengthening the global response to the threat of climate change, sustainable development, and efforts to eradicate poverty*. Intergovernmental Panel on Climate Change.

Khan, A. (2016). The global commons through a regional Lens: The Arctic council on short-lived climate pollutants. *Transnational Environmental Law*, 1–22. https://doi.org/10.1017/S2047102516000157

Khan, A., & Kulovesi, K. (2018). Black carbon and the Arctic: Global problem-solving through the nexus of science, law and space. *Review of European, Comparative & International Environmental Law, 27*, 5–14. https://doi.org/10.1111/reel.12245

McCarty et al. (2021). https://bg.copernicus.org/articles/18/5053/2021/bg-18-5053-2021.pdf

Miller, R.L. & Tegen, I. (1998). Climate response to soil dust aerosols. *Journal of Climate, 11*, 3247–3267.

Myhre, G., Shindell, D., Bréon, F.-M., Collins, W., Fuglestvedt, J., Huang, J., Koch, D., Lamarque, J.-F., Lee, D., Mendoza, B., Nakajima, T., Robock, A., Stephens, G., Takemura, T., & Zhang, H. (2013). Anthropogenic and natural radiative forcing. In T. F. Stocker, D. Qin, G.-K. Plattner, M. Tignor, S. K. Allen, J. Boschung, A. Nauels, Y. Xia, V. Bex, & P. M. Midgley (Eds.), *Climate change 2013: The physical science basis. Contribution of working group I to the fifth assessment report of the intergovernmental panel on climate change*. Cambridge University Press.

OECD. (2021). *The economic benefits of air quality improvements in Arctic council countries*. OECD Publishing.

Rogelj, J., Schaeffer, M., Meinshausen, M., Shindell, D. T., Hare, W., Klimont, Z., Velders, G. J. M., Amann, M., & Schellnhuber, H. J. (2014). Disentangling the effects of CO2 and short-lived climate forcer mitigation. *Proceedings of the National Academy of Sciences of the United States of America (PNAS), 111*(46), 16325–16330. https://doi.org/10.1073/pnas.1415631111

Sand, M., Berntsen, T. K., von Salzen, K., Flanner, M. G., Langner, J., & Victor, D. G. (2016). Response of Arctic temperature to changes in emissions of short-lived climate forcers. *Nature Climate Change, 6*, 286–290. https://doi.org/10.1038/nclimate2880

Shindell, D., Kuylenstierna, J. C. I., Vignati, E., van Dingenen, R., Amann, M., Klimont, Z., Anenberg, S. C., Muller, N., Janssens-Maenhaut, G., Raes, F., Schwartz, J., Faluvegi, G., Pozzoli, L., Kupiainen, K., Höglund-Isaksson, L., Emberson, L., Streets, D., Ramanathan, V., Hicks, K., Kim Oanh, N. T., Milly, G., Williams, M., Demkine, W., & Fowler, D. (2012). Simultaneously mitigating near-term climate change and improving human health and food security. *Science, 335*, 183–189. https://doi.org/10.1126/science.1210026

Shindell, D., Borgford-Parnell, N., Brauer, M., Haines, A., Kuylenstierna, J. C. I., Leonard, S. A., Ramanathan, V., Ravishankara, A., Amann, M., & Srivastava, L. (2017). A climate policy pathway for near- and long-term benefits. *Science, 356*, 493–494. https://doi.org/10.1126/science.aak9521

Stohl, A., Berg, T., Burkhart, J. F., Fjaeraa, A. M., Forster, C., Herber, A., Hov, Ø., Lunder, C., McMillan, W. W., Oltmans, S., Shiobara, M., Simpson, D., Solberg, S., Stebel, K., Stroem, J., Torseth, K., Treffeisen, R., Virkkunen, K., & Yttri, K. E. (2007). Arctic smoke – Record high air pollution levels in the European Arctic due to agricultural fires in Eastern Europe. *Atmospheric Chemistry and Physics, 7*, 511–534.

Stohl, A., Aamaas, B., Amann, M., Baker, L. H., Bellouin, N., Berntsen, T. K., Boucher, O., Cherian, R., Collins, W., Daskalakis, N., Dusinska, M., Eckhardt, S., Fuglestvedt, J. S., Harju, M., Heyes, C., Hodnebrog, Ø., Hao, J., Im, U., Kanakidou, M., Klimont, Z., Kupiainen, K., Law, K. S., Lund, M. T., Maas, R., MacIntosh, C. R., Myhre, G., Myriokefalitakis, S., Olivié, D., Quaas, J., Quennehen, B., Raut, J.-C., Rumbold, S. T., Samset, B. H., Schulz, M., Seland, Ø., Shine, K. P., Skeie, R. B., Wang, S., Yttri, K. E., & Zhu, T. (2015). Evaluating the climate and air quality impacts of short-lived pollutants. *Atmospheric Chemistry and Physics, 15*, 10529–10566. https://doi.org/10.5194/acp-15-10529-2015

UNEP & WMO (2011). Integrated assessment of black carbon and tropospheric ozone: Summary for decision makers. 30 p. ISBN: 978-92-807-3142-2. Available at: http://wedocs.unep.org/handle/20.500.11822/8028.

**Kaarle Kupiainen** is a Senior Specialist at the Finnish Ministry of the Environment, and has represented Finland in many Arctic Council groups working with short-lived climate pollutants. He has acted as a senior research scientist at the Finnish Environment Institute (SYKE) and as a research scholar at the International Institute for Applied Systems Analysis (IIASA) in Austria. He holds a PhD degree from the University of Helsinki.

**Mark Flanner** is an Associate Professor in the Department of Climate and Space Sciences and Engineering at the University of Michigan, since 2009. Prior to this, he was a postdoctoral scholar at the National Center for Atmospheric Research in Colorado.

**Sabine Eckhardt** received her PhD at the Technical University of Munich and is now working as a senior scientist at the Norwegian Institute for Air research. Her research interest is long-range atmospheric transport of pollutants with a focus on the polar areas.

# Chapter 10
# Permafrost Climate Feedbacks

**Benjamin W. Abbott**

## Introduction

The Earth system is sustained by interconnections. The stabilizing interactions among atmosphere, ocean, land, and biological communities have allowed life to flourish on this planet for billions of years (Schlesinger & Bernhardt, 2012). For example, when volcanic activity increases atmospheric carbon dioxide ($CO_2$), silicate and carbonate weathering accelerate, which draws down $CO_2$ via a suite of physical and biological feedbacks. Likewise, when the tropospheric temperature goes up, increased cloud formation reflects more incoming radiation due to increased albedo. While destabilizing feedbacks exist as well (for example, Milankovitch cycles and snow cover), the stabilizing feedbacks have prevailed since life began, keeping the Earth's climatic and biogeochemical conditions within an envelope that is suitable for life as we know it.

However, the Earth's stabilizing feedbacks have limits. Human greenhouse gas emissions are causing abrupt climate change that is unprecedented not only in human history, but in the history of the Earth (Tierney et al., 2020). Because of its speed and severity, anthropogenic climate change is triggering responses in ecosystems worldwide that may stabilize or further destabilize the climate system (Lenton et al., 2019). These ecosystem feedbacks are extremely complex and poorly represented in current Earth system models. This creates a risky environment for decision making, because without robust understanding of both human emissions and ecosystem feedbacks, we could trigger environmental tipping points and erode the stabilizing processes that have sustained life on Earth. The situation is even more tenuous regarding human wellbeing, because human civilization has flourished during a time of uncommonly stable climate (Steffen et al., 2015).

B. W. Abbott (✉)
Brigham Young University, Provo, UT, USA

© The Author(s), under exclusive license to Springer Nature Switzerland AG 2022
M. Finger, G. Rekvig (eds.), *Global Arctic*,
https://doi.org/10.1007/978-3-030-81253-9_10

Climate in the permafrost zone is warming at least twice as quickly as the global mean, and approximately six-fold faster over large areas of the Arctic (Huang et al., 2017; Polvani et al., 2020). Because these regions contain vast stores of organic matter, the climate feedbacks from the permafrost zone have been described as among the largest and most likely to occur. The release of carbon dioxide ($CO_2$) and methane ($CH_4$) from the permafrost zone has been termed the *permafrost carbon feedback* (Jorgenson & Osterkamp, 2005). However, I prefer the term *permafrost climate feedbacks* (E.J. Burke et al., 2012), which can include non-carbon greenhouse gases and albedo effects associated with vegetation shifts (Loranty et al., 2018; Voigt et al., 2020). Fortunately, the acronym PCF works for both terms, providing a diplomatic and concise compromise. In this chapter, I explore the ecological dynamics regulating PCF and lay out the key uncertainties that currently make our predictions of PCF range from extreme greenhouse gas release to substantial $CO_2$ uptake. I will use an approach informed by the field of ecosystem ecology, which seeks to explain spatiotemporal variation in ecosystem properties (in this case, organic matter and greenhouse gas balance) with state factors and interactive controls such as climate, hydrology, energy availability, and disturbance (Chapin et al., 2012).

## Background

One of the most common mistakes about the permafrost zone (and maybe the Earth system in general) is treating it like a single ecosystem. While there are similarities across Arctic, Boreal, and Alpine environments, the vast and remote world of permafrost is heterogeneous and context dependent. This diversity means that no single location is representative of ecological dynamics across the permafrost zone (Tank et al., 2020). This is problematic because our observations are extremely patchy, both geographically (Metcalfe et al., 2018; Mu et al., 2020) and temporally (Natali et al., 2019; Shogren et al., 2020). Sometimes a location only a few meters or minutes away will have a completely different ecological community or periglacial history, creating a distinct response to disturbance (Malone et al., 2018; Olefeldt et al., 2016; Shogren et al., 2019). To make meaningful predictions of how permafrost regions will respond to climate change, we need to appreciate the commonalities and differences of the ecosystems that constitute the permafrost zone.

A second and more pernicious mistake about the permafrost zone is to consider it as nothing but a bank of greenhouse gas precursors. These regions are home to vibrant human societies and ecological communities that depend on reliable seasons and solid permafrost. The indigenous citizens of the permafrost zone are simultaneously among the most vulnerable to climate change and the most knowledgeable about how to adapt and avoid it (Parkinson & Berner, 2009; Pearce et al., 2009; Riedlinger & Berkes, 2001). Most of the Earth's remaining terrestrial and marine wilderness occurs within the permafrost zone, increasing the urgency of protecting these ecosystems not just to avoid global climate catastrophe, but for the human and nonhuman inhabitants who call the permafrost zone home. We will need to weave both scientific and traditional ecological knowledge about the permafrost zone to sustain its diverse ecosystems and communities (Kimmerer, 2002). While I primarily rely on my lens as an ecosystem ecologist in this chapter, I remind readers that solving any of the challenges below will require going well beyond the boundaries of any research discipline and of science itself.

## *Subdividing the Permafrost Zone*

On a global scale, climate is the strongest predictor of ecosystem type and function (Chapin et al., 2012). Climate is especially influential for ecosystems in the permafrost zone because the stability of the ground depends on the persistence of subzero temperature (Tank et al., 2020; Turetsky et al., 2020). However, permafrost is not just a function of mean annual air temperature. Permafrost exists where it does because of the interaction between climate and the thermal properties of the land surface, such as albedo and thermal conductivity. This means that vegetation and snow (and hence the things that affect them) influence the extent and state of permafrost (Loranty et al., 2018; Shur & Jorgenson, 2007).

Because the permafrost zone occupies 24% of the exposed land surface in the Northern Hemisphere (22.79 × $10^6$ $km^2$), it is useful to subdivide it into more approachable units. There are many ways to divide and organize the permafrost zone, but I will focus on three approaches: permafrost coverage, biomes, and watersheds.

The percentage of the land surface underlain by permanently frozen ground is a useful criterion for subdividing the permafrost zone because this parameter often correlates with surface soil and vegetation properties (Lindgren et al., 2016; Shur & Jorgenson, 2007). There are four common permafrost categories: continuous (>90% of the land surface underlain by permafrost), discontinuous (50–90%), sporadic (10–50%), and isolated (<10%). These categories occupy 47, 19, 17, and 17% of the permafrost zone, respectively (Heginbottom et al., 2002), although the boundaries and accuracy of the categories are highly uncertain because of the physical and biological reasons addressed in other chapters in this book (Jafarov et al., 2018; Shur & Jorgenson, 2007; T. Zhang et al., 2000). These categories can also be applied to subsea permafrost, which exists under portions of the continental shelves of the Arctic Ocean and surrounding seas (Overduin et al., 2018; Sayedi et al., 2020). Subsea permafrost formed during periods when ocean levels were lower, such as during the Last Glacial Maximum (Lindgren et al., 2018). While estimates of the extent and depth of subsea permafrost are even less settled than for terrestrial permafrost, it is estimated that continuous and discontinuous permafrost underly ~3.5 × $10^6$ $km^2$ of the submerged continental shelves (Overduin et al., 2018; Sayedi et al., 2020).

The biome approach to classifying the permafrost zone uses dominant vegetation types, which are generally associated with climate and permafrost conditions

(Heginbottom et al., 2002; Shur & Jorgenson, 2007). The Arctic tundra biome covers $5.0 \times 10^6$ km$^2$ and the Boreal forest biome covers $13.7 \times 10^6$ km$^2$, although the extent of the Boreal forest depends on the definition of the southern transition to temperate forest and varies in the literature from 11.4 to $18.5 \times 10^6$ km$^2$ (Abbott, Jones, et al., 2016b; McGuire et al., 1995; Pan et al., 2011). Because most Arctic tundra is in the continuous permafrost zone (>90% permafrost cover) while most Boreal forest is in the discontinuous, sporadic, or isolated zones (0–90% cover), almost all Arctic tundra is underlain by permafrost, but most of the Boreal forest is not (Obu et al., 2019; T. Zhang et al., 2000). A third biome that exists within and beyond the boundaries of Arctic tundra and Boreal forest is Alpine permafrost. The global extent of Alpine permafrost is ~$4.0 \times 10^6$ km$^2$, including many mountainous regions worldwide (Bockheim & Munroe, 2014). The Tibetan Plateau accounts for 70% of Alpine permafrost outside of the Arctic and Boreal biomes, covering ~$1.1 \times 10^6$ km$^2$ (Bockheim & Munroe, 2014; Mu et al., 2020; Zhang et al., 2019).

Another way of classifying the permafrost zone is to consider the surface and subsurface water networks that connect land to sea. While these watersheds cut across permafrost categories and biomes, they have the advantage of grouping ecosystems based on connectivity – the ability to exchange material, energy, and organisms (Abbott, Baranov, et al., 2016a; Vonk et al., 2015; Zarnetske et al., 2018). The circumarctic watershed, defined as the drainages of the Arctic Ocean and surrounding seas, covers from 16.8 to $20.5 \times 10^6$ km$^2$ depending on whether Greenland and the Canadian Archipelago are included (Holmes et al., 2013; McClelland et al., 2012). Even for its size, this enormous drainage produces a huge amount of runoff, yielding ~3700 km$^3$ of river discharge annually (Holmes et al., 2012; McGuire et al., 2009) and an unknown amount of direct groundwater discharge (Connolly et al., 2020; Lecher, 2017). Most of the discharge from the circumarctic comes from six large rivers (listed here in descending order of discharge): the Yenisey, Ob, Lena, Mackenzie, Yukon, and Kolyma (Holmes et al., 2013). In addition to transporting globally relevant loads of dissolved and particulate organic carbon (DOC and POC, respectively), the circumarctic watershed processes and produces organic matter and inorganic nutrients, affecting overall ecosystem carbon and nutrient balance. Freshwater ecosystems play a particularly influential role in regulating carbon cycling at high latitudes, where they cover more than 50% of the landscape in some regions and account for 8% of global runoff, 36% of global lake area, and over 50% of global wetland area (Abbott et al., 2019; Aufdenkampe et al., 2011; Avis et al., 2011; Kling, 2010; McGuire et al., 2009).

## Why Do Permafrost Regions Contain So Much Carbon and Nitrogen?

Net ecosystem carbon balance (NECB) is a complete accounting of inputs and outputs of carbon in an ecosystem (Chapin et al., 2006). NECB includes vertical carbon fluxes across the surface-atmosphere boundary such as primary production, ecosystem respiration, emissions from wildfire, and trace gas flux, as well as lateral carbon fluxes such as hydrological flow and movement of organisms into and out of the system. As for your bank account, it is not the individual rate of either the inputs or the outputs that determines whether an ecosystem is a carbon sink or source, but the difference between the deposits and the withdrawals. Indeed, permafrost ecosystems are some of the least productive globally, with net primary productivity in Arctic tundra and Boreal forest typically fixing only 80 to 230 g C m$^{-2}$ year$^{-1}$, which is a tenth or less of the productivity of a tropical forest (Chapin et al., 2012).

Site its low rate of primary production, the permafrost zone has accumulated the largest pool of organic carbon on Earth (Fig. 10.1). Regions underlain by permafrost account for 16% of global soil area but contain more than 50% of the world's soil organic matter (SOM). Some of this organic matter dates to more than two million years ago, but most of it has accumulated since the Last Glacial Maximum (Lindgren et al., 2018). Permafrost ecosystems have accumulated these enormous stocks of organic carbon not because they are productive, but because they are miserly. The cold and waterlogged conditions that prevail in the permafrost zone limit the decomposition of organic matter by invertebrates, fungi, bacteria, and archaea. Because soil temperature and moisture are largely determined by climate, permafrost carbon

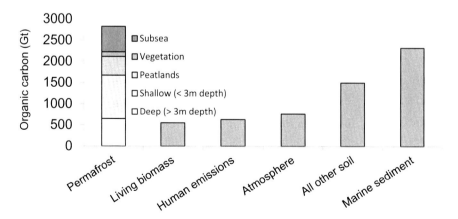

**Fig. 10.1** Permafrost organic carbon stocks compared to other global pools. Citations for the estimates for each pool are as follows: subsea permafrost (Sayedi et al., 2020), peatlands and shallow and deep permafrost (Hugelius et al., 2014; Mishra et al., 2021), vegetation (Abbott, Jones, et al., 2016b), cumulative human emissions since the Industrial Revolution (Raupach & Canadell, 2010), living biomass (Bar-On et al., 2018), atmosphere and other soil (Chapin et al., 2012), and marine sediments (Atwood et al., 2020)

has been described as climate-protected (Schmidt et al., 2011; Trumbore, 2009; Zimov et al., 2006). As climate change strips away this protection, the response of biomass, decomposition, hydrological carbon release, and disturbance such as wildfire will determine the NECB of the permafrost zone (Chen et al., 2021; Mack et al., 2004; McGuire et al., 2009; Schuur et al., 2015).

Not only do permafrost ecosystems store an uncommon amount of organic matter they distribute it in uncommon ways. Living biomass accounts for a small fraction of overall organic carbon, with plants containing ~110 gigatons (Gt) C and dead organic matter in peat, soil, and sediment containing ~2700 Gt C (Abbott, Jones, et al., 2016b; Fuchs et al., 2018; Hugelius et al., 2014; Mishra et al., 2021; Sayedi et al., 2020). This creates a biomass to necromass ratio of ~1:25 (Fig. 10.1). Another peculiarity of permafrost carbon is that, unlike most terrestrial ecosystems where SOM occurs primarily in the top meter of soil due to plant inputs, more than half of permafrost SOM is stored below 3 m (Hugelius et al., 2014; Mishra et al., 2021; Strauss et al., 2017). Three mechanisms account for this deeper distribution of SOM. First, seasonal freezing and thawing of the active-layer above the permafrost table can cause cryogenic mixing, incorporating chunks of SOM into the permafrost (Bockheim, 2007; Kaiser et al., 2007). Second, Arctic rivers transport globally relevant loads of dissolved and particulate organic carbon (Abbott, Jones, et al., 2016b; McClelland et al., 2012; Tank et al., 2016), some of which they deposit into fluvial and estuarine sediments, which can have considerable extent and depth (Fuchs et al., 2018). Third, the deposition of windblown silt during periods of glaciation and development of organic matter such as peat can raise the soil surface, incorporating plant matter into syngenetic permafrost (Bauer & Vitt, 2011; Strauss et al., 2017). This deep distribution of organic matter has implications for the vulnerability of permafrost carbon and nitrogen to surficial changes in temperature, biogeochemical conditions, and disturbance (Harden et al., 2012; Hewitt et al., 2015; Koven, Lawrence, & Riley, 2015a).

## Permafrost and Climate Response

The question of how permafrost carbon balance will respond to climate change has fueled three decades of debate (Shaver et al., 1992) and remains far from settled (McGuire et al., 2018; Sayedi et al., 2020). Because permafrost organic pools are so large (Fig. 10.1), even small changes in carbon and nitrogen balance could entrain serious consequences for regional ecosystem processes and global biogeochemical cycles (Burke et al., 2018; Koven, Lawrence, & Riley, 2015a).

## Greenhouse gases Ruled by Redox Conditions, Microbial Communities, and Substrate

Before describing the physical changes and ecological responses going on in the permafrost zone, it is worth reviewing the biogeochemistry that underlies the production and consumption of the three major greenhouse gases. Carbon dioxide ($CO_2$), methane ($CH_4$), and nitrous oxide ($N_2O$) are the most important drivers of human-caused climate change (Montzka et al., 2011), and their production and consumption in the permafrost zone will largely determine PCF. These gases are produced during decomposition of organic matter ($CO_2$ and $CH_4$) or during the transformation of nutrients released from that decomposition ($N_2O$) (Schädel et al., 2016; Voigt et al., 2020). The type, amount, and rate of production or decomposition of these gases generally depends on three factors: availability of substrate, makeup of the microbial community, and the redox state of the soil or water environment (Kolbe et al., 2019; Schlesinger & Bernhardt, 2012). The availability of SOM for decomposition following thaw depends on both inherent factors such as source and molecular composition as well as extrinsic factors such as nutrient availability and association with mineral soil (Chen et al., 2021; Schmidt et al., 2011; Wologo et al., 2021). The interaction between microbial community and redox conditions then regulate the type and amount of gas produced. For example, methanogenesis ($CH_4$ production) requires $CO_2$ or low-molecular weight organic matter for substrate and is only carried out by 50 or so species of archaea, which cannot tolerate oxygen (Sapart et al., 2017). Conversely, $N_2O$ and $CO_2$ can be produced in oxic or anoxic environments, but via different microbial processes (respectively, nitrification and denitrification for $N_2O$, and aerobic respiration and fermentation for $CO_2$) (Roy Chowdhury et al., 2015; Voigt et al., 2020). In the permafrost zone, microorganisms dominate the production and uptake of these gases, except for the uptake of $CO_2$, which is primarily accomplished by terrestrial vegetation (Table 10.1).

While warming is likely to increase the production of all three greenhouse gases in the permafrost zone (Abbott & Jones, 2015; Voigt et al., 2016), the proportions produced and consumed matter because the global warming potential of these gases varies from 1 to nearly 300 (Table 10.1). For example, the same amount of carbon converted into $CH_4$ would have an 86-fold greater effect on Earth's climate over 20 years than the same amount of carbon converted into $CO_2$ (Table 10.1). The magnitude of PCF depends on the amount of gas released to the atmosphere multiplied by its global warming potential. Waterlogged environments favor the production of $CH_4$, which is a more potent greenhouse gas but is produced much more slowly than $CO_2$ (Knoblauch et al., 2013; Schädel et al., 2016; Voigt et al., 2016). Conversely, well-drained environments favor rapid oxidation of organic matter to $CO_2$ but suppress $CH_4$ production (Sayedi et al., 2020; Zhang et al., 2019). Environments with variable soil moisture can stimulate coupled nitrification and denitrification, causing $N_2O$ production (Voigt et al., 2020). Knowing the conditions for uptake and release of these gases will equip us to make better predictions about

**Table 10.1** Characteristics of greenhouse gases associated with permafrost climate feedbacks

| Gas | Sources | Sinks | Substrate | Mean lifetime (Years)[a] | Current level (PPB)[a] | GWP (20/100 Years)[b] |
|---|---|---|---|---|---|---|
| $CO_2$ | Aerobic and anaerobic respiration in soil and water. Combustion during wildfire. | Primary productivity (photosynthesis). Silicate and carbonate weathering. | Organic matter | ~500 | 410,000 | 1/1 |
| $CH_4$ | Anaerobic metabolism of archaea in anoxic soil and water. Combustion during wildfire. | Microbial oxidation by methanotrophs in oxic waters and soils. Photochemical oxidation in troposphere. | $CO_2$ or low-molecular weight compounds (e.g. acetate) | 12.4 | 1870 | 86/34 |
| $N_2O$ | Nitrification in oxic soils and partial denitrification in anoxic soils. | Denitrification in anoxic soil and water. Photochemical breakdown in stratosphere. | Nitrate ($NO_3^-$) or ammonium ($NH_4^+$) | 121 | 333 | 268/298 |

[a]In the atmosphere
[b]Global warming potential must be defined relative to a time horizon because of the diverse atmospheric dynamics of greenhouse gases (Montzka et al., 2011; Neubauer & Megonigal, 2015)

how PCF might play out under different climate and ecosystem change scenarios (Koven, Lawrence, & Riley, 2015a; Turetsky et al., 2020).

## *How Is Climate Changing in the Permafrost Zone?*

Permafrost zone air temperature is increasing much faster than global mean temperature, largely due to feedbacks associated with sea-ice loss, decreasing snow cover, and ozone-depleting gases (Huang et al., 2017; Polvani et al., 2020). Warming has been most prevalent during the autumn and early winter in coastal areas, when sea ice is at its minimum, and in the spring at latitudes from 50 to 60 °N as snow cover decreases (AMAP, 2017). Intensification of the freshwater cycle is projected across the circumarctic watershed, including increases in precipitation, evapotranspiration (ET), storage, and discharge (Rawlins et al., 2010), although the relative magnitude of these parameters is poorly constrained (Holmes et al., 2012). Precipitation appears to have increased 5% over land north of 55° since 1950, although high interannual variability means this trend is not significant (AMAP, 2017; Peterson, 2006; Rawlins et al., 2019). While circumpolar precipitation minus ET is projected to increase by 13–25% by 2100, much of this increase is due to changes in winter precipitation and greater evaporation from the ice-free ocean

(Bintanja & Selten, 2014). Consequently, growing seasonal precipitation in many inland areas is not expected to keep up with enhanced ET, resulting in water stress (Chapin et al., 2010; Sulla-Menashe et al., 2018).

As a result of changes in temperature and precipitation, both permafrost and non-permafrost soil temperatures have warmed over the past century, causing increased active layer thickness, freeze-thaw cycling, longer duration of thaw, and widespread ground collapse or thermokarst (Biskaborn et al., 2019; Farquharson et al., 2019). Models predict widespread near-surface permafrost degradation with projections varying between 14% and 70% loss by 2100 depending on the degree of warming (Meredith et al., 2019; Saito et al., 2021; Schuur et al., 2015; Slater & Lawrence, 2013).

Permafrost degradation follows two basic trajectories. In permafrost with little or no ground ice, the soil profile can thaw from the top down without disturbing the surface (gradual active layer thickening). Alternatively, when ground ice volume exceeds soil pore space, permafrost thaw causes surface subsidence or collapse, termed thermokarst (Kokelj & Jorgenson, 2013). Thermokarst features show high morphological diversity, depending on the slope, ground ice content, and characteristics of the soil and vegetation. If thermokarst occurs on hillslopes, it can abruptly expose SOM from meters below the surface and alter soil conditions at depth, influencing substrate availability, microbial community, and redox state (Abbott & Jones, 2015; Monteux et al., 2020; Turetsky et al., 2020). Thermokarst currently affects a small portion of the landscape in most areas, although current estimates suggest it could account for between one-third and one-half of overall greenhouse gas release from the permafrost zone over the coming centuries (Abbott & Jones, 2015; Olefeldt et al., 2016; Turetsky et al., 2020).

Growing season length – historically ~100 days for Arctic tundra and ~ 150 days for Boreal forest – has increased by 2–4 days per decade from 1960 to 2000, due mostly to earlier spring thaw, and is projected to lengthen by a total of 37–60 days over pre-industrial conditions by the end of the century (Chapin et al., 2012; Euskirchen et al., 2006). Wildfire extent and severity have increased throughout the permafrost region, including in Arctic tundra (Abbott et al., 2021; Flannigan et al., 2009; Kasischke & Turetsky, 2006).

The changes in climate and disturbance described above are associated with increasing winter base flow and the seasonal contribution of ground water relative to surface water (Frey & McClelland, 2009; Meredith et al., 2019). Coupled to changes in hydrology, aquatic chemistry has experienced substantial shifts, including an increase in DOC flux in areas with peat and thick organic soils, a decrease in discharge-normalized DOC where organic soils are shallow, increases in major ion concentrations, accelerated chemical weathering, and increased inorganic nutrient concentrations (Abbott et al., 2015; Tank et al., 2020; Vonk et al., 2015).

Climate change is accelerating the thaw and erosion of Arctic coastlines due to warming air and water, in combination with increased exposure to wave action and storms due to reductions in sea ice cover (Meredith et al., 2019). Thermal collapse and erosion of coastlines delivers DOC and POC to coastal shelf waters, with collapse being most pronounced in northeastern Alaska and eastern Siberia, where coastal retreat rates now exceed 10 m year$^{-1}$ in some areas (Jones et al., 2009; Meredith et al., 2019; Vonk et al., 2012).

## *How Are These Changes Affecting Permafrost Climate Feedbacks?*

There is substantial complexity and uncertainty associated with both sides of the permafrost carbon balance equation (that is, uptake and release). This uncertainty extends beyond unknowns about the rate of warming and permafrost degradation. While simulations of the extent of permafrost thaw by 2100 vary by a factor of five, from 14% to 70%, estimates of permafrost carbon balance do not even agree in sign, ranging from 60 Gt C uptake to 135 Gt C release by 2100 (McGuire et al., 2018; Meredith et al., 2019). Several key sources of uncertainty limiting our prediction of the rate, magnitude, and type of the PCF have been identified, many of which affect both the consumption and production of greenhouse gases.

On the carbon uptake side of the equation, the following factors have been identified as key uncertainties: (1) Potential influence of disturbances such as wildfire, thermokarst, extreme weather, pests and pathogens, and invasive species; (2) uncertainties about nutrient availability, especially the fate of nitrogen and phosphorus currently stored in permafrost and active layer organic matter, (3) changes in water availability during the growing season; and (4) Sensitivity of net primary productivity to $CO_2$ fertilization (Abbott, Jones, et al., 2016b; Koven, Schuur, et al., 2015b; McGuire et al., 2018; Meredith et al., 2019). While the extended growing season, greater nutrient availability, and $CO_2$ fertilization could increase net primary production, potentially resulting in carbon accumulation in plant biomass and near-surface SOM, the disturbances and limitations listed above could preclude or offset those gains (McGuire et al., 2018; Meredith et al., 2019; Myers-Smith et al., 2020). There are also questions about the reliability of modelled carbon uptake. Increased primary productivity due to $CO_2$ fertilization accounts for most of the predicted carbon uptake in current models (Balshi et al., 2007; McGuire et al., 2018). However, $CO_2$ fertilization has not been observed or experimentally demonstrated in Arctic or Boreal environments (Abbott, Jones, et al., 2016b). Arctic tundra has experienced both spectral greening (increased greenness observed from satellites) and vegetation greening (increased growth of vegetation) at various scales (Bhatt et al., 2010; Myers-Smith et al., 2020). However, the initial response of the Boreal forest – which contains most of the permafrost zone's organic matter – shows a patchwork of increasing and decreasing primary productivity, associated with wildfire, water stress, pests, and ecosystem shifts (Alcaraz-Segura et al., 2010; Koven, 2013; Sulla-Menashe et al., 2018).

On the greenhouse gas release side of the equation, the following dynamics have been identified as major sources of uncertainty: (1) The proportion of gradual and abrupt thaw (for example, thermokarst); (2) the redox conditions and microbial community after thaw; (3) the trajectory of lateral nutrient and carbon flux through surface water networks; and (4) interactions among thawed organic matter, new organic matter inputs, and available nutrients (Mack et al., 2004; Monteux et al., 2020; Rastetter et al., 2020; Turetsky et al., 2020; Wologo et al., 2021). As for the carbon uptake uncertainties, these controlling factors respond on different spatial and temporal scales. For example, thermokarst formation, which can initially create a pulse of organic matter release, can eventually create conditions for carbon accumulation in sediment and peat (Anthony et al., 2014; Lindgren et al., 2018). Likewise, Boreal wildfire releases a large amount of ecosystem carbon as $CO_2$ and $CH_4$, but these losses can be compensated by rapid regrowth of more carbon-rich species in some environments (Mack et al., 2021). Conversely, the vertical and lateral loss of nutrients following wildfire, which could constrain the capacity of ecosystems to take up $CO_2$, could be part of a nutrient-shunting adaptation that favors recovery of permafrost-adapted vegetation (Abbott et al., 2021; Mack et al., 2011; Rastetter et al., 2020). Increased nutrient availability and high-quality carbon inputs can, in turn, accelerate or slow the decomposition of organic matter (Hartley et al., 2010; Mack et al., 2004; Wologo et al., 2021).

The permafrost climate feedback has been portrayed in popular media (and, to a lesser extent, in peer-reviewed literature) as an all-or-nothing scenario. Permafrost greenhouse gas release has been described as a tipping point, a runaway climate feedback, and, most dramatically, a time bomb. At the other extreme, some have dismissed the importance of this feedback, asserting that increases in biomass will offset any carbon losses from soil, or that changes will occur too slowly to concern current governments. Our understanding of this complex corner of the Earth is much less binary. The good news is that most studies estimate substantially less local disturbance and carbon release when human emissions are reduced compared to business as usual (Abbott, Jones, et al., 2016b; Meredith et al., 2019; Sayedi et al., 2020; Schuur et al., 2015). This means that more human emissions will result in more permafrost disruption, including all the local and global costs that go with it. In some ways, the vast and diverse permafrost ecosystems can be resistant to disturbance and slow to react to perturbation. Deep permafrost has incredible thermal inertia that only responds to changes at the surface on centennial to millennial timescales. Other aspects of this system, such as near-surface permafrost or wildfire, can be sensitive bellwethers that respond rapidly to direct human disturbance or changes in climate. In this regard, Arctic and Boreal ecosystems resemble human society.

## Conclusions

It is hard to overstate the difficulty of predicting PCF. Doing so involves trying to understand complex and diverse ecosystems over vast distances and long periods of time and trying to generate realistic and reliable predictions of how these intercoupled systems will respond to truly unprecedented global change. Add to that the fact that we are trying to have these predictions vetted and ready for negotiation in time to actually do something about the problem, and you have the perfect recipe for what ecologists call a super wicked problem (Levin et al., 2012).

Despite these challenges, I propose that we already know what we need to know about PCF. Regarding both the local effects of climate change on northern peoples and the global effects on climate, the more extreme the warming, the more extreme the consequences. Briefly consider the range of current estimates of the PCF, imperfect as they are. At one extreme, the permafrost zone could exert a substantial but not dominant influence on climate in the coming centuries. The combined emissions from terrestrial and subsea permafrost might amount to 100 Gt of carbon or more (McGuire et al., 2018; Meredith et al., 2019; Sayedi et al., 2020). While the associated amount of greenhouse gas would be enormous, it remains relatively modest compared to projected human emissions over the same time period. At the other extreme, the permafrost zone could end up removing 50 Gt of carbon or more from the atmosphere (McGuire et al., 2018). While this would be a welcome stabilizing feedback for climate, it would barely make a dent in the current and committed emissions that are already destabilizing the Earth's climate. In either case, there is only one viable path forward: stop emitting greenhouse gases to the atmosphere and remove existing greenhouse gases from it.

If we take our current understanding of the Earth system seriously, current climate targets are woefully inadequate. The most discussed goal of limiting warming to 1.5 °C by 2100 would still be a global catastrophe. A world that is 1.5 °C warmer in 2100 is likely to have millions of climate refugees, no coral reefs, and ice sheets in Greenland and West Antarctica on their way to drowning all the world's coastal cities (Lenton et al., 2019). Given the stakes and our nascent understanding of socioecological feedbacks, the only responsible choice is to get back to the Holocene as quickly as we can. Thankfully, current research in energy systems demonstrates that going back to the Holocene is both technically feasible and economically advantageous (Bogdanov et al., 2021; Brown et al., 2018; Hausfather & Peters, 2020). Whatever our specific discipline, we in the science community need to become informed and speak clearly about the urgency of a rapid and complete decarbonization of the global economy. I look to the future with faith, hope, and resolve that we can overcome this great challenge and recognize our common interest in preserving these frozen places.

## References

Abbott, B. W., & Jones, J. B. (2015). Permafrost collapse alters soil carbon stocks, respiration, CH4, and N2O in upland tundra. *Global Change Biology, 21*(12), 4570–4587. https://doi.org/10.1111/gcb.13069.

Abbott, B. W., Jones, J. B., Godsey, S. E., Larouche, J. R., & Bowden, W. B. (2015). Patterns and persistence of hydrologic carbon and nutrient export from collapsing upland permafrost. *Biogeosciences, 12*(12), 3725–3740. https://doi.org/10.5194/bg-12-3725-2015.

Abbott, B. W., Baranov, V., Mendoza-Lera, C., Nikolakopoulou, M., Harjung, A., Kolbe, T., Balasubramanian, M. N., Vaessen, T. N., Ciocca, F., Campeau, A., Wallin, M. B., Romeijn, P., Antonelli, M., Gonçalves, J., Datry, T., Laverman, A. M., de Dreuzy, J.-R., Hannah, D. M., Krause, S., … Pinay, G. (2016a). Using multi-tracer inference to move beyond single-catchment ecohydrology. *Earth-Science Reviews, 160*(Suppl C), 19–42. https://doi.org/10.1016/j.earscirev.2016.06.014.

Abbott, B. W., Jones, J. B., Schuur, E. A. G., Chapin, F. S., III, Bowden, W. B., Bret-Harte, M. S., Epstein, H. E., Flannigan, M. D., Harms, T. K., Hollingsworth, T. N., Mack, M. C., McGuire, A. D., Natali, S. M., Rocha, A. V., Tank, S. E., Turetsky, M. R., Vonk, J. E., Wickland, K. P., Aiken, G. R., … Zimov, S. (2016b). Biomass offsets little or none of permafrost carbon release from soils, streams, and wildfire: An expert assessment. *Environmental Research Letters, 11*(3), 034014. https://doi.org/10.1088/1748-9326/11/3/034014.

Abbott, B. W., Bishop, K., Zarnetske, J. P., Minaudo, C., Chapin, F. S., Krause, S., Hannah, D. M., Conner, L., Ellison, D., Godsey, S. E., Plont, S., Marçais, J., Kolbe, T., Huebner, A., Frei, R. J., Hampton, T., Gu, S., Buhman, M., Sara Sayedi, S., … Pinay, G. (2019). Human domination of the global water cycle absent from depictions and perceptions. *Nature Geoscience, 12*(7), 533–540. https://doi.org/10.1038/s41561-019-0374-y.

Abbott, B. W., Rocha, A. V., Shogren, A., Zarnetske, J. P., Iannucci, F., Bowden, W. B., Bratsman, S. P., Patch, L., Watts, R., Fulweber, R., Frei, R. J., Huebner, A. M., Ludwig, S. M., Carling, G. T., & O'Donnell, J. A. (2021). Tundra wildfire triggers sustained lateral nutrient loss in Alaskan Arctic. *Global Change Biology, 27*(7), 1408–1430. https://doi.org/10.1111/gcb.15507.

Alcaraz-Segura, D., Chuvieco, E., Epstein, H. E., Kasischke, E. S., & Trishchenko, A. (2010). Debating the greening vs. browning of the north American boreal forest: Differences between satellite datasets. *Global Change Biology, 16*(2), 760–770. https://doi.org/10.1111/j.1365-2486.2009.01956.x.

AMAP. (2017). *Snow, water, ice and permafrost. Summary for policy-makers* (p. 20).

Anthony, K. M. W., Zimov, S. A., Grosse, G., Jones, M. C., Anthony, P. M., Iii, F. S. C., Finlay, J. C., Mack, M. C., Davydov, S., Frenzel, P., & Frolking, S. (2014). A shift of thermokarst lakes from carbon sources to sinks during the Holocene epoch. *Nature, 511*(7510), 452–456. https://doi.org/10.1038/nature13560.

Atwood, T. B., Witt, A., Mayorga, J., Hammill, E., & Sala, E. (2020). Global patterns in marine sediment carbon stocks. *Frontiers in Marine Science, 7*. https://doi.org/10.3389/fmars.2020.00165.

Aufdenkampe, A. K., Mayorga, E., Raymond, P. A., Melack, J. M., Doney, S. C., Alin, S. R., Aalto, R. E., & Yoo, K. (2011). Riverine coupling of biogeochemical cycles between land, oceans, and atmosphere. *Frontiers in Ecology and the Environment, 9*(1), 53–60. https://doi.org/10.1890/100014.

Avis, C. A., Weaver, A. J., & Meissner, K. J. (2011). Reduction in areal extent of high-latitude wetlands in response to permafrost thaw. *Nature Geoscience, 4*(7), 444–448. https://doi.org/10.1038/ngeo1160.

Balshi, M. S., McGuire, A. D., Zhuang, Q., Melillo, J., Kicklighter, D. W., Kasischke, E., Wirth, C., Flannigan, M., Harden, J., Clein, J. S., Burnside, T. J., McAllister, J., Kurz, W. A., Apps, M., & Shvidenko, A. (2007). The role of historical fire disturbance in the carbon dynamics of the pan-boreal region: A process-based analysis. *Journal of Geophysical Research, 112*(G2). https://doi.org/10.1029/2006JG000380.

Bar-On, Y. M., Phillips, R., & Milo, R. (2018). The biomass distribution on earth. *Proceedings of the National Academy of Sciences, 115*(25), 6506–6511. https://doi.org/10.1073/pnas.1711842115.

Bauer, I. E., & Vitt, D. H. (2011). Peatland dynamics in a complex landscape: Development of a fen-bog complex in the sporadic discontinuous permafrost zone of northern Alberta, Canada: Development of peatland in the sporadic discontinuous permafrost zone, Canada. *Boreas, 40*(4), 714–726. https://doi.org/10.1111/j.1502-3885.2011.00210.x.

Bhatt, U. S., Walker, D. A., Raynolds, M. K., Comiso, J. C., Epstein, H. E., Jia, G., Gens, R., Pinzon, J. E., Tucker, C. J., Tweedie, C. E., & Webber, P. J. (2010). Circumpolar Arctic tundra vegetation change is linked to sea ice decline. *Earth Interactions, 14*(8), 1–20. https://doi.org/10.1175/2010EI315.1.

Bintanja, R., & Selten, F. M. (2014). Future increases in Arctic precipitation linked to local evaporation and sea-ice retreat. *Nature, 509*(7501), 479–482. https://doi.org/10.1038/nature13259.

Biskaborn, B. K., Smith, S. L., Noetzli, J., Matthes, H., Vieira, G., Streletskiy, D. A., Schoeneich, P., Romanovsky, V. E., Lewkowicz, A. G., Abramov, A., Allard, M., Boike, J., Cable, W. L., Christiansen, H. H., Delaloye, R., Diekmann, B., Drozdov, D., Etzelmüller, B., Grosse, G., … Lantuit, H. (2019). Permafrost is warming at a global scale. *Nature Communications, 10*(1), 264. https://doi.org/10.1038/s41467-018-08240-4.

Bockheim, J. G. (2007). Importance of cryoturbation in redistributing organic carbon in permafrost-affected soils. *Soil Science Society of America Journal, 71*(4), 1335. https://doi.org/10.2136/sssaj2006.0414N.

Bockheim, J. G., & Munroe, J. S. (2014). Organic carbon pools and genesis of alpine soils with permafrost: A review. *Arctic, Antarctic, and Alpine Research, 46*(4), 987–1006. https://doi.org/10.1657/1938-4246-46.4.987.

Bogdanov, D., Ram, M., Aghahosseini, A., Gulagi, A., Oyewo, A. S., Child, M., Caldera, U., Sadovskaia, K., Farfan, J., De Souza Noel Simas Barbosa, L., Fasihi, M., Khalili, S., Traber, T., & Breyer, C. (2021). Low-cost renewable electricity as the key driver of the global energy transition towards sustainability. *Energy, 120467.* https://doi.org/10.1016/j.energy.2021.120467.

Brown, T. W., Bischof-Niemz, T., Blok, K., Breyer, C., Lund, H., & Mathiesen, B. V. (2018). Response to 'Burden of proof: A comprehensive review of the feasibility of 100% renewable-electricity systems. *Renewable and Sustainable Energy Reviews, 92*, 834–847. https://doi.org/10.1016/j.rser.2018.04.113.

Burke, E. J., Hartley, I. P., & Jones, C. D. (2012). Uncertainties in the global temperature change caused by carbon release from permafrost thawing. *The Cryosphere, 6*(5), 1063–1076. https://doi.org/10.5194/tc-6-1063-2012.

Burke, E. J., Chadburn, S. E., Huntingford, C., & Jones, C. D. (2018). CO2 loss by permafrost thawing implies additional emissions reductions to limit warming to 1.5 or 2 °C. *Environmental Research Letters, 13*(2), 024024. https://doi.org/10.1088/1748-9326/aaa138.

Chapin, F. S., Woodwell, G. M., Randerson, J. T., Rastetter, E. B., Lovett, G. M., Baldocchi, D. D., Clark, D. A., Harmon, M. E., Schimel, D. S., Valentini, R., Wirth, C., Aber, J. D., Cole, J. J., Goulden, M. L., Harden, J. W., Heimann, M., Howarth, R. W., Matson, P. A., McGuire, A. D., … Schulze, E.-D. (2006). Reconciling carbon-cycle concepts, terminology, and methods. *Ecosystems, 9*(7), 1041–1050. https://doi.org/10.1007/s10021-005-0105-7.

Chapin, F. S., McGuire, A. D., Ruess, R. W., Hollingsworth, T. N., Mack, M. C., Johnstone, J. F., Kasischke, E. S., Euskirchen, E. S., Jones, J. B., Jorgenson, M. T., Kielland, K., Kofinas, G. P., Turetsky, M. R., Yarie, J., Lloyd, A. H., & Taylor, D. L. (2010). Resilience of Alaska's boreal forest to climatic change. *Canadian Journal of Forest Research, 40*(7), 1360–1370. https://doi.org/10.1139/X10-074.

Chapin, F. S., Matson, P. A., & Vitousek, P. M. (2012). *Principles of terrestrial ecosystem ecology.* New York: Springer. https://doi.org/10.1007/978-1-4419-9504-9.

Chen, L., Fang, K., Wei, B., Qin, S., Feng, X., Hu, T., Ji, C., & Yang, Y. (2021). Soil carbon persistence governed by plant input and mineral protection at regional and global scales. *Ecology Letters, 24*(5), 1018–1028. https://doi.org/10.1111/ele.13723.

Connolly, C. T., Cardenas, M. B., Burkart, G. A., Spencer, R. G. M., & McClelland, J. W. (2020). Groundwater as a major source of dissolved organic matter to Arctic coastal waters. *Nature Communications, 11*(1), 1479. https://doi.org/10.1038/s41467-020-15250-8.

Euskirchen, E. S., McGuire, A. D., Kicklighter, D. W., Zhuang, Q., Clein, J. S., Dargaville, R. J., Dye, D. G., Kimball, J. S., McDonald, K. C., Melillo, J. M., Romanovsky, V. E., & Smith, N. V. (2006). Importance of recent shifts in soil thermal dynamics on growing season length, productivity, and carbon sequestration in terrestrial high-latitude ecosystems. *Global Change Biology, 12*(4), 731–750. https://doi.org/10.1111/j.1365-2486.2006.01113.x.

Farquharson, L. M., Romanovsky, V. E., Cable, W. L., Walker, D. A., Kokelj, S., & Nicolsky, D. (2019). Climate change drives widespread and rapid thermokarst development in very cold permafrost in the Canadian High Arctic. *Geophysical Research Letters, 2019GL082187*. https://doi.org/10.1029/2019GL082187.

Flannigan, M., Stocks, B., Turetsky, M., & Wotton, M. (2009). Impacts of climate change on fire activity and fire management in the circumboreal forest. *Global Change Biology, 15*(3), 549–560. https://doi.org/10.1111/j.1365-2486.2008.01660.x.

Frey, K. E., & McClelland, J. W. (2009). Impacts of permafrost degradation on arctic river biogeochemistry. *Hydrological Processes, 23*(1), 169–182. https://doi.org/10.1002/hyp.7196.

Fuchs, M., Grosse, G., Jones, B. M., Strauss, J., Baughman, C. A., & Walker, D. A. (2018). Sedimentary and geochemical characteristics of two small permafrost-dominated Arctic river deltas in northern Alaska. *Arktos, 4*(1), 20. https://doi.org/10.1007/s41063-018-0056-9.

Harden, J. W., Koven, C. D., Ping, C.-L., Hugelius, G., David McGuire, A., Camill, P., Jorgenson, T., Kuhry, P., Michaelson, G. J., O'Donnell, J. A., Schuur, E. A. G., Tarnocai, C., Johnson, K., & Grosse, G. (2012). Field information links permafrost carbon to physical vulnerabilities of thawing. *Geophysical Research Letters, 39*(15), n/a-n/a. https://doi.org/10.1029/2012GL051958.

Hartley, I. P., Hopkins, D. W., Sommerkorn, M., & Wookey, P. A. (2010). The response of organic matter mineralisation to nutrient and substrate additions in sub-arctic soils. *Soil Biology and Biochemistry, 42*(1), 92–100. https://doi.org/10.1016/j.soilbio.2009.10.004.

Hausfather, Z., & Peters, G. P. (2020). Emissions – The 'business as usual' story is misleading. *Nature, 577*(7792), 618–620. https://doi.org/10.1038/d41586-020-00177-3.

Heginbottom, J., Brown, J., Ferrians, O., & Melnikov, E. S. (2002). *Circum-Arctic map of permafrost and ground-ice conditions, version 2* (Data set). NSIDC. https://doi.org/10.7265/SKBG-KF16.

Hewitt, R. E., Bennett, A. P., Breen, A. L., Hollingsworth, T. N., Taylor, D. L., Chapin, F. S., & Rupp, T. S. (2015). Getting to the root of the matter: Landscape implications of plant-fungal interactions for tree migration in Alaska. *Landscape Ecology*. https://doi.org/10.1007/s10980-015-0306-1.

Holmes, R. M., McClelland, J. W., Peterson, B. J., Tank, S. E., Bulygina, E., Eglinton, T. I., Gordeev, V. V., Gurtovaya, T. Y., Raymond, P. A., Repeta, D. J., Staples, R., Striegl, R. G., Zhulidov, A. V., & Zimov, S. A. (2012). Seasonal and annual fluxes of nutrients and organic matter from large rivers to the Arctic Ocean and surrounding seas. *Estuaries and Coasts, 35*(2), 369–382. https://doi.org/10.1007/s12237-011-9386-6.

Holmes, R. M., Coe, T., Fiske, G. J., Gurtovaya, T., McClelland, J. W., Shiklomanov, A. I., Spencer, R. G., Tank, S. E., & Zhulidov, A. V. (2013). Climate change impacts on the hydrology and biogeochemistry of Arctic rivers. *Climatic Change and Global Warming of Inland Waters: Impacts and Mitigation for Ecosystems and Societies*, 3–26.

Huang, J., Zhang, J., Wang, L., Hao, M., Zhang, Q., Nie, S., Zhang, X., Chen, X., Lin, Y., Yao, Y., Xu, Y., Yin, Y., Luo, Y., & Zhao, Z. (2017). Recently amplified arctic warming has contributed to a continual global warming trend. *Nature Climate Change, 7*(12), 875. https://doi.org/10.1038/s41558-017-0009-5.

Hugelius, G., Strauss, J., Zubrzycki, S., Harden, J. W., Schuur, E. A. G., Ping, C.-L., Schirrmeister, L., Grosse, G., Michaelson, G. J., Koven, C. D., O'Donnell, J. A., Elberling, B., Mishra, U., Camill, P., Yu, Z., Palmtag, J., & Kuhry, P. (2014). Estimated stocks of circumpolar permafrost carbon with quantified uncertainty ranges and identified data gaps. *Biogeosciences, 11*(23), 6573–6593. https://doi.org/10.5194/bg-11-6573-2014.

Jafarov, E. E., Coon, E. T., Harp, D. R., Wilson, C. J., Painter, S. L., Atchley, A. L., & Romanovsky, V. E. (2018). Modeling the role of preferential snow accumulation in through talik development and hillslope groundwater flow in a transitional permafrost landscape. *Environmental Research Letters, 13*(10), 105006. https://doi.org/10.1088/1748-9326/aadd30.

Jones, B. M., Arp, C. D., Jorgenson, M. T., Hinkel, K. M., Schmutz, J. A., & Flint, P. L. (2009). Increase in the rate and uniformity of coastline erosion in Arctic Alaska. *Geophysical Research Letters, 36*(3). https://doi.org/10.1029/2008GL036205.

Jorgenson, M. T., & Osterkamp, T. E. (2005). Response of boreal ecosystems to varying modes of permafrost degradation. *Canadian Journal of Forest Research, 35*(9), 2100–2111. https://doi.org/10.1139/x05-153.

Kaiser, C., Meyer, H., Biasi, C., Rusalimova, O., Barsukov, P., & Richter, A. (2007). Conservation of soil organic matter through cryoturbation in arctic soils in Siberia. *Journal of Geophysical Research, 112*(G2). https://doi.org/10.1029/2006JG000258.

Kasischke, E. S., & Turetsky, M. R. (2006). Recent changes in the fire regime across the North American boreal region—Spatial and temporal patterns of burning across Canada and Alaska. *Geophysical Research Letters, 33*(9). https://doi.org/10.1029/2006GL025677.

Kimmerer, R. W. (2002). Weaving traditional ecological knowledge into biological education: A call to action. *Bioscience, 52*(5), 432–438. https://doi.org/10.1641/0006-3568(2002)052[0432:WTEKIB]2.0.CO;2.

Kling, G. (2010). Land water interactions. In *Alaska's changing arctic: Ecological consequences for tundra, streams, and lakes.*

Knoblauch, C., Beer, C., Sosnin, A., Wagner, D., & Pfeiffer, E.-M. (2013). Predicting long-term carbon mineralization and trace gas production from thawing permafrost of Northeast Siberia. *Global Change Biology, 19*(4), 1160–1172. https://doi.org/10.1111/gcb.12116.

Kokelj, S. V., & Jorgenson, M. T. (2013). Advances in Thermokarst research: Recent advances in research investigating Thermokarst processes. *Permafrost and Periglacial Processes, 24*(2), 108–119. https://doi.org/10.1002/ppp.1779.

Kolbe, T., de Dreuzy, J.-R., Abbott, B. W., Aquilina, L., Babey, T., Green, C. T., Fleckenstein, J. H., Labasque, T., Laverman, A. M., Marçais, J., Peiffer, S., Thomas, Z., & Pinay, G. (2019). Stratification of reactivity determines nitrate removal in groundwater. *Proceedings of the National Academy of Sciences*, 201816892. https://doi.org/10.1073/pnas.1816892116.

Koven, C. D. (2013). Boreal carbon loss due to poleward shift in low-carbon ecosystems. *Nature Geoscience, 6*(6), 452–456. https://doi.org/10.1038/ngeo1801.

Koven, C. D., Lawrence, D. M., & Riley, W. J. (2015a). Permafrost carbon–climate feedback is sensitive to deep soil carbon decomposability but not deep soil nitrogen dynamics. *Proceedings of the National Academy of Sciences, 112*(12), 3752–3757. https://doi.org/10.1073/pnas.1415123112.

Koven, C. D., Schuur, E. A. G., Schädel, C., Bohn, T. J., Burke, E. J., Chen, G., Chen, X., Ciais, P., Grosse, G., Harden, J. W., Hayes, D. J., Hugelius, G., Jafarov, E. E., Krinner, G., Kuhry, P., Lawrence, D. M., MacDougall, A. H., Marchenko, S. S., McGuire, A. D., ... Turetsky, M. (2015b). A simplified, data-constrained approach to estimate the permafrost carbon–climate feedback. *Philosophical Transactions of the Royal Society A, 373*(2054), 20140423. https://doi.org/10.1098/rsta.2014.0423.

Lecher, A. L. (2017). Groundwater discharge in the Arctic: A review of studies and implications for biogeochemistry. *Hydrology, 4*(3), 41. https://doi.org/10.3390/hydrology4030041.

Lenton, T. M., Rockström, J., Gaffney, O., Rahmstorf, S., Richardson, K., Steffen, W., & Schellnhuber, H. J. (2019). Climate tipping points—Too risky to bet against. *Nature, 575*(7784), 592–595. https://doi.org/10.1038/d41586-019-03595-0.

Levin, K., Cashore, B., Bernstein, S., & Auld, G. (2012). Overcoming the tragedy of super wicked problems: Constraining our future selves to ameliorate global climate change. *Policy Sciences, 45*(2), 123–152. https://doi.org/10.1007/s11077-012-9151-0.

Lindgren, A., Hugelius, G., Kuhry, P., Christensen, T. R., & Vandenberghe, J. (2016). GIS-based maps and area estimates of northern hemisphere permafrost extent during the last glacial maximum: LGM permafrost. *Permafrost and Periglacial Processes, 27*(1), 6–16. https://doi.org/10.1002/ppp.1851.

Lindgren, A., Hugelius, G., & Kuhry, P. (2018). Extensive loss of past permafrost carbon but a net accumulation into present-day soils. *Nature, 560*(7717), 219. https://doi.org/10.1038/s41586-018-0371-0.

Loranty, M. M., Abbott, B. W., Blok, D., Douglas, T. A., Epstein, H. E., Forbes, B. C., Jones, B. M., Kholodov, A. L., Kropp, H., Malhotra, A., Mamet, S. D., Myers-Smith, I. H., Natali,

S. M., O'Donnell, J. A., Phoenix, G. K., Rocha, A. V., Sonnentag, O., Tape, K. D., & Walker, D. A. (2018). Reviews and syntheses: Changing ecosystem influences on soil thermal regimes in northern high-latitude permafrost regions. *Biogeosciences, 15*(17), 5287–5313. https://doi.org/10.5194/bg-15-5287-2018.

Mack, M. C., Schuur, E. A. G., Bret-Harte, M. S., Shaver, G. R., & Chapin, F. S. (2004). Ecosystem carbon storage in arctic tundra reduced by long-term nutrient fertilization. *Nature, 431*(7007), 440–443. https://doi.org/10.1038/nature02887.

Mack, M. C., Bret-Harte, M. S., Hollingsworth, T. N., Jandt, R. R., Schuur, E. A. G., Shaver, G. R., & Verbyla, D. L. (2011). Carbon loss from an unprecedented Arctic tundra wildfire. *Nature, 475*(7357), 489–492. https://doi.org/10.1038/nature10283.

Mack, M. C., Walker, X. J., Johnstone, J. F., Alexander, H. D., Melvin, A. M., Jean, M., & Miller, S. N. (2021). Carbon loss from boreal forest wildfires offset by increased dominance of deciduous trees. *Science, 372*(6539), 280–283. https://doi.org/10.1126/science.abf3903.

Malone, E. T., Abbott, B. W., Klaar, M. J., Kidd, C., Sebilo, M., Milner, A. M., & Pinay, G. (2018). Decline in ecosystem δ13C and mid-successional nitrogen loss in a two-century postglacial chronosequence. *Ecosystems, 21*(8), 1659–1675. https://doi.org/10.1007/s10021-018-0245-1.

McClelland, J. W., Holmes, R. M., Dunton, K. H., & Macdonald, R. W. (2012). The Arctic Ocean Estuary. *Estuaries and Coasts, 35*(2), 353–368. https://doi.org/10.1007/s12237-010-9357-3.

McGuire, A. D., Melillo, J. M., Kicklighter, D. W., & Joyce, L. A. (1995). Equilibrium responses of soil carbon to climate change: Empirical and process-based estimates. *Journal of Biogeography, 22*(4/5), 785. https://doi.org/10.2307/2845980.

McGuire, A. D., Anderson, L. G., Christensen, T. R., Dallimore, S., Guo, L., Hayes, D. J., Heimann, M., Lorenson, T. D., Macdonald, R. W., & Roulet, N. (2009). Sensitivity of the carbon cycle in the Arctic to climate change. *Ecological Monographs, 79*(4), 523–555.

McGuire, A. D., Lawrence, D. M., Koven, C., Clein, J. S., Burke, E., Chen, G., Jafarov, E., MacDougall, A. H., Marchenko, S., Nicolsky, D., Peng, S., Rinke, A., Ciais, P., Gouttevin, I., Hayes, D. J., Ji, D., Krinner, G., Moore, J. C., Romanovsky, V., … Zhuang, Q. (2018). Dependence of the evolution of carbon dynamics in the northern permafrost region on the trajectory of climate change. *Proceedings of the National Academy of Sciences, 115*(15), 3882–3887. https://doi.org/10.1073/pnas.1719903115.

Meredith, M., Sommerkorn, M., Cassotta, S., Derksen, C., Ekaykin, A., Hollowed, A., Kofinas, G., Mackintosh, A., Melbourne-Thomas, J., Muelbert, M. M. C., Ottersen, G., Pritchard, H., & Schuur, E. A. G. (2019). *Polar regions* (IPCC special report on the ocean and cryosphere in a changing climate). IPCC. https://www.ipcc.ch/site/assets/uploads/sites/3/2019/11/07_SROCC_Ch03_FINAL.pdf.

Metcalfe, D. B., Hermans, T. D. G., Ahlstrand, J., Becker, M., Berggren, M., Björk, R. G., Björkman, M. P., Blok, D., Chaudhary, N., Chisholm, C., Classen, A. T., Hasselquist, N. J., Jonsson, M., Kristensen, J. A., Kumordzi, B. B., Lee, H., Mayor, J. R., Prevéy, J., Pantazatou, K., … Abdi, A. M. (2018). Patchy field sampling biases understanding of climate change impacts across the Arctic. *Nature Ecology & Evolution, 1*. https://doi.org/10.1038/s41559-018-0612-5.

Mishra, U., Hugelius, G., Shelef, E., Yang, Y., Strauss, J., Lupachev, A., Harden, J. W., Jastrow, J. D., Ping, C.-L., Riley, W. J., Schuur, E. A. G., Matamala, R., Siewert, M., Nave, L. E., Koven, C. D., Fuchs, M., Palmtag, J., Kuhry, P., Treat, C. C., … Orr, A. (2021). Spatial heterogeneity and environmental predictors of permafrost region soil organic carbon stocks. *Science Advances, 7*(9), eaaz5236. https://doi.org/10.1126/sciadv.aaz5236.

Monteux, S., Keuper, F., Fontaine, S., Gavazov, K., Hallin, S., Juhanson, J., Krab, E. J., Revaillot, S., Verbruggen, E., Walz, J., Weedon, J. T., & Dorrepaal, E. (2020). Carbon and nitrogen cycling in Yedoma permafrost controlled by microbial functional limitations. *Nature Geoscience, 13*(12), 794–798. https://doi.org/10.1038/s41561-020-00662-4.

Montzka, S. A., Dlugokencky, E. J., & Butler, J. H. (2011). Non-CO2 greenhouse gases and climate change. *Nature, 476*(7358), 43–50. https://doi.org/10.1038/nature10322.

Mu, C., Abbott, B. W., Norris, A. J., Mu, M., Fan, C., Chen, X., Jia, L., Yang, R., Zhang, T., Wang, K., Peng, X., Wu, Q., Guggenberger, G., & Wu, X. (2020). The status and stability of permafrost carbon on the Tibetan Plateau. *Earth-Science Reviews, 211*, 103433. https://doi.org/10.1016/j.earscirev.2020.103433.

Myers-Smith, I. H., Kerby, J. T., Phoenix, G. K., Bjerke, J. W., Epstein, H. E., Assmann, J. J., John, C., Andreu-Hayles, L., Angers-Blondin, S., Beck, P. S. A., Berner, L. T., Bhatt, U. S., Bjorkman, A. D., Blok, D., Bryn, A., Christiansen, C. T., Cornelissen, J. H. C., Cunliffe, A. M., Elmendorf, S. C., … Wipf, S. (2020). Complexity revealed in the greening of the Arctic. *Nature Climate Change, 10*(2), 106–117. https://doi.org/10.1038/s41558-019-0688-1.

Natali, S. M., Watts, J. D., Rogers, B. M., Potter, S., Ludwig, S. M., Selbmann, A.-K., Sullivan, P. F., Abbott, B. W., Arndt, K. A., Birch, L., Björkman, M. P., Bloom, A. A., Celis, G., Christensen, T. R., Christiansen, C. T., Commane, R., Cooper, E. J., Crill, P., Czimczik, C., … Zona, D. (2019). Large loss of CO2 in winter observed across the northern permafrost region. *Nature Climate Change, 9*(11), 852–857. https://doi.org/10.1038/s41558-019-0592-8.

Neubauer, S. C., & Megonigal, J. P. (2015). Moving beyond global warming potentials to quantify the climatic role of ecosystems. *Ecosystems, 18*(6), 1000–1013. https://doi.org/10.1007/s10021-015-9879-4.

Obu, J., Westermann, S., Bartsch, A., Berdnikov, N., Christiansen, H. H., Dashtseren, A., Delaloye, R., Elberling, B., Etzelmüller, B., Kholodov, A., Khomutov, A., Kääb, A., Leibman, M. O., Lewkowicz, A. G., Panda, S. K., Romanovsky, V., Way, R. G., Westergaard-Nielsen, A., Wu, T., … Zou, D. (2019). Northern Hemisphere permafrost map based on TTOP modelling for 2000–2016 at 1 km2 scale. *Earth-Science Reviews, 193*, 299–316. https://doi.org/10.1016/j.earscirev.2019.04.023.

Olefeldt, D., Goswami, S., Grosse, G., Hayes, D., Hugelius, G., Kuhry, P., McGuire, A. D., Romanovsky, V. E., Sannel, A. B. K., Schuur, E. A. G., & Turetsky, M. R. (2016). Circumpolar distribution and carbon storage of thermokarst landscapes. *Nature Communications, 7*, 13043. https://doi.org/10.1038/ncomms13043.

Overduin, P. P., von Deimling, T. S., Miesner, F., Grigoriev, M. N., Ruppel, C., Vasiliev, A., Lantuit, H., Juhls, B., & Westermann, S. (2018). Submarine permafrost map in the Arctic modeled using 1-D transient heat flux (SuPerMAP). *Journal of Geophysical Research: Oceans, 0*(0). https://doi.org/10.1029/2018JC014675.

Pan, Y., Birdsey, R. A., Fang, J., Houghton, R., Kauppi, P. E., Kurz, W. A., Phillips, O. L., Shvidenko, A., Lewis, S. L., Canadell, J. G., Ciais, P., Jackson, R. B., Pacala, S. W., McGuire, A. D., Piao, S., Rautiainen, A., Sitch, S., & Hayes, D. (2011). A large and persistent carbon sink in the world's forests. *Science, 333*(6045), 988–993. https://doi.org/10.1126/science.1201609.

Parkinson, A. J., & Berner, J. (2009). Climate change and impacts on human health in the Arctic: An international workshop on emerging threats and the response of Arctic communities to climate change. *International Journal of Circumpolar Health, 68*(1), 84–91. https://doi.org/10.3402/ijch.v68i1.18295.

Pearce, T. D., Ford, J. D., Laidler, G. J., Smit, B., Duerden, F., Allarut, M., Andrachuk, M., Baryluk, S., Dialla, A., Elee, P., Goose, A., Ikummaq, T., Joamie, E., Kataoyak, F., Loring, E., Meakin, S., Nickels, S., Shappa, K., Shirley, J., & Wandel, J. (2009). Community collaboration and climate change research in the Canadian Arctic. *Polar Research, 28*(1), 10–27. https://doi.org/10.1111/j.1751-8369.2008.00094.x.

Peterson, B. J. (2006). Trajectory shifts in the Arctic and subarctic freshwater cycle. *Science, 313*(5790), 1061–1066. https://doi.org/10.1126/science.1122593.

Polvani, L. M., Previdi, M., England, M. R., Chiodo, G., & Smith, K. L. (2020). Substantial twentieth-century Arctic warming caused by ozone-depleting substances. *Nature Climate Change*, 1–4. https://doi.org/10.1038/s41558-019-0677-4.

Rastetter, E. B., Kling, G. W., Shaver, G. R., Crump, B. C., Gough, L., & Griffin, K. L. (2020). Ecosystem recovery from disturbance is constrained by N cycle openness, vegetation-soil N distribution, form of N losses, and the balance between vegetation and soil-microbial processes. *Ecosystems*. https://doi.org/10.1007/s10021-020-00542-3.

Raupach, M. R., & Canadell, J. G. (2010). Carbon and the anthropocene. *Current Opinion in Environmental Sustainability, 2*(4), 210–218. https://doi.org/10.1016/j.cosust.2010.04.003.

Rawlins, M. A., Steele, M., Holland, M. M., Adam, J. C., Cherry, J. E., Francis, J. A., Groisman, P. Y., Hinzman, L. D., Huntington, T. G., Kane, D. L., Kimball, J. S., Kwok, R., Lammers, R. B., Lee, C. M., Lettenmaier, D. P., McDonald, K. C., Podest, E., Pundsack, J. W., Rudels, B., … Zhang, T. (2010). Analysis of the Arctic system for freshwater cycle intensification: Observations and expectations. *Journal of Climate, 23*(21), 5715–5737. https://doi.org/10.1175/2010JCLI3421.1.

Rawlins, M. A., Cai, L., Stuefer, S. L., & Nicolsky, D. (2019). Changing characteristics of runoff and freshwater export from watersheds draining northern Alaska. *The Cryosphere, 13*(12), 3337–3352. https://doi.org/10.5194/tc-13-3337-2019.

Riedlinger, D., & Berkes, F. (2001). Contributions of traditional knowledge to understanding climate change in the Canadian Arctic. *Polar Record, 37*(203), 315–328. https://doi.org/10.1017/S0032247400017058.

Roy Chowdhury, T., Herndon, E. M., Phelps, T. J., Elias, D. A., Gu, B., Liang, L., Wullschleger, S. D., & Graham, D. E. (2015). Stoichiometry and temperature sensitivity of methanogenesis and CO2 production from saturated polygonal tundra in Barrow, Alaska. *Global Change Biology, 21*(2), 722–737. https://doi.org/10.1111/gcb.12762.

Saito, K., Walsh, J. E., Bring, A., Brown, R., Shiklomanov, A., & Yang, D. (2021). Future trajectory of Arctic system evolution. In D. Yang & D. L. Kane (Eds.), *Arctic hydrology, permafrost and ecosystems* (pp. 893–914). Springer. https://doi.org/10.1007/978-3-030-50930-9_30.

Sapart, C. J., Shakhova, N., Semiletov, I., Jansen, J., Szidat, S., Kosmach, D., Dudarev, O., van der Veen, C., Egger, M., Sergienko, V., Salyuk, A., Tumskoy, V., Tison, J.-L., & Röckmann, T. (2017). The origin of methane in the east Siberian Arctic shelf unraveled with triple isotope analysis. *Biogeosciences, 14*(9), 2283–2292. https://doi.org/10.5194/bg-14-2283-2017.

Sayedi, S. S., Abbott, B. W., Thornton, B. F., Frederick, J. M., Vonk, J. E., Overduin, P., Schädel, C., Schuur, E. A. G., Bourbonnais, A., Demidov, N., Gavrilov, A., He, S., Hugelius, G., Jakobsson, M., Jones, M. C., Joung, D., Kraev, G., Macdonald, R. W., McGuire, A. D., … Frei, R. J. (2020). Subsea permafrost carbon stocks and climate change sensitivity estimated by expert assessment. *Environmental Research Letters, 15*(12), 124075. https://doi.org/10.1088/1748-9326/abcc29.

Schädel, C., Bader, M. K.-F., Schuur, E. A. G., Biasi, C., Bracho, R., Čapek, P., De Baets, S., Diáková, K., Ernakovich, J., Estop-Aragones, C., Graham, D. E., Hartley, I. P., Iversen, C. M., Kane, E., Knoblauch, C., Lupascu, M., Martikainen, P. J., Natali, S. M., Norby, R. J., … Wickland, K. P. (2016). Potential carbon emissions dominated by carbon dioxide from thawed permafrost soils. *Nature Climate Change*. https://doi.org/10.1038/nclimate3054.

Schlesinger, W. H., & Bernhardt, E. S. (2012). *Biogeochemistry: An analysis of global change*. Academic.

Schmidt, M. W. I., Torn, M. S., Abiven, S., Dittmar, T., Guggenberger, G., Janssens, I. A., Kleber, M., Kögel-Knabner, I., Lehmann, J., Manning, D. A. C., Nannipieri, P., Rasse, D. P., Weiner, S., & Trumbore, S. E. (2011). Persistence of soil organic matter as an ecosystem property. *Nature, 478*(7367), 49–56. https://doi.org/10.1038/nature10386.

Schuur, E. A. G., McGuire, A. D., Schädel, C., Grosse, G., Harden, J. W., Hayes, D. J., Hugelius, G., Koven, C. D., Kuhry, P., Lawrence, D. M., Natali, S. M., Olefeldt, D., Romanovsky, V. E., Schaefer, K., Turetsky, M. R., Treat, C. C., & Vonk, J. E. (2015). Climate change and the permafrost carbon feedback. *Nature, 520*(7546), 171–179. https://doi.org/10.1038/nature14338.

Shaver, G. R., Billings, W. D., Chapin, F. S., Giblin, A. E., Nadelhoffer, K. J., Oechel, W. C., & Rastetter, E. B. (1992). Global change and the carbon balance of Arctic ecosystems. *Bioscience, 42*(6), 433–441. https://doi.org/10.2307/1311862.

Shogren, A. J., Zarnetske, J. P., Abbott, B. W., Iannucci, F., Frei, R. J., Griffin, N. A., & Bowden, W. B. (2019). Revealing biogeochemical signatures of Arctic landscapes with river chemistry. *Scientific Reports, 9*(1), 1–11. https://doi.org/10.1038/s41598-019-49296-6.

Shogren, A. J., Zarnetske, J. P., Abbott, B. W., Iannucci, F., & Bowden, W. B. (2020). We cannot shrug off the shoulder seasons: Addressing knowledge and data gaps in an Arctic headwater. *Environmental Research Letters*. https://doi.org/10.1088/1748-9326/ab9d3c.

Shur, Y. L., & Jorgenson, M. T. (2007). Patterns of permafrost formation and degradation in relation to climate and ecosystems. *Permafrost and Periglacial Processes, 18*(1), 7–19. https://doi.org/10.1002/ppp.582.

Slater, A. G., & Lawrence, D. M. (2013). Diagnosing present and future permafrost from climate models. *Journal of Climate, 26*(15), 5608–5623. https://doi.org/10.1175/JCLI-D-12-00341.1.

Steffen, W., Richardson, K., Rockström, J., Cornell, S. E., Fetzer, I., Bennett, E. M., Biggs, R., Carpenter, S. R., de Vries, W., de Wit, C. A., Folke, C., Gerten, D., Heinke, J., Mace, G. M., Persson, L. M., Ramanathan, V., Reyers, B., & Sörlin, S. (2015). Planetary boundaries: Guiding human development on a changing planet. *Science, 347*(6223), 1259855. https://doi.org/10.1126/science.1259855.

Strauss, J., Schirrmeister, L., Grosse, G., Fortier, D., Hugelius, G., Knoblauch, C., Romanovsky, V., Schädel, C., Schneider von Deimling, T., Schuur, E. A. G., Shmelev, D., Ulrich, M., & Veremeeva, A. (2017). Deep Yedoma permafrost: A synthesis of depositional characteristics and carbon vulnerability. *Earth-Science Reviews, 172*, 75–86. https://doi.org/10.1016/j.earscirev.2017.07.007.

Sulla-Menashe, D., Woodcock, C. E., & Friedl, M. A. (2018). Canadian boreal forest greening and browning trends: An analysis of biogeographic patterns and the relative roles of disturbance versus climate drivers. *Environmental Research Letters, 13*(1), 014007. https://doi.org/10.1088/1748-9326/aa9b88.

Tank, S. E., Striegl, R. G., McClelland, J. W., & Kokelj, S. V. (2016). Multi-decadal increases in dissolved organic carbon and alkalinity flux from the Mackenzie drainage basin to the Arctic Ocean. *Environmental Research Letters, 11*(5), 054015. https://doi.org/10.1088/1748-9326/11/5/054015.

Tank, S. E., Vonk, J. E., Walvoord, M. A., McClelland, J. W., Laurion, I., & Abbott, B. W. (2020). Landscape matters: Predicting the biogeochemical effects of permafrost thaw on aquatic networks with a state factor approach. *Permafrost and Periglacial Processes*. https://doi.org/10.1002/ppp.2057.

Tierney, J. E., Poulsen, C. J., Montañez, I. P., Bhattacharya, T., Feng, R., Ford, H. L., Hönisch, B., Inglis, G. N., Petersen, S. V., Sagoo, N., Tabor, C. R., Thirumalai, K., Zhu, J., Burls, N. J., Foster, G. L., Goddéris, Y., Huber, B. T., Ivany, L. C., Turner, S. K., … Zhang, Y. G. (2020). Past climates inform our future. *Science, 370*(6517). https://doi.org/10.1126/science.aay3701.

Trumbore, S. (2009). Radiocarbon and soil carbon dynamics. *Annual Review of Earth and Planetary Sciences, 37*(1), 47–66. https://doi.org/10.1146/annurev.earth.36.031207.124300.

Turetsky, M. R., Abbott, B. W., Jones, M. C., Anthony, K. W., Olefeldt, D., Schuur, E. A. G., Grosse, G., Kuhry, P., Hugelius, G., Koven, C., Lawrence, D. M., Gibson, C., Sannel, A. B. K., & McGuire, A. D. (2020). Carbon release through abrupt permafrost thaw. *Nature Geoscience, 13*(2), 138–143. https://doi.org/10.1038/s41561-019-0526-0.

Voigt, C., Lamprecht, R. E., Marushchak, M. E., Lind, S. E., Novakovskiy, A., Aurela, M., Martikainen, P. J., & Biasi, C. (2016). Warming of subarctic tundra increases emissions of all three important greenhouse gases – Carbon dioxide, methane, and nitrous oxide. *Global Change Biology*, n/a-n/a. https://doi.org/10.1111/gcb.13563.

Voigt, C., van Delden, L., Marushchak, M. E., Biasi, C., Abbott, B. W., Elberling, B., Siciliano, S. D., Sonnentag, O., Stewart, K. J., Yang, Y., & Martikainen, P. J. (2020). *Nitrous oxide fluxes from permafrost regions* (Data set). PANGAEA. https://doi.pangaea.de/10.1594/PANGAEA.919217.

Vonk, J. E., Sánchez-García, L., van Dongen, B. E., Alling, V., Kosmach, D., Charkin, A., Semiletov, I. P., Dudarev, O. V., Shakhova, N., Roos, P., Eglinton, T. I., Andersson, A., & Gustafsson, Ö. (2012). Activation of old carbon by erosion of coastal and subsea permafrost in Arctic Siberia. *Nature, 489*(7414), 137–140. https://doi.org/10.1038/nature11392.

Vonk, J. E., Tank, S. E., Bowden, W. B., Laurion, I., Vincent, W. F., Alekseychik, P., Amyot, M., Billet, M. F., Canário, J., Cory, R. M., Deshpande, B. N., Helbig, M., Jammet, M., Karlsson, J., Larouche, J., MacMillan, G., Rautio, M., Walter Anthony, K. M., & Wickland,

K. P. (2015). Reviews and syntheses: Effects of permafrost thaw on Arctic aquatic ecosystems. *Biogeosciences, 12*(23), 7129–7167. https://doi.org/10.5194/bg-12-7129-2015.

Wologo, E., Shakil, S., Zolkos, S., Textor, S., Ewing, S., Klassen, J., Spencer, R. G. M., Podgorski, D. C., Tank, S. E., Baker, M. A., O'Donnell, J. A., Wickland, K. P., Foks, S. S. W., Zarnetske, J. P., Lee-Cullin, J., Liu, F., Yang, Y., Kortelainen, P., Kolehmainen, J., … Abbott, B. W. (2021). Stream dissolved organic matter in permafrost regions shows surprising compositional similarities but negative priming and nutrient effects. *Global Biogeochemical Cycles, 35*(1), e2020GB006719. https://doi.org/10.1029/2020GB006719.

Zarnetske, J. P., Bouda, M., Abbott, B. W., Saiers, J., & Raymond, P. A. (2018). Generality of hydrologic transport limitation of watershed organic carbon flux across ecoregions of the United States. *Geophysical Research Letters, 45*(21), 11,702–11,711. https://doi.org/10.1029/2018GL080005.

Zhang, T., Heginbottom, J. A., Barry, R. G., & Brown, J. (2000). Further statistics on the distribution of permafrost and ground ice in the Northern Hemisphere [1]. *Polar Geography, 24*(2), 126–131. https://doi.org/10.1080/10889370009377692.

Zhang, Q., Yang, G., Song, Y., Kou, D., Wang, G., Zhang, D., Qin, S., Mao, C., Feng, X., & Yang, Y. (2019). Magnitude and drivers of potential methane oxidation and production across the Tibetan alpine permafrost region. *Environmental Science & Technology, 53*(24), 14243–14252. https://doi.org/10.1021/acs.est.9b03490.

Zimov, S. A., Davydov, S. P., Zimova, G. M., Davydova, A. I., Schuur, E. A. G., Dutta, K., & Chapin, F. S. (2006). Permafrost carbon: Stock and decomposability of a globally significant carbon pool. *Geophysical Research Letters, 33*(20). https://doi.org/10.1029/2006GL027484.

**Benjamin W. Abbott** is an assistant professor at Brigham Young University in Provo, Utah, USA. After receiving his Ph.D. from the University of Alaska Fairbanks, he worked as a Marie Skłodowska Curie postdoctoral fellow with the French National Centre for Scientific Research (CNRS) and then for Michigan State University.

# Chapter 11
# Impacts of Global Warming on Arctic Biota

Mathilde Le Moullec and Morgan Lizabeth Bender

## Introduction

Climate change, which is magnified toward the poles with the Arctic amplification and the convergence of atmospheric and marine currents from the southerly latitudes, affects biota through multiple mechanisms. In this chapter, we introduce three major ecological changes in the rapidly warming Arctic, highlighting how changes at the base of the food web can ripple across the ecosystem. First, we review the major ecological and evolutionary consequences of sea-ice retreat on marine and terrestrial wildlife: the disappearance of ice-associated species' habitat; the altered productivity of the ice edge ecosystem; and the loss of connectivity between populations. Second, we assess the ways in which boreal species invade and outcompete Arctic species by reducing diet quality and increasing stress through (apparent) competition, predation, and disease exposure. Third, we document an altered tundra ecosystem with changes in vegetation productivity and availability related to rainier polar winters. Finally, we discuss the importance of the spatial scale at which climate change will act on distant populations, where a variety of environmental gradients, like divergent vegetation trends, are key to sustaining biodiversity.

M. Le Moullec (✉)
Norwegian University of Science and Technology (NTNU), Trondheim, Norway

Laval University (ULAVAL), Québec, QC, Canada
e-mail: mathilde.lemoullec@ntnu.no

M. L. Bender
The Arctic University of Norway (UiT), Tromsø, Norway
e-mail: morgan.l.bender@uit.no

© The Author(s), under exclusive license to Springer Nature Switzerland AG 2022
M. Finger, G. Rekvig (eds.), *Global Arctic*,
https://doi.org/10.1007/978-3-030-81253-9_11

# Sea-Ice Retreat and Its Ecological and Evolutionary Consequences on Wildlife

Between cracks, attached underneath, walking on top, swimming along the edges, and in surrounding waters, a large diversity of organisms are associated with the sea-ice either for some or all of their life cycle (Lund-Hansen et al., 2020; Sakshaug et al., 2009). With climate change, this ecosystem is currently undergoing drastic changes as it shrinks, thins, and drifts more quickly around the Arctic. Sea-ice that lasts over multiple summers, called multi-year ice, hosts organisms that are highly adapted to subzero temperatures and wide variations in water salinity and amount of sunlight. The multi-year ice habitat is disappearing and may completely vanish by 2040, according to recently published climate scenarios (Meredith et al., 2019). First-year ice that forms during the annual fall freeze-up, melting over the summer, will likely be found farther north into the future. However, the extent of this ice will be reduced and the ice edge will move northwards, above the deep Arctic Ocean. The ice edge itself is a highly productive ecosystem when located close to the coast and above shallow seas, such as in the Bering and Barents Seas. It is the favored habitat of iconic species like narwhal *Monodon monoceros*, beluga *Delphinapterus leucas,* and bowhead *Balaena mysticetus* whales, as well as polar bears *Ursus maritimus*. Shifting the timing of the ice edge retreat in the spring, combined with the more northern location of the edge, threatens productivity (Wassmann & Reigstad, 2011).

The marine and terrestrial systems are tightly connected in the Arctic because of its geography of extensive coastline encircling the Arctic Ocean. In this way, disappearing sea-ice implies major changes on land. Sea-ice connected to land has direct influences on coastal and inner fjords climate by promoting cold stable weather. Sea-ice is also a surface that land plants and animals use to travel and disperse, thereby connecting continents like Russia and Canada, but also islands within archipelagos (Post et al., 2013, 2019). We will explain how the functioning of ecosystems linked to multi-year ice, the ice edge, and coastal areas are altered by disappearing sea-ice (Figures 11.1 and 11.2).

## *Loss of Multi-Year Ice Threatens Native Ice-Associated Communities*

The multi-year ice environment, commonly comprised of two-to-seven-year-old ice, has different physical and chemical properties than first-year ice and hosts different ecological communities. In these communities, organisms have evolved physiologies, morphologies, and life history strategies specific to a lifetime in the sea-ice. Examples of these highly adapted ice-associated organisms are the ice algae species *Melosira arctica*, *Nitschia frigida,* or the amphipod *Gammarus wilkitskii,* an iconic zooplankton species (Sakshaug et al., 2009). Unlike first-year ice, multi-year ice does not need to be recolonized by organisms each winter. Multi-year ice drifting into first-year ice promotes the dispersal of native species. Thereafter, older and thicker ice have higher abundances of these species (Sakshaug et al., 2009). Within multi-year ice, the detritivorous amphipod *G. wilkitskii* is a species with a longevity of five to six years, which inhabits the underside of the ice with a spiky large-sized body shape that would otherwise sink in open water. In contrast, the herbivorous amphipod *Apherusa glacialis* is short-lived (one to two years) and has high fecundity. Being a more motile species, it is commonly the first amphipod to colonize first-year ice (Beuchel & Lønne, 2002; Sakshaug et al., 2009). These contrasting life-history strategies will influence how the species will face an ice-free ocean. A recent study on the circumpolar distribution and life history of this amphipod suggests that it may have the ability to inhabit an ice-free Arctic Ocean (Kunisch et al., 2020). However, ice-associated amphipods completing parts of their life cycle as benthic organisms on shallow sea shelves before ascending to the sea-ice, such as several species from the *Onisimus* genus, will face difficulties when first-year sea-ice is restricted to the deep Arctic Ocean. In the near future, first-year ice may primarily be colonized by pelagic organisms and, in this way, would resemble Antarctic communities (Sakshaug et al., 2009).

These zooplankton communities can act as indicators of climate change effects on marine ecosystems because they are the link between the primary producers and higher tropic levels. Changes in the abundance and composition of these native ice fauna can ripple through the food web. Little auks *Alle alle* feeding in the multi-year ice can have 80 percent of their diet composed of *A. glacialis* (Sakshaug et al., 2009), which is also the favorite food resource of polar cod *Boreogadus saida* in the ice. The polar cod is a key ice-associated species, as their eggs and larvae use the sea-ice for protection from storms, UV damage, and predation (Gradinger & Bluhm, 2004; Spencer et al., 2020). Recruitment collapse threatens polar cod populations as sea-ice retreats and temperature increases in key nursery areas around the Arctic Ocean (Huserbråten et al., 2019). Furthermore, polar cod is the main food resource of most seabirds (such as the ivory gull *Pagophila eburnean* and the northern fulmar *Fulmarus glacialis*), seals (such as the ringed *Pusa hispida* and the harp seal *Pagophilus groenlandicus*), and whales (such as the beluga) feeding in the ice. The ringed seal, which also feeds on the amphipod *G. wilkitskii*, breed and live year-round in the ice and are the only seal species to actively maintain breathing holes, also in multi-year ice (Kovacs et al., 2011). Ringed seals are the main prey of polar bears, although polar bears are rarely present in multi-year ice, but rather hunt in more open ice at the ice edge (Lone et al., 2018). These highly specialized ice-associated species are at risk of extinction unless they can quickly cope with the disappearance of summer sea-ice (Laidre et al., 2015).

## *Loss of a Productive Habitat at the Ice Edge*

The ice edge, as well as areas of ice formation (that is, polynyas), are highly productive ecosystems, stimulated by light availability and the mixing of nutrients. Light is a limiting factor for life in Arctic waters, especially during the polar night and under sea-ice and snow. As the light returns in the spring, ice-algae starts to bloom under the ice, followed by a second major bloom of pelagic phytoplankton as the ice retreats. The timing between the two events is synchronized with the reproduction and developmental stages of planktonic grazers. The Arctic copepod *Calanus glacialis* represent 80% of Arctic zooplankton biomass and plays a major role in assimilating essential omega-3 fatty acids produced abundantly by the ice algae, and widely utilized by Arctic marine organisms. In the Arctic food-web, these native copepods are energy-rich food for secondary consumers like fish, which in turn are food for the top predators like seals, whales, and polar bears. The earlier sea-ice break-up comes with earlier phytoplankton bloom, which shortens the lag-time between the ice and pelagic algae blooms, resulting in a potential mismatch with the copepod developmental stages and their food. In mismatched years, their recruitment and growth are reduced, which alters the diet of higher trophic levels (Leu et al., 2011; Søreide et al., 2010).

Another threat is the change of location of the ice edge productivity that is moving northwards, away from the coast and shallow seas. Extensive areas of shallow oceans, approximately 200 m deep, that surround the Arctic Ocean, like the Bering or Barents Sea, are seasonally covered by ice and have a tight coupling between the bottom and the surface waters. Detritus from the large spring blooms settle to the bottom, feeding a diverse benthos community (Søreide et al., 2013). The water layers mix during storms, bringing bottom-nutrients to the surface, which in turn feed the ice and pelagic algae (Dunton et al., 2005). Such rapid advective processes do not occur above the deep Arctic Ocean of ca. 5000 m depth, and the primary production that is currently low under multi-year ice is expected to remain low with first-year ice (Lund-Hansen et al., 2020; Wassmann & Reigstad, 2011). Although the now-extensive open water areas above shallow seas have shown increased primary productivity in recent decades, the bloom onset has advanced and the planktonic communities (phyto- and zooplankton) are changing (Lewis et al., 2020).

The rich benthos communities in the shallow seas provide food for several organisms using the sea-ice as a resting platform, such as common eider ducks *Somateria mollissima* (Heath et al., 2006), bearded seals *Erignathus barbatus*, and walruses *Odobenus rosmarus*. Bottom-feeders like walruses feed in the shallow seas where they can reach the bottom. As their resting ice platform retreats northwards above the deep Arctic Ocean, walruses have now opted to stay on land, restricting their feeding environments to nearshore areas (Jay et al., 2012; Post et al., 2013). Walrus aggregation areas in northwestern Alaska already exhibit signs of resource shortage (Jay et al., 2012), and reduced access to resources can only sustain smaller populations. Conversely, although they are good swimmers, polar bear females roaming the offshore sea-ice face difficulties reaching their denning sites on land.

Furthermore, young cubs cannot swim long distances, which makes it challenging to return to the sea-ice (Lone et al., 2018).

The ice edge and polynyas in the vicinity of coastal areas are also hunting hotspots for indigenous peoples. For example, the community of Igloolik, Canada, is adjacent to a polynya. Thus, reduced productivity and access to the ice edge can influence indigenous communities' subsistence and endangered and emblematic Arctic species.

## Alteration of the Coastal Sea-Ice and Consequences for Land Organisms

When a fresh breeze flows along the ice-covered coast, it cools the nearby land (Macias-Fauria et al., 2017). Fast-land ice creates stability in winter weather. It acts as an isolation lid, preventing water masses from releasing heat near the coast, and can thereby buffer against moderate local warm-weather events. In spring, the date of sea-ice breakup along the Arctic coasts has significantly advanced in the past decades, if any fast-land ice has even formed over the winter. In turn, open water can influence local coastal weather on land, accelerating snowmelt and vegetation onset (Macias-Fauria et al., 2017; Post et al., 2013). Herbivores can benefit from earlier and greater access to their food resources. However, this requires that migratory herbivores have reached their coastal summer grounds and that their offspring production will still match the timing of peak quality and quantity of vegetation production (e.g., Doiron et al., 2015 [snow geese *Chen caerulescens*], Kerby & Post, 2013 [caribou *Rangifer tarandus*]). Reproduction is a complex process governed by many environmental and biological processes (such as gestation or breeding time) and cannot readily shift temporally in most wildlife. The evolution of such traits, requiring earlier birth or egg-laying, may not keep up with the rate of climate change.

Sea-ice is a traveling platform for numerous terrestrial organisms that connects thousands of Arctic archipelagos as one "solid ground". Numerous flora species bordering the Fram Strait (in East Greenland, Svalbard, and Iceland) dispersed passively with the help of sea-ice, where driftwood carrying seeds comes from large rivers like the Yenisey or the Yukon (Alsos et al., 2016). The Arctic fox *Vulpes lagopus* is an iconic long-distance traveler, whose fastest recorded travel was 3506 km on ice in 76 days between Svalbard and northern Canada (Fuglei & Tarroux, 2019). The presence of sea-ice predicts the genetic isolation of Arctic foxes in areas that are no longer connected to sea-ice, such as Iceland, Fennoscandia, or the Aleutians – areas that are now subject to isolation and subsequent inbreeding (Norén et al., 2017). Loss of sea-ice in recent decades was also found to have increased the genetic isolation of reindeer and caribou inhabiting high-Arctic archipelagos (Jenkins et al., 2016; Peeters et al., 2020). In contrast, dispersal limitation can restrict the spread of some diseases carried by Arctic foxes, like rabies (Simon et al., 2019) (Fig. 11.1).

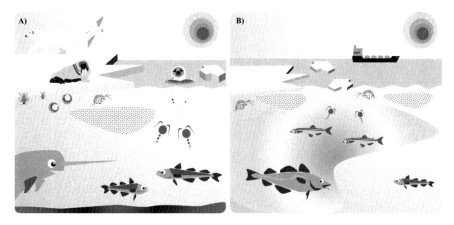

**Fig. 11.1** Schematic representation of (**a**) multi-year sea-ice Arctic ecosystems above shallow seas with the Arctic wildlife of polar bear and walrus close to the sea-ice edge; sea-ice algae, amphipods *G. wilkitzkii*, *A. glacialis*, and polar cod embryos under the ice; and ice-associated seals, narwhals, polar cod, and large-sized Arctic copepods species in the cold water; (**b**) under a possible climate change scenario, with thin first-year ice above the deep Arctic Ocean (accessible to cargo ships), incoming boreal species (Atlantic cod, capelin, and small-sized copepods) and altered primary productivity patterns with warmer waters, and few remaining Arctic species (*A. glacialis* and polar cod)

## Effects of Boreal Invasion on Arctic Species

The habitat of Arctic species is shrinking as the sea-ice retreats, the Atlantic and Pacific waters flow further North, and the temperature increases. New species, which are often better adapted to these new environmental conditions, enter the system from southern boreal regions; this process is known as the borealization of the Arctic (Polyakov et al., 2020) (Figure 11.1b and 11.2b). Incoming warm waters from the South not only open a shortcut for shipping through the Arctic, but also for wildlife. The Northwest Passage, between the Atlantic and the Pacific, is expected to be a zone of interchange for boreal fish species (Wisz et al., 2015). While the northward limits of boreal species' range are generally due to lack of adaptation to climate harshness and resource limitation, the Arctic species' southward range limit is determined by competition (Callaghan et al., 2004). Ecological changes occur when there are no escape routes for Arctic wildlife and cohabitation must take place, both at sea and on land. Generalist boreal species can outcompete the highly specialized Arctic species because of their more versatile intrinsic traits, such as a more diverse diet, higher motility, and being a less energetic prey item, but also through predation, hybridization and diseases/parasitism risks (Ims & Ehrich, 2013; Polyakov et al., 2020).

## Replacement Species of Lower Quality

Although borealization of the Arctic Ocean and shelf seas provides increasingly abundant pelagic food resources for Arctic predators, prey items are often smaller in size and have lower lipid content. These alternative prey items of lower quality result in a slower transfer of energy through the food web. For example, the Atlantic copepod *C. finmarchicus* and the pelagic fish caplin *Mallotus villosus* have been found to be alternative prey for the copepod *C. glacialis* and the polar cod, respectively, which are in turn eaten by several Arctic seabird species, such as the little auk and black-legged kittiwake *Rissa tridactyla* (Amélineau et al., 2019; Møller & Nielsen, 2020; Vihtakari et al., 2018). Little auks have been found to increase their energy expenditure in flying longer trips in years with a high inflow of Atlantic water, which decreased their body conditions, chicks' feeding rate, and their growth rate (Amélineau et al., 2019). These changes likely explained the decrease in adult and chick survival rates measured in Svalbard colonies (Hovinen et al., 2014).

The range overlap of related boreal and Arctic species can now lead to species hybridization. The native Arctic species' genetic integrity can be altered by genetic introgression from boreal species and lead to a loss of genetic biodiversity. The genetic diversity of a species represents the pool of genes where natural selection can act and thereafter determines the rate of evolution. In salmonid Arctic species, like the Arctic charr *Salvelinus alpinus,* hybrids are often less fit in warmer waters than their boreal parent, with reduced growth rates and survival (Dupont Cyr et al., 2018; Young et al., 2016). However, a recent contradictory point of view proposed that hybridization can rescue some Arctic genes from going extinct, even if the Arctic species may vanish (Charles & Stehlik, 2020). Hybridization may speed up the process of introducing warm-adapted genes into cold-adapted species. It is suggested that the pizzly – the hybrid between a polar and grizzly bear *Ursus arctos* – has both temperate and Arctic genes, yet the possible speciation into a new species is unknown (Derocher, 2012, p. 40). Hybridization may not rescue polar bears as a species, but in fact preserve part of their genetic diversity.

## Arctic Specialists Outcompeted by Boreal Generalists

In different biomes, similar ecological niches are occupied by species with comparable functional traits. The "equivalent" boreal species of an Arctic species is commonly larger and has a more diverse diet than the Arctic species. Therefore, they are often better competitors and even predators of their Arctic counterparts. Boreal fish species have functional traits characterized by large body sizes, a fish-eating diet, and high motility, and are replacing Arctic fish communities that are characterized by smaller, slower bottom-associated species (Frainer et al., 2017). This directional shift is attractive for fisheries that have intensified their activity northward. For example, large stocks of Atlantic cod *Gadus morhua* and haddock

*Melanogrammus aeglefinus* are now found off the coast of Svalbard. These generalist species, with an opportunist diet, become important competitors and predators of key Arctic fish like polar cod (Polyakov et al., 2020; Sakshaug et al., 2009). Atlantic cod is considered a new top predator of the marine Arctic system that is competing with marine mammals, and it is the likely cause of the decrease in body condition seen in harp seals (Bogstad et al., 2015). Ice-associated marine mammals also see an increase in predation risk with the northward expansion of killer whales *Orcinus orca* distribution, in addition to being exposed to new incoming diseases (Kovacs et al., 2011).

On land, species borealization is caused by the combination of both climatic and anthropogenic changes (Callaghan et al., 2004; Elmhagen et al., 2015; Festa-Bianchet et al., 2011). Northward expansions of generalists and opportunists like the raven and the red fox *Vulpes vulpes* pose a serious threat to tundra ecosystem functioning. In addition to being a better competitor, red foxes can directly predate on Arctic foxes. In northern Scandinavia, red foxes are the main limiting factor of Arctic foxes' recolonization (Hamel et al., 2013). Numerous native tundra predators, like Arctic foxes in most tundra regions, are specialized in hunting small rodents like lemmings and mostly reproduce in peak prey abundance years. These specialist predators are part of the cycling dynamics of tundra species, with their relative abundance lagging behind the cyclic abundance of rodents (Ims & Fuglei, 2005). However, generalist predators like red foxes commonly find alternative subsidies when rodent abundances are low, potentially allowing them to maintain a stable reproductive success independent of the cyclic dynamic of the tundra. At low rodent densities, red foxes preyed more intensively on ground-nesting birds such as waterfowls, ptarmigans, and waders (Elmhagen et al., 2015; Ims & Fuglei, 2005). Similarly, ravens increased predation of waterfowl-type nests when rodents were scarce (Ims et al., 2013). The number of generalist predators and their rodent catching rate can reduce prey abundance for native specialist predators, which has further consequences on their reproduction success (Henden et al., 2010). So far, empirical support of dampened rodent cycles has mainly been associated with changes in winter temperature and snowpack properties, at a time of year when rodents are still out of reach of boreal predators (Kausrud et al., 2008).

As an alternative to competition arising from a boreal and an Arctic predator sharing the same food resource, "apparent competition" happens when a boreal and an Arctic prey share the same predator. Such apparent competition is a major cause of North American caribou decline. Hence, the southern range distribution of caribou is now overlapping with the northern range expansion of moose *Alces alces* and white-tailed deer *Odocoileus virginianus*, favored by land-use changes due to forestry. Because the grey wolf *Canis lupus* is the predator of these boreal species, the northward expansion of deer and moose has resulted in an indirect increased predation pressure on caribou by the grey wolf, thereby contributing to the former's decline (Festa-Bianchet et al., 2011). In the marine system, apparent competition may be occurring in the Barents Sea between two groups of zooplankton, krill, and copepods, which share capelin as a predator. As the biomass of krill is predicted to

# 11 Impacts of Global Warming on Arctic Biota

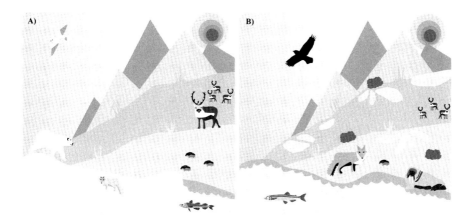

**Fig. 11.2** Schematic representation of Arctic tundra ecosystems (**a**) when sea-ice still reaches land supporting polar bears and Arctic foxes and cooling coastal terrestrial climates where Arctic herbivores like reindeer/caribou and lemmings dominate; and (**b**) under a climate change scenario where the sea-ice loss isolates populations, drives walruses ashore, and warms coastal terrestrial weather. The tundra landscape is greening with shrub expansion, but also shows patches of browning, while boreal species like red fox and raven predate/scavenge on Arctic herbivores

increase with climate change, the increased predation pressure is likely to negatively affect copepod abundance (Stige et al., 2018) (Fig. 11.2).

## Tundra Ecosystem Altered from Changes in Vegetation Productivity and Availability

The rapidly warming Arctic has led to an overall increase in vegetation growth and shrub expansion, known as the "greening" of the Arctic (Berner et al., 2020; Myers-Smith et al., 2011). Arctic greening can increase the radiative energy retention of the earth, leading to a positive feedback on global climate warming (Zhang et al., 2020). Yet, recent contrasting observations have documented patterns of vegetation browning (Berner et al., 2020; Myers-Smith et al., 2020) (Figure 11.2). Arctic "browning" refers to observations of extensive plant damage and decelerating increase in vegetation greening, deviating from expectations based on rising temperature trends (Myers-Smith et al., 2020; Phoenix & Bjerke, 2016; Vickers et al., 2016). As polar winters become rainier, one of the suggested drivers of decelerating greening is related to frost damage after total snow melt (that is, thaw-freeze events), or meltwater refreezing on the ground in a thick ice layer encapsulating the vegetation (ice-encasement events) (Phoenix & Bjerke, 2016; Vickers et al., 2016). The associated increase frequency of winter warm spells impacts the vegetation availability, quantity, and quality for Arctic herbivores, leading to ecosystem-wide impacts (Hansen et al., 2013; Treharne et al., 2019). Climate warming can homogenize the effects of weather on populations across large spatial scales by increasing population spatial synchrony (Hansen et al., 2020). However, the complex patterns emerging from diverging vegetation growth trajectories may instead enhance local heterogeneity and possibly buffer against climate change (Buchwal et al., 2020; Myers-Smith et al., 2020). The resilience of tundra species will depend on what spatial-scale and temporal-scale climate change may homogenize local weather variables, such as winter warm spells.

## *Arctic Vegetation Greening or Browning?*

Arctic vegetation greening occurs as land and soil temperature increases. Mechanistically, the soil active layer deepens and micro-organisms' activity accelerates, which promotes decomposition and greater nutrient availability for plants' roots (Natali et al., 2012). In addition, the increased global atmospheric carbon dioxide concentrations, the substrate for photosynthesis, act as a fertilizer. Consequently, below-ground roots can expand, like the unseen part of the iceberg (Iversen et al., 2015), and above-ground biomass can enlarge. The latter is the fraction that is widely monitored by satellites to reconstruct vegetation maps and plot greening trends (Berner et al., 2020).

The northward expansion of shrub density and height largely influences these greening trends and plays a crucial role in climate feedback loops (Myers-Smith et al., 2011; Zhang et al., 2020). The shrub canopy is the interface of energy flux between the atmosphere, soil, and permafrost when present. In summer, increased shrub biomass favors negative feedbacks on climate warming with increased evapotranspiration cooling the soil and increased carbon intake, also through storage in woody structures. However, in winter, positive feedback on climate warming occurs through reduced albedo and altered snowpack properties. Shrubs trap snow, which better insulates the ground. In addition, they tend to stick out above the snow cover, increasing radiative energy intake, which results in higher land and soil temperatures. The warmer land surface temperature can influence mean sea level pressure, which can modify atmospheric and oceanic circulation in a way that vegetation greening indirectly accelerates sea-ice melt (Zhang et al., 2020). Just like the albedo reduction between open water and ice-covered water, expanding shrub cover masks the high albedo properties of snow-covered surfaces and contributes to the "Arctic amplification" of climate change.

Recently, observations have attributed the contrasting phenomenon of Arctic vegetation "browning" to several causes. For instance, Arctic herbivore density increases can counteract shrub expansion through increased browsing (Christie et al., 2015). Furthermore, Arctic vegetation "likes it hot" (Van der Wal & Stien, 2014), but to what extent? Drought stress will likely increase in importance, particularly as permafrost recedes to a greater soil depth and has less influence on surface conditions, such as limiting water supply from melting (see previous chapter). Soil moisture plays an important role in limiting shrub growth as temperature increases (Gamm et al., 2017). Dryer soils also increase the frequency of tundra fires. Last but not least, extreme winter warm spells resulting in thaw-freeze or ice-encapsulation events can lead to shoot or plant death (Milner et al., 2016; Phoenix & Bjerke, 2016). The extent of browning can be broad, but mainly vary locally according to topography. Thus, considerable spatial complexity exists between Arctic greening and browning (Berner et al., 2020; Buchwal et al., 2020; Myers-Smith et al., 2020).

## *Ecosystem Consequences of Winter Warm Spells*

Winter vegetation availability is the key factor that controls the abundance of large Arctic herbivores, particularly in the high-Arctic. After warm spells that result in ice-encasement of the short-growing vegetation for several months, the vegetation available to resident herbivores decreases drastically and can therefore only sustain a small population. A high density of herbivores on ice-locked tundra has triggered massive die-offs of large herbivores like caribou and muskoxen *Ovibos moschatus*, but can also influence dynamics of rodents and ptarmigan (Forbes et al., 2016; Hansen et al., 2013; Kausrud et al., 2008; Rennert et al., 2009). The community of resident herbivores can fluctuate in synchrony across time, according to the frequency of these rain-on-snow events. Effects cascade further to scavenger predators, who thrive with abundant carcasses but crash with a one-year delay, as is the case for the Arctic fox on Svalbard (Hansen et al., 2013). The Arctic fox, in turn, can influence population dynamics of ground-nesting birds like migratory geese through increased gosling predation (Fuglei et al., 2003). Migratory geese and reindeer/caribou can, in turn, modulate their food resource, the vegetation community, depending on their densities. On the contrary, warm spells resulting in total snowmelt are more common in the low-Arctic and facilitate herbivore access to food resources. Snowpack is an insulating coat for small mammals, like lemmings that reproduce in winter under the snow, out of the reach of most predators. Similarly to tundra ice-encasement, total snowmelt will have a negative influence on rodents' survival and reproduction, with further consequences for cycling tundra ecosystems (see above) (Kausrud et al., 2008).

Winter warm spells can also alter vegetation quantity and quality for herbivores the following summer. Plant damage from thaw-freeze and ice-encasement events can reduce plant growth (Bokhorst et al., 2011; Le Moullec et al., 2020; Phoenix & Bjerke, 2016). Following a winter with thaw-freeze events, the overall capacity of the tundra carbon sink was reduced even when damage was not visible (Treharne et al., 2019). However, damaged plants often invest in shoot re-growth at the cost of other traits like flowering. This can also occur when plants grow in sub-optimal conditions like when ice-encasement has lasting cooling effects on summer soil temperature, delaying plant development (Le Moullec et al., 2019). Such compensatory plant growth was observed in several shrub species, including herbivores' favorite shrubs, and, in some cases, could even overshoot production compared with undamaged plots (Bokhorst et al., 2011; Le Moullec et al., 2019; Milner et al., 2016). In addition, leaf re-growth and delayed growth can be more digestible and nutritious for herbivores (Torp et al., 2010). However, such trade-offs, compensating vegetative growth at the cost of reproduction, is not a long-lasting strategy for plants' performance, and negative consequences are expected when acute stress from winter warm spells will become chronic. More knowledge on chronic effects is required and becomes particularly urgent as we rapidly head toward a rain-dominated Arctic winter (AMAP, 2017; Bintanja & Andry, 2017).

## *Spatial Variation as a Key to Maintaining Tundra Biodiversity*

Enhancing our understanding of climate change (both changes in long-term weather trends and frequency of extreme events) effects on the Arctic biota lies in partitioning the role of local heterogeneity and large-scale common drivers of population fluctuations. While large-scale, common climate drivers can homogenize population abundances across long distances through synchrony in their population dynamics, local heterogeneity can complexify these patterns (Hansen et al., 2020; Moran, 1953). The co-fluctuation of distant population abundances (that is, high population synchrony) can lead to a regional mass extinction, for instance following extreme events (Heino et al., 1997). For example, co-fluctuating summer temperatures have a synchronizing effect on Arctic plant growth across large areas, and co-fluctuating winter warm spell events (with vegetation ice-encasement) have a synchronizing effect on reindeer populations, especially when reindeer are at high densities (Hansen et al., 2019; Le Moullec et al., 2020).

Both summer temperature and the frequency of winter warm spells will increase with background climate warming, as well as the occurrence of climate extremes (AMAP, 2017; Meredith et al., 2019). This could ultimately increase large-scale extinction risks of some Arctic plant and animal species. However, the consequences of summer warming and winter warm spells also depend on their combined effect. Winter warm spells result in plant damage and/or changes of soil properties altering plant growth the following summer, thereafter enhancing local heterogeneity (Le Moullec et al., 2020). Furthermore, differences in the strengths of the effect of winter warm spells compared to summer temperatures can lead to divergent reindeer population trends; again, increasing local heterogeneity, which can potentially buffer against population synchrony (Hansen et al., 2019). The extent of future winter warm spells, how they will influence snowpack properties (total snowmelt, basal ice formation), and how the effects can be counteracted by summer warming are very important for determining the effects of climate change on tundra heterogeneity and subsequent species biodiversity.

## Conclusion and Future Outlooks

Climate change acts across many ecological levels of Arctic ecosystems. When effects are seen at the base of the food web, such as in marine plankton or tundra vegetation communities, effects ripple throughout the ecosystem. Responses to climate change in iconic top predators can result from critical ecosystem-wide changes. This demonstrates the tight connections in ecosystem functioning within the Arctic, increasing the need for ecosystem-based management practices.

Maintenance of biodiversity is ensured by heterogeneous – and thus complex – responses of species to their environments. The gradient of environmental effects on genes, species, and communities will determine ecosystem resilience to climate change (Ims & Ehrich, 2013). As the Arctic climate warms, environmental factors may become more synchronized across large spatial scales, accordingly homogenizing population fluctuations, which increases the risk of extensive biodiversity loss (Hansen et al., 2020). However, this may not be as simple because of the complexity of Arctic species responses to the combined effect of winter and summer warming. In some cases, these combined effects could buffer against climate change. Still, our understanding of such complex key mechanisms is insufficient and relies on detailed, long-term studies across populations, species, and ecosystems. In the remote Arctic, combining knowledge from international scientific unions with traditional ecological knowledge from indigenous peoples is a unique resource to overcome such knowledge gaps (Gagnon et al., 2020).

Across the globe, the ecological consequences of climate change results in the shrinkage and fragmentation of Arctic species' favored habitat, and the need for species to track and adapt to spatial and temporal changes in access to resources, stresses from new competitors, predators, and diseases. The Earth's ecosystems are under change. While the Arctic is a climate change "hot-spot", mechanisms described herein can serve as early warnings for widespread altered ecosystem functioning. However, climate change is only one side of the ongoing multi-factor changes the Arctic biota is facing. A combination of climate (such as sea-ice retreat) and land-use changes (such as industrialization) are fragmenting habitats. Furthermore, the "Arctic amplification" term defined by abiotic parameters could also include the magnified exposure and effect of pollutants on Arctic species. How highly specialized Arctic species will adapt to these changes is uncertain, and conservation actions are difficult to set, especially when some effects are now unmitigable, such as the multi-year ice disappearance in a couple of decades. Nevertheless, we can control our human footprint to how land-use (such as forestry and mining) and sea-use (such as shipping) contributes to climate change in the Arctic. Gaining awareness and changing practices of mass consumption in industrial regions, far away from the Arctic, is the main conservation action that can be taken to sustain Arctic biodiversity.

**Acknowledgements** The authors are grateful to Professor Sandra Hamel for her constructive comments on this chapter.

# References

Alsos, I. G., Ehrich, D., Seidenkrantz, M.-S., Bennike, O., Kirchhefer, A. J., & Geirsdottir, A. (2016). The role of sea ice for vascular plant dispersal in the Arctic. *Biology Letters, 12*(9), 20160264. https://doi.org/10.1098/rsbl.2016.0264

AMAP. (2017). Snow, water, ice and permafrost in the Arctic (SWIPA). In *Arctic*. Monitoring and Assessment Programme (AMAP).

Amélineau, F., Grémillet, D., Harding, A. M. A., Walkusz, W., Choquet, R., & Fort, J. (2019). Arctic climate change and pollution impact little auk foraging and fitness across a decade. *Scientific Reports, 9*(1), 1014. https://doi.org/10.1038/s41598-018-38042-z

Berner, L. T., Massey, R., Jantz, P., Forbes, B. C., Macias-Fauria, M., Myers-Smith, I., … Goetz, S. J. (2020). Summer warming explains widespread but not uniform greening in the Arctic tundra biome. *Nature Communications, 11*(1), 4621. https://doi.org/10.1038/s41467-020-18479-5

Beuchel, F., & Lønne, O. J. (2002). Population dynamics of the sympagic amphipods Gammarus wilkitzkii and Apherusa glacialis in sea ice north of Svalbard. *Polar Biology, 25*(4), 241–250. https://doi.org/10.1007/s00300-001-0329-8

Bintanja, R., & Andry, O. (2017). Towards a rain-dominated Arctic. *Nature Climate Change, 7*(4), 263–267. https://doi.org/10.1038/nclimate3240

Bogstad, B., Gjøsæter, H., Haug, T., & Lindstrøm, U. (2015). A review of the battle for food in the Barents Sea: Cod vs. marine mammals [review]. *Frontiers in Ecology and Evolution, 3*(29). https://doi.org/10.3389/fevo.2015.00029

Bokhorst, S., Bjerke, J. W., Street, L. E., Callaghan, T. V., & Phoenix, G. K. (2011). Impacts of multiple extreme winter warming events on sub-Arctic heathland: Phenology, reproduction, growth, and $CO_2$ flux responses. *Global Change Biology, 17*(9), 2817–2830. https://doi.org/10.1111/j.1365-2486.2011.02424.x

Buchwal, A., Sullivan, P. F., Macias-Fauria, M., Post, E., Myers-Smith, I. H., Stroeve, J. C., … Welker, J. M. (2020). Divergence of Arctic shrub growth associated with sea ice decline. *Proceedings of the National Academy of Sciences, 117*(52), 33334.

Callaghan, T. V., Björn, L. O., Chernov, Y., Chapin, T., Christensen, T. R., Huntley, B., … Zöckler, C. (2004). Biodiversity, distributions and adaptations of Arctic species in the context of environmental change. *Ambio: A Journal of the Human Environment, 33*(7), 404–417. https://doi.org/10.1579/0044-7447-33.7.404

Charles, K. M., & Stehlik, I. (2020). Assisted species migration and hybridization to conserve cold-adapted plants under climate change. *Conservation Biology*. https://doi.org/10.1111/cobi.13583

Christie, K. S., Bryant, J. P., Gough, L., Ravolainen, V. T., Ruess, R. W., & Tape, K. D. (2015). The role of vertebrate herbivores in regulating shrub expansion in the Arctic: A synthesis. *Bioscience, 65*(12), 1123–1133. https://doi.org/10.1093/biosci/biv137

Derocher, A. E. (2012). *Polar bears: A complete guide to their biology and behavior*. The Johns Hopkins University Press.

Doiron, M., Gauthier, G., & Lévesque, E. (2015). Trophic mismatch and its effects on the growth of young in an Arctic herbivore. *Global Change Biology, 21*(12), 4364–4376. https://doi.org/10.1111/gcb.13057

Dunton, K. H., Goodall, J. L., Schonberg, S. V., Grebmeier, J. M., & Maidment, D. R. (2005). Multi-decadal synthesis of benthic–pelagic coupling in the Western Arctic: Role of cross-shelf advective processes. *Deep Sea Research Part II: Topical Studies in Oceanography, 52*(24), 3462–3477. https://doi.org/10.1016/j.dsr2.2005.09.007

Dupont Cyr, B.-A., Dufresne, F., Christen, F., Desrosiers, V., Proulx, É., Le François, N. R., … Blier, P. U. (2018). Hybridization between char species (*Salvelinus alpinus* and *Salvelinus fontinalis*): A fast track for novel allometric trajectories. *Biology Open, 7*(10), bio033332. https://doi.org/10.1242/bio.033332

Elmhagen, B., Kindberg, J., Hellström, P., & Angerbjörn, A. (2015). A boreal invasion in response to climate change? Range shifts and community effects in the borderland between forest and tundra. *Ambio, 44*(1), 39–50. https://doi.org/10.1007/s13280-014-0606-8

Festa-Bianchet, M., Ray, J. C., Boutin, S., Côté, S. D., & Gunn, A. (2011). Conservation of caribou (*Rangifer tarandus*) in Canada: An uncertain future. *Canadian Journal of Zoology, 89*(5), 419–434. https://doi.org/10.1139/z11-025

Forbes, B. C., Kumpula, T., Meschtyb, N., Laptander, R., Macias-Fauria, M., Zetterberg, P., … Bartsch, A. (2016). Sea ice, rain-on-snow and tundra reindeer nomadism in Arctic Russia. *Biology Letters, 12*(11). https://doi.org/10.1098/rsbl.2016.0466

Frainer, A., Primicerio, R., Kortsch, S., Aune, M., Dolgov, A. V., Fossheim, M., & Aschan, M. M. (2017). Climate-driven changes in functional biogeography of Arctic marine fish communities. *Proceedings of the National Academy of Sciences, 114*(46), 12202. https://doi.org/10.1073/pnas.1706080114

Fuglei, E., Øritsland, N. A., & Prestrud, P. (2003). Local variation in arctic fox abundance on Svalbard, Norway. *Polar Biology, 26*, 93–98. https://doi.org/10.1007/s00300-002-0458-8

Fuglei, E., & Tarroux, A. (2019). Arctic fox dispersal from Svalbard to Canada: One female's long run across sea ice. *Polar Research, 38*. https://doi.org/10.33265/polar.v38.3512

Gagnon, C. A., Hamel, S., Russell, D. E., Powell, T., Andre, J., Svoboda, M. Y., & Berteaux, D. (2020). Merging indigenous and scientific knowledge links climate with the growth of a large migratory caribou population. *Journal of Applied Ecology, 57*(9), 1644–1655. https://doi.org/10.1111/1365-2664.13558

Gamm, C. M., Sullivan, P. F., Buchwal, A., Dial, R. J., Young, A. B., Watts, D. A., … Cornelissen, H. (2017). Declining growth of deciduous shrubs in the warming climate of continental western Greenland. *Journal of Ecology, 106*, 640–654. https://doi.org/10.1111/1365-2745.12882

Gradinger, R. R., & Bluhm, B. A. (2004). In-situ observations on the distribution and behavior of amphipods and Arctic cod (Boreogadus saida) under the sea ice of the High Arctic Canada Basin. *Polar Biology, 27*(10), 595–603. https://doi.org/10.1007/s00300-004-0630-4

Hamel, S., Killengreen, S. T., Henden, J. A., Yoccoz, N. G., & Ims, R. A. (2013). Disentangling the importance of interspecific competition, food availability, and habitat in species occupancy: Recolonization of the endangered Fennoscandian arctic fox. *Biological Conservation, 160*, 114–120. https://doi.org/10.1016/j.biocon.2013.01.011

Hansen, B. B., Grotan, V., Aanes, R., Saether, B. E., Stien, A., Fuglei, E., … Pedersen, A. O. (2013). Climate events synchronize the dynamics of a resident vertebrate community in the High Arctic. *Science, 339*(6117), 313–315. https://doi.org/10.1126/science.1226766

Hansen, B. B., Grøtan, V., Herfindal, I., & Lee, A. M. (2020). The Moran effect revisited: Spatial population synchrony under global warming. *Ecography, 43*(11), 1591–1602. https://doi.org/10.1111/ecog.04962

Hansen, B. B., Pedersen, Å. Ø., Peeters, B., Le Moullec, M., Albon, S. D., Herfindal, I., … Aanes, R. (2019). Spatial heterogeneity in climate change effects decouples the long-term dynamics of wild reindeer populations in the High Arctic. *Global Change Biology, 25*(11), 3656–3668. https://doi.org/10.1111/gcb.14761

Heath, J. P., Gilchrist, H. G., & Ydenberg, R. C. (2006). Regulation of stroke pattern and swim speed across a range of current velocities: Diving by common eiders wintering in polynyas in the Canadian Arctic. *Journal of Experimental Biology, 209*(20), 3974. https://doi.org/10.1242/jeb.02482

Heino, M., Kaitala, V., Ranta, E., & Lindström, J. (1997). Synchronous dynamics and rates of extinction in spatially structured populations. *Proceedings of the Royal Society of London, Series B, 264*(1381), 481–486.

Henden, J.-A., Ims, R. A., Yoccoz, N. G., Hellström, P., & Angerbjörn, A. (2010). Strength of asymmetric competition between predators in food webs ruled by fluctuating prey: The case of foxes in tundra. *Oikos, 119*(1), 27–34. https://doi.org/10.1111/j.1600-0706.2009.17604.x

Hovinen, J. E. H., Welcker, J., Descamps, S., Strøm, H., Jerstad, K., Berge, J., & Steen, H. (2014). Climate warming decreases the survival of the little auk (Alle alle), a high Arctic avian predator. *Ecology and Evolution, 4*(15), 3127–3138. https://doi.org/10.1002/ece3.1160

Huserbråten, M. B. O., Eriksen, E., Gjøsæter, H., & Vikebø, F. (2019). Polar cod in jeopardy under the retreating Arctic Sea ice. *Communications Biology, 2*(1), 407. https://doi.org/10.1038/s42003-019-0649-2

Ims, R. A., & Ehrich, D. (2013). Chapter 12. Terrestrial ecosystems. In *Arctic biodiversity assessment. Status and trends in Arctic biodiversity. Conservation of Arctic Flora and Fauna (CAFF)* (pp. 384–441).

Ims, R. A., & Fuglei, E. (2005). Trophic interaction cycles in tundra ecosystems and the impact of climate change. *Bioscience, 55*(4), 311–322. https://doi.org/10.1641/0006-3568(2005)055[0311:ticite]2.0.co;2

Ims, R. A., Henden, J.-A., Thingnes, A. V., & Killengreen, S. T. (2013). Indirect food web interactions mediated by predator–rodent dynamics: Relative roles of lemmings and voles. *Biology Letters, 9*(6), 20130802. https://doi.org/10.1098/rsbl.2013.0802

Iversen, C. M., Sloan, V. L., Sullivan, P. F., Euskirchen, E. S., McGuire, A. D., Norby, R. J., … Wullschleger, S. D. (2015). The unseen iceberg: Plant roots in arctic tundra. *New Phytologist, 205*(1), 34–58. https://doi.org/10.1111/nph.13003

Jay, C. V., Fischbach, A. S., & Kochnev, A. A. (2012). Walrus areas of use in the Chukchi Sea during sparse sea ice cover. *Marine Ecology Progress Series, 468*, 1–13. https://www.int-res.com/abstracts/meps/v468/p1–13

Jenkins, D. A., Lecomte, N., Schaefer, J. A., Olsen, S. M., Swingedouw, D., Côté, S. D., … Yannic, G. (2016). Loss of connectivity among island-dwelling Peary caribou following sea ice decline. *Biology Letters, 12*(9) http://rsbl.royalsocietypublishing.org/content/12/9/20160235.abstract

Kausrud, K. L., Mysterud, A., Steen, H., Vik, J. O., Østbye, E., Cazelles, B., … Stenseth, N. C. (2008). Linking climate change to lemming cycles. *Nature, 456*(7218), 93–97. https://doi.org/10.1038/nature07442

Kerby, J. T., & Post, E. (2013). Advancing plant phenology and reduced herbivore production in a terrestrial system associated with sea ice decline. *Nature Communications, 4*(1), 2514. https://doi.org/10.1038/ncomms3514

Kovacs, K. M., Lydersen, C., Overland, J. E., & Moore, S. E. (2011). Impacts of changing sea-ice conditions on Arctic marine mammals. *Marine Biodiversity, 41*(1), 181–194. https://doi.org/10.1007/s12526-010-0061-0

Kunisch, E. H., Bluhm, B. A., Daase, M., Gradinger, R., Hop, H., Melnikov, I. A., … Berge, J. (2020). Pelagic occurrences of the ice amphipod Apherusa glacialis throughout the Arctic. *Journal of Plankton Research, 42*(1), 73–86. https://doi.org/10.1093/plankt/fbz072

Laidre, K. L., Stern, H., Kovacs, K. M., Lowry, L., Moore, S. E., Regehr, E. V., … Ugarte, F. (2015). Arctic marine mammal population status, sea ice habitat loss, and conservation recommendations for the 21st century. *Conservation Biology, 29*(3), 724–737. https://doi.org/10.1111/cobi.12474

Le Moullec, M., Isaksen, K., Petit Bon, M., Jónsdóttir, I. S., Varpe, Ø., Hendel, A.-L., … Hansen, B. B. (2019). *Towards rainy Arctic winters: effects of experimental icing on tundra plants and their soil conditions* MET Report, Issue 08. https://www.met.no/publikasjoner/met-report/met-report-2019

Le Moullec, M., Sandal, L., Grøtan, V., Buchwal, A., & Hansen, B. B. (2020). Climate synchronises shrub growth across a high-arctic archipelago: Contrasting implications of summer and winter warming. *Oikos, 129*(7), 1012–1027. https://doi.org/10.1111/oik.07059

Leu, E., Søreide, J. E., Hessen, D. O., Falk-Petersen, S., & Berge, J. (2011). Consequences of changing sea-ice cover for primary and secondary producers in the European Arctic shelf seas: Timing, quantity, and quality. *Progress in Oceanography, 90*(1), 18–32. https://doi.org/10.1016/j.pocean.2011.02.004

Lewis, K. M., van Dijken, G. L., & Arrigo, K. R. (2020). Changes in phytoplankton concentration now drive increased Arctic Ocean primary production. *Science, 369*(6500), 198. https://doi.org/10.1126/science.aay8380

Lone, K., Merkel, B., Lydersen, C., Kovacs, K. M., & Aars, J. (2018). Sea ice resource selection models for polar bears in the Barents Sea subpopulation. *Ecography, 41*(4), 567–578. https://doi.org/10.1111/ecog.03020

Lund-Hansen, L. C., Søgaard, D. H., Sorrell, B. K., Gradinger, R., & Meiners, K. M. (2020). *Arctic Sea ice ecology: Seasonal dynamics in algal and bacterial productivity.* Springer Polar Sciences.

Macias-Fauria, M., Karlsen, S. R., & Forbes, B. C. (2017). Disentangling the coupling between sea ice and tundra productivity in Svalbard. *Scientific Reports, 7*(1), 8586. https://doi.org/10.1038/s41598-017-06218-8

Meredith, M., Sommerkorn, M., Cassotta, S., Derksen, D., Ekaykin, A., Hollowed, A., … Schuur, E. A. G. (2019). Polar regions. In H.-O. Pörtner, D. C. Roberts, V. Masson-Delmotte, P. Zhai, M. Tignor, E. Poloczanska, K. Mintenbeck, A. Alegría, M. Nicolai, A. Okem, J. Petzold, B. Rama, & N. M. Weyer (Eds.), *IPCC special report on the ocean and cryosphere in a changing climate*.

Milner, J. M., Varpe, Ø., Van der Wal, R., & Hansen, B. B. (2016). Experimental icing affects growth, mortality, and flowering in a high arctic dwarf shrub. *Ecology and Evolution, 6*(7), 2139–2148. https://doi.org/10.1002/ece3.2023

Møller, E. F., & Nielsen, T. G. (2020). Borealization of Arctic zooplankton—smaller and less fat zooplankton species in Disko Bay, Western Greenland. *Limnology and Oceanography, 65*(6), 1175–1188. https://doi.org/10.1002/lno.11380

Moran, P. A. P. (1953). The statistical analysis of the Canadian lynx cycle. *Australian Journal of Zoology, 1*(3), 291–298.

Myers-Smith, I. H., Forbes, B. C., Wilmking, M., Hallinger, M., Lantz, T., Blok, D., … Hik, D. S. (2011). Shrub expansion in tundra ecosystems: Dynamics, impacts and research priorities. *Environmental Research Letters, 6*(4), 045509. https://doi.org/10.1088/1748-9326/6/4/045509

Myers-Smith, I. H., Kerby, J. T., Phoenix, G. K., Bjerke, J. W., Epstein, H. E., Assmann, J. J., … Wipf, S. (2020). Complexity revealed in the greening of the Arctic. *Nature Climate Change, 10*(2), 106–117. https://doi.org/10.1038/s41558-019-0688-1

Natali, S. M., Schuur, E. A. G., & Rubin, R. L. (2012). Increased plant productivity in Alaskan tundra as a result of experimental warming of soil and permafrost. *Journal of Ecology, 100*(2), 488–498. https://doi.org/10.1111/j.1365-2745.2011.01925.x

Norén, K., Dalén, L., Flagstad, Ø., Berteaux, D., Wallén, J., & Angerbjörn, A. (2017). Evolution, ecology and conservation – Revisiting three decades of Arctic fox population genetic research. *Polar Research, 36*(sup1), 4. https://doi.org/10.1080/17518369.2017.1325135

Peeters, B., Le Moullec, M., Raeymaekers, J. A. M., Marquez, J. F., Røed, K. H., Pedersen, Å. Ø., … Hansen, B. B. (2020). Sea ice loss increases genetic isolation in a high Arctic ungulate metapopulation. *Global Change Biology, 26*(4), 2028–2041. https://doi.org/10.1111/gcb.14965

Phoenix, G. K., & Bjerke, J. W. (2016). Arctic browning: Extreme events and trends reversing arctic greening. *Global Change Biology, 22*(9), 2960–2962. https://doi.org/10.1111/gcb.13261

Polyakov, I. V., Alkire, M. B., Bluhm, B. A., Brown, K. A., Carmack, E. C., Chierici, M., … Wassmann, P. (2020). Borealization of the Arctic Ocean in response to anomalous advection from sub-Arctic seas. *Frontiers in Marine Science, 7*(491). https://doi.org/10.3389/fmars.2020.00491

Post, E., Alley, R. B., Christensen, T. R., Macias-Fauria, M., Forbes, B. C., Gooseff, M. N., … Wang, M. (2019). The polar regions in a 2°C warmer world. *Science Advances, 5*(12), eaaw9883. https://doi.org/10.1126/sciadv.aaw9883

Post, E., Bhatt, U. S., Bitz, C. M., Brodie, J. F., Fulton, T. L., Hebblewhite, M., … Walker, D. A. (2013). Ecological consequences of sea-ice decline. *Science, 341*(6145), 519–524. https://doi.org/10.1126/science.1235225

Rennert, K. J., Roe, G., Putkonen, J., & Bitz, C. M. (2009). Soil thermal and ecological impacts of rain on snow events in the circumpolar Arctic. *Journal of Climate, 22*(9), 2302–2315. https://doi.org/10.1175/2008JCLI2117.1

Sakshaug, E., Johnsen, G., & Kovacs, K. (2009). *Ecosystem Barents Sea*. Tapir Academic Press.

Simon, A., Tardy, O., Hurford, A., Lecomte, N., Bélanger, D., & Leighton, P. (2019). Dynamics and persistence of rabies in the Arctic. *Polar Research, 38*(0). 10.33265/polar.v38.3366.

Søreide, J. E., Carroll, M. L., Hop, H., Ambrose, W. G., Hegseth, E. N., & Falk-Petersen, S. (2013). Sympagic-pelagic-benthic coupling in Arctic and Atlantic waters around Svalbard revealed by stable isotopic and fatty acid tracers. *Marine Biology Research, 9*(9), 831–850. https://doi.org/10.1080/17451000.2013.775457

Søreide, J. E., Leu, E. V. A., Berge, J., Graeve, M., & Falk-Petersen, S. (2010). Timing of blooms, algal food quality and Calanus glacialis reproduction and growth in a changing Arctic. *Global Change Biology, 16*(11), 3154–3163. https://doi.org/10.1111/j.1365-2486.2010.02175.x

Spencer, M. L., Vestfals, C. D., Mueter, F. J., & Laurel, B. J. (2020). Ontogenetic changes in the buoyancy and salinity tolerance of eggs and larvae of polar cod (*Boreogadus saida*) and other gadids. *Polar Biology, 43*(8), 1141–1158. https://doi.org/10.1007/s00300-020-02620-7

Stige, L. C., Kvile, K. Ø., Bogstad, B., & Langangen, Ø. (2018). Predator-prey interactions cause apparent competition between marine zooplankton groups. *Ecology, 99*(3), 632–641. https://doi.org/10.1002/ecy.2126

Torp, M., Olofsson, J., Witzell, J., & Baxter, R. (2010). Snow-induced changes in dwarf birch chemistry increase moth larval growth rate and level of herbivory. *Polar Biology, 33*(5), 693–702. https://doi.org/10.1007/s00300-009-0744-9

Treharne, R., Bjerke, J. W., Tømmervik, H., Stendardi, L., & Phoenix, G. K. (2019). Arctic browning: Impacts of extreme climatic events on heathland ecosystem $CO_2$ fluxes. *Global Change Biology, 25*(2), 489–503. https://doi.org/10.1111/gcb.14500

Van der Wal, R., & Stien, A. (2014). High-arctic plants like it hot: A long-term investigation of between-year variability in plant biomass. *Ecology, 95*(12), 3414–3427. https://doi.org/10.1890/14-0533.1

Vickers, H., Høgda, K. A., Solbø, S., Karlsen, S., Tømmervik, H., Aanes, R., & Hansen, B. B. (2016). Changes in greening in the high arctic: Insights from a 30 year AVHRR max NDVI dataset for Svalbard. *Environmental Research Letters, 11*, 105004. https://doi.org/10.1088/1748-9326/11/10/105004

Vihtakari, M., Welcker, J., Moe, B., Chastel, O., Tartu, S., Hop, H., … Gabrielsen, G. W. (2018). Black-legged kittiwakes as messengers of Atlantification in the Arctic. *Scientific Reports, 8*(1), 1178. https://doi.org/10.1038/s41598-017-19118-8

Wassmann, P., & Reigstad, M. (2011). Future Arctic Ocean seasonal ice zones and implications for pelagic-benthic coupling. *Oceanography, 24*(3), 220–231. http://www.jstor.org/stable/24861317

Wisz, M. S., Broennimann, O., Grønkjær, P., Møller, P. R., Olsen, S. M., Swingedouw, D., … Pellissier, L. (2015). Arctic warming will promote Atlantic–Pacific fish interchange. *Nature Climate Change, 5*(3), 261–265. https://doi.org/10.1038/nclimate2500

Young, M. K., Isaak, D. J., McKelvey, K. S., Wilcox, T. M., Pilgrim, K. L., Carim, K. J., … Schwartz, M. K. (2016). Climate, demography, and zoogeography predict introgression thresholds in salmonid hybrid zones in Rocky Mountain streams. *PLoS One, 11*(11), e0163563. https://doi.org/10.1371/journal.pone.0163563

Zhang, W., Döscher, R., Koenigk, T., Miller, P. A., Jansson, C., Samuelsson, P., … Smith, B. (2020). The interplay of recent vegetation and sea ice dynamics—Results from a regional earth system model over the Arctic. *Geophysical Research Letters, 47*(6), e2019GL085982. https://doi.org/10.1029/2019GL085982

**Dr. Mathilde Le Moullec** is a postdoctoral researcher at the Centre for Biodiversity Dynamics of the Norwegian University of Science and Technology (NTNU) in Trondheim, Norway, and at Laval University (ULAVAL), Quebec, Canada.

**Dr. Morgan Lizabeth Bender** is a postdoctoral researcher at UiT The Arctic University of Norway in Tromsø, Norway.

# Chapter 12
# Pollution and Monitoring in the Arctic

Tatiana Yu. Sorokina

## Introduction

The Arctic is a huge sea area, as well as territories. The Arctic region has long been considered as one of the cleanest and least polluted regions in the world. Is this really the case? This question can be answered by analyzing the distinctive features of the northernmost region.

For a long time, the Arctic areas were among the least explored and least populated in the world (McCannon, 2012). Even now, this remains true to some extent. The research focus in the Arctic was, for many years, related to the economy and possible benefits from the exploitation of northern nature (Avango et al., 2018). Therefore, researchers were primarily interested in finding the reserves of resources and developing ways to extract them in harsh climatic conditions. The associated environmental pollution remained unnoticed. At the same time, even the development of industry, the increase in the extraction of natural resources, and the associated modernization of northern settlements are not enough to keep people in the Far North. A massive outflow of population from northern cities and towns has been recorded in all Arctic States (Heleniak, 2014). Therefore, the number and density of population in the Arctic remains small today, even in the light of the general increase in the world's population and the overpopulation of many southern cities.

The previous point is associated with the relatively late appearance of industrial and military facilities in the region, which are the main pollutants of nature. It is important to pay attention to one particularly interesting issue. In the more southern regions of the Arctic states, industrial enterprises and military bases appeared earlier

---

T. Y. Sorokina (✉)
Northern Arctic Federal University, Arkhangelsk, Russia
e-mail: t.sorokina@narfu.ru

© The Author(s), under exclusive license to Springer Nature Switzerland AG 2022
M. Finger, G. Rekvig (eds.), *Global Arctic*,
https://doi.org/10.1007/978-3-030-81253-9_12

than they did in the north (for objective reasons); however, this does not mean that they have not yet managed to pollute the Arctic to the same scale. In some Arctic regions, certain pollutants were detected at an unacceptably high level. There are several reasons for this, but the main reason is that, for a long period, many States (represented by the chief decision-makers) considered the Arctic to be too distant, which led to neglect of the environmental pollution and a complete failure of waste management. Moreover, some Arctic territories and water areas have become testing grounds for new types of weapons.

The harsh northern climate still significantly limits the anthropogenic load on the region, even though the entire global community, not just the Arctic States, has taken a course for enhanced development of the Arctic. Low average annual temperatures and permafrost prevent pollutants from spreading widely over the territory and penetrating deep into the soil. Long-term glaciers, as well as the freezing of some marine areas in winter, do not allow the Arctic seas to be used all year round and at full capacity (CBS news, 2021). Extremely low air temperatures, strong winds, and snowstorms sometimes require the suspension of certain types of human activity (Canadian Center, 2019). However, these same factors significantly aggravate the implementation of environmental measures, such as the cleaning up territories and water areas after natural and man-made emergencies.

The world is now recognizing that all the large-scale environmental problems that humanity faces in the twenty-first century affect the Arctic region. Prime among these, of course, is global climate change. The changes taking place on the planet are clearly visible in the Arctic. Even space satellites record the rapid decrease of glaciers. In this chapter, we will not deal with the issues of the impact of climate change on the Arctic ecosystems. We will discuss the main types of pollution in the Arctic, their sources, as well as provoking factors. Special attention is paid to environmental and biological monitoring systems, which are the main mechanisms for assessing the extent of pollution, its consequences, as well as risks to humans and other inhabitants of the northernmost region.

The study represents the current understanding of the environmental monitoring activities in the Arctic, which are related to pollution control and health risk management associated with environmental pollution. We investigated environmental monitoring activities from three perspectives, examining: (1) what types of pollution are of greatest concern in the Arctic, (2) who are the stakeholders in the monitoring activities, and (3) how to access the monitoring data. This paper provides an overview of environmental monitoring conducted in the Arctic by reviewing existing information on state of the Arctic environment in terms of pollution by various types of pollutants, including toxic elements, radiation, and paying special attention to POPs and microplastic. We summarize recently published information about environmental pollutants found in the human blood of Arctic inhabitants. We then explore various monitoring initiatives and obligations identifying the main stakeholders in this process. Finally, we discuss possible challenges when gaining access to monitoring data referring to adopted legal regulations and practices.

## Background

Environmental observations in the Arctic started from the moment it was settled by people. They began as individual observations of the weather, habitat, flora and fauna by the first settlers. This knowledge helped people to survive in the harsh northern conditions. Today, we refer to such monitoring data as the traditional knowledge of the indigenous peoples of the North (Arctic Center, 2021). Later, the massive invasion of the Arctic started. The first Arctic expeditions took place quite recently, from the overall perspective of human history. The development of Arctic territories and the Arctic Ocean have always been accompanied by regular environmental observations. During this relatively short period of time, each of the Arctic countries has collected a huge amount of environmental data. However, this is not enough. The development of industry and new technologies is changing our world at a rapid pace, and the Arctic, with its fragile ecosystem, is changing along with the rest of the world. Constant monitoring of the state of the environment (environmental monitoring) is perhaps the most reliable and correct tool for obtaining adequate information for making strategically important decisions and determining the risks that these decisions may entail. Many of the international environmental agreements contain requirements for the party states to conduct monitoring activities in particular spheres. Although there are not many specific Arctic agreements at the moment, several other existing international obligations of the states are applicable to the Arctic. All Arctic states are interested in gaining more environmental information, and thus developing national systems of monitoring. Some other states and international organizations are also active actors in this area. Of particular concern today is the increasing anthropogenic impact on the Arctic and the increase in levels and trends of environmental pollution associated with it. At the same time, the huge number of agreements, regulations, published monitoring data, reports, and initiatives make it difficult to see the full picture of pollution and monitoring in the Arctic. This paper aims to fill this knowledge gap and to describe environmental monitoring in the Arctic in terms of objects (which are natural objects and various types of pollution), subjects (stakeholders), and the handling of the monitoring data obtained.

## Materials and Methods

We reviewed and analyzed the peer-reviewed publications on conducted monitoring in the Arctic, as well as international agreements and national regulations. We had worked with the scientific articles and reviews using the Web of Science platform (webofknowledge.com), and had also monitored the official websites of the United Nations Organization, the World Health Organization (WHO), the Arctic Council and its working groups, and national authorities of the Russian Federation and other Arctic States. We also consulted experts on Arctic issues (including its pollution). These included researchers from Russia (Northern (Arctic) Federal University named after M.V. Lomonosov, Northern State Medical University, N. Laverov Federal Center for Integrated Arctic Research of the Ural Branch of the Russian Academy of Sciences, North-West Public Health Research Center, and many others), Norway (National Institute of Occupational Health, UiT the Arctic University of Norway, Akvaplan-niva, etc.), Finland (University of Lapland, etc.), Canada (Environmental Health Science and Research Bureau), and other experts from non-Arctic states. Guided by the previous scientific experience of these expert groups, four topics were prioritized to draw an adequate picture of pollution and monitoring in the Arctic. The chosen topics are outlined in the text as follows: (1) pollution in the Arctic, where two types of pollution (microplastic and POPs) were specially highlighted, (2) human biological monitoring, (3) stakeholders of the environmental monitoring in the Arctic, and (4) access to monitoring data.

## Pollution in the Arctic

Arctic ecosystems are very vulnerable. However, the anthropogenic load on the region increases every year, which is always associated with large-scale pollution and other adverse environmental changes. It is not only modern local human activity that causes adverse environmental consequences. Some contaminants enter the Arctic through cross-border pollution. Some elements are formed in nature itself. In this section, we seek to understand what types of pollution the fragile Arctic ecosystems are exposed to and what are the possible sources of such pollution.

The Arctic has large reserves of mineral resources (European Environment Agency, 2017). The region has many open deposits of hydrocarbons, unique deposits of diamonds, rare earth metals and other raw materials that are in demand on world markets. Active anthropogenic activity, including the extraction of these resources, is one of the main causes of serious environmental problems that may restrain the further development of the region (Gosudarstvennii doklad o sostoyanii i ob okhrane okruzhaushei sredi Rossiiskoi Federatsii, 2019). The main sources of pollution in the Arctic zone are fuel and energy enterprises, the metallurgical enterprises of the mining and mineral processing, chemical industry, timber, pulp and paper industry, and transport (Dauvalter, 2020; Svetlov et al., 2020; Yaraghi et al., 2020).

Many Arctic cities have high levels of atmospheric pollution (Brozovsky et al., 2021; Thorp et al., 2021). The increase in the concentration of carbon dioxide recorded at the Russian and American background stations is mainly due to the overall increase in anthropogenic $CO_2$ emissions. In the Arctic, the concentrations of pollutants in the atmospheric air, including gas and aerosol impurities such as heavy metals, are monitored. The highest concentrations of pollutants are found in the air of cities, especially near industrial enterprises (Gosudarstvennii doklad o sostoyanii i ob okhrane okruzhaushei sredi Rossiiskoi Federatsii, 2019). Some indicators have exceeded the established standards. Most often, sulfur dioxide, carbon monoxide, formaldehyde, nitrogen oxide, and benz(a)pyrene are found in the atmospheric air (Gosudarstvennii doklad o sostoyanii i ob okhrane okruzhaushei sredi Rossiiskoi Federatsii, 2019).

An important distinguishing feature of the Arctic region is the presence of significant water resources, represented in the form of vast basins of the largest rivers. These rivers are part of the resource base of the region. At the same time, they directly affect the overall ecological situation, climate, and the state of the Arctic flora and fauna. Water reservoirs in the Arctic are extremely vulnerable to pollutants due to slow self-cleaning processes. The main water pollution is associated with the following sources: mining and metallurgical plants, oil and gas complexes, pulp and paper mills, gold mining enterprises, wastewater discharges from housing and communal services in settlements, as well as transport and fishing fleets (Dauvalter, 2020; Dinu et al., 2020; Yaraghi et al., 2020). Nickel, manganese, copper, molybdenum, iron, and mercury compounds make a significant contribution to surface water pollution. Hydrobiological observations of the state of freshwater ecosystems in the Arctic zone of the Russian Federation have been carried out for the main ecological communities: phytoplankton, zooplankton and zoobenthos (Gosudarstvennii doklad o sostoyanii i ob okhrane okruzhaushei sredi Rossiiskoi Federatsii, 2019). Each of these communities is observed for a number of parameters that allow us to obtain information about the quantitative and qualitative composition of the ecosystems of surface water bodies.

An important feature of the Arctic soils is the large proportion of permafrost territories characterized by low temperature and a small layer of seasonal thawing. In some places, the depth of freezing of rocks reaches 1500 m. However, the economic development of the Arctic territories (infrastructure construction, geological exploration, mining, etc.) causes serious damage to the soils, destroying them (Ji et al., 2017).

The beginning of the development of the Northern Sea Route and the development of the Arctic territories with the extensive use of the nuclear icebreaker fleet and nuclear power plants, including floating ones, makes it relevant to systematically assess the radiation situation in the Arctic region. The radiometric network on the territory of the Arctic zone of the Russian Federation includes 94 points of observations for the exposure dose rate of gamma radiation, 43 points of observations for radioactive fallout, and eight points of observations for radioactive aerosols of the spirit, as well as five points of observations on the White Sea and one point-on the Barents Sea for coastal water pollution. According to official data, radioactive contamination of the Arctic air and marine environments from 2016 to 2019 was at a low level and there is a tendency to reduce it. Potential sources of technogenic radionuclides entering the waters of the Barents Sea are the nuclear submarines "Komsomolets" sunk as a result of accidents in the Norvezh Sea, whose waters are exchanged with the waters of the Barents Sea, and "K-159" in the Barents Sea. In the areas where they are located, periodic (most recently in 2014), expeditionary comprehensive surveys of the marine environment are carried out (water, bottom sediments and marine organisms are monitored). There were no detected leaks of radioactive materials from sunken submarines (Gosudarstvennii doklad o sostoyanii i ob okhrane okruzhaushei sredi Rossiiskoi Federatsii, 2019).

The disposal of by-products of past economic activities, preserved military and civilian objects, disturbed land requiring reclamation, as well as abandoned objects in the waters of the seas and rivers, are of great importance for the environmental safety of the territories in the region. In terms of the type of economic activity, more than 97% of all waste generated in the Arctic region was produced in mining, construction, transportation, and storage (Gosudarstvennii doklad o sostoyanii i ob okhrane okruzhaushei sredi Rossiiskoi Federatsii, 2019).

## *Microplastic*

Plastic pollution in the Arctic is now recognized as a growing problem. Plastic is currently the most widespread type of garbage in the world (EcoWatch, 2018; Ocean conservancy, 2018). At the beginning of the twentieth century, polymer synthesis technologies were optimized, which led to the mass production of light, strong, inert, and corrosion-resistant plastics in the middle of the last century (Cole et al., 2011). The annual demand for plastics is increasing, even though many societies have already recognized the problem of plastic pollution and are committed to reducing the production and consumption of this type of material (Andrady, 2011). However, the problem of plastic pollution depends not only on the volume of production and consumption. The lifetime of a plastic product can be a thousand years, which significantly complicates its disposal and natural processing. Microplastics are plastic fragments and plastic particles up to 5 mm in size. Microplastics can be of two types. Primary microplastics are microgranules of plastic that are specially produced and added to various products (for example, to cosmetics, to fillers of furniture, toys, etc.). Secondary microplastics are plastic fragments with a size of no more than 5 mm, which are formed as a result of the destruction of larger plastic objects.

Five main factors play major roles in the entry of plastic into the Arctic environment: (a) irrational management of solid household waste management, (b) pollution of the sea and river coastal strip by vacationers, (c) direct discharge of household wastewater into Arctic rivers and seas, (d) incoming garbage from sea and river vessels, and (e) fishing equipment. Large plastic waste is slowly but inevitably destroyed under the influence of solar radiation, mechanical, and biological factors (Lassen et al., 2015). Thus, a large number of macro-, micro-, and nanoparticles of plastic appear in the fragile environment of the Arctic. Plastic has a negative impact on the northern nature. Plastic can contribute to the body formation disorders of marine animals, fish, and birds and can cause death by suffocation or ingestion. It can block animals' gastrointestinal tracts and small particles of plastic can penetrate deep into living organisms. In some animals, plastic was found in the blood, lymph, and liver. Some data indicate the ability of plastic particles to accumulate persistent organic pollutants on their surface (Rodrigues et al., 2019).

Today, plastic is found in all corners of the world, even in the most remote ones where there is no human activity. The Arctic is no exception. Studies have shown the presence of microplastics in surface and underground waters (Cozar et al., 2017; Kanhai et al., 2019; Lusher et al., 2015; Morgana et al., 2018; Mu et al., 2019). Some authors claim that microplastic particles are present in the entire sea water column in the Arctic, accumulating in the bottom sediments (Kanhai et al., 2018; Tekman et al., 2020). Moreover, sea ice can serve as a temporary storage of microplastics (Kanhai et al., 2019; Kanhai et al., 2020; O'Hara et al., 2004; Peeken et al., 2018). The detected concentrations of microplastics in ice cores from the central part of the Arctic Ocean significantly exceed those obtained in more polluted regions of the planet (Obbard et al., 2014). Researchers have discovered the presence of microplastics in the snow on the territory of the Svalbard archipelago (Bergman et al., 2019), while others examined the Barents Sea in the Svalbard archipelago and found plastic particles in 95% of seawater samples (Lusher et al., 2015). Disheartening results were obtained during the research works in the White, Kara and Greenland Seas (Tosic et al., 2020). The authors of the circumpolar expedition "Tara Oceans 2013" even talked about the existence of a "sixth garbage spot" (Cozar et al., 2017). The microplastic is found in the northern fish, crustaceans, and other representatives of the Arctic fauna (Fang et al., 2021; Moore et al., 2020; Morgana et al., 2018).

## *Persistent Organic Pollutants*

The Arctic region is negatively affected by persistent organic pollutants (POPs), which are heterogeneous carbon-based compounds that can be of natural or anthropogenic origin (Mouly et al., 2016). According to the Stockholm Convention on Persistent Organic Pollutants, 35 POPs are currently included in the list of the most harmful substances (Stockholm Convention on Persistent Organic Pollutants, 2001). There is minimal direct use of POPs chemicals in the Arctic region. Nevertheless, POPs, their metabolites and decomposition products are transported to the Arctic from more southern regions by atmosphere and by ocean currents and rivers (De Wit et al., 2010; De Wit et al., 2006). POPs can also enter the Arctic with migratory species of birds, fish, and wild animals. The physical and chemical properties of POPs allow them to bioaccumulate and biomagnify in the adipose tissues of living organisms. That is, POPs are lipophilic pollutants. Consumption of products contaminated with POPs can disrupt the endocrine, reproductive, and immune systems, and can potentially cause cancer and other adverse effects. Monitoring of persistent organic pollutants in various biological and environmental matrices became an urgent task. The Stockholm Convention on Persistent Organic Pollutants obliges Member States to

> …promote and implement, both at the national and international levels, relevant research, monitoring and cooperation… (Stockholm Convention on Persistent Organic Pollutants, 2001).

Indigenous peoples of the Arctic are particularly at risk of adverse health effects (AMAP, 2004, 2009, 2015). The traditional way of life and nutrition is the main reason. In different regions of the Arctic, 30–50% of the local diet consists of northern fish (for example, pink salmon, Arctic char, navaga, humpback whitefish, northern pike, and others). Published data indicate that fish accumulate POPs (Dudarev et al., 2019; AMAP, 2018; Braune et al., 2005). The results obtained show that the main POPs in fish are the decomposition products of dichlorodiphenyltrichloroethane (DDT) and congeners of polychlorinated biphenyls (PCBs). Studies conducted in the Arctic confirm the relationship between the fat content in fish and the concentration of POPs. In all of the studied fish with a high fat content (especially in pink salmon), the content of POPs remains significantly higher (two or more times higher) than in less-fat fish species. This dependence is particularly noticeable for $\sum$DDT, trans-nonachlor, trans-chlordane, and cis-chlordane (Lakhmanov et al., 2020). Persistent toxic substances affect humans through traditional food and form a significant risk factor for the health of people living in the Arctic (Bae et al., 2018; Dudarev et al., 2019; Fiore et al., 2019; Lee et al., 2014; Lind et al., 2019; Rosen et al., 2018; Yegambaram et al., 2015; Zheng et al., 2016; Gray et al., 2017; Mouly et al., 2016).

Not only Arctic fish contain POPs; gray and bowhead whales are part of the traditional diet of the indigenous peoples of the Eastern Arctic. This diet exists for hundreds of years and is a prerequisite for human survival and adaptation in the harsh conditions of the Arctic. Hunting for marine mammals is allowed in almost all Arctic States: Russian Federation (Chukotskiy Autonomus Okrug), the United States (Alaska), Canada, Norway, Iceland, and Denmark (Greenland and the Faroe Islands). The fat, muscle, liver, and skin of whales are healthy food for people living in the Arctic climate (Mouly et al., 2016; Reynolds et al., 2006). However, the regular consumption of such foods can lead to high levels of toxic fat-soluble pollutants, such as POPs. The same researchers found POPs ($\Sigma$PCB) in blubber and liver of gray and bowhead whales in quantities significantly exceeding the EU consumption limits. The consumption of $\Sigma$PCB may exceed the acceptable level of consumption (WHO recommend less than 20 ng/kg of body weight per day), as on average, indigenous people consume up to 49 kg of marine mammal tissue annually (Chukmasov et al., 2019). Other marine mammals consumed by the Arctic natives may include bowhead whales (*Balaena mysticetus*), belugas (*Delphinapterus leucas*), walruses (*Odobenus rosmarus*), and some phocid species, which also may contain POPs.

Concentrations of some of the POPs listed in the 2001 Stockholm Convention have tended to decrease in Arctic marine biota over the past two decades. This can be explained by the effective implementation of international and national agreements and regulations that have established bans and restrictions on the use and release of POPs. At the same time, the concentrations of polychlorinated biphenyls (PCBs) and chlordanes remained relatively constant and are still found in nature (AMAP, 2016). Moreover, thousands of new synthetic chemicals are produced each year, and recently there have been reports of the presence of brominated flame retardants (BFRs) and polybrominated diphenyl ethers (PBDEs) (De Wit et al., 2010) in humans and marine biota (AMAP, 2017).

## Human Biological Monitoring

Environmental conditions have a great effect on the health of people. However, until recently, issues of human health protection in the Arctic and environmental protection have been separated in terms of time, space, and the area of competence of the responsible institutions. At the end of the twentieth century, the global community began to discuss the need to combine the efforts of specialists from different spheres of knowledge in order to achieve qualitatively new results in the field of public health. This is how the "One Health" approach appeared. Areas where this approach is particularly relevant include food safety, the fight against zoonoses (diseases that can spread between animals and humans), and the fight against antibiotic resistance (World Health Organization, 2017).

Various types of pollution of the Arctic nature have a negative impact on the health of northern residents. In this regard, the main task of human biological monitoring in the Arctic was to identify public health risks associated with the impact of environmental pollutants (heavy metals, persistent organic pollutants, etc.). Special attention in this effort is given to the Arctic indigenous peoples. Indigenous peoples living in the North are at the greatest risk, as their traditional way of life and diet is based on close interaction with the Arctic wildlife.

The main difference between the Russian Federation and other countries where northern indigenous communities are also found is that in Russia there are a large number of nationalities (ethnic groups) as such (Minority Rights Group International, 2021). Nearly each region of the Russian Federation represents its own nationality, and many of them even represent more than two nationalities. Most of the people living in Russia lead the same post-soviet way of life and their main language is Russian (the Soviet State had somehow united all of them together – Russians, Tatars, Udmurts, Chechens, etc.). Indigenous small-numbered peoples of the Russian North are some of the last who continue to lead a relatively traditional lifestyle. The list of these peoples is set by the Government (Postanovlenie Pravitel'stva RF, 2000). The main idea of the Federal Law on indigenous rights' guarantees (Federal'nij Zakon Rossiiskoi Federatsii, 1999) and this list of indigenous peoples of the North is to ensure their access to the traditional fishing, hunting, herding, and so on (for example, they have privileges to access the natural resources of the North, including the ability to catch restricted species listed in the Red Book).

Russia has developed a special legislation on the legal status of the indigenous small-numbered peoples of the North and approved a list of national groups that belong to them (Kryazhkov, 2013). Russian legislation identifies peoples as indigenous small-numbered peoples of the Russian North if they: (i) have historically lived in the North; (ii) are a small group (up to 50,000 people); and (iii) lead a traditional lifestyle. One additional and important issue is that such people should identify themselves as indigenous for this particular region (North). When conducting monitoring studies, it is important to conduct a questionnaire that, in addition to other mandatory sections, includes the possibility of answering about nationality (ethnicity). This makes it possible to determine which indigenous groups of peoples live in the Arctic area chosen for the study, and which residents are newcomers to these territories. For example, if there are Nenets, Russians, Komi, Udmurts, and Ukrainians among the respondents, then only the Nenets are considered to be indigenous by law. Since the rest identify themselves as other peoples (and taking into account the areas of traditional habitation of these nationalities), they do not identify themselves as indigenous inhabitants of the Arctic (self-identification is one of the main criteria). In the given case, if it is necessary to select the indigenous peoples of the Arctic for the purpose of the study, only the Nenets would be chosen. The other groups mentioned above are non-indigenous to the Arctic.

The Russian legislative provisions are nearly the same as international agreements. For example, WHO identifies indigenous peoples as follows (World Health Organization, 2020):

> Indigenous populations are communities that live within, or are attached to, geographically distinct traditional habitats or ancestral territories, and who identify themselves as being part of a distinct cultural group, descended from groups present in the area before modern states were created and current borders defined. They generally maintain cultural and social identities, and social, economic, cultural and political institutions, separate from the mainstream or dominant society or culture.

Large-scale and regular biomonitoring studies have been conducted in Norway, Greenland, and Canada, where this type of monitoring is supported by the government (Government of Canada, 2020). An active player in this type of activity is the AMAP (Arctic Monitoring and Assessment Program of the Arctic Council), which initiates such studies. Based on the results of its activities, AMAP publishes relevant reports that present not only quantitative indicators of pollutants detected in human organisms, but also contain forecasts and developed recommendations for reducing environmental health risks of the Arctic population (Abass et al., 2018; AMAP, 2009, 2015). The international biomonitoring map of the Arctic has long had gaps in the data on the Russian Arctic. Taking this into account, as well as the positive experience of other Arctic States, the Russian Federation initiated a project to develop a methodology for conducting national Arctic biomonitoring in 2017 (Sorokina, 2019).

Studies conducted for more than 20 years have revealed the following data. Persistent organic pollutants have been found in the blood of Arctic residents, although the serum concentrations of POPs are generally low and similar in all Arctic countries. Significant differences were found between localities, even within the same geographical area, as well as between indigenous and non-indigenous inhabitants. As a rule, the concentrations of POPs are directly related to traditional fatty foods (AMAP, 2009; Bjerregaard et al., 2001). The differences in the concentrations of POPs in blood serum observed in various studies can be explained, among other things, by differences in the structure of nutrition (Araujo-León et al., 2019). The concentrations of most POPs have decreased in recent decades (Varakina et al., 2021).

Mercury (Hg) is naturally present in the environment. People are exposed to Hg mainly as a result of the consumption of fish (marine and freshwater) and seafood, as well as marine mammals. However, although the traditional diet of Arctic populations often includes a lot of seafood and fish, the detected concentrations of mercury indicate that no more than 10% of northern people have mercury consumption above the EFA recommendations (Sobolev et al., 2021).

Lead is considered a good indicator of Pb exposure. A spatial trend of higher B-Pb content among the coastal population was identified, which may be due to the use of Pb for hunting and fishing. Lead contamination of drinking water sources near populated areas from mining activities may also contribute to an increase in the amount of B-Pb (Sobolev et al., 2021).

Ongoing studies on the determination of toxicants in the blood of northern residents have forced scientists to pay attention to essential elements. The lack of these elements can increase the harmful health effects caused by environmental pollutants. Essential elements deficiency may also reduce the body's ability to adapt to the harsh climatic conditions of the Arctic. Iron (Fe) deficiency is an important risk factor for adverse health outcomes and is a diagnostic problem, especially in rural areas. For the purpose of determining iron deficiency, the concentration of S-ferritin is measured. Recent studies in the European part of the Russian Arctic have indicated critically low concentrations of Fe in the blood of local residents (Sobolev et al., 2021).

Iodine (I) is an important element involved in the regulation of numerous metabolic processes. I-deficiency disorders (IDS) include endemic goiter, hypothyroidism, mental retardation, and other adverse effects (Korobitsyna et al., 2020). The results indicate that the iodine status of many residents of the Arctic territories is below the recommended levels (Sobolev et al., 2021). These findings show that iodine deficiency is a recurring public health problem. This requires urgent public health measures at the national, regional, and individual levels.

## Stakeholders of the Environmental Monitoring in the Arctic

Environmental monitoring is a system of long-term observations of the state of the environment, assessment, and forecast of changes under the influence of natural and anthropogenic factors. The environment as a whole and its individual components (land, water, atmospheric air, and other natural objects) are monitored. Environmental monitoring itself is extremely important. Based on the results of monitoring, it is possible to draw conclusions about the ongoing natural processes, make forecasts, identify risks to the well-being of ecosystems and public health, and calculate economic risks, challenges, and advantages. Various stakeholders conduct environmental monitoring in the Arctic.

***International Level*** The main work on monitoring in the Arctic today is entrusted to the Arctic Council. The Arctic Council was established in 1996 to promote (Ottawa Declaration, 1996):

> …cooperation, coordination and interaction among the Arctic States, with the involvement of the Arctic Indigenous communities and other Arctic inhabitants on common Arctic issues, in particular issues of sustainable development and environmental protection in the Arctic.

The Arctic Council's six working groups (AMAP, SDWG, CAFF, ACAP, EPPR, and PAME) are involved in conducting various types of monitoring in the Arctic. The Arctic Council is currently implementing more than 40 projects that are somehow related to monitoring in the Arctic, including environmental monitoring. The system of organizing work in the form of working groups and sufficient project management has already proven its effectiveness. The results of monitoring conducted under the auspices of the Arctic Council have an international authority and are used by other participants of international relations (for example, WHO, UNEP, and the Global Monitoring Plan under the Stockholm Convention on POPs).

***Governmental Environmental Monitoring*** Taking into account the importance of environmental monitoring for the proper statecraft, it has already become an integral state's responsibility. Governmental monitoring is carried out on behalf of the Arctic States by various national institutions, laboratories, scientists, and so on. The following specific features can characterize governmental monitoring:

(a) In some countries, such as Russia, governmental monitoring is the most extensive and expensive. Russia has established a special state hydrometeorological service (Roshydromet), which is subordinate to the Ministry of Natural Resources and Ecology. Its main task is to organize and conduct environmental monitoring in the country, including the Arctic region. In addition, in accordance with the state task, monitoring works are carried out by various scientific organizations. Many different state bodies are involved in these processes and allocate significant budget funds to this. Even the Ministry of Science and Higher Education and the Ministry for the Development of the Far East and the Arctic finance environmental monitoring in the form of grants or subsidies. Information on the results of state monitoring is usually published in the public domain, with the exception of information that has a restricted access mode (these exceptions are discussed below).

(b) Governmental monitoring covers all natural objects and all spheres of anthropogenic impact on nature. This is due to the states' interest to have as much data as possible on the state of the environment and the processes occurring in it. This is necessary to ensure the ability to make prompt management decisions.

(c) The governments conduct environmental monitoring even where other entities are prohibited from doing so. This refers to regime objects (nuclear power facilities, military facilities, etc.) and specific types of pollution (such as radiation pollution). Such restrictions are set for various reasons, such as the increased danger of the object to "ordinary" people or to ensure state security.

(d) Officially published governmental monitoring data are considered a reliable source of environmental information. These data are believed to be verified and authorized by the state. Governmental environmental monitoring data are published on the official websites of responsible state bodies and organizations. Official governmental reports on the state of the environment are also published annually. These published data (together with some secret data) form the basis of management decisions made by the Arctic states.

(e) Governmental environmental monitoring is always limited to the territory of a particular country, even in the Arctic. Any sovereign state has the right to monitor only the processes occurring within its state borders. Therefore, the Russian Federation conducts environmental monitoring only on the territory of its Arctic regions (Murmansk and Arkhangelsk Regions, Nenets, Yamalo-Nenets, Chukotka Autonomous Okrugs, Krasnoyarsk Kray, the Republics of Karelia, Komi, and Sakha-Yakutia). Three exceptions come to mind. The first is the Svalbard Archipelago, where, according to the Svalbard Treaty, not only Norway, but also Russia can conduct scientific research. The second is the high seas. According to the 1982 UN Convention on the Law of the Sea:

> the high seas are open to all States…
> freedom of the high seas … comprises, inter alia, … freedom of scientific research.

It should be kept in mind that within the high seas, not only Arctic States can conduct observations, but also other countries, private companies and other entities. The provisions of UNCLOS also apply to the Arctic Seas, although the implementation of any research projects in the Arctic seas is extremely complex, requires enormous infrastructure, financial and other investments, and is still carried out with the sanction and support of some of the Arctic States. The third exception is that states may agree to conduct joint monitoring observations. This is either done by international (bilateral or multilateral) agreements between countries or foreign specialists are involved in the implementation of specific projects on the territory of the Arctic State within the framework of an employment or other civil law contract. Anything that does not apply to the three cases listed above is usually a spy story.

Industrial environmental monitoring. Environmental monitoring is carried out not only by the state, but also by industrial companies. The companies themselves conduct observations of natural processes on their own and adjacent territories. This makes it possible to: (i) assess the environmental consequences caused by their activities; (ii) prevent or minimize the damage caused to nature or individual natural objects; (iii) assess the effectiveness of the introduction of new eco-friendly technologies; and (iv) assess the effectiveness of measures to eliminate the environmental damage caused. For these purposes, special divisions are created in the structure of companies, and specialists with appropriate training are hired. If laboratory tests are required, some companies create their own laboratories and purchase expensive equipment. Others invite independent labs and their experts. Today, all business companies operating in the Arctic are engaged in industrial environmental monitoring. A sound environmental policy of the company has become a prerequisite for its investment attractiveness. This was not always that the case. Industrial giants have long been considered as the main destructors of natural ecosystems and polluters of the environment (Arctic cleanup program, 2013; NBC news, 2020), and they still are. They are actively assisted by military facilities, which have always been numerous in the Arctic (especially in the Soviet Russia). When, in the mid-twentieth century, the world community realized the scale of the environmental problems created by the Industrial Revolution, some suggested limiting economic development in order to avoid more disastrous consequences (Meadows et al., 1972).

However, humanity decided to follow the path prescribed in the concept of sustainable development (Report of the World Commission on Environment and Development, 1987; Sustainable Development Goals, 2020) and tried to "make friends" between the economy, ecology, and society. Thus, environmental obligations and industrial environmental monitoring have become an integral part of business in the twenty-first century. Today, all industrial companies in all Arctic states conduct monitoring, but the reliability of their own monitoring data published by the companies is not so obvious. On one hand, business is an interested party in this relationship. Companies do not want to compromise themselves. Moreover, if environmental standards are exceeded, there is a risk of large fines and holding the perpetrators legally responsible. On the other hand, the success of a business today depends largely on reputation, and companies understand this. The publication of deliberately false information can lead to even more negative consequences. Only a few will wish to take such risks, especially in the Arctic, where it is impossible to implement any major project without large-scale investments.

***Public Environmental Monitoring*** The name implies that the public carries out this type of monitoring. The first thing that comes to mind is personal observations of people living, working and coming to the Arctic. Such observations can be made from time to time or constantly. Environmental monitoring, especially climate monitoring, is popular among northerners who are engaged in agriculture for their own needs (animal husbandry, vegetable growing, gardening, etc.). However, the most prominent actor in public environmental monitoring is public environmental organizations (environmental NGOs). Many well-known and not so well-known environmental organizations are implementing research projects in the Arctic. Based on the results of their work, they publish relevant reports and commentaries. However, their work does not always involve scientific research. They also conduct actions to attract public attention to some of the most acute problems and risks (Permanent Court of Arbitration, 2017). In this regard, governments are not always ready to take the work of public activists seriously. However, we believe this resource is highly promising in terms of efficiency and responsiveness to possible incidents or dangers, when the state monitoring machine may be too clumsy. In addition, public initiatives have always been a deterrent mechanism against possible arbitrariness of the state. However, we have not yet found any publications in highly rated scientific journals based on the results of independent public research work.

## Access to Monitoring Data

Arctic environmental monitoring data form environmental information. Therefore, when discussing access to monitoring data, it is necessary to take into account the regime of environmental information. This regime is regulated by international norms, but primarily by the national legislation of the Arctic States.

So what is environmental information? Perhaps the most complete definition is given in the UNECE Convention on Access to Information, Public Participation in Decision-Making and Access to Justice in Environmental Matters (Aarhus Convention, 1998) (World Health Organization, 2019). We refer to this document for one more reason. Most of the Arctic States are parties to this agreement (Norway, Sweden, Iceland, and Finland). Denmark joined the Convention, but without the Faroe Islands and Greenland, explaining that "full implementation of the Convention in these areas may imply unnecessary and inadequate bureaucratization". Article 2 of the 1998 Aarhus Convention states that:

> 'Environmental information' means any information in written, visual, aural, electronic or any other material form on:
> (a) The state of elements of the environment, such as air and atmosphere, water, soil, land, landscape and natural sites, biological diversity and its components, including genetically modified organisms, and the interaction among these elements;
> (b) Factors, such as substances, energy, noise and radiation, and activities or measures, including administrative measures, environmental agreements, policies, legislation, plans and programmes, affecting or likely to affect the elements of the environment within the scope of subparagraph (a) above, and cost-benefit and other economic analyses and assumptions used in environmental decision-making;
> (c) The state of human health and safety, conditions of human life, cultural sites and built structures, inasmuch as they are or may be affected by the state of the elements of the environment or, through these elements, by the factors, activities or measures referred to in subparagraph (b) above.

This definition shows that not only the results, but any information related to environmental monitoring, is treated as environmental. This applies to information about the activities of government agencies, organizations, people who are engaged in monitoring the Arctic, regulations governing monitoring, developed monitoring programs, results obtained, identified risks to the environment and people, forecasts, and so on. In general, access to information about regulations that establish the rights and obligations of citizens, the powers of public authorities, as well as about the emergencies and accidents affecting the safety and health of citizens, their consequences, about natural disasters, and their official forecasts cannot be classified. However, the availability of environmental monitoring results is not so obvious. The 1998 Aarhus Convention (Article 4, sub-paragraphs 3 and 4) contains a fairly broad list of possibilities for refusing to provide environmental information. All of these points are also applicable in the case of requesting access to the results of environmental monitoring of the Arctic. The agreement makes the following reservation:

> the anticipated grounds for refusal shall be interpreted in a restrictive way, taking into account the public interest served by disclosure and taking into account whether the information requested relates to emissions into the environment.

Still, the agreement leaves quite broad opportunities for refusal. One may have difficulty obtaining access to monitoring data if they form (a) a state or commercial secret, (b) relate to personal data of a person, or (c) relate to intellectual property, and in a number of other cases specified both in the Convention and in the national legislation of the Arctic States.

In the Arctic States that are not parties to the Convention, the procedure for access to environmental information is similar to the Convention provisions. In Russia, for example, according to the Federal Law "On State Secrets", information about the state of the environment is classified as intelligence that is not subject to be referred to state secret and classified (Law of the Russian Federation No 5485-1, 1993). At the same time, this law contains a large list of intelligence making up state secrets; this refers to data on the activities of military facilities and weapons, and other data whose disclosure could affect the national security of the state. However, what is clearly described in the law is not so easy to implement in practice. In this regard, the incident that occurred in the North of the Russia on August 8, 2019, sometimes called the Nenoksa accident (Bulletin of Atomic Scientists, 2019), is illustrative. This situation is interesting precisely from the point of view of access of Russian citizens, the international community and the scientific community to environmental information (in particular, environmental monitoring data). Residents of nearby localities received information about the incident not from local administrations or the media, but from the federal media at least 2 h later. The report contained only brief information about a certain explosion, but there was no data on radiation levels or other factors that could negatively affect the well-being of a person. Therefore, this news could have gone unnoticed among the population. Only a few politicians tried to raise public awareness about the possible risks associated with the presence of a large number of restricted-access, especially important, and military facilities nearby. This was followed by a series of brief and contradictory speeches and publications by officials, who did not bring more clarity to the picture of what happened. Some publications were even subsequently removed from the official websites of government agencies. It became known that, on the day of the incident, a short-term rise in the level of radiation was recorded in the nearest city of Severodvinsk. However, according to a report by two specialized state services responsible for environmental and social-hygienic monitoring, this level was estimated as "a negligible risk level, therefore, measures for emergency response and normalization of the radiation situation, as well as measures for radiation protection of the population are not required" (Roshydromet, 2019). However, this publication was dated August 12, 2019 (now deleted from the official website meteorf.ru). Four days had passed since the incident, which meant it was too late to alert people about possible risks. The same happened in Chernobyl in 1986 (World Nuclear Association, 2020). It seemed that the Chernobyl disaster had exposed unacceptable gaps in the Soviet system of warning about the risks associated with radiation emissions, so it would be fair to expect that a new informing system had been developed over the intervening three decades; unfortunately, confusion reigned. There is reason to believe that this was due to the inability of officials to clearly distinguish where open information about the state of the environment and public health risks ends, and where state secrets begin. It is noteworthy that the official website of the Northern Branch of Hydrometeorology and Environmental Monitoring, responsible for conducting and publishing the results of monitoring of radiation pollution of the environment on the territory of the Arkhangelsk region and the Nenets Autonomous Okrug, had not yet published open access data for August 2019 (Northern Branch of Hydrometeorology and Environmental Monitoring, 2020).

This case showed the difficulties in implementing the legislation on state secrets, which specifically prescribes list of classified information. In cases where there is no such list, or it is more abstract (such as commercial secrets), there may occur even more confusion with access to monitoring data.

## Discussion

Pollution in the Arctic can be both local and transboundary. Cross-border pollutants enter the Arctic with air masses, sea currents, and migratory species of animals, birds, and fish. These are the ways in which the Far North receives microplastics and some types of chemicals that have never been used or produced in the Arctic. Climate change is associated with an increase in the amount of incoming pollutants: the ice is melting, exposing huge sea areas; the amount of precipitation and storm winds is increasing; and the migration routes of fauna are changing. However, the most significant effect is caused by local pollution in the Arctic, which is closely related to anthropogenic activities. We acquired this data with the help of environmental monitoring, which has been carried out in the Arctic for a long time. The state of atmospheric air, water bodies, soils, flora and fauna are monitored. In addition, specific types of pollution are monitored, such as the radiation situation, the presence of persistent organic pollutants in the environment and the human body, or the relatively new problem of plastic in the Arctic. As a rule, the highest concentrations of pollutants and substances are found near settlements or industrial facilities. This allows us to draw a reasonable conclusion that the pollution of the northern nature largely depends on the pace of industrial and economic development of territories and urbanization. In this regard, it is particularly important to rely on environmental monitoring data and other environmentally relevant information when making management decisions about the development of territories and the implementation of Arctic megaprojects. Even today, we see cases of refusal to finance projects due to the negative impact on the Arctic environment or lack of information about possible environmental risks (Washington Post, 2019). We can assume that this may become a global trend due to the increasing environmental concern in modern society.

The present study has specifically highlighted two types of pollution: persistent organic pollutants and microplastics. This is due to two reasons: (i) the special concern of the Arctic States and the global community about these types of pollution (only the working groups of the Arctic Council are implementing several projects at once in relation to the monitoring and study of these pollutants, not to mention other international and national initiatives); and (ii) the extent of the spread of these pollutants in the Arctic, which is partly due to transboundary transport routes. The presence of POPs in the environment and human samples has been of concern to scientists and the rest of the global community for several decades. Despite the measures taken (for example, the prohibition of certain types of substances, and strict regulation of the production and use of others), POPs are still detected in various natural objects and in human blood and breast milk. However, there is also good news. Recent publications report that the concentrations of these substances have decreased significantly. This trend is fair for the entire Arctic, as it was revealed by scientists from Canada, Russia, Norway, and other Arctic States. However, it is still too early to breathe a sigh of relief. The need for research on new chemicals of particular concern is already being actively discussed, both in the scientific community and in political circles (AMAP, 2017). In this regard, there is reason to believe that new additions to the Stockholm Convention (or the achievement of new international agreements on new toxic substances) will be initiated in the near future.

In September 2019, WHO declared that microplastic is not harmful to human health (World Health Organization, 2019). Nevertheless, research in this area is not only continuing, but also gaining momentum. Today, national and international research teams are working in the Arctic region. The results of their work indicate that plastic has already reached even the most remote areas of the planet. The Arctic region and its most inaccessible areas are no exception and are filled with microplastic particles. Unfortunately, we find these plastic particles even in the Arctic fauna, which is a matter of concern. Some states have already announced a change of political course towards limiting or even completely banning the use of certain plastic products (CTV News, 2021). At the same time, there is no clear answer to the question of what effect microplastics can have on human health (especially with the continuing trend of increasing its quantity and accumulation in nature), there are also supporters of the further development of the plastic industry. For example, in 2019, Russia, the largest Arctic state, announced its intention to increase the production of plastics and become one of the leaders in this market. The COVID-19 pandemic changed the behavior of world's society towards plastics. Plastic products like gloves and shields became everyday essentials for fighting the disease. The demand for plastics will continue to grow. Unfortunately, the peer-reviewed publications we found on the topic of plastic in the Arctic are limited to stating the fact of the discovery of microplastics in various samples. Data on its impact on humans and biota is still too small to draw any certain conclusions.

Any environmental pollution affects human health to some degree, since humans are part of nature. The most major sources of environmental pollutants entering the human body are air, drinking water and nutrition. Many environmental toxicants cause significant negative effects. The list of possible health disorders is quite large, ranging from disorders in the gastrointestinal tract, endocrine, immune or reproductive systems, to different types of cancer (AMAP, 2004, 2009, 2015). With the increasing level of environmental pollution in the Arctic, which is caused by an increase in anthropogenic pressure on the region and ongoing climate change, the risks to the health of the population of the northern territories associated with exposure to environmental toxicants increase significantly as well. For a long time, environmental monitoring and human health monitoring were carried out independently of each other. To a certain extent, they still are. However, the One Health approach has significantly changed the perception in this area in favor of combining the efforts of different specialists for the purposes of preserving the health of nature and humans, and achieving sustainable development goals. In our opinion, this approach is one of the most promising in terms of further development of scientific research and monitoring.

Today, environmental monitoring in the Arctic is carried out by various stakeholders. Environmental problems cannot be purely local. Therefore, the global community is highly concerned about the environmental situation in the Arctic region. This is also facilitated by the general trend towards the globalization of the modern world. Various monitoring and research initiatives are being implemented, not only within the framework of the Arctic Council, but also by other international stakeholders. Many international environmental agreements provide the obligations of member states to organize and conduct monitoring. For example, the Stockholm Convention on Persistent Organic Pollutants provides the need for party states to conduct systematic observations of changes in the concentrations of POPs, the extent of their distribution, and the sources of their release to the environment (Stockholm Convention on Persistent Organic Pollutants, 2001). The Global Monitoring Plan is designed within this agreement for the entire world, including the Arctic. The example of this Convention clearly shows the international collaboration on monitoring, not only between different States, but also between international communities. Thus, the main contributor of data on the Arctic region to the Global Monitoring Report, which is compiled and published under the Convention, are the results of the research work of the Arctic Monitoring and Assessment Program (AMAP) and other working groups of the Arctic Council (Second Global Monitoring Report, 2017).

However, each Arctic State implements its own national monitoring system in the Arctic. Despite the globalization processes that we have already mentioned above, and the increase in international environmental obligations, it is the national monitoring systems that has always been the most extensive, both in terms of coverage of the Arctic territories and natural objects, and in terms of types of pollution studied and the methods used. The Arctic States have full access to environmental monitoring data collected within their state borders and jurisdictions. Here again we need to refer to the regulation of access to monitoring data, which, as our study has shown, remains contradictory. On one hand, we are guaranteed free access and openness of information about the state of the environment, as well as information related to our health, rights, and freedoms. On the other hand, international agreements and national norms contain many reservations that do not allow us to clearly define the border where the national interests of a particular sovereign state in the field of security can legitimately give the monitoring data the status of secret intelligence. Therefore, we are of the opinion that regulation on this issue requires significant improvement at both the international and national levels.

## Conclusions

Monitoring in the Arctic is large-scale work. Arctic and non-Arctic States and other players are interested in it; for example, international entities such as the Arctic Council and its working groups, non-governmental organizations, residents of the northern territories themselves and so on. Various specialists are involved in the implementation of monitoring activities. The condition and pollution of individual natural objects (soils, atmospheric air, water reservoirs, etc.) and natural complexes are monitored in the Arctic. Monitoring activities also include monitoring of certain types of pollution (for example, POPs, plastic and microplastic, radiation pollution, etc.). Monitoring data form the basis for government management decisions and forecasts, as well as economic and investment decisions. All this prompted us to formulate the following basic principles of monitoring activities in the Arctic.

1. *Interdisciplinary approach.* Specialists from different fields of knowledge (analytical chemistry, biology, ecology, law, statistical analysis, health care, etc.) should be involved in the monitoring of Arctic ecosystems on an ongoing basis.
2. *Integrated approach.* To make competent and reasonable management decisions, actors should not limit themselves to dry data of environmental and/or biological monitoring. It is necessary to assess the existing risks based on the information received and develop recommendations for minimizing the identified risks. A special role in this work is played by the process of informing stakeholders and other interested parties about the identified problems and developed recommendations.
3. *International and interregional cooperation.* Today, research work in the Arctic is not possible without the exchange of experience and competencies between specialists from different countries. The Arctic is a global region and it is impossible to divide it by state borders. Also, state borders cannot protect against the region's environmental problems and pollution. In this regard, the Arctic States, as well as other participants of the Arctic Council and the world community, need to join efforts in protecting the Arctic environment and the health of the population living on this territory.

The implementation of these principles will make it possible to most effectively address emerging and worsening environmental problems in the Arctic, as well as to make more scientifically sound and balanced management decisions related to the sustainable development of the northern territories.

## References

Abass, K., Emelyanova, A., & Rautio, A. (2018). Temporal trends of contaminants in Arctic human populations. *Environmental Science and Pollution Research, 25*, 28834–28850.

AMAP. (2004). *Persistent toxic substances, food security and indigenous peoples of the Russian North.* Final Report. Arctic Monitoring and Assessment Programme. Oslo, Norway. 192 p.

AMAP. (2009). *AMAP assessment 2009: Human health in the Arctic*. Arctic Monitoring and Assessment Programme. Oslo, Norway. Xiv+254 pp.
AMAP. (2015). *AMAP assessment 2015: Human health in the Arctic*. Arctic Monitoring and Assessment Programme. Oslo, Norway. vii + 165 pp.
AMAP. (2016). *Temporal trends in persistent organic pollutants in the Arctic*. AMAP assessment 2015. Arctic Monitoring and Assessment Programme. Oslo, Norway.
AMAP. (2017). *AMAP assessment 2016: Chemicals of Emerging Arctic Concern*. Arctic Monitoring and Assessment Programme. Oslo, Norway.
AMAP. (2018). *Biological effects of contaminants on Arctic wildlife and fish*. Arctic Monitoring and Assessment Programme (AMAP). Tromsø, Norway.
Andrady, A. L. (2011). Microplastics in the marine environment. *Marine Pollution Bulletin, 62*(8), 1596–1605.
Araujo-León, J. A., Mena-Rejón, G. J., Canché-Pool, E. B., & Ruiz-Piña, H. A. (2019). Biomonitoring Organochlorine Pesticides in Didelphis virginiana from Yucatan, Mexico by GC-ECD. *Bulletin of Environmental Contamination and Toxicology, 102*, 836–842.
Arctic Center. University of Lapland. Traditional Knowledge. Retrieved February 20, 2021., from https://www.arcticcentre.org/EN/arcticregion/Arctic-Indigenous-Peoples/Traditional-knowledge
Arctic cleanup program. (2013). Retrieved December 17, 2020, from https://www.rgo.ru/en/projects/arctic-cleanup-program
Avango, D., Hogselius, P., & Nilsson, D. (2018). Swedish explorers, in-situ knowledge, and resource-based business in the age of empire. *Scandinavian Journal of History, 43*(3), 324–347.
Bae, J., Kim, S., Barr, D. B., & Buck Louis, G. M. (2018). Maternal and paternal serum concentrations of persistent organic pollutants and the secondary sex ratio: A population-based preconception cohort study. *Environmental Research, 161*, 9–16.
Bergman, M., Mutzel, S., Primke, S., Tekman, M. B., Trachsel, J., & Gerdts, G. (2019). White and wonderful? Microplastics prevail in snow from the Alpsto the Arctic. *Science Advances, 5*(8), Eaax1157.
Bjerregaard, P., Dewailly, E., Ayotte, P., Pars, T., Ferron, L., & Mulvad, G. (2001). Exposure of Inuit in Greenland to organochlorines through the marine diet. *Journal of Toxicology and Environmental Health, 62*, 69–81.
Braune, B. M., Outridge, P. M., Fisk, A. T., Muir, D. C. G., Helm, P. A., Hobbs, K., Hoekstra, P. F., Kuzyk, Z. A., Kwan, M., Leteher, R. J., et al. (2005). Persistent organic pollutants and mercury in marine biota of the Canadian Arctic: An overview of spatial and temporal trends. *Science of the Total Environment, 351–352*, 4–56.
Brozovsky, J., Gaitani, N., & Gustavsen, A. (2021). A systematic review of urban climate research in cold and polar climate regions. *Renewable & Sustainable Energy Reviews, 138*, 110551.
Bulletin of Atomic Scientists. (2019). *The Nenoksa accident: A timeline of confusing and conflicting reports*. 75 Years and Counting. Retrieved September 20, 2020, from https://thebulletin.org/2019/08/the-nenoksa-accident-a-timeline-of-confusing-and-conflicting-reports/
Canadian Center for Occupational Health and Safety. (2019). *Cold Environments – Working in the Cold*. Retrieved January 17, 2021, from https://www.ccohs.ca/oshanswers/phys_agents/cold_working.html
CBS News. (2021). Odynova, A. *Russian tanker cuts a previously impossible path through the warming Arctic*. Retrieved February 23, 2021, from https://www.cbsnews.com/news/russian-tanker-cuts-a-previously-impossible-path-through-the-warming-arctic/
Chukmasov, P., Aksenov, A., Sorokina, T., Varakina, Y., Sobolev, N., & Nieboer, E. (2019). North Pacific baleen whales as a potential source of Persistent Organic Pollutants (POPs) in the diet of the indigenous peoples of the Eastern Arctic Coasts. *Toxics, 7*, 65.
Cole, M., Lindeque, P., Halsband, C., & Galloway, T. S. (2011). Microplastics as contaminants in the marine environment: A review. *Marine Pollution Bulletin, 62*(12), 2588–2597.
Cozar, A., Marti, E., Duarte, C. M., Garcia-de-Lomas, J., van Sebille, E., Ballatore, T. J., Eguiluz, V. M., Gonzales-Gordillo, J. I., Pedrotti, M. L., Echevarria, F., Trouble, R., & Irigoien,

X. (2017). The Arctic Ocean as a dead end for floating plastics in the North Atlantic branch of the Thermohaline Circulation. *Science Advances, 3*(4), e1600582.

CTV News. (2021). Aiello, R. *Canada banning plastic bags, straws, cutlery and other single-use items by the end of 2021*. Retrieved November 7, 2020, from https://www.ctvnews.ca/climate-and-environment/canada-banning-plastic-bags-straws-cutlery-and-other-single-use-items-by-the-end-of-2021-1.5135968

Dauvalter, V. A. (2020). Geochemistry of lakes in a zone impacted by an Arctic Iron-producing Enterprise. *Geochemistry International, 58*(8), 933–946.

De Wit, C. A., Alaee, M., & Muir, D. C. G. (2006). Levels and trends of brominated flame retardants in the Arctic. *Chemosphere, 64*, 209–233.

De Wit, C. A., Herzke, D., & Vorkamp, K. (2010). Brominated flame retardants in the Arctic environment –trends and new candidates. *Science of the Total Environment, 408*, 2885–2918.

Dinu, M. I., Shkinev, V. M., Moiseenko, T. I., Dzhenloda, R. K., & Danilova, T. V. (2020). Quantification and speciation of trace metals under pollution impact: Case study of a subarctic Lake. *Water, 12*(6), 1641.

Dudarev, A., Yamin-Pasternak, S., Pasternak, I., & Chupakhin, V. (2019). Traditional diet and environmental contaminants in coastal Chukotka I: Study design and dietary patterns. *International Journal of Environmental Research and Public Health, 16*, 702.

EcoWatch. (2018). *Chow, L. 10 Most Common Types of Beach Litter Are All Plastic*. Retrieved April 19, 2021, from https://www.ecowatch.com/beach-litter-plastics-ocean-conservancy-2581760475.html

European Environment Agency. (2017). Arctic resources. Retrieved February 10, 2021, from https://www.eea.europa.eu/data-and-maps/figures/arctic-resources

Fang, C., Zheng, R. H., Hong, F. K., Jiang, Y. L., Chen, J. C., Lin, H. S., Lin, L. S., Lei, R. B., Bailey, C., & Bo, J. (2021). Microplastics in three typical benthic species from the Arctic: Occurrence, characteristics, sources, and environmental implications. *Environmental Research, 192*, 110326.

Federal'nij Zakon Rossiskoi Federatsii o Garantiakh Prav Korennikh Malochislennikh Narodov Rossiiskoy Federatsii [Federal Law of Russian Federation (RF) on Guarantees of Rights of Indigenous Small-Numbered People (Law on Guarantees of Rights)] art. 1 (1999), available in Russian on the legal information portal of the Russian Federation, at http://pravo.gov.ru/proxy/ips/?docbody=&nd=102059473

Fiore, M., Oliveri Conti, G., Caltabiano, R., Buffone, A., Zuccarello, P., Cormaci, L., Cannizzaro, M. A., & Ferrante, M. (2019). Role of emerging environmental risk factors in thyroid Cancer: A brief review. *International Journal of Environmental Research and Public Health, 16*, 1185.

Gosudarstvennii doklad o sostoyanii i ob okhrane okruzhaushei sredi Rossiiskoi Federatsii v 2019 godu (available only in Russian). Retrieved 1 March, 2021 from https://www.mnr.gov.ru/docs/gosudarstvennye_doklady/proekt_gosudarstvennogo_doklada_o_sostoyanii_i_ob_okhrane_okruzhayushchey_sredy_rossiyskoy_federat2019/

Government of Canada. National biomonitoring initiatives. Retrieved October 17, 2020., from https://www.canada.ca/en/health-canada/services/chemical-substances/chemicals-management-plan/monitoring-surveillance/national-biomonitoring-initiatives.html

Gray, J. M., Rasanayagam, S., Engel, C., & Rizzo, J. (2017). State of the evidence 2017: An update on the connection between breast cancer and the environment. *Environmental Health, 16*, 94.

Heleniak, T. (2014). *Migration in the Arctic. Arctic Yearbook*. Available at: https://arcticyearbook.com/images/yearbook/2014/Scholarly_Papers/4.Heleniak.pdf. Accessed on 27 Oct 2020.

Ji, X. W., Abakumov, E., Chigray, S., Saparova, S., Polyakov, V., Wang, W. J., Wu, D. S., Li, C. L., Huang, Y., & Xie, X. C. (2021). Response of carbon and microbial properties to risk elements pollution in Arctic soils. *Journal of Hazardous Materials, 408*, 124430.

Kanhai La Daana, K., Johanson, C., Frias, J. P. G. L., Gardfeldt, K., Thompson, R. C., & O'Connor, I. (2018). Deep sea sediments of the Arctic Central Basin. *Marine Pollution Bulletin, 130*, 8–18.

Kanhai La Daana, K., Kardfeldt, K., Lyashevska, O., Hassellov, M., Thompson, R. C., & O'Connor, I. (2019). Microplastics in sub-surface waters of the Arctic Central Basin: A potential sink for microplastic. *Deep Sea Research Part I: Oceanographic Research Papers, 145*, 137–142.

Kanhai La Daana, K., Katarina, G., Krumpen, T., & Thompson, R. C. (2020). Microplastics in sea ice and seawater beneath ice floes from the Arctic Ocean. *Scientific Reports, 10*(1), 1–11.

Korobitsyna, R., Aksenov, A., Sorokina, T., Trofimova, A., Sobolev, N., Grjibovski, A. M., Chashchin, V., & Thomassen, Y. (2020). Iodine status of women and infants in Russia: A systematic review. *International Journal of Environmental Research and Public Health, 17*(22), 8346.

Kryazhkov, V. A. (2013). Development of Russian legislation on northern indigenous peoples. *Arctic Review on Law and Politics, 4*(2), 140–155. ISSN 1891-6252.

Lakhmanov, D., Varakina, Yu., Aksenov, A., Sorokina, T., Sobolev, N., Kotsur D., Plakhina E., Chashchin, V., & Thomassen, Y. (2020). Persistent Organic Pollutants (POPs) in fish consumed by the indigenous peoples from Nenets autonomous Okrug. *Environments, 7*. ISSN 2076-3298.

Lassen, C., Hansen, S. F., Magnusson, K., Hartmann, N. B., Jensen P. R., Nielsen, T. G., & Brinch, A. (2015). *Microplastics: occurrence, effects and sources of releases to the environment in Denmark.* Available at: http://mst.dk/service/publikationer/publikationsarkiv/2015/nov/rapport-om-mikroplast. Accessed 7 Nov 2020.

Law of the Russian Federation No. 5485-1 of July 21, 1993 *On state secrets.* Available online: https://www.wto.org/english/thewto_e/acc_e/rus_e/WTACCRUS58_LEG_125.pdf. Accessed on 20 Sept 2020.

Lee, D. H., Porta, M., Jacobs, D. R., & Vandenberg, L. N. (2014). Chlorinated persistent organic pollutants, obesity, and type 2 diabetes. *Endocrine Reviews, 35*, 557–601.

Lind, P. M., Salihovic, S., Stubleski, J., Kärrman, A., & Lind, L. (2019). Association of Exposure to persistent organic pollutants with mortality risk: An analysis of data from the Prospective Investigation of Vasculature in Uppsala Seniors (PIVUS) study. *JAMA Network Open, 2*, 193070.

Lusher, A. V., Tirelli, V., O'Connor, I., & Officer, R. (2015). Microplastics in Arctic polar waters: The first reported values of particles in surface and sub-surface samples. *Scientific Reports, 5*, 14947.

McCannon, G. (2012). *A history of the Arctic: Nature, exploration and exploitation.* Reaktion Books.

Meadows, D. H., Meadows, D. L., Jorgen, R., & Behrens III, W. W. (1972). *The limits to growth.* A Report for the Club of Rome's Project on the Predicament of Mankind. Universe Books.

Minority Rights Group International. (2020). *Russian Federation.* Retrieved September 27, 2020, from https://minorityrights.org/country/russian-federation/

Moore, R. C., Loseto, L., Noel, M., Etemadifar, A., Brewster, J. D., MacPhee, S., Bendell, L., & Ross, P. S. (2020). Microplastics in beluga whales (Delphinapterus leucas) from the Eastern Beaufort Sea. *Marine Pollution Bulletin, 150*, 110723.

Morgana, S., Ghigliotti, L., Estevez-Calvar, N., Stifanese, R., Wieckzorek, A., Doyle, T., Christiansen, J. S., Faimali, M., & Garaventa, F. (2018). Microplastics in the Arctic: A case study with sub-surface water and fish samples off Northeast Greenland. *Environmental Pollution, 242*, 1078–1086.

Mouly, T. A., & Toms, L.-M. L. (2016). Breast cancer and persistent organic pollutants (excluding DDT): A systematic literature review. *Environmental Science and Pollution Research, 23*, 22385–22407.

Mu, J., Zhang, S., Qy, L., Jin, F., Fang, C., Ma, X., Zhang, W., & Wang, J. (2019). Microplastics abundance and characteristics in surface waters from Northwest Pacific, the Bering Sea, and the Chukchi Sea. *Marine Pollution Bulletin, 143*, 58–65.

NBC news. (2020). Bodner, M. *Russia launches major clean-up operation after huge Arctic fuel spill.* Retrieved January 23, 2021, from https://www.nbcnews.com/news/vladimir-putin/russia-launches-major-clean-operation-after-huge-arctic-fuel-spill-n1224581

Northern Branch of Hydrometeorology and Environmental Monitoring. Radiation situation (available only in Russian). Retrieved June 27, 2020., from http://www.sevmeteo.ru/monitoring/radiation/

O'Hara, T. M., Hoekstra, P., Hanns, C., Muir, D., Wetzel, D., & Reynolds, J. A. (2004). *Preliminary assessment of the nutritive value of select tissues from the bowhead whale based on suggested nutrient daily intakes.* Retrieved February 27, 2019, from http://www.north-slope.org/assets/images/uploads/SC-56-E2.ohara_food_value.pdf

Obbard, R. W., Sadri, S., Wong, Y. Q., Khitun, A. A., Baker, I., & Thompson, R. C. (2014). Global warming releases microplastic legacy frozen in Arctic Sea ice. *Earth's Future, 2*(6), 315–320.

Ocean conservancy. (2018). *International Coastal Cleanup.* Retrieved April 19, 2021, from https://oceanconservancy.org/wp-content/uploads/2018/06/FINAL-2018-ICC-REPORT.pdf

Ottawa Declaration. (1996). Retrieved February 20, 2021, from https://oaarchive.arctic-council.org/handle/11374/85

Peeken, I., Primke, S., Beyer, B., Gutermann, J., Katlein, C., Krumpen, T., Bergmann, M., Hehemann, L., & Gerdts, G. (2018). Arctic sea ice is an important temporal sink and means of transport for microplastic. *Nature Communications, 9*(1), 1–12.

Permanent Court of Arbitration. (2017). *The Arctic Sunrise Arbitration (Netherlands v. Russia).* Retrieved July 21, 2020, from https://pca-cpa.org/en/cases/21/

Postanovlenie Pravitel'stva RF ot 24 Marta 2000 goda N 255 O Edinnom Perechne Korennikh Malochislennykh Narodov Rossiskoi Federatsii [Resolution of the Government of the RF from March 24, 2000, N 255 on the Unified Registry of Indigenous Small-Numbered Peoples of the RF], available in Russian on the website of the legal information portal Garant, at http://base.garant.ru/181870/

Report of the World Commission on Environment and Development: Our Common Future. Retrieved July 20, 2020., from http://www.un-documents.net/our-common-future.pdf

Reynolds, J. E., Wetzel, D. L., & O'Hara, T. M. (2006). Human health implications of Omega-3 and Omega-6 fatty acids in blubber of the bowhead whale (Balaena mysticetus). *Arctic, 59*, 155–164.

Rodrigues, J., Duarte, A. C., Santos-Echeandia, J., & Rocha-Santos, T. (2019). Significance of interactions between microplastics and POPs in the marine environment: A critical review. *TrAC Trends in Analytical Chemistry, 111*, 252–260.

Rosen, E. M., Muñoz, M. I., McElrath, T., Cantonwine, D. E., & Ferguson, K. K. (2018). Environmental contaminants and preeclampsia: A systematic literature review. *Journal of Toxicology and Environmental Health, Part B Critical Reviews, 21*, 291–319.

Roshydromet. (2019). Retrieved November 17, 2020, from http://www.meteorf.ru/

Second Global Monitoring Report. (2017). *Conference of the parties to the Stockholm Convention on persistent organic pollutants* (p. 129). Eighth meeting.

Sobolev, N., Ellingsen, D. G., Belova, N., Aksenov, A., Sorokina, T., Trofimova, A., Varakina, Y., Kotsur, D., Grjibovski, A., Chashchin, V., Bogolitsyn, K., & Thomassen, Y. (2021). Essential and non-essential elements in biological samples of inhabitants residing in Nenets Autonomous Okrug of the Russian Arctic. *Environment International., 152*(29), 106510.

Sorokina, T. Y. (2019). A national system of biological monitoring in the Russian Arctic as a tool for the implementation of the Stockholm Convention. *International Environmental Agreements: Politics, Law and Economics, 19*, 341–355.

Stockholm Convention on Persistent Organic Pollutants was adopted on 22 May 2001 in Stockholm, Sweden. Available online: http://chm.pops.int/Portals/0/Repository/convention_text/UNEP-POPS-COP-CONVTEXT-FULL.English.PDF. Accessed on 19 Sept 2020.

Sustainable Development Goals. Retrieved July 20, 2020., from https://sdgs.un.org/

Svetlov, A. V., Pripachkin, P. V., Masloboev, V. A., & Makarov, D. V. (2020). Classification of low-grade copper-nickel ore and mining waste by ecological Hazard and hydrometallurgical processability. *Journal of Mining Science, 56*(2), 275–282.

Tekman, M. B., Wekerle, C., Loretnz, C., Primke, S., Hasemann, C., Gerdts, G., & Bergmann, M. (2020). Tying up loose ends of microplastic pollution in the Arctic: Distribution from the sea surface through the water column to deep-sea sediments at the Hausgarten observatory. *Environmental Science & Technology, 54*(7), 4079–4090.

Thorp, T., Arnold, S. R., Pope, R. J., Spracklen, D. V., Conibear, L., Knote, C., Arshinov, M., Belan, B., Asmi, E., Laurila, T., Skorokhod, A. I., Nieminen, T., & Petaja, T. (2021). Late-spring and summertime tropospheric ozone and $NO_2$ in western Siberia and the Russian Arctic: Regional model evaluation and sensitivities. *Atmospheric Chemistry and Physics, 21*(6), 4677–4697.

Tosic, T. N., Vruggink, M., & Vesman, A. (2020). Microplastics quantification in surface waters of the Barents, Kara and White seas. *Marine Pollution Bulletin, 161*, 111745.

UNECE Convention on Access to Information, Public Participation in Decision-Making and Access to Justice in Environmental Matters done on 25 June 1998 in Aarhus, Denmark. Available online: https://unece.org/DAM/env/pp/documents/cep43e.pdf. Accessed on 19 Sept 2020.

United Nations Convention on the Law of the Sea of 10 December 1982. (UNCLOS). Retrieved April 27, 2021, from https://www.un.org/depts/los/convention_agreements/texts/unclos/unclos_e.pdf

Varakina, Y., Lakhmanov, D., Aksenov, A., Trofimova, A., Korobitsyna, R., Belova, N., Sobolev, N., Kotsur, D., Sorokina, T., Grjibovski, A., Chashchin, V., & Thomassen, Y. (2021). Concentrations of persistent organic pollutants in women's serum in the European Arctic Russia. *Toxics, 9*(6).

Washington Post. (2019). *Energy 202: Goldman Sachs rules out financing for Arctic drilling. Will other U.S. banks follow?* Retrieved August 2, 2020 from https://www.washingtonpost.com/news/powerpost/paloma/the-energy-202/2019/12/17/the-energy-202-goldman-sachs-rules-out-financing-for-arctic-drilling-will-other-u-s-banks-follow/5df7d2c2602ff125ce5b503b/

World Health Organization. (2017). *One Health*. Retrieved January 27, 2021, from https://www.who.int/news-room/q-a-detail/one-health

World Health Organization. (2019). *Microplastics in drinking-water*. Geneva. Licence: CC BY-NC-SA 3.0 IGO. 124 p.

World Health Organization. (2020). *Indigenous populations*. Retrieved December 27, 2020, from https://www.who.int/topics/health_services_indigenous/en/

World Nuclear Association. (2020). Chernobyl Accident (1986). Retrieved June 27, 2020, from https://www.world-nuclear.org/information-library/safety-and-security/safety-of-plants/chernobyl-accident.aspx

Yaraghi, N., Ronkanen, A. K., Haghighi, A. T., Aminikhah, M., Kujala, K., & Klove, B. (2020). Impacts of gold mine effluent on water quality in a pristine sub-Arctic river. *Journal of Hydrology, 589*(125170).

Yegambaram, M., Manivannan, B., Beach, T. G., & Halden, R. U. (2015). Role of environmental contaminants in the etiology of Alzheimer's disease: A review. *Current Alzheimer Research, 12*, 116–146.

Zheng, T., Zhang, J., Sommer, K., Bassig, B. A., Zhang, X., Braun, J., Xu, S., Boyle, P., Zhang, B., Shi, K., et al. (2016). Effects of environmental exposures on Fetal and childhood growth trajectories. *Annals of Global Health, 82*, 41–99.

**Tatiana Yu. Sorokina** is an Associate Professor (since 2011) at the Northern (Arctic) Federal University (NArFU), Arkhangels, Russia and Head of NArFU's Arctic Biomonitoring Laboratory (since 2017).

# Part III
# Economics and Geopolitics

# Chapter 13
# The Quest for the Ultimate Resources: Oil, Gas, and Coal

**Andrey Krivorotov**

## Introduction

Natural resources, whether they are timber, fur, gold or fossil fuels, have traditionally been among the crucial drivers of Arctic exploration and economic development. Coal and oil deposits have been known in the Arctic for some 300 years, but their large-scale development only began in the twentieth century and it is only in the past two decades that the search for Arctic oil and gas has become one of the key policy and industry priorities worldwide.

The issue became highly politicized and attracted widespread publicity in Arctic and non-Arctic states, which has somehow blurred the real investment considerations. This chapter seeks to give a brief practical account of the actual status and prospects for Arctic oil, gas, and coal development against the background of both global and national political and market trends.

## *Analytical Framework*

There are at least two key stakeholders in each extractive industry project: the investor, who does the actual extraction business; and the host government, which protects the public interest, either as a resource owner or as a regulator.

---

A. Krivorotov (✉)
MGIMO University, Odintsovo, Russia

Russian International Affairs Council (RIAC), Moscow, Russia

International Arctic Social Science Association (IASSA), Cedar Falls, IA, USA
e-mail: krivorotov@starlink.ru

Both stakeholders assess each investment project according to their specific sets of criteria, weighing up the eventual pros and cons. For the investor, the equation is rather simple: if the feasibility study demonstrates that the project may yield margins above the established corporate threshold, it is deemed commercially attractive.

The government makes a similar comparison, but it involves a wide array of considerations. A successful extractive industry project may bring about numerous positive effects for the host country, including:

- Enhanced security of energy supply for national consumers
- Tax payments and other relevant budgetary revenues
- Industrial and regional policy effects
- Strengthening the country's international positions.

On the other hand, project implementation is also associated with certain administrative costs (and direct losses for the budget in case of subsidizing), plus major environmental impacts.

The authorities influence the commercial attractiveness of each project by managing the national investment climate, essentially with two principal sets of levers:

- Legal and administrative framework (referred to as 'Laws' in our figures)
- Direct and indirect taxation and other fiscal mechanisms, including various forms of tax relief and subsidizing ('Taxes').

If both parties deem a project attractive, the investor may implement it, delivering the resources to the market and generating, among other things, a cash flow to the national budget. Alternatively, a project may be politically attractive, but commercially non-viable. In this case, the government may encourage investments by introducing tax incentives and thus getting the project done, albeit with a negative budgetary effect (Fig. 13.1).

The approaches of both the investors and, even more so, the governments are influenced by a number of overarching considerations and underlying global trends.

## *Policy Considerations*

From politicians' perspectives, although Arctic resources are more expensive than those produced in Persian Gulf, they have strong advantages as a secure source of energy supply, found in politically stable areas, relatively close to the key northern hemisphere markets.

These considerations were always of prime importance for the former Soviet Union with its highly autonomous economy, but also for Western nations after the 1973 Oil Crisis. The sudden fuel shortage in the EU and US markets spurred intensified exploration and development of the North American and North Sea potential, plus the entry of Soviet (Arctic-produced) oil and gas into the West European market.

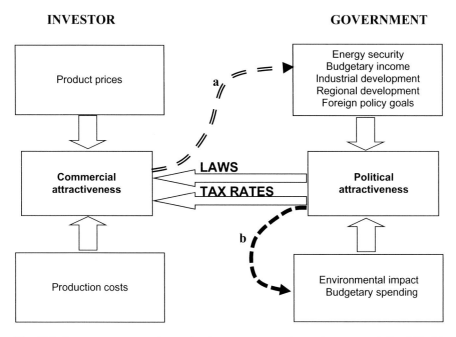

**Fig. 13.1** Investor-government interaction in resource extraction under commercially viable (**a**) and subsidized (**b**) projects. (Source: Author)

Good geological prospects also play a role. The proven petroleum reserves[1] north of the Arctic Circle are rather limited so far. Some 400 oil and gas fields were known by 2007, with total reserves of about 40 bbl (5.5bt) of oil and 1136 tcf (31tcm) of gas, which represented only about 2.3% and 0.4% of the global reserves, respectively (Gautier et al., 2009). Despite a few major Arctic discoveries in recent years, the overall share remains low, but there are high chances of finding new major fields as the area is largely underexplored.

The increased attention to Arctic sovereignty and prosperity has also highlighted the importance of industrial activities in remote regions. Large-scale extractive projects, including the related transportation infrastructure, may generate strong ripple effects in the form of research and technology development, new businesses, jobs, fiscal incomes to local budgets, etc. At the same time, they also establish a visible national presence in the Arctic and create opportunities for cooperation with foreign investors.

---

[1] Arctic countries use several resource classifications, but they all make a distinction between resources (potential volumes of fuel, which may exist, with some probability, in a territory) and reserves (deposits that are proven to be there upon the results of exploratory drilling).

## Technologies

Technological development has a dual impact on the attractiveness of Arctic oil and gas fields. On the one hand, it is a great help in terms of unbundling the Arctic's resource potential. Unmanned offshore technologies are a brilliant example, like seabed completion and compression, ultra-deep-sea development at water depths up to 3000 m, multiphase flows to shore, remotely operated vehicles (ROVs), etc. Such solutions make it possible to avoid the costly surface installations, thus cutting the capital and operational expenditures dramatically, and to enhance the operational safety by eliminating the risks related to the human factor, floating ice, and iceberg impacts.

On the other hand, the same progress in extractive technologies increases competition in the market, opening new horizons for non-Arctic resource development as well. Besides, the progress of technologies also takes place in other relevant areas like energy efficiency and renewable energy sources, which has an adverse overall effect on the oil and gas industry, diminishing the global demand for hydrocarbons.

## Health, Safety, and Environment (HSE)

Concerns for the fragile Arctic environment have traditionally been the key constraint of resources development in the region (Conley et al., 2013). Meanwhile, the approaches to the Arctic petroleum HSE issues have undergone a serious evolution over decades.

In the twentieth century, the discussions focused on contamination risks in the actual up-, mid- and downstream operations. There were major accidents at the time, accompanied by large oil spills and sometimes human casualties, such as the crashes of the *Piper Alpha* and *Alexander Kielland* oil rigs in the North Sea, coal mine accidents, a major oil pipeline leakage in Komi, northern Russia, or *Exxon Valdez* tanker running aground in Prince William Sound in Alaska.

This approach still dominates in Russia, due to the complicated HSE heritage from the Soviet era and the lower focus on climate change. Meanwhile, the public debate in the other Arctic nations in the early twenty-first century concentrated heavily on greenhouse gases. Reflecting the global trends, governments and the general public viewed the Arctic petroleum industry predominantly as a source of $CO_2$ emissions, during both the production and the burning of hydrocarbons.

The accident on the Macondo well offshore the Gulf of Mexico in April 2010, which resulted in human casualties, a loss of the *Deepwater Horizon* drilling rig, and an unparalleled oil spill, became an important turning point, delivering a heavy blow on the HSE image of big oil and gas business. Within a few months, governmental agencies all over the world introduced much more stringent environmental protection regimes and practices for offshore oil exploration and production. The Arctic attracted a special attention due to its extraordinary challenges of harsh climatic conditions, dark seasons, insufficient infrastructure, and the presence of sea ice. Numerous Arctic offshore projects either stalled or were postponed til indefinite in the subsequent years due to the increased public awareness, coupled with the fall in oil and gas prices.

However, several major onshore fields were put on stream and licensing of new acreages went on, albeit not always successfully. Industrial organizations, regional Arctic authorities, and NGOs lobbied continued operations. Norway made remarks in the Arctic Council in 2016 (supported by Canada, Denmark, and Russia) stressing the economic development value of oil and gas, as opposed to merely 'source of environmental risk' (Wilson Rowe, 2018).

Besides, new energy phenomena are in the making in both the North American and North Sea shelves, such as offshore wind and wave power generation (including conversion of elder petroleum platforms), extended platform lifetimes as a viable alternative to their expensive and risky decommissioning, etc. These trends remove the traditional 'oil *vs* environment' dichotomy, although the petroleum industry as a whole is subject to an increasing ecological scrutiny.

## Brief History

The commercial and political attractiveness of Arctic fuels, as presented in our scheme, is not fixed. Apart from the abovementioned long-term trends, it also varies strongly in line with the price fluctuations and the overall international situation. While it may sound paradoxical, the petroleum business, where field lifecycles stretch over decades, is often opportunistic and influenced heavily by the current commodity and stock markets, as well as by the changing government priorities. Therefore, we feel it is necessary to first provide a brief general outline of the global trends over the past decades, and then to analyze how each Arctic nation reacted to these, reflecting its unique domestic and foreign policy agendas. The international interest in fossil fuels, like other Arctic resources, has gone through several boom and bust periods.

Industrial exploration and production of Arctic coal started in the early twentieth century in the Norwegian archipelago of Spitsbergen (Svalbard). Companies from Norway, Sweden, Netherlands, Russia, Great Britain, USA etc. staked their claims and opened coal mines there, but only Norwegian and Soviet mines continued operations after the Great Depression. Major coal production was also established in various parts of the Soviet Arctic, such as Komi, Yakutia, and Chukotka, while in the West it suffered strongly from the competition of cheaper coal from Australia and developing countries.

The first large-scale Arctic petroleum development started in the 1960s and 1970s, receiving strong impetus from the Oil Crisis. Many giant fields were developed in that period in the US (Alaska) and the USSR (Western Siberia). However, the oil price fall in 1986 and the subsequent low prices through the end of the century with a new collapse in 1998, plus the continued trade liberalization, made new

Arctic developments unattractive. The oil business started to show renewed interest to the Arctic in the mid-2000s, reflecting new major trends:

- Oil prices rallied after the 2003 US military operation in Iraq, which made Arctic fields commercial again.
- The 2004 legal case against Shell overstating its booked reserves, in the aftermath of which oil companies realized that the global reserves were finite and largely occupied.
- Deliberate rise of the national champions (state-owned oil companies in producing countries) on the account of Western majors.
- The 2003 forecast by the US Energy Information Administration predicting a growing shortage of natural gas in the US after 2007, which boosted LNG industry worldwide.

The political attention to the Arctic was growing in parallel. The Russian private expedition, which planted the national titanium flag on the North Pole seabed in 2007, became the true trigger. Within a short time, the Arctic turned into a popular and 'trendy' global issue, rising high on politicians' lists of priorities.

The US Geological Survey (USGS) added to this by issuing its Circum-Arctic Resource Appraisal in 2008, where it estimated undiscovered resources north of the Arctic Circle at 89.983 bbl (12.329 bmt) of oil and 1669 tcf (45,054 tcm) of natural gas. Some 64% of this potential was concentrated in three provinces: West Siberian Basin (Russia), Arctic Alaska (USA), and East Barents Basin (Russia and Norway) (Bird et al., 2008). This represented approximately one-fourth of the global *estimated undiscovered* resources, while in political and media language, this appraisal transformed soon into the popular maxim that "the Arctic contains 25 percent of the world's oil and gas". Combined with the burst in offshore technologies and the intense ice melt, this created expectations of a rapid industrial development, a resulting 'Arctic (resource) race', and increased international tension, given the unsettled legal issues in the area (Fairhall, 2010; Howard, 2009).

Within a few years, however, the situation changed. The Arctic remained high on the political agenda, with the regional states adopting their official strategies, ambitious civil and military plans for the area. Meanwhile, Canada, Denmark/Greenland, Norway, Russia, and the US signed the Ilulissat Declaration in 2008, in which they committed to resolve any Arctic territorial claims based on international law. Russia and Norway set a good example by settling the 40-year-long Barents Sea dispute peacefully in 2010, whereby they divided some 175,000 sq. km of water and continental shelf. The cooperation in the Arctic Council was flourishing and brought about the first legally binding accords, including notably the 2013 Agreement on Cooperation on Marine Oil Pollution Preparedness and Response in the Arctic.

In economic terms, the global downturn broke out in 2008, followed by a heavy drop in oil and gas prices, which undermined the Arctic projects economy (Lasserre, 2021). While oil prices recovered within a few months, the gas market received a fundamental blow from the US shale revolution, which turned the United States into a net exporter of gas. Simultaneously, new offshore oil and gas provinces were emerging in southern latitudes like deep-sea Gulf of Mexico, Brazil subsalt, northern Australia, and East Africa, which represented strong competitors to the Arctic shelves. Scenarios of an oil-inspired Arctic tension seemed no more feasible.

The period since 2014 reinforced these trends by adding up major global projects and systemic shocks, which distracted the politicians' and companies' attention further from the Arctic. These included:

- The situation in the world oil market, where prices fell again in 2014 and remained moderate for years, followed by the 2020 slump.
- Acute international crises across Europe and Asia (from Brexit in the West to North Korean missile tests in the East).
- Attempts to create trade and investment megablocks (US-sponsored TTIP and TPP, Russia-led Eurasian Economic Union, and China's Belt and Road Initiative), etc.

The Arctic petroleum development, especially in the shelf, suffered further from the growing climate awareness (highlighted by the 2015 Paris Agreement) and international tensions (Morgunova & Westphal, 2016). One result of the 2014 Ukraine crisis was sectoral sanctions imposed by the US, EU, and Norway against supply of equipment and services for Arctic and deep-sea drilling to Russia. Russia was thereby deprived of the badly needed offshore technologies and Western industries lost their biggest potential market.

Many companies and government agencies remain committed to the Arctic. Overall, however, the 'Arctic petroleum euphoria' that dominated the speeches of politicians and industry leaders in the 2000s has essentially gone, replaced by reduced expectations, scaled down exploration activities, and relatively few field developments.

## Country-by-Country Analysis

The aim of this section is to deliver concise sketches of national experiences of all Arctic Council member states, except for Finland and Sweden, which have no coal or petroleum resources. We will focus primarily on the country-specific issues, closely related to the domestic political and economic conditions. The presentations follow roughly the same patterns: to describe a country's Arctic possessions, their proven reserves and resource expectations, the governmental policies (including strategy papers where appropriate) and their key drivers, the actual exploration and production activities, and mid-term prospects.

## *Canada*

Canada is the second-largest Arctic nation in terms of territory, only behind Russia, and the fourth biggest oil and gas producer globally. However, its enormous Arctic petroleum potential is virtually the least developed. The Canadian North consists of three administrative territories: Yukon, Northwest Territories, and Nunavut, which also includes the vast Canadian Arctic Archipelago. The country has also claimed a huge sector of the sea and shelf areas between its northern coast and the North Pole. Governing oil and gas operations is a shared responsibility of the federal and provincial governments and subject to mandatory consultations with the First Nations. Canadian petroleum production currently concentrates on two key areas: onshore the western Provinces of Saskatchewan, Alberta and British Columbia, with a big influx of heavy bitumen tar sands; and offshore Newfoundland and Labrador in Atlantic. Operating the North Atlantic fields, including notably Hibernia and Hebron (commissioned in 1997 and 2016, respectively) has provided Canada with unique Arctic-relevant competences, such as the construction of the world's heaviest ice-resistant concrete gravity base platform or experience in iceberg management. Exploration in the Arctic itself took place between the 1960s and the early 1980s and from about 2005 to 2015, with over 500 wells drilled. The largest number of oil and gas fields, 52 in total, were discovered in Mackenzie Delta and the adjacent Beaufort Sea shelf. They contain 172.75 mcm (over 800 mbl) of recoverable crude oil and condensate, and 254.7 bcm of marketable conventional natural gas, plus immense gas hydrate resources. Potential resources are some 10 times higher (Osadetz et al., 2005). However, both times companies quit the area, citing lower oil prices, a lack of pipelines, and regulatory hurdles (Dawson, 2015).

The environmental pressure on the industry also plays a role. Inuit from a small Nunavut settlement in Baffin Island, supported by Greenpeace, appealed the decision by Canada's National Energy Board in 2014 and obtained a court ruling prohibiting 2D seismic surveys in the Baffin Bay and Davis Strait. Similarly, environmentalists had earlier challenged 3D seismic surveys offshore Beaufort Sea as potentially harmful for bowhead whales.

Contemporary approaches by Canadian authorities to Arctic oil and gas seem very restrictive, reflecting, among other things, the high priority of climate change in the energy policy (Snow, 2016). In March 2016, Prime Minister Justin Trudeau released a joint statement with US President Barack Obama, which allowed for commercial activities in the Arctic "only when the highest safety and environmental standards are met, including national and global climate and environmental goals, and Indigenous rights and agreements", plus specific, science-based standards for oil and gas development and exploration (US-Canada Joint Statement, 2016). Soon after, Trudeau introduced a five-year long moratorium on Arctic offshore operations. Characteristically, federal authorities seemed also somewhat reluctant in 2020 to help the Newfoundland offshore industry, which had been badly hit by low oil prices and the COVID-19 pandemic (Fleming, 2020).

The Arctic and Northern Policy Framework, issued by Canada's government in 2019, contains no specific goals or funding and is more of a general invitation to a co-development process with the northern territories than a finalized Arctic strategy (Kikkert and Lackenbauer, 2019). It does not mention oil and gas, and neither does the Pan-Territorial Chapter to this Framework adopted jointly by the three territories, while Nunavut indicated that petroleum activities could potentially supplement the ongoing ore mining in future (Canada's Arctic and Northern Policy, 2020).

The expiration of the above moratorium in 2021 revived this debate. However, given the predominantly negative attitudes of the authorities and the Indigenous Peoples, any large-scale petroleum activities are unlikely.

## *Denmark/Greenland*

The Danish Kingdom consists of three asymmetric elements: continental Denmark, Faroe Islands and the only Arctic part, Greenland, which geographically belongs to North America. It is the world's biggest island, with some 2.1 million sq. km of land area (mostly ice-covered) and merely 57,000 inhabitants, predominantly Inuit.

Since 2009, Greenland has enjoyed self-government in nearly all issues except foreign, security policy, and currency emission. According to opinion polls, two-thirds of the population favor secession from Denmark, but the Greenlandic economy is not self-sufficient, relying heavily on seafood (over 90% of exports) and the annual transfer from Copenhagen. Developing its abundant natural resources, like metal ore and potentially hydrocarbons, is the island's main hope for a rapid economic development and prosperity (Krivorotov, 2020). Thus, the oil and gas activities, which started in the 1970s, have for decades evolved in close connection to the nation-building (Poppel, 2018).

The USGS estimated Greenland's oil and gas resources at 31 bbl off the coast of Northeast Greenland and 17 bbl in areas between Greenland and Canada. Both the Danish Arctic Strategy adopted in 2011 by all the three parts of the realm (Kingdom of Denmark, 2011) and the first Greenlandic Hydrocarbon Strategy of 2009 were reasonably optimistic in this respect. Keen to see a rapid petroleum development, Greenland pursued an 'open doors' policy to attract investors with liberal licensing rules and modest tax rates.

Foreign companies' interest in the Greenlandic shelf was going up and down and peaked in 2011. At that moment, many industry leaders, like ExxonMobil, ConocoPhillips, Shell, DONG from Denmark, and GDF Suez, held exploration licenses for offshore blocks off the east or west coasts. NUNAOIL A/S, the Greenlandic national oil company established in 1985, also had a 12.5% share in each license. However, Cairn Energy, a medium-sized UK-based independent producer, made the principal practical efforts by drilling eight exploration wells in 2010–2011 in the Davis Strait (out of the 15 ever drilled offshore Greenland). Cairn Energy discovered only minor, non-commercial gas resources and finally withdrew from the island with losses of over $570 m.

Many companies followed suit and left Greenland soon after. Although the 2014 Hydrocarbon Strategy expected to drill one well every 2 years, actual exploration has not gone beyond seismic shooting. The number of active exploration licenses off Greenland decreased from 46 in 2014 to 11 in 2019, and exploration costs fell from roughly €600 m in 2011 to €10 m in 2018.

Until 2021, the Greenlandic Government continued its petroleum-related efforts. Premier Kim Kielsen promoted the island in Houston in early 2020. The new Hydrocarbon Strategy, approved in January 2020, implied gathering more geological data to reduce investors' risks, additional tax incentives, and a stronger focus on inviting smaller companies to onshore exploration (Naalakkersuisut, 2020). However, experts doubted whether Greenlandic resources, which were likely to take decades to develop, would ever find their niche in the future oil market (Kristensen, 2020). Ultimately, under the new Government, which came to power in 2021, Greenland halted further oil exploration citing climate concerns, decided to join Paris Agreement and became a co-founder of Beyond Oil And Gas Alliance of nations favoring phase-out of oil and gas production.

## *Iceland*

Situated on the Mid-Atlantic ridge, Iceland has almost no sedimentary basins on- or offshore. The only exception is the Dreki area in its north-eastern shelf, which in geological terms may be similar to the North Sea. The Icelandic authorities opened Dreki for exploration after the 2008 bank crisis, which nearly ruined the national economy. However, the first licensing round, held in 2009 by Orkustofnun, the energy authority, failed as there were merely two bids submitted and later revoked by three medium-sized Norwegian and Swedish companies. The main reasons were, apparently, the 2008 oil price shock, unclear legislation, and insufficient marketing (Gilja, 2009).

During the second and the third rounds in 2013–2014, Orkustofnun issued three licenses to consortia of Icelandic, British, Norwegian, and Chinese companies (Iceland divvys… 2014; Kliewer, 2014). However, no sizeable exploration followed, the licenses were relinquished, and future prospects remain doubtful.

## *Norway*

The Norwegian Arctic includes the two northernmost counties, Nordland and Troms og Finnmark, the high latitude archipelago of Spitsbergen (Svalbard) and continental shelves and economic zones, which are several times bigger than Norway's land territory. While the national petroleum production history is 50 years long, its Arctic piece only started in the twenty-first century. All of Norway's oil and gas resources are found in the continental shelf, which consists of three distinct provinces: the North, Norwegian, and Barents Seas. The North Sea south of 62°N was opened for exploration and production by a single resolution in 1965, but any movement further north was subject to careful impact assessments. The first well in the Norwegian Sea was drilled in 1981 and in the Barents Sea in 1987 only. Norne, the first field developed off the coast of Nordland (still south of the Arctic Circle), was launched in 1997.

In 2001, Kjell Magne Bondevik's minority government established an environmentally motivated moratorium on Arctic oil and gas operations. The moratorium was lifted 3 years later, but the valuable Lofoten-Vesterålen area in the Norwegian Sea is still protected by the powerful fishers' lobby. Since then, the Norwegian Arctic offshore operations have concentrated in the Barents Sea, plus a few licenses south of Lofoten.

Contemporary Norwegian petroleum policy seeks to strike a delicate balance between industrial and environmental goals. On one hand, there is a strong desire across the party lines to make Norway a global forerunner in meeting the Paris climate goals and promoting green energy like wind power and hydrogen. Influential environmental NGOs criticize the petroleum industry as the main source of greenhouse gas emissions. They have even challenged, albeit unsuccessfully, the results of the 23rd licensing round in 2016, when 40 blocks in the Barents Sea were awarded, with the Supreme Court (so called "climate lawsuit") (Supreme Court of Norway, 2020).

On the other hand, the petroleum sector has created over 200,000 jobs (including the service industries) and brought enormous budgetary revenues, unparalleled among the developed nations (Noreng, 2018). The national sovereign wealth fund has about $1 trn on its accounts, and seven out of 10 Norwegians want the oil age to continue.

Moving into the Arctic is an important tool to reach this goal. As Table 13.1 demonstrates, while the North Sea dominates heavily in the daily production and proven reserves, nearly two-thirds of the undiscovered resources are in the Barents Sea. The 2010 delimitation treaty with Russia increased these hopes by adding 87,600 sq. km to the Norwegian shelf.

The Arctic petroleum issue also has strong foreign and regional policy undertones. Norway was the first country to launch a High North strategy in 2006, aiming

**Table 13.1** Key figures on Norwegian continental shelf (end 2020)

|  | North Sea | Norwegian Sea | Barents Sea | NORWAY TOTAL |
|---|---|---|---|---|
| Number of fields in operation | 67 | 21 | 2 | 91 |
| Number of fields under development | 6 | 2 | 1 | 9 |
| Production in 2020 | 166.1 | 54.3 | 5.6 | 226.0 |
| Cumulative production | 6280.3 | 1403.2 | 92.6 | 7776.1 |
| Remaining reserves | 2022.6 | 428.0 | 276.4 | 2727.0 |
| Contingent resources | 637.5 | 342.1 | 184.8 | 1428.4 |
| Undiscovered resources | 665.0 | 665.0 | 2505.0 | 3835.0 |

*Sources*: Resource Accounts as of 31 December 2020 https://www.npd.no/en/facts/resource-accounts-and-analysis/resource-accounts-as-of-31-december-2020/; Fields https://www.norskpetroleum.no/fakta/felt/

*Notes*: Production and resources are indicated in million standard cubic meters (scm) oil equivalent, 1 scm = 6.29bl = 0.84 t crude oil = 1000 scm natural gas at a global leadership in Arctic research, technologies, and governance (Government of Norway, 2006). With the Snøhvit gas field under development and the unique Norwegian experience of operations under the harsh northern climate, the oil and gas industry came naturally into focus. The center-left coalition cabinet of Jens Stoltenberg, a strong proponent of accelerated petroleum development, declared the Arctic the key national priority and created expectations of a new "Persian Gulf in the Barents Sea". Several editions of the strategy followed, but the principal approaches have remained unchanged despite the power shift in late 2013.

The role attributed to Arctic oil and gas activities as a regional driver has varied somewhat over the years, reflecting the progress in field development and licensing, varying exploration results and respective revisions of resource estimates. The latest edition of the strategy, submitted by Erna Solberg's conservative cabinet in late November 2020, underlines the important ripple effects of the recent petroleum development in the North. The government has pledged to facilitate a profitable oil and gas production by maintaining predictable framework conditions and holding regular licensing rounds on new acreages, plus to continue petroleum relevant R&D and competence development in and for the High North (Government of Norway, 2020).

Every field development in the Norwegian Arctic has been a technological and industrial breakthrough, reinforcing the unmatched national leadership in cold climate offshore solutions. Characteristically, all but one of these developments are operated by Equinor (previously Statoil), the 67% state-owned national champion (Krivorotov & Finger, 2019). The Norwegian authorities also used their roles as the regulators and the key shareholder in Equinor to maximize the regional effects for creating jobs and local businesses in Northern Norway (Arbo & Hersoug, 2010).

Snøhvit, the only Norwegian field to get earmarked tax breaks, was commissioned in 2007 despite many technical and business troubles, becoming the first-ever Arctic deposit developed far from the shoreline. It uses a subsea completion and holds the world record in multiphase flow to shore (147 km). Snøhvit gas is liquefied under a proprietary technology at the Melkøya facility, the first LNG plant in Europe and the first north of the Arctic Circle. The project shareholders notably include Neptune Energy, a Chinese-controlled producer. Goliat oilfield, commissioned by Italian Eni in 2016, applies a unique Norwegian-designed cylindrical platform, which is the world's largest floating production, storage, and offloading unit (FPSO).

Aasta Hansteen project was launched in 2018 in the northern part of the Norwegian Sea at water depths of 1270 m. It includes a SPAR type platform, which became the first in Norway and the world's biggest, and the new Polarled pipeline, the first-ever to bring Norwegian Arctic gas to Europe. The large Johan Castberg oilfield (about 560 mbl), to be delivered in 2024, is set to become the northernmost field in Norway. Norway's first-ever FPSO ship and a large subsea production system will be deployed. By applying advanced technologies, Equinor managed to cut the project cost by half and the breakeven oil price from 80 to $35/bl within 3 years (Equinor, 2017). Johan Castberg shall produce oil in 30 years according to the plan, while the Snøhvit lifecycle may be extended till 2050. However, Johan Castberg is, so far, the last ongoing Barents Sea development. Contrary to expectations, the massive seismic shooting and exploratory drilling in the area have resulted in few Arctic discoveries like Wisting, Alta and Gohta. Their development prospects remain unclear, particularly under low oil prices.

In 2020, a year marked by oil market shocks, the authorities delivered clear signals of their desire to enhance exploration and production, like major tax breaks introduced in June and the controversial 25th licensing round announcement in November. While the Norwegian Government was still awaiting the Supreme Court ruling on the "climate lawsuit", it invited companies to bid for a record number of 136 Arctic blocks (Staalesen, 2020).

Given the positions of all the leading political parties and the majority of the voters, Norway will continue to seek opportunities for Arctic petroleum development despite the environmental lobby. The new Cabinet which took the office in October 2021 confirms this commitment. The uncertain resource base and the market situation will be the main exogenous parameters. While there is also Norwegian and Russian coal mining in Spitsbergen, the output is decreasing, as both nations focus more on developing alternative businesses like tourism.

## Russia

The Arctic Zone of the Russian Federation (AZRF), first defined in 2014, (President of Russian Federation, 2014) covers over 5 million sq. km, or some 30%, of the Russian landmass, plus its vast economic zones and continental shelf claimed up to the North Pole. Contrary to other Arctic countries, AZRF is economically self-sufficient. Inhabited by 2.5 million people (1.8% of the national population), it represents over 6% of the GDP and about 20% of the budgetary incomes, thanks to its enormous petroleum and mineral resources. According to the official estimates as of late 2019, the Russian Arctic reserves amounted to 7.3 bt (53.3 bbl) of oil and 55 tcm (1845 tcf) of natural gas, therein 41% offshore (TASS, 2019).

Despite being a party to the Kyoto Protocol and now to Paris Agreement, Russia has, until recently, not been active in climate debates, since the massive closedowns in its heavy industry after 1990 left the country with large unfilled emissions quotas. Now, it pays growing attention to climate issues, such as by including them into the priorities of its Arctic Council chairmanship and adopting a national low-carbon development strategy through 2050. Nevertheless, the authorities maintain that an environmentally safe exploitation of Arctic resources is both achievable and compatible with sustainable development goals. The Russian national Energy and Arctic Strategies through 2035, both adopted in 2020, envisage a stabilized, eventually a lower oil output over the period, while coal production shall grow between 10% and 52%, natural gas extraction between 18% and 35%, and gas liquefaction by a factor of 4–7.2, with a major AZRF contribution (Table 13.2). Western Siberia, the main petroleum producing area since the 1970s, has entered into a mature phase, the exploration and production activities move gradually to Eastern Siberia, the Far East, and the Arctic. The Government supports this development with tax incentives. The AZRF

**Table 13.2** Resources production goals, Russian Arctic and Energy Strategies through 2035

|  | 2018 (base level) | 2024 | 2035 |
|---|---|---|---|
| Oil incl. gas condensate, mt | 555.9 | 555–560 | 490–555 |
| Therein in AZRF, percent | 17.3 | 20 | 26 |
| Natural gas, bcm | 727.6 | 795.1–820.6 | 859.7–1000.7 |
| Therein in AZRF, percent | 82.7 | 82 | 79 |
| Liquefied natural gas (LNG), mt | 18.9 | 46–65 | 80–140 |
| Therein in AZRF, mt | 8.6 | 43 | 91 |
| Coal, mt | 439.3 | 448–530 | 485–668 |

*Sources*: President of the Russian Federation, 2020; Government of the Russian Federation, 2020

share in the national oil and LNG production shall increase significantly through 2035. Natural gas production is already concentrated in the Arctic region of Yamal Nenets Autonomous Area, where it also moves even further north.

The last decade has witnessed several breakthrough projects onshore Russian Arctic, often in cooperation with foreign investors. Large-scale field developments are accompanied by a massive construction of new pipelines, railways, ports, airports, housing, icebreakers, and ice-class tankers.

Gazprom, the Russian natural gas company, has launched its Yamal Megaproject starting with the unique Bovanenkovo field with proven reserves of 4.9 tcm. Commissioned gradually in 2012–2018, this field alone has a capacity of delivering 115 bcm/y, which is one-fifth of Gazprom's entire production. In 2019, Gazprom made the next step north and started developing the Kharasavey gas field, with a lifecycle to stretch through the twenty-first century. The total potential gas output in Yamal is estimated at 310–360 bcm/y (Gazprom, 2020). Besides supplying Arctic gas to Central Russia and Europe, Gazprom negotiates constructing a pipeline from Yamal Nenets Area to China (the so-called Western route, to complement the existing Power of Siberia line).

Further north of these fields, Yamal LNG company commissioned a large gas liquefaction plant in Sabetta. The plant reached its full capacity of 16.5 mt/y of LNG in late 2018, a year ahead of schedule. Yamal LNG is a joint venture of Novatek, Russian private gas producer (50.1%), Total from France (20%), and two Chinese entities, China National Petroleum Corporation (20%) and Silk Road Foundation (9.9%).

Gazprom Neft, Gazprom's oil subsidiary, delivered its Novy Port project in the east coast of the Yamal Peninsula in 2014, which became the first in Russia to win the prestigious Excellence in Project Integration award. There are large expansion plans; the production license for Novy Port field has been extended till 2150.

Within a decade, these projects created a brand-new petroleum province of global scale in Yamal, which will produce for decades to come and is still expanding. A similar development has taken place in the adjacent Gydan Peninsula, with the northernmost oilfield in operation in Russia, Vostochno-Messoyakhskoye, commissioned by Gazprom Neft and Rosneft, the Russian national oil company, in 2016, and Novatek's ongoing Arctic LNG 2 project. The massive construction under these projects and the export of their products have contributed to achieving another national goal, a dramatic increase of navigation along the Northern Sea Route (NSR) (Grigoryev, 2019). Its cargo turnover grew from 2 mt in 2010 to 33 mt in 2020, with plans to reach 90 mt in 2030 and 130 mt in 2035.

Rosneft implements another Arctic megaproject in the Taimyr Peninsula, thus establishing a new oil province in the Eastern Siberia. It started with Vankor oilfield commissioned in 2009, with an annual production above 22 mt. Oil from Vankor is transported south, to ultimately feed into the East Siberia-Pacific Ocean (ESPO) trunk pipeline to China. Indian companies hold 49.9% of Vankor shares.

Rosneft works now on a massive expansion, the Vostok Oil project, to develop several fields north and east of Vankor. Their resource potential exceeds 5 bt (36.5 bbl) of oil, and total investments are estimated at 10 trn rubles, or about €115 bn. These fields will facilitate production and shipment of up to 100 mt/y (2 mb/d). Rosneft has pledged to deliver 30 mt (220 mb) of oil from Taimyr via the NSR by 2024 (Rosneft, 2020). Trafigura, a Singapore-based trading company, purchased 10% of Vostok Oil in late 2020, and a consortium of two more traders, Vitol (Netherlands) and Mercantile & Maritime Energy (Singapore), secured a purchase of another 5% in October 2021. Indian and Chinese companies have also indicated their interest. The new Arctic oil and gas projects are some of the biggest post-Soviet industrial developments in entire Siberia (Kryukov et al., 2020).

Meanwhile, the status and prospects for the Russian Arctic shelf are modest, despite its enormous proven reserves and potential resources estimated at 113.9 mtoe. There are several reasons for that, including heavy ice conditions, insufficient infrastructure along the Arctic coast, a lack of proprietary technologies and experience in developing continental shelf, legal restrictions on foreign involvement, and Western sectoral sanctions.

The Russian laws, as amended in 2008, confine the potential Arctic shelf operators to state-controlled enterprises, Gazprom (therein Gazprom Neft) and Rosneft, which hold some 140 licenses for 85% of the nation's Arctic shelf. In 2016, however, the Russian Subsoil Agency introduced a moratorium on further awards until the companies comply with the obligations under their existing licenses.

Prirazlomnoe is the only Russian Arctic offshore field in operation,[2] developed with a unique ice-resistant gravity base platform in the shallow waters of Pechora Sea. It was commissioned by a Gazprom Neft subsidiary in 2013, after major technical delays. The field contains over 70 mt of oil and may produce up to 5.5 mt/y, the actual figure for 2019 was 3.14 mt.

All other major projects remain either frozen or in a preparatory phase. Gazprom, Total, and Statoil had an extensive cooperation in 2008–2015 on the unique Shtokman field in the Barents Sea with proven gas reserves of 3.9 tcm. The project

---

[2] Formally speaking, Novatek operates Yurkharovskoe nearshore shelf field in Yamal, with peak production of 38.9 bcm in 2014, but the entire development was done by horizontal wells drilled from land.

stalled in 2012 after the shale gas revolution in the USA, which had originally been the target market. Rosneft and ExxonMobil, which discovered the giant Pobeda (Victory) oil and gas field in the Kara Sea in 2014, had to stop the exploration and ultimately dissolve their JV due to the newly imposed US sanctions.

At present, Gazprom plans to develop several nearshore fields in the Ob bay, plus Russian companies continue Arctic shelf exploration onboard Russian and Chinese vessels. Gazprom and Rosneft discovered six new deposits offshore the Barents and Kara Seas in 2019–2020, including the unique V.A. Dinkov (391 bcm) and Marshall Rokossovsky (514 bcm) gas and condensate fields. Still, the companies are cautious about their production plans and acknowledge the need for international cooperation. To this end, the Russian government finalized a bill in late 2020 to lift the bar for private (both Russian and foreign) companies to work in the Arctic shelf.

Russia also continues Arctic coal mining. Komi Republic government plans to maintain the present production level of 12–15mt/y through 2030 (MacKenzie, 2020), new projects are mulled in Chukotka and Taimyr. In the most optimistic case, however, Arctic mines will represent just 5–6% of the nationwide production.

## *United States*

Alaska is the only Arctic part of the USA and one of its principal oil and gas producers. The large-scale development started in the North Slope in 1968 with the discovery of Prudhoe Bay, the largest field in the North America with initial reserves of over 25 bbl of oil and condensate. Prudhoe Bay was commissioned in 1977 together with the newly built 1287 km-long Trans Alaska Pipeline System (TAPS) to the ice-free port of Valdez in the south coast.

Some 16 bbl of oil were produced at Prudhoe Bay in the first 40 years of operation. The field alone has provided about 80% of all Alaskan oil and is poised to work for years ahead. Besides, there is (or has been) oil and gas production at several other fields in the North Slope and in the Cook Inlet area with among other Kenai LNG, the world's eldest liquefaction plant in operation (since 1969). The first-ever Arctic offshore field developments have also taken place in the Beaufort Sea in shallow waters several miles off Alaska's north coast, using artificial gravel islands like Endicott (1987) and Northstar (2001).

The North Slope production peaked in 1988 at 1.97 mbl/d, when Alaska supplied about one-quarter of the entire US crude. This enormous production changed the entire life of Alaska, providing dozens of billion US dollars in public revenues to the state budget and Indigenous Peoples' corporations. Since then, however, it started decreasing to the present-day plateau of some 500,000 b/d (Alaska Department of Revenue, 2017; Ragsdale, 2008).

There has been strong lobbying both in favor and against an expansion of oil and gas operations in Alaska, for financial, political, and technical reasons. The discussions tend to concentrate around three main, rather controversial topics.

The first regards licensing (leasing in US terms) new onshore acreage in the North Slope, where most lands lie within the National Petroleum Reserve-Alaska (NPR-A) and Alaska National Wildlife Refuge (ANWR), strongly protected by federal laws and local environmentalists. The other relates to new offshore lease sales. The third concerns Alaska LNG, a megaproject estimated at $43.4 bn, which includes development of a big gas field at Point Thompson in the North Slope, constructing a 1300 km long gas pipeline to the Cook Inlet and a large LNG plant in Nikiski with a total output capacity of 20 mt/y.

All three issues have long histories, reflecting both the market fluctuations, progress in exploration, and changes in the federal government's approaches, which are more partisan in the United States than in any other Arctic nation. George W. Bush (US President in 2000–2008), who had close ties with the Texan oil business and whose presidency had a major impact of the September 11, 2001 terrorist attacks, was a strong proponent of leasing more of Alaskan OCS[3] acreage and drilling in the ANWR, quoting the national energy security needs. Barack Obama (president from 2008–2016), by contrast, was critical to the oil and gas industry. He set ambitious climate goals, promoted renewable energy, and eliminated numerous tax breaks for oil companies. The *Deepwater Horizon* accident in 2010 reinforced these approaches strongly. US Secretary of Interior Ken Salazar welcomed the 2009 court ruling, which vacated the entire 2007–2012 OCS leasing program, ruling that Bush Administration officials had not conducted sufficient scientific and environmental analysis before scheduling the lease sales off Alaska. Under President Obama, most of the Alaskan shelf was exempted from leasing.

The unsuccessful corporate efforts also discouraged investors. Shell had invested about $7 bn in purchasing lease rights and preparing the fleet in the Beaufort Sea, but its 2012 drilling was cut short, and in 2013 the *Kulluk* drilling rig came ashore north of Alaska, illustrating the high HSE risks. Shell finally withdrew from Alaska and closed down the rep office there in 2015, and most oil companies followed the path within the next few months. Onshore, Alaska LNG stalled as well. ExxonMobil, the operator, left the project and the state in 2016.

Donald Trump (president 2016–2020) excluded the US from the Paris Agreement and was a strong lobbyist of unbundling the Alaskan petroleum potential. During his presidency, Eni began oil exploration drilling in the Beaufort Sea in 2017, Shell indicated that it considered resuming its operations, and ConocoPhillips' plan to develop Willow prospect in the NPR-A received approval. Trump's first visit to Beijing in 2017 was marked by a signing of an MoU between the Alaskans (the state government and the operator) and a group of Chinese companies on participation in Alaska LNG. He also fought to resume lease sales in ANWR and arranged an auction in January 2021, days before leaving the office.

---

[3] OCS stands for the US Outer Continental Shelf, a federal jurisdiction, which covers all shelf areas beyond the states' territorial waters.

Not all of these efforts were successful, however, due to low oil prices and the refusal by many leading banks to extend financing to ANWR drilling (Drill music, 2020). BP, the operator of Prudhoe Bay, withdrew from the field in 2019, and the Alaska LNG deal with the Chinese suffered from the US–China trade war.

President Joe Biden halted any ANWR drilling just hours after his inauguration on January 20, 2021 (Schreiber & Rosen, 2021). In broader terms, his climate-oriented agenda and $1.8 bn clean energy plan may complicate oil and gas development nationwide (Blackmon, 2020).

Under the circumstances, it would be reasonable to expect new attempts to revitalize Alaskan petroleum business in the longer run, but the political and market uncertainties will complicate the development.

## Conclusions and Future Outlook

As our analysis demonstrates, while Arctic oil and gas have remained the focus of the political and industry debates since early 2000s, the actual progress has been rather limited and uneven, particularly in the shelf. What seemed to be a new era in Arctic development around 2005 now looks more like yet another 'tidal wave', which is generally descending, with few tangible results left. Russia is the only country to have approved field development plans beyond 2024.

There are serious reasons for that, including oil price fluctuations, mixed exploration results, growing environmental concerns, changes in the banks' policies, government-imposed moratoria, and negative court rulings. Overall, the Arctic oil and gas has recently become significantly less attractive from both political and commercial viewpoints. The COVID-19 pandemic has further exacerbated the problems by reducing the global petroleum demand and intensifying international tensions. The environmental impacts of the 2020 lockdown, when the atmosphere improved dramatically over the EU, Russia, or China, is also sure to enhance the global climatic focus on fossil fuels.

This being said, the present wave of renewed interest to the Arctic has already brought about substantial changes in the public awareness of the region and of its resource potential, in the respective political perceptions, technological approaches, and business practices. The contemporary Arctic oil and gas industry has a number of distinct structural features, which will eventually shape its future development in the medium term.

In the area of public attitudes, we witness an unprecedented, environmentally motivated pressure on the industry. The past decade witnessed the IMO ban on use of heavy fuel oil in the Arctic, increased focus on black carbon emissions, and massive protests by environmentalist NGOs (including Greenpeace attempts to board platforms offshore Greenland, Russia and Norway) and by Indigenous Peoples who have gone the legal way to stop petroleum exploration in Arctic Canada and Alaska. The debates on climate change, which is twice as intensive in the Arctic as it is for the world overall, have made petroleum industry no longer 'trendy' in most Arctic nations – even in Norway, where it is the key taxpayer and employer.

European oil and gas companies adapt to this challenge by transforming into broadly defined energy businesses with a strong focus on renewables, which often includes corporate renaming and massive rebranding campaigns. What seems the key goal for the petroleum industry, however, is to restore its positive image and, in more general terms, to protect the very idea of a sustainable Arctic resources development. Milestone megaprojects, like those underway in Norway and Russia, serve as highly relevant model cases. Advanced technologies, including smart wells, carbon capture and storage, and diesel/renewable cogeneration solutions for remote Arctic settlements, are likely to gain increasing popularity.

In the corporate structure, there is a strong trend towards a broader variety of actors, both in terms of their size and nationality. The global majors, which dominated the Arctic oil and gas industry until the late 1990s, are now successfully challenged by national state-owned champions and medium-sized companies. The recent legal amendments suggested in Russia to attract private investors to its Arctic shelf fit well into this trend.

Companies from Asian countries, especially from China, are increasingly active in the Arctic, backed strongly by their governments. The Chinese investors, which are predominantly interested in delivering the extracted resources to their home country, establish new value chains and logistic lines to connect the Arctic to Asia. The Western sectoral sanctions, which are likely to be long-lasting, have given additional impetus to this development by making Russia rely more on Asian countries as potential markets and as sources of financing and technologies.

On the government side, we also see a broad spectrum of approaches. The policies pursued by Arctic nations vary widely depending largely on two key factors: the importance of petroleum industry for the national economy and the country's interpretation of climate policy goals.

Norway and Russia have demonstrated a strong political will to continue large-scale Arctic exploration and production. In the US, Alaskan oil and gas development has evolved into a complicated domestic policy issue, with clear division lines between the two parties. This is negative for the industry, given its lengthy investment cycles and need for a long-term regulatory stability. Canadians remain predominantly negative to Arctic oil and gas, while Denmark/Greenland and Iceland suffer from discouraging exploration results.

Summing up, the circumpolar petroleum development is likely to be increasingly heterogeneous, with multifaceted investment regimes, stringent sustainability requirements, a stronger emphasis on technological and managerial innovations, a growing involvement of non-regional actors and continued environmental debates. The new actors may make the industry more robust by contributing with their unique technologies, know-how, and marketing channels.

In the longer run, the ongoing low-carbon transition will eventually become the single most important fundamental trend to affect the industry development. For decades, oil business relied on M. King Hubbert's 'peak oil' theory, which predicted that the global production would reach a maximum at some point and then start to fall, for geological reasons (King Hubbert, 1956). Together with the projected permanent growth in the oil demand, this created expectations of a continuous rise in oil prices that would justify the development of marginal resources, like those in the Arctic.

In the last decade, however, the rapid growth in solar and wind generation, hydrogen energy, smart grids and homes, etc. has stabilized oil consumption in the EU and North America. Global oil demand is powered mainly by China and India. Instead of the traditional 'peak oil', energy researchers have started talking about an upcoming 'peak oil demand', arguing more about its specific timing (most predictions range between 2028 and 2040). The world oil trade will increasingly turn into a buyer's market, with a harsher competition among suppliers and a general downward price trend. The market will also change in geographic terms. The massive increase of electric and hybrid cars in Europe and North America may result in a preemptive export of used gasoline- and diesel-driven cars to developing countries. The center of gravity of the global oil demand will squeeze respectively to southern latitudes, further away from the Arctic.

These circumstances jeopardize future resource development. When asked by the author in November 2020 if the new market trends would ruin Arctic petroleum projects, Per Magnus Nysveen, senior partner and head of analysis of the authoritative Norwegian Rystad Energy consulting company, responded: "Yes, this is the supply source that we expect would be first out and most at risk." The long-term prospects of the Arctic coal industry, given its heavy carbon footprint, are even riskier, while natural gas may be used longer as the cleanest hydrocarbon, among other things for LNG bunkering.

This is in no way a final judgement. An eventual discovery of large commercial resources in the Arctic, politically motivated decisions, an oil price hike caused by a major international crisis, progress in technologies, etc. may seriously change the industry prospects. On the whole, however, it appears that it will no longer be the principal economic pillar.

While Arctic fuels production is, or may become, very important for some countries, subnational regions, and companies, it is not going to play any material role in the global markets. Even under the most optimistic scenarios, the entire circumpolar production will represent merely a few per cent of the world total. Arctic oil and gas industry by all proximity will concentrate on specific geographic areas and serve as a marginal niche producer for target markets.

# References

Alaska Department of Revenue. (2017). *Production history and forecast by production area from Fall 2017 RSB*. Retrieved from http://www.tax.alaska.gov/sourcesbook/AlaskaProduction.pdf

Arbo, P., & Hersoug, B. (Eds.) (2010). *Oljevirksomhetens inntog i nord. Næringsutvikling, politikk og samfunn* [Petroleum industry on the march to the north. Business development, politics and society]. Gyldendal, Oslo.

Bird, K., Charpentier, R., Gautier, D., Houseknecht, D., Klett, T., Pitman, J., Moore, T., Schenk, C., Tennyson, M., & Wandrey, C. (2008). Circum-Arctic resource appraisal: Estimates of undiscovered oil and gas North of the Arctic Circle. *United States Geological Survey, Fact Sheet* 2008–3049.

Blackmon, D. (2020, December). *In a Biden/Harris Presidency, any fracking ban is least of industry's worries*. World Oil (pp. 43–44.)

Canada's Arctic and Northern Policy. (2020). Retrieved from https://www.rcaanc-cirnac.gc.ca/eng/1562782976772/1562783551358

Conley, H., et al. (2013). *Arctic economics in the 21st century. The benefits and costs of cold*. CSIS, Rowman & Littlefield.

Dawson, C. (2015). Exxon Mobil, BP Suspend Canadian Arctic Exploratory Drilling Program in Beaufort Sea. *The Wall Street Journal*, June 26.

Drill Music. (2020). *The economist*, August 22: 34.

Equinor. (2017). *How we cut the break-even price from USD 100–USD 27 per barrel*. Retrieved from https://www.equinor.com/en/magazine/achieving-lower-breakeven.html

Fairhall, D. (2010). *Cold front: Conflict ahead in Arctic waters*. Tauris, L.

Fleming, C. (2020). Canadian offshore industry in crisis mode. *World Oil*, June 2020: 33–36.

Gazprom. (2020). *Yamal*. Retrieved from https://www.gazprom.com/projects/yamal/

Gautier, D., Bird, K., Charpentier, R., et al. (2009). Assessment of undiscovered oil and gas in the Arctic. *Science*, *324*, 1175–1179.

Gilja, A. (2009). *Islandske overraskelser* [Icelandic surprises]. Retrieved from http://www.offshore.no/nyheter/sak.aspx?id=25132

Government of Norway. (2006). *The Norwegian Government's High North strategy*.

Government of Norway. (2020). *Meld. St. 9 (2020–2021) Mennesker, muligheter og norske interesser i nord* [Report to the Storting (white paper) People, opportunutites and Norwegian interests in the North].

Government of Russia. (2020). Энергетическая стратегия Российской Федерации на период до 2035 года [Energy Strategy of the Russian Federation through 2035]. Approved by Directive of June 9, 2929 No. 1523-r.

Grigoryev, M. N. (2019). Logistical schemes for mineral commodities year-round shipment in the water area of the Northern Sea route. *Arctic Herald, 2*(27), 84–95.

Howard, R. (2009). *The Arctic gold rush: The new race for tomorrow's natural resources*. Continuum, L.

Iceland divvys final Dreki license. (2014). Retrieved from https://www.oedigital.com/news/456856-iceland-divvys-final-dreki-license

Kikkert, P., & Whitney Lackenbauer, P. (2019). Canada's Arctic and Northern policy framework: A roadmap for the future? In L. Heininen, H. Exner-Pirot, & J. Barnes (Eds.), *Redefining Arctic security: Arctic yearbook 2019* (pp. 322–331). Arctic Portal.

King Hubbert, M. (1956). *Nuclear energy and the fossil fuels*. Retrieved from https://web.archive.org/web/20080527233843/http://www.hubbertpeak.com/hubbert/1956/1956.pdf

Kingdom of Denmark. (2011). *Denmark, Greenland and the Faroe Islands: Kingdom of Denmark Strategy for the Arctic 2011–2020*.

Kliewer, G. (2014). CNOOC wins Iceland license. *Offshore, 74*(3), 24.

Kristensen, K. (2020). Grønland er for sent ude [Greenland runs late]. *Sermitsiaq, 9*, 15.

Krivorotov, A., & Finger, M. (2019). State-owned enterprises in the Arctic. In M. Finger & L. Heininen (Eds.), *The GlobalArctic handbook* (pp. 45–62). Springer.

Krivorotov, A. (2020). North Atlantic in China-U.S. relations. In E. Safronova (Ed.), *China in world and regional politics (History and modernity)* (pp. 168–183). Issue XXV. Institute of Far Eastern Studies, RAS.

Kryukov, V. A., Lavrovskii, B. L., Seliverstov, V. E., Suslov, V. I., & Suslov, N. I. (2020). Siberian development vector: Based on cooperation and interaction. *Studies on Russian Economic Development, 31*(5), 495–504.

Lasserre, F. (2021). Exploitation des ressources naturelles dans l'Arctique. Une évolution contrastée dans les soubresauts du marché mondial. *Études du CQEG* n°3.

Mackenzie, R. (2020). *Komi intends to maintain coal production at the level of 12–15 million tons until 2030*. Retrieved from http://www.rusmininfo.com/news/22-12-2020/komi-intends-maintain-coal-production-level-12-15-million-tons-until-2030

Morgunova, M., & Westphal, K. (2016). *Offshore hydrocarbon resources in the Arctic. From cooperation to confrontation in an era of geopolitical and economic turbulence?* (SWP research paper RP 3). German Institute for International and Security Affairs.

Naalakkersuisut. (2020). Redegørelse til Inatsisartut vedrørende olie/gas-aktiviteter i Grønland. Per 23. juni 2020. [Report to the Inatsisartut (Parliament of Greenland) regarding oil and gas activities in Greenland. As of June 23, 2020]. Retrieved from https://naalakkersuisut.gl/da/Publikationer?pn=2

Noreng, Ø. (2018). *Oljeboblen. Enestående muligheter, forspilte sjanser* [The oil bubble. Outstanding opportunities, lost chances]. Gyldendal.

Osadetz, K. G., Morrell, G. R., Dixon, J., et al. (2005). Beaufort Sea–Mackenzie Delta basin: A review of conventional and nonconventional (gas hydrate) petroleum reserves and undiscovered resources. *Geological Survey of Canada Bulletin, 585*, 1–19.

Poppel, B. (2018). Arctic oil & gas Development: The case of Greenland. In L. Heininen & H. Exner-Pirot (Eds.), *Arctic yearbook 2018* (pp. 328–360). Northern Research Forum.

President of the Russian Federation. (2014). О сухопутных территориях Арктической зоны Российской Федерации [On Land Territories of the Arctic Zone of the Russian Federation]. Decree of May 2, 2014 No. 296.

President of the Russian Federation. (2020). Стратегия развития Арктической зоны Российской Федерации и обеспечения национальной безопасности на период до 2035 года [Strategy for the Development of the Russian Federation Arctic Zone and Protecting National Security through 2035]. Approved by Decree of October 26, 2020 No. 645.

Ragsdale, R. (Ed.). (2008). *Harnessing a Giant: 40 years at Prudhoe Bay*. Petroleum News.

Rosneft. (2020). *Igor Sechin Reports to President on Implementation of Promising Projects Vostok Oil and Zvezda Shipyard*. Retrieved from https://www.rosneft.com/press/today/item/204069/

Schreiber, M., & Rosen, Y. (2021). *Biden halts oil and gas development in Arctic refuge hours after inauguration*. Retrieved from https://www.arctictoday.com/biden-halts-oil-and-gas-development-in-arctic-refuge-hours-after-inauguration/

Snow, N. (2016). Canada's NEB announces changes as it issues energy future 2016 report. *Oil & Gas Journal, 14*, 17–19.

Staalesen, A. (2020). *Despite a looming climate lawsuit, Norway offers oil companies a fresh set of blocks*. Retrieved from https://www.arctictoday.com/despite-looming-climate-lawsuit-norway-offers-oil-companies-a-fresh-set-of-blocks/

Supreme Court of Norway. (2020). *Judgment in the "climate lawsuit"*. Retrieved from https://www.domstol.no/en/Enkelt-domstol/supremecourt/nyhet/2020/judgment-in-the-climate-lawsuit/

TASS. (2019). *Russia's Arctic oil reserves estimated at 7.3 bln tonnes*. Retrieved from https://tass.com/economy/1088516

U.S.-Canada Joint Statement on Climate, Energy, and Arctic Leadership. (March 10, 2016). *The White House, President Barack Obama*. Retrieved from https://obamawhitehouse.archives.gov/the-press-office/2016/03/10/us-canada-joint-statement-climate-energy-and-arctic-leadership

Wilson Rowe, E. (2018). *Arctic governance. Power in cross-border cooperation*. Manchester University Press.

**Andrey Krivorotov** is an assistant professor in Odintsovo branch, Moscow State Institute of International Relations (MGIMO University), an advisor (since 2008) to the CEO of Gazprom, an expert (since 2012) of the Russian International Affairs Council (RIAC) and a member of the International Arctic Social Science Association (IASSA) since 2016.

# Chapter 14
# Arctic Fisheries in a Changing Climate

**Franz J. Mueter**

## Introduction

Changes in sea ice cover, warmer ocean temperatures, and associated changes in ocean currents have important impacts on biological and chemical processes in the Arctic Ocean and its marginal seas (Mueter et al., 2021; Wassmann, 2011). Impacts are especially pronounced on the major Arctic inflow shelves, where northward currents from the Pacific and Atlantic transport nutrients and plankton into the Arctic. These regions, including the Bering and Chukchi seas, the Barents Sea/Fram Strait region, and the West Greenland shelf, are the major gateways into the Arctic and form a transition zone between Subarctic and Arctic marine ecosystems. They are also highly productive and provide important ecosystem services by supporting large gadid (cod-like fishes), flatfish, and crab stocks. In addition to commercially important stocks, they support seabirds and marine mammals that are critical to food security for Indigenous and local communities.

Approximately 15% of the world's marine fishes are caught in the Subarctic and Arctic (Zeller et al., 2016), and a large majority of these catches occur in the southern portions of the Arctic inflow shelves (Fig. 14.1). By contrast, the interior Arctic shelves and the Central Arctic Ocean (CAO) support mostly small but important subsistence catches (Zeller et al., 2011). The inflow shelves are also the most likely locations for southern or 'boreal' species expanding into the Arctic (Carmack & Wassmann, 2006), but the permanent expansion of fish into the Arctic depends on the availability of suitable habitats for spawning and feeding (Hollowed et al., 2013).

Changing sea ice conditions, such as reduced sea ice extent, earlier sea ice melt in the spring, and later sea ice formation in the fall, may already be ushering in a profound transformation of some Arctic regions such as the Bering Strait system

---

F. J. Mueter (✉)
University of Alaska Fairbanks, Juneau, AK, USA
e-mail: fmueter@alaska.edu

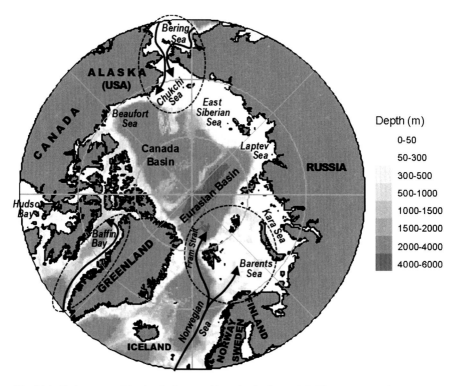

**Fig. 14.1** Bathymetry of the Arctic Ocean with major basins and shelf seas. Arrows denote major inflows and dashed ovals highlight the three inflow shelves

(Huntington et al., 2020). Impacts of these changes on the biological system include changes in the timing of important biological processes, shifts in the distribution of marine organisms ranging from plankton to whales, and changes in the amount and quality of food available for fish, seabirds, and mammals. These ongoing changes present significant challenges to coastal communities, commercial fishers, and others who depend on living marine resources. In response to these challenges, national governments and the international community have taken some proactive measures to protect vulnerable Arctic ecosystems. These include provisional prohibitions on commercial fishing in US waters north of Bering Strait (NPFMC, 2009) and in the Norwegian Arctic north and west of Svalbard (Jørgensen et al., 2020), as well as a 16-year moratorium on commercial fishing in the CAO under the 2018 international *Agreement to Prevent Unregulated High Seas Fisheries in the Central Arctic Ocean.*

These prohibitions are not permanent bans on commercial fishing; instead, they are intended as interim measures to allow for adequate research and data collection to assess the ability of Arctic marine ecosystems to support commercial fisheries. Although none of the agreements has a built-in funding mechanism to support research needs or develop management approaches, the Arctic research community, Arctic nations, Indigenous peoples, and international organizations are developing frameworks for monitoring and research and are taking steps towards some form of governance of potential CAO fisheries (Grebmeier et al., 2021; Prip, 2019). Fishery experts consider it unlikely that the CAO could support a sustainable commercial fishery anytime soon, but conditions in the gateways have changed more rapidly than anticipated, making any predictions regarding fisheries in the CAO highly uncertain.

To assess the current state and the future of Arctic fisheries, I briefly describe the relevant marine environment, summarize recent and expected future changes in Arctic marine ecosystems, and discuss the implications of these changes for fisheries resources in the Arctic. Finally, I discuss some of the ongoing efforts to develop research and management frameworks in support of the conservation and sustainable use of Arctic marine ecosystems.

## Arctic Marine Ecosystems and Their Fishery Resources

The Arctic Ocean is one of the least explored ocean regions due to its remoteness and the challenges of conducting research in ice-covered waters. These factors, as well as the low economic potential for fisheries, have historically limited oceanographic and fisheries research to coastal zones, areas where sea ice melted during summer, and the surface ocean within the summer pack ice. Consequently, our knowledge about Arctic fishes, particularly under perennial sea ice and in deeper areas, is severely limited. One of the consequences of a longer ice-free season has been easier access to Arctic waters and their natural resources. This has spurred interest in oil and gas exploration, shipping, tourism, and potential new fisheries. Along with these activities, research on Arctic marine ecosystems increased sharply beginning in the 1990s (Wassmann et al., 2011), as have efforts by fisheries agencies to effectively survey more areas.

### *Physical Setting and Biological Dynamics*

The Arctic Ocean consists of several deep basins separated by seafloor ridges and surrounded by broad continental shelves (typically <200 m, Fig. 14.1). Some of its defining and biologically relevant physical characteristics are extremely cold waters, low light conditions and seasonal or permanent ice cover. Near-freezing waters limit the distribution of many organisms that lack appropriate adaptations. The long polar night and the presence of sea-ice, especially when covered by snow, limit photosynthesis (Tremblay et al., 2015) and the ability of zooplankton or fish to feed visually for much of the year (Varpe et al., 2015). In addition to low temperatures and low light conditions, much of the Arctic Ocean experiences extreme seasonal fluctuations due to the seasonal advance and retreat of sea ice (Fig. 14.2). Species living in these waters have adapted to seasonal extremes and include endemic Arctic species

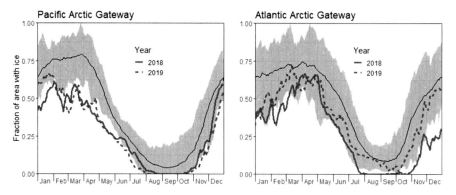

**Fig. 14.2** Daily sea ice concentrations on the two major Arctic inflow shelves (Fig. 14.1) based on data from the NOAA National Snow and Ice Data Center (Meier et al., 2017). Thin black line and grey bands denote the mean and range of observed sea-ice concentrations from 1988–2010. Daily concentration for the most recent years shown as separate lines

with physiological adaptations to cold winter conditions, short-lived species that complete their life cycle during the short production season, species with dormant overwintering stages, and seasonal migrants such as some fish, seabirds, and whales that capitalize on high summer production, but overwinter south of the Arctic. The challenges of living in cold, often dark, and extremely variable conditions result in a relatively low diversity of zooplankton (Kosobokova et al., 2011) and fish (Mecklenburg et al., 2011) in Arctic waters compared to more temperate regions, and a decrease in diversity from the inflow shelves to the interior shelves and to the CAO (Mueter et al., 2013). In contrast, benthic invertebrates living in and on the sediment are highly diverse and comprise over 90% of the Arctic marine fauna (Piepenburg et al., 2011; Sirenko, 2001). The reasons for the high benthic diversity are not fully understood, but almost certainly include a high rate of export of living and dead organic material from surface waters to the seafloor.

Biological processes and fish communities in the Arctic, especially on the inflow shelves, are also strongly impacted by ocean currents. The Arctic Ocean is connected to the Pacific via the shallow (50 m) and narrow (85 km) Bering Strait between Alaska (US) and Russia and to the Atlantic via the broad Barents Sea shelf, the deep Fram Strait between Greenland and Svalbard, Davis Strait west of Greenland, and numerous shallow straits in the Canadian Arctic Archipelago (Fig. 14.1). Pacific waters entering the Arctic through Bering Strait transport heat, freshwater, nutrients, and plankton into the Chukchi Sea (Hunt Jr. et al., 2016). Similarly, Atlantic waters entering across the Barents Sea shelf and via the east side of Fram Strait are an important source of nutrients and plankton for the Arctic. Atlantic waters also enter the Arctic along the west coast of Greenland, but largely recirculate within Hudson Bay rather than flowing into the Arctic Basin. The Atlantic inflows are much larger than flow through Bering Strait and their source waters are saltier and therefore heavier than Pacific waters. These denser waters sink below the surface layer and make up the intermediate and deep layers of the

Arctic, whereas the lighter Arctic surface waters are primarily of Pacific origin. Inflows are compensated by outflows along the east coast of Greenland through western Fram Strait and through the Canadian Arctic Archipelago. As warmer waters from the south flow into the Chukchi Sea and the Barents Sea, respectively, they encounter and mix with Arctic water masses that are advected southward onto these shelves. In these transition zones, species with primarily Pacific or Atlantic affinities mix and interact with true Arctic species.

Biological dynamics in much of the Arctic Ocean are governed by a short and intense production season. Much of the seasonal production is attributed to sea ice algae that live in or are associated with the ice and are released into the water when the ice melts. On many seasonally ice-covered shelf regions, especially shallow shelves such as in the northern Bering and Chukchi seas, only a small fraction of the intense, ice-associated spring bloom is consumed in the water column (Campbell et al., 2009), while much of the production is transported horizontally to be consumed elsewhere or is deposited on the seafloor to support a rich benthic community (Grebmeier et al., 2006). The transport of nutrients and plankton from the highly productive Subarctic into the Arctic across the Arctic inflow shelves provides both an important nutrient subsidy for primary production in the Arctic, as well as prey for upper-trophic-level consumers in the water column and on the seafloor (Wassmann et al., 2015). Thus, Arctic marine ecosystems are strongly influenced by and benefit from upstream processes.

## *Fish Communities in the Arctic*

The fish fauna in the Arctic Ocean comprises at least 242 species (Mecklenburg et al., 2011), but has a lower species diversity than adjacent areas of the North Pacific and North Atlantic oceans. Fish faunas in the North Pacific and North Atlantic evolved separately prior to the opening of the Bering Land Bridge 3.5 million years ago, with a much larger diversity of fishes in the North Pacific compared to the North Atlantic due to the much shorter geological history of the Atlantic. The opening of the Bering Land Bridge resulted in the great trans-Arctic Interchange (Vermeij, 1991), with many families of fishes that had evolved in the North Pacific contributing species to the Atlantic fauna during this period (for example, smelts, eelpouts, sculpins, poachers, snailfish, and gunnels). However, the cod-like fishes (family Gadidae) evolved in the Atlantic (Briggs, 2003) and contributed two species to the North Pacific – walleye pollock (*Gadus chalcogrammus*) and Pacific cod (*G. macrocephalus*) – both of which support important commercial fisheries today. The intensification of northern hemisphere glaciation that began about 2.9 million years ago reduced sea surface temperatures in the Arctic Ocean to the freezing point, resulting in the elimination of boreal species and the development of the modern Arctic fish fauna.

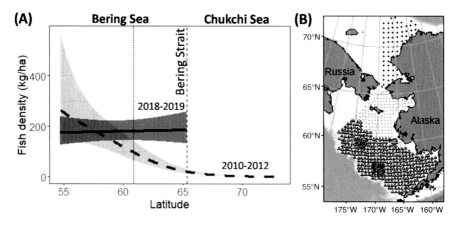

**Fig. 14.3** (**a**) Average latitudinal trends in the density of fish from the southeast Bering Sea to the northern Chukchi Sea based on trawl surveys conducted during the cool period 2010–2012 (dashed line) and during recent warm years (solid line) with 95 percent confidence bands. The location of Bering Strait is indicated by a vertical dashed line. No samples were available for the Chukchi Sea during the recent warm period. Dotted vertical line denotes northern extent of the historical survey area. Data from the Alaska Fisheries Science Center, NOAA. (**b**) Station locations sampled during trawl surveys in the standard survey area (triangles), the northern Bering Sea (grey crosses) and the Chukchi Sea (circles)

These evolutionary processes resulted in a strong gradient of decreasing diversity and abundance of high-latitude fishes from the Subarctic to the High Arctic (Fig. 14.3). Very high densities of gadids, flatfish, and crab in the Subarctic seas support some of the largest fisheries in the world, including walleye pollock in the eastern Bering Sea, at times the largest single-species fishery worldwide, and the Northeast Arctic stock of Atlantic cod (*Gadus morhua*) in the Norwegian and Barents seas. Large commercial fisheries on these species operate at the doorstep to the Arctic and follow their targets as they migrate into seasonally ice-covered waters to feed. In contrast to high abundances in the Subarctic seas, fish communities in the Arctic are characterized by low abundances of relatively small fish species.

The Arctic fish fauna is dominated by polar cod (*Boreogadus saida*), a small gadid (typically <30 cm) with a circumpolar distribution. Polar cod serve an important ecosystem role in the Arctic, supporting many seabirds and marine mammals that are resident in or migrate to the Arctic to feed on juveniles or adult cod, or on their predators. Polar cod are adapted to low light conditions and have a high concentration of salt and antifreeze glycoprotein in their blood to cope with freezing temperatures. They are often closely associated with sea ice, where they feed on ice-associated fauna and seek shelter in ice crevices, but also occur on the seafloor and in large mid-water aggregations to a depth of at least 1200 m. They have historically supported a commercial fishery in the Barents Sea, but the fishery is currently closed due to low abundances. Although often caught in local subsistence fisheries, they typically make up a small portion of these catches compared to typical coastal species such as whitefish (Coregonidae) and Arctic char (*Salvelinus alpinus*).

Despite the ecological importance of polar cod, their life history and biology remain poorly understood in most regions of the Arctic, although research on this key forage species has accelerated in recent years (Mueter et al., 2020). Polar cod are not only sensitive to warming and loss of sea ice (Gjøsæter et al., 2020; Laurel et al., 2017), but are also impacted by increased anthropogenic activity associated with the reduction of sea ice, such as increased ship traffic (Ivanova et al., 2020) and increased oil and gas activity (Nahrgang et al., 2016). In general, Arctic taxa appear to be highly sensitive to habitat changes, and polar cod in particular can serve as a sentinel species that responds quickly to changes in temperature and ice extent at the transition zone between the Subarctic and Arctic (Marsh & Mueter, 2020).

## Climate Change Impacts on Arctic Marine Environments

The world has been warming for at least a century due to increasing levels of atmospheric $CO_2$, with temperatures in the Arctic region increasing roughly twice as fast as the rest of the world (IPCC, 2013). Increased $CO_2$ levels have resulted in changes in ice cover, water temperatures, ocean currents, and ocean acidity. These changes to ocean habitats have important impacts on biological and biogeochemical processes, particularly along the edges of the Arctic in the marginal ice zone (Wassmann, 2011). Although temperature and ice conditions in the Arctic have high natural variability associated with extreme seasonal and interannual fluctuations, recent marine heatwaves exceed natural variability and have been attributed to global warming with high certainty (Walsh et al., 2018). Such heatwaves can be highly disruptive for coastal communities (e.g. Huntington et al., 2020) and their frequency and magnitude are likely to increase globally (Frölicher et al., 2018).

Expected changes in the ocean environment will affect how fish and their food will grow, how they will be distributed in the ocean, and which species will prosper or decline. Anticipated impacts of climate change on marine organisms have been qualitatively assessed both globally (Pörtner et al., 2014) and for the Arctic (Mueter et al., 2021). Quantitative projections based on numerical ecosystem models show likely increases in fish biomass in the Arctic using a variety of models (Lotze et al., 2019; Tittensor et al., 2018). However, such projections are particularly challenging for high-latitude shelf regions, where sea-ice dynamics regulate much of the primary production and where complex connections between the water column and the seafloor often dominate the transfer of energy from primary producers to fish and other consumers. Based on our current understanding, I provide a brief synthesis of expected changes in physical and biological processes and how they will affect upper trophic levels in the Arctic, particularly the species that support commercial and subsistence harvests. Some of the most relevant biological effects are changes in overall primary production, changes in the size and species composition of planktonic organisms that form the basis of the food web, changes in the timing of sea-ice retreat and associated biological processes (phenology), shifting distributions of biological communities (borealization), and effects from ocean acidification.

Primary production in the Arctic includes production by sea-ice algae embedded in the ice, phytoplankton production in the water column below the ice, and open-water phytoplankton production. This production provides the basis for any existing and future fisheries in the Arctic Ocean and is largely controlled by light and nutrient availability. Total production and the balance between under-ice production and open-water production is expected to change as a consequence of global warming, but these changes will not be uniform across the Arctic (Mueter et al., 2021; Tremblay et al., 2015). It is likely that primary production over the main inflow shelves will continue recent increasing trends because a longer ice-free season can enhance late season production, given sufficient nutrients. However, several factors may limit any increases in production in Arctic seas. First, future changes in advection through the Arctic gateways are highly uncertain and reduced advection would decrease nutrient supplies to the Arctic. Second, light availability may limit additional fall production in the Barents Sea due to its higher latitude compared to the northern Bering and Chukchi seas. Third, increased primary production will not always translate into increased production at higher trophic levels due to changes in phytoplankton composition, potential mismatches between phytoplankton and zooplankton grazers, and changes in the abundance and quality of zooplankton prey available for upper trophic levels (Mueter et al., 2021).

Primary production along most of the Arctic interior shelf and slope regions has also been increasing and will likely continue to increase in the future due to more ice-free areas and a longer ice-free season. Less ice increases light availability and promotes enhanced wind mixing and upwelling of deep waters along the continental slope, fueling additional production in surface waters. Enhanced production could support a higher biomass of resident fish such as polar cod or new migrants such as Pacific cod or Atlantic cod. However, it is unlikely that boreal species will be able to permanently expand into these regions as long as continued cold winter temperatures limit their presence. Moreover, distances from southern overwintering areas may be too large to allow boreal fish to capitalize on any increases in production in the high Arctic by extending their summer feeding migrations.

Future trends in primary production in the CAO are highly uncertain, despite increases in light availability associated with the loss of sea ice. Primary production is expected to decrease in the Canada Basin because increased precipitation and freshwater runoff strengthens water column stratification, preventing deep nutrients being mixed into the surface layer (Polyakov et al., 2020). By contrast, there is a potential for substantial increases in primary production in the Eurasian Basin because the inflow of warmer Atlantic water at depth weakens stratification, allowing for enhanced vertical mixing. For the Arctic Ocean overall, a plausible scenario is that future increases in primary production will lead to increased fishery production, particularly in the Bering Strait region, on the West Greenland shelf, and in northern portions of the Barents Sea slope, possibly extending into the Eurasian basin.

In addition to changes in the amount of primary production, changes in the timing of production and in the size and species composition of plankton prey affect the production of fish and other consumers. Ecological theory predicts that warming results in a shift to smaller average body sizes due to smaller individual sizes and/or replacement of large species by smaller species (Daufresne et al., 2009). Smaller average sizes result in a less efficient transfer of energy to fish and other consumers because food chains become longer and energy is lost at each step. Changes in the size structure of plankton communities have already been documented in the Arctic and are supported by laboratory studies and models (Mueter et al., 2021). The size and species composition of plankton are also impacted by the timing of the spring bloom because the life cycle of some key species is tied to the timing of sea ice melt. Such changes at the bottom of the food web can have cascading effects on upper trophic levels. For example, earlier sea ice retreat or the absence of sea ice on the eastern Bering Sea shelf results in the replacement of the large, lipid-rich copepod *Calanus glacialis*, an important prey species for juvenile pollock and many other consumers, with smaller zooplankton that lack lipid stores. Consequently, young pollock are unable to store sufficient energy prior to their first winter and suffer high overwinter mortality (Heintz et al., 2013), resulting in substantial decreases in pollock biomass in subsequent years. Projections of pollock biomass through the end of the century suggest future declines in pollock biomass under a variety of harvest scenarios and temperature changes (Holsman et al., 2020). Similar shifts from large, lipid-rich zooplankton to smaller zooplankton during warm years occur on the West Greenland shelf (Møller & Nielsen, 2020) and in the Barents Sea (Aarflot et al., 2018), but their impacts on fish have not yet been quantified.

One of the most obvious and best documented consequences of increasing ocean temperatures is a poleward shift in the distribution of terrestrial and marine organisms (Pecl et al., 2017). These shifts are especially evident on the Arctic inflow shelves as plankton, forage fish, and large mobile predators expand northward into the Arctic (Fossheim et al., 2015; Stevenson & Lauth, 2019). The rates and directions of these expansions vary among species (Mueter & Litzow, 2008), resulting in a changing mix of species and a re-organization of food webs (Kortsch et al., 2015). The consequences for commercial fishers and subsistence users are only beginning to emerge and may involve a profound transformation of the social-ecological-economic systems on the inflow shelves (Huntington et al., 2020).

In addition to warming, rising $CO_2$ levels in the atmosphere are associated with an increased uptake of $CO_2$ by the ocean, making ocean waters more corrosive (ocean acidification). The effects of ocean acidification on marine organisms are variable and are generally a larger concern for juveniles and for shell-building organisms. Ocean acidification negatively affects some important components of the food web, such as shelled pteropods, mussels, and crab (Falkenberg et al., 2018). Although the broader implications for commercial species are uncertain, shellfish such as red king crab (*Paralithodes camtschaticus*) are generally more susceptible to ocean acidification than fish (Hurst et al., 2013; Long et al., 2013). Any effects of ocean acidification are likely to be more severe in the Pacific Arctic, where waters have naturally high acidity, and these effects carry substantial risks for fishery-dependent communities in the region (Mathis et al., 2015).

## Fisheries Research and Management for a Sustainable Arctic

Arctic fisheries management faces two distinct challenges. First, existing fisheries in Arctic marginal seas, particularly on the inflow shelves (Fig. 14.1), are changing rapidly due to shifting distributions and changes in abundance of commercially exploited species. These changes occur largely within the Exclusive Economic Zone (EEZ) of individual Arctic nations (<200 miles from shore). Therefore, adaptive management measures are needed at the national level, but bilateral cooperation will be needed where shifting stocks cross national boundaries. Second, the potential for future fisheries in the Central Arctic Ocean beyond national jurisdictions requires enhanced international cooperation, expanded monitoring efforts in remote Arctic waters, and an appropriate governance framework.

### *Changing Fisheries in Arctic Shelf Regions*

Successful fishery management requires an understanding of the biology of the target species (growth, maturity, fecundity) and its geographical distribution, along with estimates of current abundance, past abundance trends, and past catches. For commercial stocks in temperate and high-latitude seas, this information is typically collected during annual research surveys that encompass much or all of the distributional range of major target stocks. These surveys collect samples using bottom or mid-water trawls, along with oceanographic and other ecosystem information (such as nutrient concentrations and zooplankton abundances). Stock assessment scientists then use information on age composition, growth, and maturity, along with estimated densities (weight per unit area) in the survey region to estimate both current and historical abundances and to simulate future abundances under different harvest scenarios. Outputs from these analyses provide the basis for managers to specify annual quotas, which are typically constrained by harvest control rules that translate current abundances into acceptable maximum catch levels (e.g., Kvamsdal et al., 2016). Arctic nations have used various forms of this type of annual assessment and management successfully to manage commercial fisheries in the Bering Sea (Russia, USA), the Northeast Atlantic (Canada, Greenland, Iceland), and the Barents Sea (Norway, Russia).

The assessment and management of high-latitude fisheries faces challenges arising from shifts in distribution and changes in the abundance of major stocks. Arctic fishing nations are well positioned to meet these challenges due to well-established research programs and strong management institutions. Historically, large changes in the abundance of commercial stocks have been associated with both over-harvesting and climate regime shifts, but causes are difficult to untangle and may interact in complex ways (Möllmann & Diekmann, 2012). Although appropriate harvest control rules provide a robust tool to address changes in abundance, given adequate enforcement and accountability measures, such changes are accompanied by considerable socio-economic disruptions (Barbeaux et al., 2020; Hamilton, 2007).

Like changes in abundance, shifts in distribution of fish stocks have occurred in the past, but have accelerated under recent climate warming (Stevenson & Lauth, 2019). As stocks have shifted or expanded beyond historic survey areas, national agencies have adapted by adding additional survey stations, both in response to and in anticipation of further northward shifts. For example, the US now conducts annual surveys in the northern Bering Sea (Stevenson & Lauth, 2019) and Russia has expanded its surveys into the western Chukchi Sea on the Pacific side (Orlov et al., 2020) and into the Kara Sea on the Atlantic side. These surveys require substantial resources but have become a priority in the face of changing distributions to support accurate assessments.

Management responses to the northward expansion of fish stocks differ substantially among countries. Russian fishery managers recently authorized the first commercial fishery for pollock north of Bering Strait (Arctic Today, June 25th, 2020) based on recent survey results. By contrast, the US specified precautionary catch limits set to zero for all potential target species in federally managed waters north of Bering Strait (NPFMC, 2009). These limits remain in place until sufficient information is available for establishing scientifically based reference points. Similarly, Norway established a Fisheries Protection Zone in the northern Barents Sea and Fram Strait region that provisionally closes about 445,000 km$^2$ to fishing in order to protect sensitive habitat (Jørgensen et al., 2020).

The northward movement of migratory stocks such as pollock, Pacific cod, and Atlantic cod has also resulted in stocks crossing national boundaries, raising concerns about catch allocation and appropriate forms of cooperation (Koubrak & VanderZwaag, 2020). For example, eastern Bering Sea pollock has long been a transboundary stock that migrates between US and Russian waters. However, its recent expansion into the northern Bering Sea suggests a change in migratory behavior that could pose increased risks to this stock, given the lack of coordination between Russia and the US. An even larger challenge may be the redistribution of Pacific cod in the eastern Bering Sea, which was previously limited to the US EEZ, but now likely mixes with Russian cod in the northern Bering Sea (Thompson & Thorson, 2019). Similar challenges exist in the Barents Sea, but a long history of collaborative ecosystem surveys between Norway and Russia and shared management under the Joint Norwegian-Russian Fisheries Commission may facilitate a coordinated response.

As commercial stocks re-distribute to more northern waters, they have also shifted away from existing ports, favoring sectors of the fishery that are independent of shore-based processing facilities, such as catcher processors. In the northern Bering Sea, these vessels are likely to overlap with local subsistence fishermen and hunters, resulting in safety concerns and a potential for resource conflicts.

## *Fisheries Monitoring and Research in the Central Arctic Ocean*

Researchers and managers have long recognized the need for a proactive approach to fishery assessment and management in the CAO. Concerns over the potential harm from unregulated fishing activities in the CAO has led to a series of meetings among the Arctic nations, other nations with a strong interest in the Arctic, Indigenous groups, and non-governmental organizations. These meetings culminated in the 2018 *Agreement to Prevent Unregulated High Seas Fisheries in the Central Arctic Ocean*, signed by Canada, Denmark/Greenland, the European Union, Iceland, Japan, the People's Republic of China, the Republic of Korea, Norway, the Russian Federation, and the United States of America. The parties agreed to refrain from commercial fishing in the CAO for 16 years and called for a joint program of scientific research and monitoring to establish a scientific basis for the conservation and management of fisheries resources in the CAO.

In parallel with the diplomatic negotiations, *Scientific Experts on Fish Stocks in the Central Arctic Ocean* (FiSCAO) met to develop a draft plan for monitoring and research in the CAO (NOAA, 2018). Building on these efforts, Grebmeier et al. (2021) proposed a step-wise approach toward enhanced monitoring and research in the CAO that builds on existing and planned efforts. The approach would begin by synthesizing available information to agree on a standard set of measurements along strategically placed shelf to basin transects. Building on existing programs, coordinated sampling along these transects would be conducted annually and would add a fishery component to all surveys. Initial analyses would establish fishery-relevant thresholds such as the abundance of key fish species, their prey, and potential predators. If and when these thresholds are exceeded, additional sampling effort would be triggered to improve the assessment of fishery resources and new thresholds would be established.

This approach requires efficient and cost-effective tools to assess the abundance of fish in the CAO. Standard fisheries trawl surveys require dedicated ships and are likely cost-prohibitive, but alternative tools are available for initial monitoring efforts. For example, echo sounders are routinely used to monitor fish in the water column. While such acoustic surveys are challenging when ice is present, they have been used in the CAO to demonstrate the existence of a wide-spread, deep scattering layer consisting of low abundances of unidentified fish (Snoeijs-Leijonmalm et al., 2021). When fish biomass is known to be dominated by a single species, such as polar cod, autonomous saildrones equipped with echo sounders can provide a practical and cost-effective approach to assess mid-water biomass in remote regions (Levine et al., 2021). Similarly, a small number of stationary echo sounders moored to the seafloor have the potential to provide reliable indices of abundance in areas where fish consistently aggregate for spawning or feeding (De Robertis et al., 2018). A relatively recent approach to assessing the presence of fish species is to match genetic material obtained by filtering water samples (environmental DNA or eDNA) against known baselines (Hansen et al., 2018). Although the approach shows promise for assessing the relative abundances of species in a given area, further validation and better baselines will be needed before it can be applied to monitor abundances of fish in the CAO. A drawback of these methods is that no samples of fish are obtained to measure biological parameters needed for management. However, these approaches can provide initial information on the presence and relative abundance of fishes in the CAO before additional monitoring efforts are put in place, consistent with the step-wise approach proposed by Grebmeier et al. (2021).

## Conclusions

As Arctic marine environments continue to be transformed by increasing atmospheric $CO_2$ concentrations, biological systems are responding in both predictable and unpredictable ways. Among the predictable responses is the continued northward shift and expansion of southern species into the Arctic, disrupting existing patterns of resource use. In contrast, the potential for southern species to expand into, or for resident species to increase in the CAO and support future commercial fisheries is highly uncertain, but is widely considered to be low at this time.

Arctic nations are well positioned to respond to and manage ongoing changes in the shelf regions under their jurisdiction. Nevertheless, challenges associated with fish stocks expanding across national boundaries and commercial fleets expanding into areas that are critical to the food security of Arctic Indigenous peoples will require improved coordination and cooperation among local, national, international, and Indigenous governments. The perceived low fisheries potential in international waters of the CAO, combined with existing frameworks for cooperation, provide a window of opportunity for the international community to develop the science and the governance structures needed to ensure the conservation and sustainable stewardship of Arctic marine resources.

## References

Aarflot, J. M., Skjoldal, H. R., Dalpadado, P., & Skern-Mauritzen, M. (2018). Contribution of Calanus species to the mesozooplankton biomass in the Barents Sea. *ICES Journal of Marine Science, 75*(7), 2342–2354.

Barbeaux, S. J., Holsman, K., & Zador, S. (2020). Marine heatwave stress test of ecosystem-based fisheries management in the Gulf of Alaska Pacific Cod Fishery. *Frontiers in Marine Science, 7*, 703.

Briggs, J. C. (2003). Marine centres of origin as evolutionary engines. *Journal of Biogeography, 30*, 1–18.

Campbell, R. G., Sherr, E. B., Ashjian, C. J., Plourde, S., Sherr, B. F., Hill, V., & Stockwell, D. A. (2009). Mesozooplankton prey preference and grazing impact in the Western Arctic Ocean. *Deep Sea Research II, 56*(17), 1274–1289.

Carmack, E., & Wassmann, P. (2006). Food webs and physical–biological coupling on pan-Arctic shelves: Unifying concepts and comprehensive perspectives. *Progress in Oceanography, 71*(2–4), 446–477.

Daufresne, M., Lengfellner, K., & Sommer, U. (2009). Global warming benefits the small in aquatic ecosystems. *Proceedings of the National Academy of Sciences, 106*(31), 12788.

De Robertis, A., Levine, R., & Wilson, C. D. (2018). Can a bottom-moored echo sounder array provide a survey-comparable index of abundance? *Canadian Journal of Fisheries and Aquatic Sciences, 75*(4), 629–640.

Falkenberg, L. J., Jelmert, A., Mark, F. C., Rost, B., Schulz, K. G., & Thor, P. (2018). Biological responses to ocean acidification. In *AMAP assessment 2018: Arctic Ocean acidification* (pp. 15–28). Arctic Monitoring and Assessment Programme (AMAP).

Fossheim, M., Primicerio, R., Johannesen, E., Ingvaldsen, R. B., Aschan, M. M., & Dolgov, A. V. (2015). Recent warming leads to a rapid borealization of fish communities in the Arctic. *Nature Climate Change, 5*(7), 673–677.

Frölicher, T. L., Fischer, E. M., & Gruber, N. (2018). Marine heatwaves under global warming. *Nature, 560*(7718), 360–364.

Gjøsæter, H., Huserbråten, M., Vikebø, F., & Eriksen, E. (2020). Key processes regulating the early life history of Barents Sea polar cod. *Polar Biology, 43*(8), 1015–1027.

Grebmeier, J. M., Cooper, L. W., Feder, H. M., & Sirenko, B. I. (2006). Ecosystem dynamics of the Pacific-influenced Northern Bering and Chukchi Seas in the Amerasian Arctic. *Progress in Oceanography, 71*(2–4), 331–361.

Grebmeier, J. M., Huntington, H. P., Mueter, F. J., Pulsifer, P. L., & Davies, J. (2021). A stepwise progression to fisheries ecosystem science in the Central Arctic Ocean. *Marine Policy*. In Review.

Hamilton, L. C. (2007). Climate, fishery and society interactions: Observations from the North Atlantic. *Deep Sea Research II, 54*, 2958–2969.

Hansen, B. K., Bekkevold, D., Clausen, L. W., & Nielsen, E. E. (2018). The sceptical optimist: Challenges and perspectives for the application of environmental DNA in marine fisheries. *Fish and Fisheries, 19*(5), 751–768.

Heintz, R. A., Siddon, E. C., Farley, E. V., & Napp, J. M. (2013). Correlation between recruitment and fall condition of age-0 pollock (Theragra chalcogramma) from the eastern Bering Sea under varying climate conditions. *Deep Sea Research II, 94*, 150–156.

Hollowed, A. B., Planque, B., & Loeng, H. (2013). Potential movement of fish and shellfish stocks from the sub-Arctic to the Arctic Ocean. *Fisheries Oceanography, 22*(5), 355–370.

Holsman, K. K., Haynie, A. C., Hollowed, A. B., Reum, J. C. P., Aydin, K., Hermann, A. J., Cheng, W., Faig, A., Ianelli, J. N., Kearney, K. A., & Punt, A. E. (2020). Ecosystem-based fisheries management forestalls climate-driven collapse. *Nature Communications, 11*(1), 4579.

Hunt, G. L., Jr., Drinkwater, K. F., Arrigo, K., Berge, J., Daly, K. L., Danielson, S., Daase, M., Hop, H., Isla, E., Karnovsky, N., Laidre, K., Mueter, F. J., Murphy, E. J., Renaud, P. E., Smith, W. O., Jr., Trathan, P., Turner, J., & Wolf-Gladrow, D. (2016). Advection in polar and subpolar environments: Impacts on high latitude marine ecosystems. *Progress in Oceanography, 149*, 40–81.

Huntington, H. P., Danielson, S. L., Wiese, F. K., Baker, M., Boveng, P., Citta, J. J., De Robertis, A., Dickson, D. M. S., Farley, E., George, J. C., Iken, K., Kimmel, D. G., Kuletz, K., Ladd, C., Levine, R., Quakenbush, L., Stabeno, P., Stafford, K. M., Stockwell, D., & Wilson, C. (2020). Evidence suggests potential transformation of the Pacific Arctic ecosystem is underway. *Nature Climate Change, 10*(4), 342–348.

Hurst, T. P., Fernandez, E. R., & Mathis, J. T. (2013). Effects of ocean acidification on hatch size and larval growth of walleye pollock (Theragra chalcogramma). *ICES Journal of Marine Science, 70*(4), 812–822.

IPCC. (2013). Climate change 2013: The physical science basis. In T. F. Stocker, D. Qin, G.-K. Plattner, M. Tignor, S. K. Allen, J. Boschung, A. Nauels, Y. Xia, V. Bex, & P. M. Midgley (Eds.), *Contribution of Working Group I to the Fifth Assessment Report of the Intergovernmental Panel on Climate Change*. Cambridge University Press.

Ivanova, S. V., Kessel, S. T., Espinoza, M., McLean, M. F., O'Neill, C., Landry, J., Hussey, N. E., Williams, R., Vagle, S., & Fisk, A. T. (2020). Shipping alters the movement and behavior of Arctic cod (Boreogadus saida), a keystone fish in Arctic marine ecosystems. *Ecological Applications, 30*(3), e02050.

Jørgensen, L. L., Bakke, G., & Hoel, A. H. (2020). Responding to global warming: New fisheries management measures in the Arctic. *Progress in Oceanography, 188*, 102423.

Kortsch, S., Primicerio, R., Fossheim, M., Dolgov Andrey, V., & Aschan, M. (2015). Climate change alters the structure of arctic marine food webs due to poleward shifts of boreal generalists. *Proceedings of the Royal Society B: Biological Sciences, 282*(1814), 20151546.

Kosobokova, K. N., Hopcroft, R. R., & Hirche, H.-J. (2011). Patterns of zooplankton diversity through the depths of the Arctic's central basins. *Marine Biodiversity, 41*(1), 29–50.

Koubrak, O., & VanderZwaag, D. L. (2020). Are transboundary fisheries management arrangements in the Northwest Atlantic and North Pacific seaworthy in a changing ocean? *Ecology and Society, 25*(4), 42.

Kvamsdal, S. F., Eide, A., Ekerhovd, N. A., Enberg, K., Gudmundsdottir, A., Hoel, A. H., Mills, K. E., Mueter, F. J., Ravn-Jonsen, L., Sandal, L. K., Stiansen, J. E., & Vestergaard, N. (2016). Harvest control rules in modern fisheries management. *Elementa, 4*, 000114.

Laurel, B. J., Copeman, L. A., Spencer, M., & Iseri, P. (2017). Temperature-dependent growth as a function of size and age in juvenile Arctic cod (Boreogadus saida). *ICES Journal of Marine Science, 74*(6), 1614–1621.

Levine, R. M., De Robertis, A., Grünbaum, D., Woodgate, R., Mordy, C. W., Mueter, F., Cokelet, E., Lawrence-Slavas, N., & Tabisola, H. (2021). Autonomous vehicle surveys indicate that flow reversals retain juvenile fishes in a highly advective high-latitude ecosystem. *Limnology and Oceanography, 66*(4), 1139–1154.

Long, W. C., Swiney, K. M., & Foy, R. J. (2013). Effects of ocean acidification on the embryos and larvae of red king crab, Paralithodes camtschaticus. *Marine Pollution Bulletin, 69*(1), 38–47.

Lotze, H. K., Tittensor, D. P., Bryndum-Buchholz, A., Eddy, T. D., Cheung, W. W. L., Galbraith, E. D., Barange, M., Barrier, N., Bianchi, D., Blanchard, J. L., Bopp, L., Büchner, M., Bulman, C. M., Carozza, D. A., Christensen, V., Coll, M., Dunne, J. P., Fulton, E. A., Jennings, S., … Worm, B. (2019). Global ensemble projections reveal trophic amplification of ocean biomass declines with climate change. *Proceedings of the National Academy of Sciences, 116*(26), 12907.

Marsh, J. M., & Mueter, F. J. (2020). Influences of temperature, predators, and competitors on polar cod (*Boreogadus saida*) at the southern margin of their distribution. *Polar Biology, 43*(8).

Mathis, J. T., Cooley, S. R., Lucey, N., Colt, S., Ekstrom, J., Hurst, T., Hauri, C., Evans, W., Cross, J. N., & Feely, R. A. (2015). Ocean acidification risk assessment for Alaska's fishery sector. *Progress in Oceanography, 136*, 71–91.

Mecklenburg, C. W., Møller, P. R., & Steinke, D. (2011). Biodiversity of arctic marine fishes: Taxonomy and zoogeography. *Marine Biodiversity, 41*(1), 109–140.

Meier, W. N., Fetterer, F., Savoie, M., Mallory, S., Duerr, R., & Stroeve, J. (2017). NOAA/NSIDC climate data record of passive microwave sea ice concentration, version 3. [Merged GSFC NASA Team/Bootstrap daily sea ice concentrations]. https://doi.org/10.7265/N59P2ZTG

Møller, E. F., & Nielsen, T. G. (2020). Borealization of Arctic zooplankton—Smaller and less fat zooplankton species in Disko Bay, Western Greenland. *Limnology and Oceanography, 65*(6), 1175–1188.

Möllmann, C., & Diekmann, R. (2012). Chapter 4 – Marine ecosystem regime shifts induced by climate and overfishing: A review for the northern hemisphere. In U. J. Guy Woodward & J. O. Eoin (Eds.), *Advances in ecological research* (Vol. 47, pp. 303–347). Academic.

Mueter, F. J., Bouchard, C., Hop, H., Laurel, B., & Norcross, B. (2020). Arctic gadids in a rapidly changing environment. *Polar Biology, 43*, 945–949.

Mueter, F. J., & Litzow, M. A. (2008). Sea ice retreat alters the biogeography of the Bering Sea continental shelf. *Ecological Applications, 18*(2), 309–320.

Mueter, F. J., Planque, B., Hunt, G. L., Jr., Alabia, I. D., Hirawake, T., Eisner, L., Dalpadado, P., Drinkwater, K. F., Harada, N., Arneberg, P., & Saitoh, S.-I. (2021). Possible future scenarios in the Gateways to the Arctic for Subarctic and Arctic marine systems: II. Prey resources, food webs, fish, and fisheries. *ICES Journal of Marine Science*, fsab122.

Mueter, F. J., Reist, J. D., Majewski, A. R., Sawatzky, C. D., Christiansen, J. S., Hedges, K. J., Coad, B. W., Karamushko, O. V., Lauth, R. R., Lynghammar, A., MacPhee, S. A., & Mecklenburg, C. W. (2013). Marine fishes of the Arctic. In M. O. Jeffries, J. A. Richter-Menge, & J. E. Overland (Eds.), *Arctic report card 2013*. NOAA. http://www.arctic.noaa.gov/reportcard

Nahrgang, J., Dubourg, P., Frantzen, M., Storch, D., Dahlke, F., & Meador, J. P. (2016). Early life stages of an arctic keystone species (Boreogadus saida) show high sensitivity to a water-soluble fraction of crude oil. *Environmental Pollution, 218*, 605–614.

NOAA. (2018). *Final report of the fifth meeting of scientific experts on fish stocks in the Central Arctic Ocean*. https://www.afsc.noaa.gov/Arctic_fish_stocks_fifth_meeting/pdfs/Final_report_of_the_5th_FiSCAO_meeting.pdf

NPFMC. (2009). *Fishery management plan for fish resources of the Arctic management area*. North Pacific Fishery Management Council, 605 W. 4th Ave., Suite 306, Anchorage, AK 99501. https://www.npfmc.org/wp-content/PDFdocuments/fmp/Arctic/ArcticFMP.pdf

Orlov, A. M., Benzik, A. N., Vedishcheva, E. V., Gafitsk, S. V., Gorbatenko, K. M., Goryanina, S. V., Zubarevich, V. L., Kodryan, K. V., Nosov, M. A., Orlova, S. Y., Pedchenko, A. P., Rybakov, M. O., Sokolov, A. M., Somov, A. A., Subbotin, S. N., Taptygin, M. Y., Firsov, Y. L., Khleborodov, A. S., & Chikilev, V. G. (2020). Fisheries research in the Chukchi Sea at the *RV Professor Levanidov* in August 2019: Some preliminary results. *Trudy VNIRO, 178*, 206–220. https://doi.org/10.36038/2307-3497-2019-178-206-220

Pecl, G. T., Araujo, M. B., Bell, J. D., Blanchard, J., Bonebrake, T. C., Chen, I. C., Clark, T. D., Colwell, R. K., Danielsen, F., Evengard, B., Falconi, L., Ferrier, S., Frusher, S., Garcia, R. A., Griffis, R. B., Hobday, A. J., Janion-Scheepers, C., Jarzyna, M. A., Jennings, S., ... Williams, S. E. (2017). Biodiversity redistribution under climate change: Impacts on ecosystems and human well-being. *Science, 355*(6332).

Piepenburg, D., Archambault, P., Ambrose, W. G., Blanchard, A. L., Bluhm, B. A., Carroll, M. L., Conlan, K. E., Cusson, M., Feder, H. M., Grebmeier, J. M., Jewett, S. C., Lévesque, M., Petryashev, V. V., Sejr, M. K., Sirenko, B. I., & Włodarska-Kowalczuk, M. (2011). Towards a pan-Arctic inventory of the species diversity of the macro- and megabenthic fauna of the Arctic shelf seas. *Marine Biodiversity, 41*(1), 51–70.

Polyakov, I. V., Alkire, M. B., Bluhm, B. A., Brown, K. A., Carmack, E. C., Chierici, M., Danielson, S. L., Ellingsen, I., Ershova, E. A., Gårdfeldt, K., Ingvaldsen, R. B., Pnyushkov, A. V., Slagstad, D., & Wassmann, P. (2020). Borealization of the Arctic Ocean in response to anomalous advection from Sub-Arctic Seas. *Frontiers in Marine Science, 7*, 491.

Pörtner, H.-O., Karl, D. M., Boyd, P. W., Cheung, W. W. L., Lluch-Cota, S. E., Nojiri, Y., Schmidt, D. N., & Zavialov, P. O. (2014). Ocean Systems. In C. B. Field, V. R. Barros, D. J. Dokken, K. J. Mach, M. D. Mastrandrea, T. E. Bilir, M. Chatterjee, K. L. Ebi, Y. O. Estrada, R. C. Genova, B. Girma, E. S. Kissel, A. N. Levy, S. MacCracken, P. R. Mastrandrea, & L. L. White (Eds.), *Climate change 2014: Impacts, adaptation, and vulnerability. Part A: Global and sectoral aspects. Contribution of Working Group II to the Fifth Assessment Report of the Intergovernmental Panel on Climate Change* (pp. 411–484). Cambridge University Press.

Prip, C. (2019). Arctic Ocean governance in light of an international legally binding instrument on the conservation and sustainable use of marine biodiversity of areas beyond national jurisdiction. *Marine Policy, 103768*.

Sirenko, B. I. (2001). List of species of free-living invertebrates of Eurasian Arctic seas and adjacent deep waters. *Explorations of the Fauna of the Seas, 51*, 1–129.

Snoeijs-Leijonmalm, P., Gjøsæter, H., Ingvaldsen, R. B., Knutsen, T., Korneliussen, R., Ona, E., Rune Skjoldal, H., Stranne, C., Mayer, L., Jakobsson, M., & Gårdfeldt, K. (2021). A deep scattering layer under the North Pole pack ice. *Progress in Oceanography, 102560*.

Stevenson, D. E., & Lauth, R. R. (2019). Bottom trawl surveys in the northern Bering Sea indicate recent shifts in the distribution of marine species. *Polar Biology, 42*(2), 407–421.

Thompson, G. G., & Thorson, J. T. (2019). Assessment of the Pacific Cod Stock in the Eastern Bering Sea. In *Stock assessment and fishery evaluation report for the groundfish resources of the Bering Sea/Aleutian Islands regions*. North Pacific Fisheries Management Council.

Tittensor, D. P., Eddy, T. D., Lotze, H. K., Galbraith, E. D., Cheung, W., Barange, M., Blanchard, J. L., Bopp, L., Bryndum-Buchholz, A., Büchner, M., Bulman, C., Carozza, D. A., Christensen, V., Coll, M., Dunne, J. P., Fernandes, J. A., Fulton, E. A., Hobday, A. J., Huber, V., … Walker, N. D. (2018). A protocol for the intercomparison of marine fishery and ecosystem models: Fish-MIP v1.0. *Geosci Model Dev, 11*(4), 1421–1442.

Tremblay, J.-É., Anderson, L. G., Matrai, P., Coupel, P., Bélanger, S., Michel, C., & Reigstad, M. (2015). Global and regional drivers of nutrient supply, primary production and $CO_2$ drawdown in the changing Arctic Ocean. *Progress in Oceanography, 139*, 171–196.

Varpe, Ø., Daase, M., & Kristiansen, T. (2015). A fish-eye view on the new Arctic lightscape. *ICES Journal of Marine Science, 72*(9), 2532–2538.

Vermeij, G. J. (1991). Anatomy of an invasion: The trans-Arctic interchange. *Paleobiology, 17*, 281–307.

Walsh, J. E., Thoman, R. L., Bhatt, U. S., Bieniek, P. A., Brettschneider, B., Brubaker, M., Danielson, S., Lader, R., Fetterer, F., Holderied, K., Iken, K., Mahoney, A., McCammon, M., & Partain, J. (2018). The high latitude marine heat wave of 2016 and its impacts on Alaska. *Bulletin of the American Meteorological Society, 99*(1), S39–S43.

Wassmann, P. (2011). Arctic marine ecosystems in an era of rapid climate change. *Progress in Oceanography, 90*(1–4), 1–17.

Wassmann, P., Duarte, C. M., Agustí, S., & Sejr, M. K. (2011). Footprints of climate change in the Arctic marine ecosystem. *Global Change Biology, 17*(2), 1235–1249.

Wassmann, P., Kosobokova, K. N., Slagstad, D., Drinkwater, K. F., Hopcroft, R. R., Moore, S. E., Ellingsen, I., Nelson, R. J., Carmack, E., Popova, E., & Berge, J. (2015). The contiguous domains of Arctic Ocean advection: Trails of life and death. *Progress in Oceanography, 139*, 42–65.

Zeller, D., Booth, S., Pakhomov, E., Swartz, W., & Pauly, D. (2011). Arctic fisheries catches in Russia, USA, and Canada: Baselines for neglected ecosystems. *Polar Biology, 34*, 955–973.

Zeller, D., Palomares, M. L. D., Tavakolie, A., Ang, M., Belhabib, D., Cheung, W. W. L., Lam, V. W. Y., Sy, E., Tsui, G., Zylich, K., & Pauly, D. (2016). Still catching attention: Sea around us reconstructed global catch data, their spatial expression and public accessibility. *Marine Policy, 70*, 145–152.

**Franz J. Mueter** is the President's Professor of Quantitative Fisheries and Ecosystems at the College of Fisheries and Ocean Sciences, University of Alaska Fairbanks, U.S.A.

# Chapter 15
# Infrastructure Projects in the Global Arctic

**Alexander Pilyasov**

## Introduction

When I am asked what the "ideal" Arctic is, I usually answer that it means the vastness of spaces and resources as in the Russian Arctic; the art of distributing natural rent in favor of local residents, as in the American Arctic (Alaska); local indigenous entrepreneurship and community economic development, as in the Canadian Arctic; a sense of their land and belonging to it, as in Greenland; and infrastructure development, as in the Scandinavian Arctic (in the north of Norway, Sweden, and Finland). Indeed, a significant part of the global Arctic is deprived of land transport infrastructure, and the level of road infrastructure (for example, per 100 km$^2$) is several times less than in the developed regions of the world and in the non-Arctic parts of the Arctic countries and regions (Shpak, 2012).

The infrastructural arrangement of the Arctic is the same challenge for humanity, as the exploration of space or the depths of the ocean, as the conquest of the peaks of mountainous countries. Examples of the construction of new Arctic roads or railways will forever remain in the global annals of mankind, in the "information field of the Earth" as acts of unparalleled courage and heroism, often paid for by numerous human lives. It is always more expensive, more difficult, more dramatic, and more dangerous than the construction of transport infrastructure facilities of a similar length in the developed regions of the world, where the first builders are always backed by nearby cities and towns, alternative road routes, and a constant close human presence. Not so in the Arctic. Here the roads are lonely in space, they have no alternative, there are no closely located cities and towns and other places of concentrated residence of human communities to which one can "shout" in case of danger.

A. Pilyasov (✉)
Lomonosov Moscow State University, Moscow, Russia

© The Author(s), under exclusive license to Springer Nature Switzerland AG 2022
M. Finger, G. Rekvig (eds.), *Global Arctic*,
https://doi.org/10.1007/978-3-030-81253-9_15

A person in infrastructure projects in the Arctic is left alone with natural extremes, to which he or she must adapt in order to withstand and triumphantly complete construction. The newest realities of climate change in the Arctic, faster than in the rest of the world (Climatic changes in the Arctic ..., 2014), have not radically changed the situation: the Arctic is still a challenge, naturally unpredictable (and perhaps now even more due to increased temperature amplitudes), and is risky for human life.

Therefore, each infrastructural project for the transport arrangement of the Arctic in any country is an event of national importance, in which geopolitical (military-strategic) and economic (to ensure the availability of strategically important and demanded resources in the world) tasks merge.

For a significant part of the twentieth century, the Soviet Union, concentrating the main territory of the global North and the Arctic, naturally acted as a leader in infrastructure development in the process of developing the natural resources of these territories. In the classic works of Soviet Northern scholar S. V. Slavin (1961, 1982), an inextricable link between the pioneering extraction of natural resources of uninhabited spaces and their transport arrangement was established.

Among the numerous Russian Northern development research schools that emerged in many scientific centers in the post-war period, I should especially note the Irkutsk school of economic geographers and its leader K. Kosmachev, in whose works the idea of the territorial structures of the regions of new development was first formulated (Kosmachev, 1974). This idea was subsequently developed and deepened by Kosmachev's students (Mosunov et al., 1990), who formed the idea of the territorial structures of the development process as a unity of the *trassa* – a transport channel for the penetration of "developed into the undeveloped" and numerous multi-level bases of development (rear, outpost, local).

In the developed regions of the country during the same period, economic geographers developed the concept of a linear-nodal territorial structure of highways and nodes of the settlement system (Khorev, 1981; Lappo, 1983). The main difference between the northern "trassas" of the stage of pioneering development of new spaces and "lines" in the inhabited areas was the monopoly lack of alternative transport channels in the Arctic and in the North. Northern and Arctic city bases of colonization were located on uncontested transport channels, and the "mainland" urban centers of the settlement system were always at the crossroads of several highways.

In the same post-war period, the first broad (from a global perspective) works on socio-economic development, including the transport infrastructure of the global Arctic, appeared (The Arctic Frontier, 1966; Armstrong et al., 1978), in which the major transport megaprojects of the global Arctic had been discussed. All these research efforts can be considered as important prerequisites for modern work on infrastructure projects of the global Arctic.

## Features and Drivers of Infrastructure Megaprojects in the Arctic

It is traditionally believed that large infrastructure megaprojects in the Arctic are caused by abrupt changes in the world prices for its natural resources. This is partly true; indeed, it was the Great Depression in the West in the 1930s when prices for machine tools and equipment were ultra-low, and the price of gold rose relative to the products of investment engineering, desperately needed by agrarian Soviet Russia, which was embarking on a large-scale industrialization. This led to a megaproject of a brave road arrangement in the North-East of Russia, the Kolyma highway, built in an unprecedentedly short time: 600 km in total off-road terrain, dotted with rivers and streams, in nine years (1932–1940) (Navasardov, 2004).

However, this is not the whole truth. An equally important role was played by the state's ability to carry out large-scale mobilization of financial, labor, and material resources, without which not a single Arctic infrastructure megaproject can move from the discussion phase to the implementation phase. For densely developed areas, it is fashionable to talk about public–private partnerships, concession schemes for the implementation of large infrastructure projects. Indeed, when the size of the potential market is large, and the risks of unpredictability of demand in the future are minimal, it is natural to count on the involvement of corporate structures in the business of infrastructure development.

The fundamental difference in the Arctic is that initially there is no market here and its appearance in the future is far from obvious. That is why the first step in the infrastructural development of undeveloped and sparsely populated territory is always made by the state: on its will, energy, faith, the first road trail is paved. It is always the main actor in any pioneering infrastructure megaproject in the Arctic. Further, the efforts of the state are picked up and privatized by the following corporate mining structures. Regardless of the formal legal dress of the superstructures for the pioneering development of the uninhabited territories of the Arctic, the main actor and driver of the first infrastructural arrangement is the state and its ability to quasi-military mobilization of resources in the road arrangement of a new territory.

It may seem surprising that the implementation of Arctic infrastructure megaprojects depends on the specific economic model of public policy in the Arctic country. However, the recent history of Russia and other Arctic countries shows that the ability to make super-efforts, for intense mobilization of human, financial, and material resources does not exist constantly in time, but arises during periods of weakening or oblivion of the market paradigm, the ideas of liberalism of the Anglo-Saxonian economic school of Smith-Ricardo (Smith, 2017; Ricardo, 2016) and their followers, and, conversely, the triumph of the ideas of state protectionism, state regulation (dirigism) of the German school of national political economy by F. List (List, 2017). It turns out that the very possibility of implementing a large infrastructure megaproject in the Arctic depends on the ideological position, the economic worldview of those who make economic decisions in a particular Arctic country, their enthusiasm for the ideas of one or another economic school. The fact is that infrastructural megaprojects of the Arctic are almost always unattractive when judged by the market criteria of short-term profitability. Furthermore, only a long-term state approach ensures their viability. That is why they "come to life" precisely during periods of dominance of the ideology of dirigism, and not liberalism, in public economic policy.

Without the creation of the famous Kolyma highway, it would have been impossible to begin a large-scale resource development of gold-placer deposits in the Kolyma River basin in the North-East of Russia in the 1930s. However, as our calculations have shown, the main economic structure of the development of this territory, the Dalstroy trust, was profitable only during the first period of its existence, from 1932 to 1940, when the world price of mined gold was higher than the entire volume of state capital investments in the form of cash and material resources, technical equipment that came here from the center of the country. In the subsequent periods of its existence, from 1941 to 1956, the activities of the trust were no longer profitable, and geopolitical considerations came to the fore to "hold" the vast northeastern outskirts of the country, while retaining the influence of Soviet power on it (Pilyasov, 1996).

Another interesting feature of the Arctic infrastructure megaprojects is that many of them take a long time to prepare. As a rule, the discussion in society about the need to create a new transport corridor has been going on for decades and even centuries. However, the scale of the necessary mobilization of financial, labor, material assets that are required for this is such that it is never possible to move the project from the dead center of a simple expert discussions at once. Only after centuries and decades has the discussed megaproject finally received the rights of citizenship in a specific area of the Arctic. The trigger for this is usually a war and the threat of cutting off part of the country's territory due to the impossibility of a quick redeployment of military forces, or the threat of annexing part of the territory by a neighboring unfriendly Arctic country; an overly acute need for foreign exchange resources, which are provided by the strategic resource of the Arctic periphery supplied to foreign markets; and, much less often, the country's own internal need for a rare metal, an energy carrier, a substance that is no longer present in any other part of it.

The need to build the Murmansk railway was discussed in Tsarist Russia for decades, but the First World War unprecedentedly accelerated the implementation of this Arctic megaproject. Russia needed access to the ocean instead of Petrograd, cut off by the theater of military operations, with its access to Europe through the Baltic Sea. Likewise, the railway from the center of Russia to the coal-mining Vorkuta, which was built with unprecedented speed during the Second World War: Russia badly needed coal as it had just lost the occupied Donetsk coal basin.

The fact that many long-discussed Arctic infrastructure megaprojects were eventually initiated for practical implementation by military disasters and wars means that they have, "from birth," a dual purpose and dual use: both civilian and military. Depending on the specific geopolitical situation, they can be used for peaceful purposes of exporting strategically important resources from the Arctic to world markets or for conducting military exercises, military mobilizations, etc.

There are also specific, purely Arctic institutional (intangible) factors that are capable of temporarily closing or slowing down the construction of linear infrastructure facilities. The road infrastructure of the Arctic always affects vast uninhabited areas. It would also seem, from the point of view of the problem of property rights, that this is more a blessing than a burden: uninhabited, it means there is no claimant for them; that is, they are kind of ownerless or totally state-owned.

In reality, however, this is not the case. A significant part of the territory of the Arctic federations (Russian, American, and Canadian) is the place of traditional residence and economic activity – reindeer husbandry and traditional crafts – of the indigenous small peoples of the North. This means that it is necessary to settle land claims, informal rights to these lands and territories from their first inhabitants.

The practice of such settlement differs in different countries. In Alaska, the construction of the Trans-Alaska oil pipeline was simply stopped for several years to legislatively settle the land claims of indigenous peoples, as a result of which a federal law was adopted, according to which 12 national corporations with land ownership rights were created, with an initial budget contribution from the federal government and the right of every indigenous person to be their shareholders. In Russia, large-scale infrastructure construction in the Arctic is accompanied by the conclusion of agreements between the state and corporations with local indigenous communities.

In all cases, local indigenous organizations act as a significant actor in the implementation of any spatially extended and territorially "greedy" infrastructure project. The most exemplary solution for such infrastructural construction is not just agreeing with the indigenous people about laying a new ground highway, allocating compensation money and land to them in exchange for lost territories, but also about attracting their traditional knowledge and using their experience in new construction. The works of Russian classics on the theory of Northern development (Kosmachev, 1974) provide examples of the use of traditional knowledge and sledding deer and dogs of indigenous peoples at the pioneering stage of infrastructural and economic development of the territories of Eastern Siberia.

Permafrost in a significant part of the global Arctic (Atlas of Population, Society and Economy in the Arctic, 2019) creates specific obstacles to the creation of new infrastructure facilities. In order for the road route or landfill to remain in its original form and not "meander" under the influence of melting permafrost, special engineering efforts are required in the form of thermosiphons, forced cooling of the soil, etc. The practice of building the Trans-Alaska oil pipeline confirms that this leads to higher construction costs and an increase in its terms.

There is one more feature of the Arctic infrastructure megaprojects. The concentration and mobilization of labor, financial and material resources during the peak periods of construction is so great – many times higher than the usually necessary resources for the operation stage of this facility – that it causes a socio-psychological atmosphere of "fever", with an unprecedented concentration of people in the Arctic and resources. Numerous studies confirm (Goldsmith, 1989) that these extremely eventful periods generate numerous economic costs in the form of high inflation, traditional cost overruns of the originally planned budget funds, and social diseases such as alcoholism, crime, and violence. Regardless of the particular country or particular economic model the Arctic super-project is being implemented in, economic and social costs from it are inevitable.

The infrastructural arrangement of the Arctic, with all the generality of the peripheral remoteness and climatic extremity of the global Arctic, significantly depends on the specific country in which it is carried out, its size, economic and geographical position within the continent (primarily the number of neighboring countries and the degree of geopolitical tension in relations with them) and a specific economic model of state policy (liberal or dirigistic), which the authorities of this country are currently implementing.

The size of an Arctic country and its Arctic part is important for infrastructure megaprojects. The large Arctic countries (Russia, USA, and Canada), as follows from the analysis of the strategies of socio-economic development of all countries of the global Arctic (Heininen, 2011; Pilyasov, 2014), are usually more inward-facing, towards their infrastructure problems of ensuring transport accessibility, development of large deposits of natural resources onshore and offshore. By contrast, small Arctic countries (Sweden, Norway, Finland, Iceland) are more interested in the international agenda, issues of international cooperation, global climate change, and ecology. Therefore, the topic of large infrastructure megaprojects is primarily in demand by large Arctic federations – Russian, American, and Canadian.

However, there are important differences here too, which are determined by the originality of the economic and geographical position of each Arctic federation. Located in Eurasia, inhabited by many countries, the Russian Federation is forced to have a significant number of neighbors. Therefore, in addition to economic content, every major economic or infrastructure project in the bordering Arctic also has geopolitical significance. As the former chairman of the All-Russian State TV and Radio Company, Oleg Poptsov, joked, "even if we build a bathhouse on our Arctic islands, our foreign colleagues will still reproach us for creating a new defense facility in the Arctic."

Hence, the gigantic and even zealous attention of other polar countries to how Russia is developing its infrastructure in its Arctic is evident. For the simple fact of geography, Russia, which occupies about half of the land area of the global Arctic, is destined to be the leader in terms of initiating large infrastructure projects there. And they always have both economic and geopolitical significance.

On the other hand, there are few countries in the North American continent, and infrastructural megaprojects in the Arctic usually have an economic component here, and geopolitical ones are loaded to a much lesser extent and only in force majeure periods. During the Second World War, for example, part of the transboundary Alaska highway on Canadian territory was temporarily occupied by the American army. In quiet periods, even super-large construction projects such as the Trans-Alaska pipeline across Alaska from the Prudhoe Bay field in the north to the ice-free port of Valdez in the south of the state were a purely American phenomenon and did not create any geopolitical friction and tensions from other Arctic neighboring countries. Also, for decades, the project for the construction of a gas pipeline in the Mackenzie River Delta of Canada (Boland, 2016) has always been the subject of an exclusively Canadian domestic agenda, which has never caused international friction and tensions.

## Technology and Arctic Infrastructure Megaprojects

Of course, a particular economic era and its technological and technical power, the organizational shell of economic structures, has an impact on the process of infrastructure development in the Arctic. A map of the Russian Arctic shows two groups of territories: those with a solid, well-developed network of highways, and those with only sections of road infrastructure.

What is the reason for this? In the era of the early industrial development of the Soviet Arctic by the forces of integral combines (that is, the 1930s–1950s) (Slavin, 1961), using the manual labor of the GULAG prisoners and with minimal technical equipment, completely self-sufficient socio-economic centers, which were in fact mining corrective labor camps were created and often later transformed into single-industry cities. They were connected to each other by a single network of roads or railways. These are the Murmansk region, the Komi Republic, the Magadan region, and the Norilsk industrial district.

On the other hand, the territories and regions, the development of which had already started in the era of the domination of technologically more armed Soviet departmental trusts, in the 1960s–1980s, were influenced by the narrow economic paradigm of "resources without comprehensive development of the territory"; therefore, transport construction pursued the purpose of ensuring the export of the extracted resource to the nearest seaport-input development base or multimodal transport and logistics center. Therefore, an integral system of roads or railways in the Chukotka, Yamal-Nenets, Nenets Autonomous Okrugs, where pioneering economic development began already in the 1960s and later, did not arise.

The common difference between Arctic transport projects and similar projects in the developed zone is always a relatively high concentration of resources: per kilometer of the constructed road in the Arctic there are more financial, material, and human resources than on the mainland. However, the specific ratio of these types of resources varies with the technological era.

In the early industrial era (1930–1950s), the main resource on Soviet Arctic construction sites was the inexpensive physical strength of hundreds and thousands of workers relocated here voluntarily or forcibly, and the role of technical means was minimal. However, already in the late industrial era, the ratio of labor and material assets changed in favor of the latter: tractors and bulldozers came to replace the shovel and pickaxe. In the postindustrial era, as the recent history of the Arctic exploration shows, there is a widespread gravitation towards uncrowded, robotic, partially, or fully automated and mobile solutions in road infrastructure, using artificial intelligence technologies. This is also manifested in the pioneering road infrastructure in the Arctic (Table 15.1).

**Table 15.1** Comparison of the features of the Arctic transport megaproject in the industrial and post-industrial eras

| Features | Industrial | Post-industrial |
|---|---|---|
| Dynamics | Immobile | Mobile, fly-in fly-out (FIFO model) |
| Location | Predominantly onshore | Many coastal or offshore |
| Claims | Native land claims | Nation offshore claims |
| Dissipation in space | Areawide | Platform, island-like |
| Institutions | National law and regulation only | National, international law and regulation. Arctic Council regulation involvement |
| Age and time | Predominantly frontier character (first economic layer) | Predominantly already colonized character (second economic layer) |
| Umbrella structure | Comprehensive state program | Network of pilot projects/segments as PPP |

Source: Author's estimation

The previous pioneering development process in the industrial era resulted in numerous single-industry resource settlements along a new rail or road. Due to the predominant use of the rotational method of infrastructure construction and the operation of resource facilities, now, during pioneering development, new stationary single-industry settlements (almost always problematic in the long term, after the development of their resource facility) hardly ever arise.

For example, the project for the construction and operation of the Obskaya-Polunochnoye railway along the eastern slope of the Urals provided for a rotational method of organizing work with the resettlement of workers in modular, fully equipped residential premises; the use of mobile manufacturing facilities for servicing equipment. To accommodate shifts in areas of significant work (construction of medium, large and small bridges, small culverts), at the construction sites of railway sidings, it was envisaged to create temporary camps for builders and helipads.

The temporal organization of construction work is not subordinated to the principles of a linear (stage of pioneering, mature, old development) conveyor process – from the past to the future – but various works go in parallel. Instead of the previous stage sequence, a tight temporary conjugation is envisaged, the simultaneity of several stages of construction at different sites at once, which sharply reduces the period of pioneering development.

Initially, during the industrial development of gold-placer deposits in the Upper Kolyma basin in the 1930s, two equivalent input bases remained, through the Northern Sea Route and through Magadan, located on the coast of the Sea of Okhotsk. Then only one base survived: Magadan, through which the main supplies of goods and people to the mines of the Upper Kolyma were carried out. The peculiarity of the organization of modern infrastructure projects in the Arctic is that the system of several equivalent outpost bases for the area of pioneering development is preserved (none of them decays over time). The capabilities of new information and communication technologies make it possible to maintain reliable coordination between them from the area of new development and from the "mainland", which ensures their synchronous and effective, complementary work and interaction.

## Modern Features of Arctic Infrastructure Megaprojects: Globalization, Climate Change, and Attraction to the Sea

Modern processes of globalization taking place all over the world and stopped by the global COVID-19 pandemic gave rise to the hope that large infrastructure megaprojects in the Arctic could be implemented by the resources of the entire world community, not just the country in which they are deployed. And, it would seem, there were solid grounds for such a conviction in the form of the revived project of a tunnel through the Bering Strait, discussed in the 1990s, connecting the American and Russian Arctic, the American continent and Eurasia, the Belkomur project, etc.

The project of a railway tunnel through the Bering Strait, from American Alaska to Russian Chukotka, with subsequent integration with the Trans-Siberian Railway, was first discussed at the beginning of the twentieth century. To implement it, the tsarist government of Russia even created the Northeast Siberian Society, but then it stopped the design work, frightened by the potential American colonization of the entire territory of northeastern Asian Russia (Slavin, 1961). This again confirms that, in the Russian Arctic, any large transport mega-project necessarily, in addition to the economic component, also has a geopolitical dimension.

The project for the construction of the Belkomur railway line (White Sea–Komi–the Urals, 1252 km in total, including 712 km of new construction of two sections) involves reducing the shoulder of cargo delivery from the Urals and Siberia to the seaports of Arkhangelsk to 850 km. This is only one of the projects for connecting inland railways from Russian and foreign Asia, Russian Pacific ports with seaports on the Northern Sea Route (depending on the specific project, options for docking in the seaports of Arkhangelsk, Indiga, Ust-Kara, etc. are being considered). The investor, the Chinese company Poly International Holding Co., Ltd., was ready to invest up to US$5.5 billion[1] in Belkomur, but again the project did not take place. A new project for connecting the southern railways with the Northern Sea Route is currently being discussed via Sabetta as a newly built port of the Yamal LNG project.

The infrastructural power of China, which has proven to the world its ability to implement the most complex infrastructure projects in its country, aroused hope among some Arctic countries that, based on its experience and financial resources, it would be possible to dramatically activate the infrastructure development of the Arctic, create a new network of roads to resources. It seemed that China, which had put forward an ambitious Belt and Road Initiative of multiple infrastructural connections between Asia and Europe by sea and land railways, both in the Arctic and in the tropics, should take an interest in implementing large-scale capital-intensive infrastructure projects within the Eurasian Arctic.

However, the reality turned out to be different. China's interest in acquiring ownership of key elements of the already created infrastructure system of the Arctic in

---

[1] http://www.belkomur.com/press/index.php?ELEMENT_ID=2956. Accessed on December 29, 2020.

the form of ports, terminals, and other multimodal transport and logistics centers is evident. However, the desire to implement large infrastructure megaprojects in partnership with the Arctic country, without guarantees of full control over them, did not appear. The Arctic countries, as in the pre-globalization era, must implement these projects on their own. The Chinese investor is ready to come to the already created infrastructure facilities, but not to become a leader in their pioneering implementation.

The tendencies towards mitigation of the Arctic climate, which are more active and more powerful in the Russian sector than in the Canadian and American ones, cause many infrastructure megaprojects of the Arctic to naturally gravitate towards the formerly icy, and now more and more ice-free Northern Sea Route and seaports.

The Northern Sea Route is actually multiple routes that fluctuate in space and time depending on the climate and the level of ice coverage. The Northern Sea Route can go around the Novaya and Severnaya Zemlya archipelagos from the north, bypassing the straits that separate them from the coast (high-latitude route) or close to the coast in conditions of increased ice coverage of the high-latitude route: "Previously, we went along a more southern route, which is close to the coast, now since 2011 more to the north, where the depths allow ships with a draft of more than 11 meters and therefore a large displacement to come."[2] However, the difference between the present time is that this amplitude itself has increased significantly: in the Arctic Ocean, there is now significantly more light, rapidly migrating first-year ice, which, although it breaks more easily than long-term ice, is more insidious due to its high speed of movement.

Compared to the Soviet past, the general rhythm of the Northern Sea Route is also changing. Earlier, the main production user of the route, Norilsk Nickel, had discrete transportation of mined ore (accumulated over months of work – taken out). Now new users, such as oil and gas companies NovaTEK, Gazpromneft, Lukoil, have a more frequent work rhythm, hundreds of sea voyages a year. With such regularity, one can understand the comparison of Deputy Prime Minister Yu.P. Trutnev gas carriers and supply vessels that go from Murmansk and Arkhangelsk to Sabetta with electric trains.[3]

In the last decade, intensified use of the transport capabilities of the Northern Sea Route has formed the preconditions for reanimating another long-term project for the creation of the Northern Latitudinal Transport Corridor (Kryukov et al., 2016): the Pacific port of Vanino – the Baikal-Amur Mainline – and the Arctic ports of Arkhangelsk, Indiga, or Sabetta. The point is to implement the long-cherished idea of a railway transport corridor between the Pacific and the Arctic Ocean.

The main participants reached agreement regarding the construction of specific sections of this corridor at the end of the tenth years. We refer to the construction of two sections: Nadym-Salekhard, which will connect two separate branches of the

---

[2] From interviews with respondents along the NSR.
[3] Trutnev Yu.P. Oral communication. International Forum "Arctic – Territory of Dialogue". April 2019.

Northern and Sverdlovsk railways, and the Bovanenkovo-Sabetta railroad, which will give a new impetus to the development of the Bovanenkovo gas field, turning a dead-end point on the railroad into a transit point (and here, difficult negotiations are underway between the government of the Yamal-Nenets autonomous okrug, NovaTEK and Gazprom on the sharing of construction costs). So far, a much greater understanding has been reached on the construction of the first section. In 2018, a Concession Agreement was concluded for the financing, construction, and operation of the Nadym-Salekhard section, including the construction of a bridge across the Ob from Salekhard to Obskaya, between the project company Northern Latitudinal Passage LLC and the Federal Agency of Railway Transport.

## General Types and Special Cases of Arctic Infrastructure Megaprojects

Because the global Arctic is so large and diverse and the history of its large-scale industrial development goes back more than a hundred years, the presence of different types and cases of infrastructure projects is inevitable. The main types are determined by the latitudinal or meridian strike of the transport routes. Special cases are determined by the interaction of Arctic transport megaprojects with time: temporary roads, deferred infrastructure projects, and implemented but subsequently dismantled roads.

For example, it is known that the development of Siberia, which includes the southern, steppe, and taiga zones, and the zone of the northern forest-tundra and the Arctic tundra, took place along the system of meridian river systems (in the south-north-Arctic direction) and latitudinal portages (then tracts) that formed the territorial framework of the development process. During periods of active development of Siberia, latitudinal communication usually dominated along the created highways and railways and latitudinal transport channels, and during periods of pauses in development (compression of development), "natural" (physical–geographical) communication along the channels (basins) of the great Siberian rivers (Pelyasov, 2017). The modern period of the development of Russian Siberia and the Arctic is characterized by the simultaneous discussion of both meridian (for example, the Ural Industrial – Ural Polar megaproject) and latitudinal infrastructure projects (for example, the Northern Siberian Mainline, Belkomur, etc.).

A separate interesting case of a large infrastructure facility in the Arctic is temporary roads. These are not the seasonal winter roads that dominate year-round roads in many Arctic regions of the world, but railways and highways prepared for all-season operation, albeit built according to a "temporary low-cost scheme"; one might say, "light". Such schemes were typical for the large-scale and rapid oil and gas development of Western Siberia. Our respondents in the south of Yamal recalled: "tracks were laid, dumping, only one freight train passed a day. They moved at a speed of 10–15 km per hour. No more than 20 cars were allowed on the train.

And the railway – it was built in the swamp. And no matter what kind of dumping they did, after this freight train there was a repair staff … Railroad tracks – they were driving the shift teams. It was … 1979."[4]

Among the large infrastructure megaprojects in the Arctic, a separate (and numerous) class is made up of actively discussed but not implemented construction projects. In the global Arctic over the past decades, several such well-known transport projects have accumulated, emerging from time to time as if out of nowhere due to the new world situation, and then plunging back into the abyss of oblivion transport projects. This is a transcontinental railway tunnel through the Bering Strait, and the project to create a pipeline in the Mackenzie River Valley in Canada, and the already mentioned Belkomur.

Let us dwell in more detail on one more meridian transport megaproject, which was actively discussed in Russia at the beginning of the twenty-first century: "The Industrial Urals – The Polar Urals". The core of the transport component of the project was the Polunochnoye–Obskaya–Salekhard railway and the Tyumen–Urai–Agirish–Salekhard highway, which were routed along the eastern slope of the Urals Mountains. The idea of the project was to connect the southern industrial Urals represented by the Sverdlovsk and Chelyabinsk regions with the polar Urals with its underdeveloped natural resource potential and provide access to the Northern Sea Route by a transport corridor along the shortest line.

The megaproject aims to involve in the economic turnover the mineral and raw materials and fuel and energy resources of the Polar and Subpolar Urals (these are the Arctic Yamal-Nenets and northern Khanty-Mansiysk autonomous okrugs), in order to replace with them the long-distance delivery of raw materials for old industrial enterprises of the southern Urals, whose raw material base is depleted centuries-old exploitation. Its natural integrator is the Ural mountain system, its united ore belt, the integrity of which is emphasized by the laying of the Obskaya–Polunochnaya railway line along the eastern slope of the Ural Mountains. The original idea of the megaproject was the integration of the railroad and the highway with the formation of a single poly-highway along the eastern slope of the Ural Mountains.[5] It was assumed that most of the mineral raw materials transported by the Polunochnoye–Obskaya railroad will be the products of new mining enterprises of the Polar and Subpolar Urals.

A powerful research study was carried out (Transport Corridor …, 2009b; Spatial Paradigm …, 2009a), a detailed project implementation plan was developed, and a management organizational structure was created: a state corporation that became an integral part of all strategic documents of the country and received official status. However, despite all these positive circumstances, it was eventually postponed. The official explanation consisted of insufficient geological knowledge of the mineral resource base of the Polar and Subpolar Urals, but my personal explanation was the

---

[4] Report on the transport system of Siberia and the Far East. Russian Geographical Society. M. Institute of Regional Consulting. 2018.

[5] Transport strategy of the Russian Federation for the period up to 2030 (as amended by orders of the Government of the Russian Federation of 06/11/2014 N 1032-r, of 05/12/2018 N 893-r).

continued influence of the liberal lobby in the Russian government at that time, which opposed any state mobilization of material and financial resources for large infrastructure megaprojects, seeing in them "ineffective spending of budgetary funds". Of course, with narrow accounting calculations, any transport megaproject of national importance turns out to be unprofitable in the short term, according to market criteria. However, the same megaproject in the long term, with a comprehensive account of the economic and geopolitical component, often turned out, as the economic history of the global Arctic shows, to be both economically profitable and politically feasible.

In the economic history of the Arctic of the twentieth century, there is another class of infrastructure projects: implemented, but later abandoned or dismantled projects. Many of them, after playing the role of "breakthrough to resources" during the boom in economic development, then turned out to be too expensive to operate and were therefore abandoned. First of all, polar railways fall into this class. For example, narrow-gauge railways in the Magadan Region were dismantled, passenger traffic on the Dudinka–Norilsk railroad was stopped, and Gazprom stopped operating its Novy Urengoy–Yamburg railroad.[6]

The most tragic and most striking example in this class of Arctic infrastructure megaprojects is, of course, the abandoned Vorkuta (Chum)–Salekhard-Igarka railway, the "dead road", the 501/503 Stalinist construction site from 1947 to 1953. Its idea was to link the latitudinal Arctic territories of the Komi Republic, the Yamal-Nenets Autonomous Okrug and the north of the Krasnoyarsk Territory. There is no doubt that, under the conditions of modern economic development of these territories, it would have already paid off many times over.

After the death of J. Stalin, construction of the railway was stopped, and the railway itself was abandoned. Meanwhile, only 10 years remained until the large-scale oil and gas development of Western Siberia, in which the railway could be in high demand. In the mid-1960s, when the creation of the West Siberian oil and gas complex began, many prominent scientists and Communist party and economic leaders proposed to restore the abandoned railway and continue its construction. However, these proposals have not received practical implementation.

In 1979, academician A. G. Aganbegyan, assessing the economic effect that could be obtained as a result of the implementation of the recommendations of scientists, wrote: "It is now clear to economists that the decision of the ministries of transport construction and communication.

of the USSR, which refused to restore the road from Salekhard to Nadym and further to Urengoy ten years ago, turned out to be at least a billion-dollar loss for the country" (Lamin, Timoshenko, 2012).

On the example of the "Stalinist" 501/503 construction site, the polar railway line, we see that the fate of Arctic infrastructure megaprojects, like the fate of people, sometimes turns out to be very dramatic and unexpected.

---

[6] https://www.znak.com/2015-06-04/proekt_na_kotoryy_byli_potracheny_milliardy_koncernu_ne_nuzhen_foto. Accessed December 30, 2020.

## Conclusion

In infrastructure projects in the global Arctic, natural extremity is often "overcome" by geopolitical and economic extremes (war or new pioneering economic development); that is, circumstances that suddenly emerged as a result of global military and economic cataclysms that simply force the state to undertake large-scale and expensive transport infrastructure for its unpopulated spaces.

It would be wrong to expect a uniform new infrastructure development in the global Arctic in the future. All historical experience testifies to the opposite: it is always development by leaps and bounds, under the pressure of extraordinary circumstances that necessitate volitional mobilization of material, human, and financial resources. The infrastructural development of the Arctic has never been planned; it has always been the result of extraordinary efforts by the state and the partners it mobilizes in response to the challenge of radical changes in the external geopolitical and economic environment.

Large transport projects in the Arctic are always inscribed in a specific economic era and its philosophy and are firmly integrated with it. Therefore, in order to predictably assess the activity of their deployment, the main destinations, one must first determine the general context of the emerging new era.

What factors will determine the deployment of large infrastructure projects in the global Arctic in the next 10–15 years? In multiscale logic, three groups of such drivers can be identified: global, macroregional (intercontinental), and national (intracountry), which primarily refer to large Arctic federations: Russian, American, and Canadian.

In the group of planetary factors, the factors that depend on the evolution of the technological mode, the growing role of innovation and knowledge are distinguished: dynamics of globalization/de-globalization; the rapidly growing middle class of Asian countries (primarily China) as a new consumer of resource products and recreational and other services of the Arctic. Global warming can be considered as non-economic compulsion for humanity to quickly switch to a new technological mode based on "green" energy, resources (such as rare earth metals), nature-like solutions (for example, constructive use of cold to liquefy natural gas).

How this can affect transport megaprojects in the Arctic has already been clearly demonstrated by Russia. Here, the boom of NovaTEK's Arctic LNG projects has led to the implementation of a large-scale program of radical redesign of the format for the use of the Arctic seas and the Northern Sea Route in terms of a significant reduction in the load on the icebreaker fleet and support on new specialized vessels (gas carriers) of the reinforced ice class, which are much more maneuverable and provide significant economies of scale, but require the creation of special transshipment facilities for reloading fuel to conventional ships that supply European and Asian markets. New transport megaprojects in the Arctic will undoubtedly respond to the values and priorities that are asserted by the emerging new technological order based on artificial intelligence, networked forms of economic organization, and significantly greater environmental friendliness.

It can be argued that in new infrastructural projects in the Arctic, a significantly greater role than before will be played not by the "routes" themselves – channels of penetration of the developed into the undeveloped – but completely new vehicles and a high-tech service for accompanying the transport process. This is where the fastest progress will be in the coming years. However, the very environment for the deployment of new transport projects in the Arctic will change and become more probabilistic and non-stationary: there is already a clear trend away from land to water, sea, and river environments in new transport routes in the global Arctic. In this sense, the comparison of the exploration of the Arctic with the exploration of space and the depths of the oceans, which is fashionable today, effectively conveys the nature of the new environment for the deployment of economic activity in the Arctic, which is gradually beginning to prevail.

Global changes in the natural environment and climate are pushing to the top infrastructure megaprojects that either run in the Arctic seas, the Arctic Ocean, the Northern Sea Route, or the Northwest Passage, or have access, are confined to the Arctic seas and seaports. This will become a characteristic trend in the near future in the global Arctic.

Globalization in the former Anglo-Saxonian version, asserted in recent decades, has exhausted itself. The conflicts and contradictions it generates between countries are too great. For the Arctic, this means that new mega-projects will be implemented either in a new, Chinese, version of globalization, as an integral part of the infrastructure initiative "One Belt, One Road", or as an internal agenda of large Arctic federations, as the fruit and result of their national efforts, with self-reliance and resources. The tendencies of liberal globalization and intercountry economic integration that emerged in the 1990s will be replaced by processes of national sovereignty and dissipation of the former large interstate economic and political blocs. For Arctic transport projects, this will mean strengthening the role in their deployment at the national level.

A clarification is in order here. New consumers of Arctic goods and services from Asian countries – China, India, presumably South Korea, Singapore, Vietnam, etc. – are already physically and virtually present in the Arctic as investors, tourists, researchers, and politicians. This trend, due to the rapid growth of the middle class in these countries, will intensify; they will become carriers of (or have an impact on) new priorities, values, and even tastes (demand for cold, snow, and winter) in many Arctic projects, including transport. The rise of the Arctic Asia will be inevitable and it (primarily the Russian Taimyr) will become the fastest growing part of the circumpolar North. This means that the various transport projects "from Asia to Asia" (sea, air, river) – that is, from the Arctic to South and Southeast Asia – have a high chance of being realized.

Macro-regional drivers of the development of Arctic transport projects are associated primarily with the intercontinental transit of goods and people. The Russian realities of the last 20 years show that the expectations of the 2000s that transit will provide the main load of the Northern Sea Route did not come true. In the medium term, in view of the emerging trend towards the autonomization of markets and the sovereignty of the Arctic countries, a departure from economic liberalism to dirigism, it is more logical to expect a faster development of coastal, sectional, and in-country transportation in the Arctic than the rapid development of international transit.

It is the national drivers, the strength of desire for the transport arrangement of their Arctic of large Arctic federations that will be the main factors in the deployment of new and modernization of existing routes and development bases. There are already numerous confirmations of this, such as newly built port of Sabetta in the Russian Arctic (a potentially connecting hub between inland railways and the Northern Sea Route), or plans to turn the settlement of Nome in Alaska into a major strategic US seaport in the Arctic. These trends indicate the desire of the large Arctic countries in the next 10–15 years to "reassemble" those elements of the national transport infrastructure that were previously postponed. They also clearly indicate the trend towards sovereignty, which is indicated in the behavior of the Arctic federations.

Therefore, one can see a dual nature in the processes already emerging today and undoubtedly having significant potential for prolongation. On one hand, transport projects will strengthen the international ties of the Arctic with the main consumers in Asia, Europe, America; on the other hand, there will be strong tendencies towards sovereignty, strengthening national control and integrity over their Arctic territories. In Russia, this will manifest itself in new transport projects that bind municipalities of the Arctic façade (for example, regular or charter pan-Arctic flights will be possible), in Alaska in the formation of new seaports, in northern Canada in an effort to reduce the American "tutelage" of the Northwest Passage. In most cases, land transport projects in the Arctic will not be "greenfield", but "brownfield"; that is, they will be like "plug-in sections", the completion of existing highways and railways to ensure their new through transit, previously interrupted by off-road sections.

It is imperative to add to this assessment an element of surprise and unpredictability, which is always great in any project, including transport, in the Arctic. Therefore, it is impossible to exclude the sudden emergence of an absolutely new "surprise" transport project, which will have a significant impact on the entire context of the development of the global Arctic.

Infrastructure megaprojects of the global Arctic are always implemented in an atmosphere of discussion: to develop or to preserve, keep these extreme areas of the Earth intact? I am convinced that the Arctic was given to Humanity not only to admire it, but also to train their courage and intelligence in the resource development of this last frontier of the Earth.

# References

Armstrong, T., Rogers, G., & Rowley, G. (1978). *The Circumpolar North. A political and economic geography of the Arctic and Sub-Arctic* (303p). Methuen & Co Ltd.

Atlas of Population, Society and Economy in the Arctic, Jungsberg, L., Turunen, E., Heleniak, T., Wang, S., Ramage, J., & Roto, J. (2019). *Nordregio Working Paper, 3*(Map 7), 65. Accessibility and infrastructure, pp. 64–70.

Boland, K. (2016). Unexpected possibilities. Arctic ports and northern corridors in transition. In J. Higginbotham & J. Spence (Eds.), *North of 60. Toward a renewed Canadian Arctic Agenda. Special Report* (pp. 27–34).

Goldsmith, O. S. (1989). *Analyzing economic impact in Alaska*. ISER.

Heininen, L. (2011). *Arctic strategies and policies: Inventory and comparative study* (95 p). Northern Research Forum. University of Lapland.

Khorev, B. S. (1981). *Territorial organization of society: Actual problems of regional management and planning in the USSR*. Mysl Publ. (in Russian).

Kosmachev, K. P. (1974). *Pioneering development in the taiga* (144 p). Nauka Publ. (in Russian)

Katsov, V. M., & Porfiriev, B. N. (2014). Climate change in the Arctic: Implications for the environment and the economy. In A. I. Tatarkin (Ed.), *Russian Arctic: a modern development paradigm* (pp. 197–222). Nestor istoriya Publ. (in Russian).

Kryukov, V. A., Malov, V. Yu., Tokarev, A. N., Blam, Yu. Sh., and Churashev, V. N. (2016). Assessment of the prospects for the creation of the Northern latitudinal transport corridor. ECO. No. 5. (in Russian)

Lamin, V. A., & Timoshenko, A. I. (2012). The role of transport routes in the development of Siberia. *Ural Historical Bulletin, 2*(35), 37–47. (in Russian).

Lappo, G. M. (1983). The concept of the supporting frame of the territorial structure of the national economy: Development, theoretical and practical significance, Izvestiya AN SSSR. [Bulletin of the Soviet Academy of sciences]. *Ser. Geographic, 5*, 16–28. (in Russian).

List F. (2017). *National system of political economy* (451 p). Socium Publ. (in Russian).

MacDonald, R. S. J. (Ed.). (1966). *The Arctic frontier* (310 p). University of Toronto Press.

Mosunov, V. P., Nikulnikov, Y. S., & Sysoev, A. A. (1990). Territorial structures of areas of new development. In K. P. Kosmachev (Ed.), (p. 149). Nauka Publ. (in Russian).

Navasardov, A. S. (2004). *The development of the North-East of the USSR in the 30s. XX century* (211 p). North-Eastern Research Center, Far Eastern Branch, Russian Academy of Sciences. (in Russian).

Pelyasov, A. (2017). Siberia: In search of new model of development. *Journal of Siberian Federal University. Humanities & Social Sciences, 11*(10), 1754–1778.

Pilyasov, A. N. (1996). *Regularities and specificness of the development of the North-East of Russia (retrospective and forecast)* (145 p). North-Eastern Comprehensive Research Institute, Far Eastern Branch, Russian Academy of Sciences. (in Russian).

Pilyasov, A. N. (2014). Comparative analysis of National Arctic Strategies. In V. S. Yagya, M. L. Lagutina, & T. S. Nemchinova (Eds.), *Actual problems of world politics in the XXI century: Collection of articles* (pp. 308–341) (in Russian).

Ricardo, D. (2016). *Principles of political economy and taxation*. EKSMO Publ. (in Russian).

Shpak, A. V. (2012). On the development of a transport and logistics system in the Arctic zone of the Russian Federation and the Murmansk region. *The North Industrial, 1*, 18–25. (in Russian).

Slavin, S. V. (1961). *Industrial and transport development of the North of the USSR*. Economizdat Publ. (in Russian).

Slavin, S. V. (1982). *Development of the North of the Soviet Union*. Nauka Publ. (in Russian).

Smith, A. (2017). *Research on the nature and causes of the wealth of peoples*. Lenand Publ. (in Russian).

Tatarkin, A. I. (Ed.). (2009a). *The spatial paradigm of the development of little-studied territories Experience, problems, solutions. Volume 1 and 2*. Institute of Economics, Ural Branch of RAS. (in Russian).

Tatarkin, A. I. (Ed.). (2009b). *Transport corridor "Industrial Ural – Ural Polar": Results and prospects* (298 p). . Institute of Economics, Ural Branch of RAS. (in Russian).

**Alexander Pilyasov** is a full Professor at Lomonosov Moscow State University, Moscow, Russia, general Director of the ANO "Institute of Regional Consulting", Moscow, and chief research associate at the Kola Luzin Institute of economic problems, Russian Academy of Sciences, Apatity, Russia.

# Chapter 16
# New Arctic Seaways and the Role of China in Regime Formation

**Mariia Kobzeva**

## Introduction

Arctic seaways are likely to be ice-free by the 2030s (National Centers for Environmental Information, 2020). The possibility of beneficial shipping of hydrocarbons from Arctic land and shelf via the sea routes has now been proven, with non-Arctic states making a remarkable contribution. China, which is one of the most active Asian actors in the region, chose shipping as the point of application for its Arctic politics. Specifically, the PRC elaborated the initiative of the Ice Silk Road construction and introduced several ideas to international discourse on how to improve the existing regime (Council of the People's Republic of China, 2018). Although China's involvement in shipping has been discussed extensively, China's role in regime making is understudied (Huang et al., 2015). In this regard, our key research question is: What is the role of the new circumpolar seaways for China's involvement in the Arctic regime-making?

The Arctic regime is gradually moving from liberal order based on dialogue to a more realist one (Bertelsen and Gallucci, 2016; Nilsen, 2020). China joined the Arctic development in the era of liberal norms when non-regional actors were considered as positive contributors to the Arctic regime formation (Young, 2012b). Geographically unable to become an equally important actor in the region as Arctic states, China has adopted the rules of cooperation. At the same time, China became a key challenger to the US and the liberal order – a process that was not related to the Arctic. This created the following paradox: while initially establishing its Arctic behavior according to liberal norms, China turned out to be a challenger that provokes key actors to accelerate the change in the Arctic regime towards real politics (Kobzeva, 2020).

M. Kobzeva (✉)
UiT the Arctic University of Norway, Tromsø, Norway
e-mail: mariia.kobzeva@uit.no

In this chapter, we examine these major political and economic shifts in the Arctic through the regime theory perspective. According to the classical definition, a regime presents a set of principles, norms, rules, and decision-making procedures (Krasner, 1982). It also reflects a system condition after each essential change of its components. Specifically, it shows how the new law adoption, or including new actors in the decision-making process, etc., influence other components and how a regime forms a more dense or conversely loose structure.

The regime is often fragmented into sub-regimes that address specific areas and develop on different international platforms. The Arctic regime also has such a complex structure, where instead of universal law, it has regulations on environmental security, navigation rules, etc. (Young, 2012a). The capacity of each actor to influence the regime depends on its position in the sub-regime, its own intention to change or abide by the current regulations, and the actor's power to initiate and promote the change. Great powers hold more capacities to changing the regime in their favor by having more human capital, better technologies, and finances (Drezner, 2009). For this reason, the procedure of states' participation in the regime formation illustrates mechanisms of power distribution in the regime framework.

The current Arctic regime is based on liberal norms and international law addressed to all states. At the same time, the existing legal framework is not solid enough and certain rules remain debatable (Young, 2014). This means that discontented actors must either adapt to or change the most disappointing aspects. Among the key spheres of Arctic development, shipping provides one of the best agendas for discussions with non-Arctic states (Havnes, 2020). First, it does not necessarily engage territories of the sovereign countries compared to oil and gas extraction. Second, navigation rules have already become a subject of discrepancy between the Arctic major powers, the US, and Russia. China also has its own vision regarding shipping (Kobzeva, 2020). This point emphasizes the need for other states to elaborate a response to such a challenge. In addition, navigation is one of the most dynamic sub-regimes in the Arctic since most of the adopted regulations address it, either explicitly or implicitly.

We shall first examine China's engagement in key Arctic seaways development and identify the opportunities that China can obtain there in terms of regime-making. Second, we discuss China's position in the Arctic shipping regime, its key thesis, and tools for influence. The research data include publications of official speeches and documents on official websites of Arctic states and China, as well as publications of the state media, expert publications, and analytical reports in English, Russian, and Chinese.

## China's Participation in Arctic Seaways Development

In 2018, China published its Arctic policy and voiced its interests in the region. Since then, the idea of developing Arctic seaways officially reached the top of China's Arctic agenda (Council of the People's Republic of China, 2018). As one of the world-leading sea powers, China plans to develop 'blue corridors'; that is, new global seaways that include the Northeast, the Northwest, and the Central Arctic passages (Cyberspace Administration of China, 2015; Xinhua, 2017c). To use opportunities in the Arctic, China developed the idea of the 'Ice' Silk Road' (ISR), the northern branch of China's key global initiative of the Belt and Road (BRI) (Xinhua, 2017b).

The ISR expects to provide a synergetic effect of development of the new seaways and to support a new energy system. These prospects motivated China to pursue being a competitive Arctic actor. By 2020, China's contribution to extracting hydrocarbons was comparable to those of leading European non-Arctic states like France and Germany. At the same time, Arctic shipping remains a high-maintenance item for China (Huang et al., 2015). The critical evaluation of China's readiness for Arctic shipping shows a moderate level of development in all areas. In concrete terms, the risk-bearing voyages of COSCO and other companies on the NEP remain rare and are perceived as achievements rather than routine. China has no naval ice-class ships (aside from supposed use of submarines under the Arctic ice). The Chinese contribution to Arctic cruise tourism is made by visitors, but not vessels. Finally, icebreaker shipbuilding continues to focus on research vessels. The first and second Xue Longs do not have a relevant ice-class to conduct year-round navigation in the region. The expected nuclear icebreaker will also be a "trial balloon" for Chinese polar navigation. In fact, all the vessels serve to collect information and obtain experience in navigation, but not to support the sea trade (Kobzeva, 2019).

Further, we will evaluate China's cooperation with Arctic states in the shipping area and opportunities for regime-making that China gets due to the development of three Arctic seaways.

## *The North-East Passage*

The North-East Passage (NEP) flows around the Russian and Scandinavian Arctic coastline and is the new seaway connecting China (Asia) and Europe. The NEP is the most developed regarding Arctic navigation. It also encompasses the largest ongoing and future extraction projects. Naturally, for China, the NEP is a priority among all three seaways. It attracts the most dynamic part of China's Arctic politics and becomes a focus area and a testing ground for all the ideas that China elaborates to promote its positions in the Arctic, including the Ice Silk Road.

China's cooperation here is situated in the two-folded political framework. One part falls into bilateral relations with the key shipping regulator nation – Russia, which adheres to its status-quo but is the most amiable partner regarding China's investments (Sergunin and Konyshev, 2016). Another part involves the Scandic nations. They have fewer coastal areas involved in Arctic shipping besides Norway and Iceland, but they are important contributors to the elaboration of norms and rules of shipping. Each of these states has its own developed relations with China. At the same time, as US allies, they are becoming increasingly cautious in terms of consuming Chinese investments. This situation has led to a presentiment of the new dilemma when the Arctic will be cleaved into two parts: pro-Chinese and pro-American (Bertelsen, 2020).

The cooperation with the Russian Federation is fairly successful (Kobzeva, 2020). Amid China's modest current capacities to become a full-fledged Arctic shipping nation, its commitment to cooperate is precious for Russia, which requires investments into the NSR development. Against the backdrop of "comprehensive partnership and strategic cooperation", this field falls into the scope of a large-scale bilateral political agenda. It is already recognized as having become a method of interaction between the Chinese ISR and the Russian plan of the Northern Sea Route (NSR) development that were called consonant initiatives (President of Russia, 2019b). Potentially, it seems a possible complement to the so-called docking between the Eurasian Economic Union (EAEU) being led by Moscow and the Belt and Road Initiative being guided from Beijing (President of Russia, 2019c).

Such political benevolence is firmed by direct leadership support that sees raising global markets connectivity via the Arctic as a contribution to multipolarity (President of Russia, 2018). The cooperation is framed by a series of bilateral statements and various working mechanisms that cover both the law and economic aspects of Arctic shipping. Both states support the organizations involved (TASS, 2018), which gives China free rein to put its efforts into shipping development along the Russian coast. All types of shipping, including scientific expeditions, oil exploration on the shelf, and cargo shipping, take place here. In addition, Chinese visitors take a significant part in tourist cruise voyages in this part of the Arctic (Russian Arctic National Park, 2018).

The key feature of Chinese cooperation with Russia here is a clear linkage with extracting projects. While Arctic shipping shapes into a new lane for the Arctic energy system, China becomes an investor, supplier, and consumer of it. Chinese vessels remain rare guests on the NSR, however Chinese companies contribute to infrastructure projects. Russian adjacent railway logistics and new LNG hubs are under discussion, and Sabetta port is already operating. Sabetta united the logistics of the two LNG enterprises with Chinese participation on Yamal and Gydan and became a main shipping destination in the Russian Arctic. The long-term agreement between the Russian (Sovkomflot, Novatek) and Chinese companies (COSCO, Silk Road Fund) outlined the willingness for joint work on logistics and vessel construction for cargo flow to the Asia Pacific as well as transit from Europe to Asia (Neftegaz.RU, 2019).

Despite the short- and mid-term political and economic advantages being obvious for both China and Russia, it is not the same in the security field. There are assumptions that ISR, in the case of going via Russian territory, may be a workaround for the limitations of the First and the Second Island Chains (Li et al., 2014). However, it is worth mentioning that the Bering strait, in its narrowest part, is only twice the width of Malacca and presents a similarly complicated bottleneck in direct proximity to the US and its allies in Asia-Pacific. Thus, the strategic role of shipping via NEP for China is highly dependent on Russia's position in any conflict.

All things considered, China's weaker position in the Arctic amid the opportunities to cooperate have led to the fact that its attitude to a regime here is the best compromise. Instead of promoting claims made in the White Paper, China officially recognized its readiness to take Russian regulations on the NSR into account (President of Russia, 2019a). Thus, China has turned out to be a supporter of the current regime set along the Russian Arctic coast.

Cooperation with Scandinavia is developing amid the uneasy political situation. Since Nordic states are acting out of concord with each other, it has become difficult for China to harmonize the controversial bilateral agenda with its Arctic ideas. Thus, one option for China is to overlap these vectors with a comprehensive initiative, which is again the construction of the Belt and Road. Certain projects in continental Nordic states have already fallen into this mixed reality and serve to promote the BRI (China Briefing News, 2018). An example is the idea of constructing the Nordic Arctic corridor: the railroad from Kirkenes in Norway via Finland and then the tunnel to Estonia (Kallio, 2017).

Shipping is a well-established area for cooperation between China and, independently, Denmark, Norway, Iceland, and Finland. In recent decades, the Nordic states have contributed significantly to the technological development of Chinese shipbuilding and navigation (Sverdrup-Thygeson, 2017; Tulupov, 2013). Specifically, the Finnish Aker Arctic became a solid partner for the modernization and design of Chinese scientific icebreakers, whereas Iceland has purchased ice-class vessels produced in China (Eimskip, 2020; Forsby, 2017).

Although the direct Arctic dimension of relations with China is quite weak, the BRI trail is becoming clearer. It worth mentioning that some states have shown interest in a more active partnership. For instance, Finland supported both the initiative and the idea of cooperation in the Arctic (Embassy of the PRC in the Republic of Finland, 2017; Embassy of the PRC in the Republic of Finland, 2020). Another example is Iceland, which has expressed interest in cooperation, has established the free-trade zone, and is still considering becoming a partner in the BRI (Embassy of the PRC in the Republic of Iceland, 2012; The American Presidency Project, 2019).

In the same way as for Russia, infrastructure and shipping development around the Scandinavian Arctic relates to extraction projects. China is an old partner (supplier and customer) in many Nordic enterprises, including oil and gas production. It worth mentioning the collaboration between Statoil and key Chinese corporations in extraction, engineering, and mining on the Norwegian shelf, as well as purchasing seismic data, LPG, and stakes in Norwegian extracting companies (equinor.com, 2001; ISDP, 2016). Another significant Chinese presence is in Greenland, mostly with projects for extracting minerals, and in Iceland with oil and gas exploration in the offshore Dreki area (Government of Greenland, 2015).

At the same time, any attempt by China to anchor itself in the region via buying land or infrastructure (Norway, Iceland, Greenland cases) has triggered extreme alarmism and killed any initiative at the grassroots level (H. R. Nilsen and Ellingsen, 2015). Currently, one of the most debated topics among Nordic experts in Arctic affairs is the role of the so-called GIUK (Greenland-Iceland-UK) gap for China's Arctic shipping, trade, and military strategy (Forsby, 2017). The progress on the independence of Greenland and Faroes from Denmark plays to fears of China's presence at the core of the NATO region. This places any Chinese project in the area into the framework of the Sino-American rivalry (Lanteigne and Shi, 2020).

China has shown itself as an ambitious actor and has tried not to miss opportunities to raise the profile as a valuable partner. Playing the card of Iceland's independent state, China supported the small economy in crises (Komissina, 2015). In 2019, China lobbied hard for the interests of its business in Faroese (Satariano, 2019). China's participation in the Arctic regime development yields some dividends when all the Nordic states supported the granting of the permanent observer status to China (however, they have not been the key objectors). At the same time, some of the expectations of Chinese experts, that it will be possible to get more supporters of the small Northern states regarding a more inclusive regime or promoting freedom of navigation, look far too idealistic now (Yu, 2016).

To sum up, in terms of regime, the NEP area is the best-elaborated among the three seaways: the functioning combination of the international and national laws and norms set by the Arctic Council (UNCLOS, the Polar Code, and the national legislation, mainly on the Russian side). Currently, the NEP's framework is more rigid than ever. Russia and Norway strictly prioritize its Arctic sovereignty, which China lacks. The chance to find a supporter of China's idea of a shared future in the Arctic and freedom of navigation among the small Nordic countries has failed due to anti-Chinese political sentiments. The long-discussed possible interest of parties like Iceland and Greenland to support China's involvement to balance their own positions among powerful coastal Arctic states currently seems unlikely. Thus, China's role as an innovator of the Arctic regime in this part of the High North is now on the decline side.

## *The North-West Passage (NWP)*

It is now safe to say that the Ice Silk Road will not go via the NWP. However, it is also worth noting that the opposite scenario had all the conditions to become real. For many years, Chinese business worked extensively to invest extracting industry in Canada and considered the US Alaska. These oil, gas, and mineral projects could provide a reason for the cost-intensive shipping development in this area with a lack of infrastructure (China.org.cn, 2016). However, the North American energy system via the Arctic to Asia was nipped in the bud.

Currently, amid raised tensions, it may seem hard to believe that, just a couple of years ago, Arctic history could have been changed here by establishing the Sino-American enterprise Alaska LNG (Kobzeva, 2018). This project was supposed to involve Sinopec, CIC Capital Corporation, and Bank of China, with strong financial support from the PRC. It was expected to deliver 15 million tons of LNG to China by tankers, which is five times more than from Russian Yamal LNG (the whole Alaskan project required 40 billion USD investments; 20 million tons of LNG annually could have been delivered to the port terminal in Prudhoe Bay) (DeMarban,

2017). However, neither preliminary high-level agreements nor the active position of Bill Walker's administration in Alaska or China General Chamber of Commerce-USA lobbying were able to succeed (China Daily, 2015). By the end of 2020, the cooperation disappeared completely and politicization took over economic considerations. This decision also made Russia the main and the only one key partner of China in the Arctic.

Cooperation in the Canadian Arctic that promised the establishment of the Free Trade Zone has also experienced difficulties (Fekete and Kennedy, 2012). Beijing has not become a trustworthy partner for Canadian establishment in spite of Chinese abidingness to Russian national regulations in the NSR and its poor capacities for Arctic shipment. The one attempt by the COSCO vessel to navigate along the Canadian coast fueled Ottawa's fears of China as a possible security threat (Lackenbauer et al., 2018), despite interest from Canadian entrepreneurs in such a shipping development to Asia (DRY CARGO International, 2014).

The practical ground for further cooperation with the US and Canada on the NWP remains, thanks to the business ties. A good example is the cooperation agreement of 2016 between the American Bureau of Shipping and COSCO regarding development of technologies and construction of ice-class vessels for Arctic navigation (the similar kind of cooperation as the COSCO established with Sovcomflot) (ABS, 2016).

Nevertheless, China's influence on the Arctic regime here is nullified, both due to the reasons mentioned above and, not least, due to China's demonstrative behavior (Ryan and Lamothe, 2015). In theory, the NWP has a similar regime basis as the NEP, but the US position against the UNCLOS ratification has complicated the dialogue on legal issues, even in favorable situations. There have been some Chinese expert discussions that the PRC might cooperate with the US to improve the legal mechanisms for navigation in the Bering Strait (Li et al., 2014). Right now, the hostile political juncture has malformed the regime framework and pushed China out of the legal regime-making.

## *The Central Passage (CP)*

Although the Central Passage is still hardly navigable, it could, according to recent assessments, be free of ice during navigation by around 2040. The passage has certain advantages. It makes it possible to reach Europe from China almost five days faster than via the NEP. It also crosses the so-called 'area' where China has a legal right to exploration and development (along with all other states). In addition, the CP has no fixed channels and may make it possible to avoid the Arctic states' EEZs, which usually have more strict regulations (Østreng et al., 2013). Thus, the navigation here may provide a beneficial alternative for China compared to two passages along the coast of Russia and Canada (Bennett et al., 2020).

China mentioned its interest in this passage in the White Paper and made two symbolic voyages: the eighth expedition of the Xue Long and the first Arctic voyage of the Xue Long 2 (Xinhua, 2017a). The latest voyage included sampling from the Gakkel Ridge sea bottom, which will likely be part of an 'Area' after the final delimitation of the Arctic shelf (Xinhua, 2020).

The CP regime is the loosest among the Arctic passages. This provides China with unique opportunities in terms of regime-making and places the Asian state as one of the key contributors to it. Since there are no legal reasons to limit China's participation in international discussions on the CP regime, future shipping along the passage is likely to be inextricably linked with Beijing Arctic politics. Currently, China has no technical capacities and knowledge to conduct such voyages; however, two decades seem to be enough time for working on such a task. In this regard, Chinese experts consider opportunities for China to work in advance and promote its ideas in the UN organizations, taking UNCLOS as a basis for argumentation (Li and Hu, 2015).

## China's Capacity to Influence the Arctic Regime

It has been argued that the PRC tends not to respect international rules (Kroenig and Cimmino, 2020). However, China's approach towards the Arctic regime is principally different. From the first decade of the twenty-first century, when Chinese experts still ventured opinions regarding the internationalization of the Arctic, China has made efforts to show all Arctic states and especially Russia – its key counterpart in the Arctic – its respect for existing law (Todorov, 2019). China's basic step to prove its reliability as a partner was the adoption of the Arctic Council's rules in 2013 to obtain a symbolic permanent observer status and then with publication of the White Paper (Council of the People's Republic of China, 2018). The one explanation for China's approach is that a regime supported by great powers (in the Arctic by the US and Russia) makes it possible to detect disobedient parties, casting them in a negative light (Drezner, 2009). This is contrary to China's interests in the Arctic, where the state strives to sustain a positive image of a responsible power.

The consequences of China's give-and-take approach is two-fold. The first is that since China joined the regime framework, it has had legal rights in the region. According to the letter of international law, this includes freedom of navigation and marine activities in the Area. Here comes the main discrepancy field, where China could become a challenger to Russia and Canada. The two Arctic states insist on the primacy of national law in areas covered by ice for more than six months, which is possible according to Article 234 of UNCLOS (United Nations, 1982). They also prefer national control over any passages in the EEZ. China interprets these issues in a different way (which can be learned only by reading Chinese expert publications since there was no official statement on this matter) (Yang, 2014).

The second consequence is that, as a legal Arctic actor, China has the right to participate in the governance of the region and to suggest ideas for regime improvement. Such a perspective clarifies that China's self-description as a "near-Arctic state" is not a claim for sovereignty, but for participation in Arctic development and regime improvement. This serves as an extension to the CPC belief that the international system reflects only Western needs, is unequal and unfair, and should therefore be improved. At that, to bolster its profile, China has turned the Western states' accusations of discrimination of human rights domestically into the idea of discrimination of states' rights in the international arena (Denisov and Adamova, 2017).

Regarding the Arctic, this approach is accumulated in China's key thesis on the Arctic as a common heritage of all mankind – a reflection of the concept that Hu Jintao introduced to the 18th National Congress in 2012 (China Current Affairs, 2012). Another working thesis is on the role of China as a "rights campaigner" of other "discriminated" states who do not have enough rights to voice and are excluded from the legal debates (Lanteigne, 2020). Thus, China's position becomes yet another factor that shifts the Arctic regime agenda from the internal focus to the global one.

There are regulatory lacunas that afford both short-term and long-term benefits for China. The first group may help to cut down expenses on technical requirements and services for shipping. It includes China having more active participation in the further development of the Polar Code (Liu and Hossain, 2017) and lobbying for the relaxation of the navigation rules along the NSR that Russia is able to agree with (Kobzeva, 2020). For instance, it applies further discussions regarding mandatory Russian icebreaker assistance and ice pilotage.

The second part involves ensuring China's rights in the region in the future. This is about China's contribution to the governance of the 'Area', including after the new delimitation approved by the UN (Liao, 2012). It also involves rules for fishing quotas in the Central Arctic – the issue that has already attracted China's support during elaboration of the Agreement of 2018 (Government of Canada, 2018). Finally, the elaboration of regulations for shipping in the Central Arctic Ocean and the possible promotion of the idea of the innocent passage rights remain in the White Paper text. Thus, abiding the Russian norms, China leaves itself a space to maintain its positions in the Arctic in case if circumstances change (Rainwater, 2015).

In the course of active Chinese scientific and diplomatic work in the Arctic, two directions crystallized for involvement in the current regime development. The first way is to join and remain active on all discussion platforms. For this reason, China has become a member of many Arctic fora. The most effective for China's politics are UN organizations, including the International Seabed Authority. China has proved itself as an active contributor in this structure by providing human capital, financial support, and technological experience of licensed enterprise able to do the exploration of ferromanganese crusts (Magnússon, 2015; International Seabed Authority, 2014).

The second way is for China to shape its own Arctic identity and promote ideas for an international audience (Ding and Zhang, 2016). Amid the worsened rivalry with the US, the Chinese leadership has paid great attention to new political tactics that have to be effective but not fuel controversies (He and Song, 2013). One of such approaches is the development of China's international discourse power (for an international audience, China also uses the term "right to voice" as an alternative translation). This kind of power, in the eyes of the CPC ideologists, becomes the key instrument for changing rules to affect the regime or to promote ideas for other countries (Zhao, 2016). The indicator of the discourse power effectiveness is the new international standards or narrations adopted by others, at the level of the UN at best (Zhang, 2020).

The Arctic has become a foothold for China's discourse power development. This instrument helps to design a new image of China as a new world center of inclusiveness and universalism. In this regard, the ISR as an Arctic branch of the BRI becomes a centerpiece for China's Arctic politics and accumulates the new narratives and principles of cooperation (Liu, 2020). The discourse power implies various instruments, and the most relevant of them for China's Arctic policy are diplomacy, lobbying, low politics, and soft law (Zhang and Huang, 2015).

Thus, China's capacities in Arctic regime-making are expanding due to better knowledge of the region, clarified vision, and using new political instruments. The improvement of the current regime may bring sufficient economic and political benefits to China. However, it worth noting that the PRC is not interested in seismic changes to the rules and regulations, but in moving the regime to better inclusiveness and recognition of China's legitimate place in it.

## Conclusion

China is a significant contributor to the development of the Arctic seaways, both as a market that inspires economic rise in the region and as an important operating force. China has rapidly elaborated on navigation technologies and shown a persistence to build up as an Arctic shipping nation. However, an equally important contribution is due to China's involvement in the regime-making of Arctic shipping. China develops diplomatic instruments to promote its ideas for the international arena and focuses on "discourse power" as a new way to promote its ideas in the Arctic. All these components pave the way for China to the times of an ice-free Arctic.

The role of the three Arctic passages in terms of regime-making is not the same for China. The most active regime-making appears around the North-East Passage. An extensive economic network with Arctic states here provides China with opportunities to hold a dialogue among politicians and business circles on various international and bilateral platforms. Although the NEP regime is well-established, China keeps a compromise approach and develops variable diplomacy with Arctic actors using opportunities to improve the regime. This ranges from indulgence in rules for arranging the voyages to the promotion of the idea for more inclusive of cooperation in the Arctic.

The North-West Passage falls out of China's Arctic politics due to aggravated antagonism with the US and Canada. However, the reanimation of the dialogue holds some validity thanks to the business ties. In case of a more favorable situation, it is reasonable to assume that China's tactics in regime-making here will be close to that practiced on the NEP.

Central Passage development is a convenient target for increasing China's role in Arctic regime formation. If China, by its own efforts, is able to conduct shipping on the CP, it will prove the PRC's status as an Arctic shipping nation. On this waterway, China has the best condition to participate in regime-making, including law, regulations, and principles of cooperation. In any case, China will have to consider relations with Arctic states that control straits from the Arctic Ocean into the Pacific and Atlantic. In this regard, the future success of China's involvement in regime-making around the CP depends on the results of China's activities in the NEP area.

The Arctic is becoming a narrative and situational complement for the BRI. However, the in-fact symbolic meaning of the ISR idea should not downplay its value for regime-making in the Arctic. Any successful project in the High North with Chinese participation will acquire a label of the ISR. As long as shipping remains a challenging economic area, the ISR will be focused on technological development from the establishment of the new energy system, informational logistics, and port infrastructure, to railroads in Scandinavia and Russia. Thus, the ISR will become an envelope for China's Arctic politics and its international network among Arctic states.

The Arctic regime-making in the field of shipping illustrates the new distribution of power in the region. This is currently dominated by Arctic states with a stronger systemic push from the US. In this regard, China's scope of cooperation is decreasing. China lost the opportunity to hold the dialogue on the NWP and its involvement is currently oriented to the NEP as a nexus between Russian extracting industries, the European market, and their shared attempt to link with Asia. However, the political rivalry with the US shifts the cooperation of European states with China into a question of being "with" or "against" Washington. Amid rising contradictions in EU–Russia relations, this situation has gradually narrowed China's participation to shipping in the Russian Arctic as the most available option.

The further escalation may bring two scenarios. The first one is Beijing's closer rapprochement with Russia. This scenario is not favorable for China as a great power striving for global influence. The reason is that Beijing's alliance with Moscow will irrevocably repel European partners and China may lose in terms of the regime-making. China's participation in Arctic regime formation will be narrowed down to getting here-and-now benefits in projects located in Russia. The scenario will also be even worse for Russia, which will have to heavily rely on China's investments in a crucially important region. This will deprive Russia of its strategic advantage over China in the Arctic. In general, such a progression of events will split the Arctic into two blocs and destroy the regime based on a universal approach.

The second scenario is China's political equidistance from Washington and Moscow in the Arctic. This scenario is possible in the case of the fast-aggravating conflict between the US and Russia in the Arctic with minimal involvement from China. The rivalry will inevitably lead to a decay of the current regime. This juncture will mean that Arctic states will suddenly be unable to satisfy demands through the established system of cooperation. Since the US and Russia will drive hard and will require loyalty from partners, China may act as an alternative partner that is able to tolerate different views if it helps pragmatic solutions. In this case, China as a great power will get an advantage in regime formation. The state will suggest adaptation of rules for economic cooperation (including shipping), which is now formulated as an idea of the Ice Silk Road construction. Even if they are not adopted instantly, they will strengthen China's position in the Arctic regime formation.

# References

ABS. *ABS, COSCO Sign Cooperation Agreement*. (2016, February 2). https://ww2.eagle.org/en/news/press-room/abs-cosco-sign-cooperation-agreement.html
Bennett, M. M., Stephenson, S. R., Yang, K., Bravo, M. T., & De Jonghe, B. (2020). The opening of the Transpolar Sea Route: Logistical, geopolitical, environmental, and socioeconomic impacts. *Marine Policy, 121*, 104–178. https://doi.org/10.1016/j.marpol.2020.104178.
Bertelsen, R. G. (2020, June 10). *Arktisk orden i verdensordenen*. www.nordnorskdebatt.no. https://www.nordnorskdebatt.no/5-124-12792
Bertelsen, R. G., & Gallucci, V. (2016). The return of China, post-Cold War Russia, and the Arctic: Changes on land and at sea. *Marine Policy, 72*, 240–245. https://doi.org/10.1016/j.marpol.2016.04.034.
China Briefing News. (2018, December 13). *Norway and Sichuan Build Belt and Road Arctic Bridge*. https://www.china-briefing.com/news/norway-sichuan-build-belt-road-arctic-bridge/
China Current Affairs. (2012, November 20). *Hujintao shiba da baogao (quanwen) [Hu Jintao's 18th National Congress Report (Full Text)]*. http://news.china.com.cn/politics/2012-11/20/content_27165856.htm
China Daily. (2015, September 24). *Full text: President Xi's speech on China-US ties*. https://www.chinadaily.com.cn/world/2015xivisitus/2015-09/24/content_21964069.htm
China Daily. (2016, April 20). *China charting a new course navigating the Arctic*. http://www.china.org.cn/china/2016-04/20/content_38285595.htm
Council of the People's Republic of China. (2018, January 26). *Full text: China's Arctic Policy. The White Paper*. The State http://english.www.gov.cn/archive/white_paper/2018/01/26/content_281476026660336.htm
Cyberspace Administration of China. (2015). *Shouquan fabu: Zhonghua renmin gongheguo guojia anquan fa [Authorized release: National Security Law of the People's Republic of China]*. http://www.cac.gov.cn/2015-07/01/c_1115787841.htm
DeMarban, A. (2017, November 23). Governor details China-Alaska LNG deal, but skeptics still have questions. *Anchorage Daily News*. https://www.adn.com/business-economy/energy/2017/11/22/governor-reveals-china-alaska-lng-deal-but-skeptics-still-have-questions/
Denisov, I., & Adamova, D. (2017). Formuly vneshney politiki Si TSzin'pina: osnovnyye osobennosti i problemy interpretatsii [Xi Jinping's foreign policy formulas: main features and problems of interpretation]. *Kitay v Mirovoy i Regional'noy Politike. Istoriya i Sovremennost' [China in World and Regional Politics. History and Modernity], 22*, 76–89.
Ding, H., & Zhang, C. (2016). Fan beiji gongtongti de shexiang yu zhongguo shenfen de suzao – Yi zhong jiangou zhuyi de jiedu [The Vision of Pan-Arctic Community and the Molding of Chinese Identity – An Interpretation of Constructivism]. *Jiangsu xingzheng xueyuan xuebao [The Journal of Jiangsu Administration Institute], 4*, 76–83.
Drezner, D. W. (2009). The Power and Peril of International Regime Complexity. *Perspectives on Politics, 7*(1), 65–70.
Dry Cargo International. (2014, November 4). *First arctic cargo shipped through the Northwest Passage*. https://www.drycargomag.com/first-arctic-cargo-shipped-through-the-northwest-passage
Eimskip. (2020, May 2). *Eimskip takes delivery of new Dettifoss*. https://www.eimskip.com/about-eimskip/news/general-news/eimskip-takes-delivery-of-new-dettifoss/

Embassy of the PRC in the Republic of Finland. (2017, May 15). *Ambassador Chen Li's Article on The Belt and Road Initiative.* http://www.chinaembassy-fi.org/eng/ztxw/ydyleng/t1461886.htm

Embassy of the PRC in the Republic of Finland. (2020, October 13). *China-Finland Ties Show Strong Resilience Even in Hard Times.* http://www.chinaembassy-fi.org/eng/xwdt/t1823701.htm

Embassy of the PRC in the Republic of Iceland. (2012, April 25). *Chinese Premier Wen Jiabao Pays Official Visit to Iceland.* http://is.china-embassy.org/eng/xwdt/t926273.htm

Equinor. (2001, November 9). *China buys Norwegian LPG.* https://www.equinor.com/content/statoil/en/news/archive/2001/11/09/ChinabuysNorwegianLPG.html

Fekete, J., & Kennedy, M. (2012, February 9). Multibillion-dollar deals 'new level' for Canada-China relationship. *National Post.* https://nationalpost.com/news/canada/china-and-canada-reach-deals-on-air-travel-oil-and-uranium-and-pandas

Forsby, A. B. (2017). Denmark's relationship with China: An odd couple's quest for bilateral harmony. In *China and Nordic Diplomacy.* Bjørnar Sverdrup-Thygeson, Wrenn Yennie Lindgren, Marc Lanteigne (Edit.) (pp. 27–44). Routledge. https://doi.org/10.4324/9781315144702-3

Government of Canada. (2018, October 3). *Agreement to prevent unregulated high seas fisheries in the central Arctic Ocean.* https://www.dfo-mpo.gc.ca/international/agreement-accord-eng.htm

Government of Greenland. (2015, January 8). *New strong force behind London Mining Greenland.* https://naalakkersuisut.gl//en/Naalakkersuisut/News/2015/01/080115-London-Mining

Havnes, H. (2020). The Polar Silk Road and China's role in Arctic governance. *Journal of Infrastructure, Policy and Development, 4*(1), 121–138. https://doi.org/10.24294/jipd.v4i1.1166.

He, G., & Song, X. (2013). Chuangzaoxing jieru: Zhongguo canyu beiji diqu shiwu de tiaojian yu lujing tansuo [Creative Involvement: The Conditions and Approach Explorations of China's Involvement in Arctic Region Affairs]. *Ping yang xuebao [Pacific Journal], 21*(3), 51–58.

Huang, L., Lasserre, F., & Alexeeva, O. (2015). Is China's interest for the Arctic driven by Arctic shipping potential? *Asian Geographer, 32*(1), 59–71. https://doi.org/10.1080/10225706.2014.928785.

International Seabed Authority. (2014, April 29). *China Ocean Mineral Resources Research and Development Association (COMRA) and ISA Sign Exploration Contract.* https://www.isa.org.jm/news/china-ocean-mineral-resources-research-and-development-associationcomra-and-isa-sign

ISDP. (2016). *Sino-Nordic Relations: Opportunities and the Way Ahead. Institute for Security and Development Policy.* http://isdp.eu/publication/sino-nordic-relations-opportunities-way-ahead/

Kallio, J. (2017). Finland and China: Pragmatism prevails. In *China and Nordic Diplomacy.* Bjørnar Sverdrup-Thygeson, Wrenn Yennie Lindgren, Marc Lanteigne (Edit.) (pp. 45–59). Routledge. https://doi.org/10.4324/9781315144702-4

Kobzeva, M. (2018). Kitay kak vozmozhnyy partnor SSHA v Arktike: nastoyashcheye i budushcheye [China as the possible partner of the USA in the Arctic: Today and tomorrow]. *Natsional'nyye interesy: prioritety i bezopasnost' [National Interests: Priorities and Security], 14*(9), 1762–1778.

Kobzeva, M. (2019). China's Arctic policy: Present and future. *The Polar Journal, 9*(1), 94–112. https://doi.org/10.1080/2154896X.2019.1618558.

Kobzeva, M. (2020). Strategic partnership setting for Sino-Russian cooperation in Arctic shipping. *The Polar Journal,* 1–19. https://doi.org/10.1080/2154896X.2020.1810956.

Komissina, I. (2015). Arkticheskiy vektor vneshney politiki Kitaya [The Arctic Vector of China's Foreign Policy]. *Problemy Natsional'noy Strategii [Problems of the National Strategy], 1*(28) https://www.elibrary.ru/item.asp?id=23415138.

Krasner, S. D. (1982). Structural Causes and Regime Consequences: Regimes as Intervening Variables. *International Organization, 36*(2), 185–205.

Kroenig, M., & Cimmino, J. (2020). *Global strategy 2021: An allied strategy for China.* Atlantic Council. https://www.atlanticcouncil.org/global-strategy-2021-an-allied-strategy-for-china/

Lackenbauer, P. W., Lajeunesse, A., Manicom, J., & Lasserre, F. (2018). China's Arctic Ambitions and What they Mean for Canada (1st). *University of Calgary Press.* https://www.jstor.org/stable/j.ctvf3w20h

Lanteigne, M. (2020, April 28). Identity and Relationship-Building in China's Arctic Diplomacy. *The Arctic Institute.* https://www.thearcticinstitute.org/identity-relationship-building-china-arctic-diplomacy/

Lanteigne, M., & Shi, M. (2020, May 25). Greenland in the Middle: The Latest Front in a Great Power Rivalry. *The Polar Connection.* https://polarconnection.org/greenland-in-the-middle/

Li, Z., & Hu, M. (2015). "Beiji hangdao" kaitong yu zhongguo ji qi shou yingxiang quyu de maoyi zengzhang qianli fenxi [Analysis of the trade potential for the development of the "Arctic seaway" and China and adjacent areas]. *Jidi yanjiu [Polar Research], 27*(4), 429–438.

Li, J., Zhan, L., & Ma, P. (2014). Zhongguo Kaifa Haishang Dongbei Hangdao De Zhanlue Tuijin Gouxiang [China's Strategic Vision for the Development of the Northeast Passage]. *Dongbei Caijing Daxue Xuebao [Journal of Dongbei University of Finance and Economics], 2*, 43–51.

Liao, Q. (2012). Beiji Dalujia Falu Zhidu Yanjiu [Research on the Legal System of the Arctic Continental Shelf] Liaoning Xingzheng Xueyuan Xuebao. *Journal of Liaoning Administration Institute, 12*(2012), 63–65.

Liu, R. C. (2020). China's "Great Overseas Propaganda" Under the Belt and Road Initiative. In J. T. Jacob & T. A. Hoang (Eds.), *China's Search for 'National Rejuvenation': Domestic and Foreign Policies under Xi Jinping* (pp. 169–183). Springer. https://doi.org/10.1007/978-981-15-2796-8_12.

Liu, N., & Hossain, K. (2017). China and the Development of International Law on Arctic Shipping. In *Arctic Law and Governance: The role of China, Finland and the EU.* Timo Koivurova, Qin Tianbao, Tapio Nykänen and Sébastien Duyck (eds). Hart Publishing. https://doi.org/10.5040/9781474203302

Magnússon, B. M. (2015). China as the Guardian of the International Seabed Area in the Central Arctic Ocean. *The Yearbook of Polar Law Online, 7*(1), 83–101. https://doi.org/10.1163/2211-6427_004.

National Centers for Environmental Information. (2020, February 28). Predicting the Future of Arctic Ice. http://www.ncei.noaa.gov/news/arctic-ice-study

Neftegaz, R. U. (2019, June 7). *NOVATEK, COSCO, Sovkomflot i Fond Shelkovogo Puti podpisali soglasheniye v otnoshenii kompanii Morskoy arkticheskiy transport [NOVATEK, COSCO, Sovcomflot and the Silk Road Fund signed an agreement on Arctic Marine Transport].* https://neftegaz.ru/news/partnership/453166-novatek-cosco-sovkomflot-i-fond-shelkovogo-puti-podpisali-soglashenie-v-otnoshenii-kompanii-morskoy-/

Nilsen, T. (2020, September 11). London calling to the faraway north, leads largest NATO task force into the Barents Sea since last Cold War. *The Independent Barents Observer.* https://thebarentsobserver.com/en/security/2020/09/london-calling-faraway-north-leads-largest-nato-task-force-barents-sea-last-cold

Nilsen, H. R., & Ellingsen, M.-B. (2015). The power of environmental indifference. A critical discourse analysis of a collaboration of tourism firms. *Ecological Economics, 109*, 26–33. https://doi.org/10.1016/j.ecolecon.2014.10.014.

Østreng, W., Eger, K. M., Fløistad, B., Jørgensen-Dahl, A., Lothe, L., Mejlænder-Larsen, M., & Wergeland, T. (2013). *Shipping in Arctic Waters: A Comparison of the Northeast, Northwest and Trans Polar Passages* (2013th ed.). Berlin/Heidelberg: Springer. https://doi.org/10.1007/978-3-642-16790-4

President of Russia. (2018, June 8). *Sovmestnoye zayavleniye Rossiyskoy Federatsii i Kitayskoy Narodnoy Respubliki [Joint statement of the Russian Federation and the People's Republic of China].* http://kremlin.ru/supplement/5312

President of Russia. (2019a, June 5). *Sovmestnoye zayavleniye Rossiyskoy Federatsii i Kitayskoy Narodnoy Respubliki ob ukreplenii global'noy strategicheskoy stabil'nosti v sovremennuyu epokhu [Joint statement by the Russian Federation and the People's Republic of China on strengthening global strategic stability in the modern era].* http://kremlin.ru/supplement/5412

President of Russia. (2019b, June 5). *Zayavleniya dlya pressy po itogam rossiysko-kitayskikh peregovorov [Press statements following Russian-Chinese talks].* http://kremlin.ru/events/president/news/60672

President of Russia. (2019c, September 27). *Zasedaniye kruglogo stola foruma "Odin poyas, odin put'" [The Round Table Discussion of the Forum "One Belt – One Road"].* http://kremlin.ru/events/president/news/60393

Rainwater, S. (2015). *International Law and the 'Globalization' of the Arctic: Assessing the Rights of Non-Arctic States in the High North.* Emory University School of Law. Atlanta, GA. http://law.emory.edu/eilr/content/volume-30/issue-1/comments/international-law-globalization-arctic-rights-high-north.html

Russian Arctic National Park. *Statistika [Statistics]* (2018). http://www.rus-arc.ru/ru/Tourism/Statistics

Ryan, M., & Lamothe, D. (2015, September 4). Chinese naval ships came within 12 nautical miles of American soil. *Washington Post.* https://www.washingtonpost.com/world/national-security/chinese-naval-ships-came-within-12-nautical-miles-of-american-soil/2015/09/04/dee5e1b0-5305-11e5-933e-7d06c647a395_story.html

Satariano, A. (2019, December 20). At the Edge of the World, a New Battleground for the U.S. and China. *The New York Times.* https://www.nytimes.com/2019/12/20/technology/faroe-islands-huawei-china-us.html

Sergunin, A., & Konyshev, V. (2016). *Russia in the Arctic. Hard or Soft Power?* ibidem-Verlag. https://publications.hse.ru/en/books/201496493

Sverdrup-Thygeson, B. (2017). The Norway-China relationship: For better, for worse, for richer, for poorer. In B. Sverdrup-Thygeson, W. Y. Lindgren, & M. Lanteigne (Eds.), *China and Nordic Diplomacy* (pp. 77–100). Routledge. https://doi.org/10.4324/9781315144702-6.

TASS. (2018, May 15). *Rossiya i Kitay razrabatyvayut memorandum o sovmestnom osvoyenii Arktiki [Russia and China are Developing a Memorandum on Joint Development of the Arctic].* https://tass.ru/mezhdunarodnaya-panorama/5200505

The American Presidency Project. (2019, September 4). *Remarks by the Vice President and Prime Minister Katrín Jakobsdóttir of Iceland Before a Bilateral Meeting in Keflavík, Iceland.* https://www.presidency.ucsb.edu/documents/remarks-the-vice-president-and-prime-minister-katrin-jakobsdottir-iceland-before-bilateral

Todorov, A. (2019, March 12). Kuda vedet Severnyy morskoy put'? [Where Does the Northern Sea Route Lead?]. *Russian Council of International Affairs.* https://russiancouncil.ru/analytics-and-comments/analytics/kuda-vedet-severnyy-morskoy-put/

Tulupov, D. (2013). Skandinavskiy vektor arkticheskoy politiki Kitaya [Scandinavian Vector of China's Arctic Policies]. *Mirovaya ekonomika i mezhdunarodnyye otnosheniya [World Economy and International Relations], 9*, 61–68.

United Nations. (1982, December 10). *United Nations Convention on the Law of the Sea.* https://treaties.un.org/Pages/showDetails.aspx?objid=0800000280043ad5

Xinhua. (2017a, July 20). *China's ice breaker sets sail for Arctic rim expedition.* http://www.xinhuanet.com//english/2017-07/20/c_136458956.htm

Xinhua. (2017b, June 20). *Full text: Vision for Maritime Cooperation under the Belt and Road Initiative.* http://www.xinhuanet.com/english/2017-06/20/c_136380414.htm

Xinhua. (2017c, November 4). *Full text of Xi Jinping's report at 19th CPC National Congress.* https://www.chinadaily.com.cn/china/19thcpcnationalcongress/2017-11/04/content_34115212.htm

Xinhua. (2020, September 17). *China's polar icebreaker heading home from Arctic expedition.* http://www.xinhuanet.com/english/2020-09/17/c_139376451.htm

Yang, J. (2014). Beiji hangyun yu zhongguo beiji zhengce dingwei [Arctic shipping and China's Arctic policy positioning]. *Guoji guancha [International Review], 01*, 123–137.

Young, O. R. (2012a). Arctic Tipping Points: Governance in Turbulent Times. *Ambio, 41*(1), 75–84. https://doi.org/10.1007/s13280-011-0227-4.

Young, O. R. (2012b). Building an international regime complex for the Arctic: Current status and next steps. *The Polar Journal, 2*(2), 391–407. https://doi.org/10.1080/2154896X.2012.735047.

Young, O. R. (2014). The Effectiveness of International Environmental Regimes: Existing Knowledge, Cutting-edge Themes, and Research Strategies. In M. M. Betsill, K. Hochstetler, & D. Stevis (Eds.), *Advances in International Environmental Politics* (pp. 273–299). Palgrave Macmillan UK. https://doi.org/10.1057/9781137338976_11.

Yu, S. (2016, September 25). Zhangdejiang dui fenlan jinxing zhengshi youhao fangwen [Zhang Dejiang pays an official goodwill visit to Finland]. *Xinhua*. http://www.xinhuanet.com/world/2016-09/25/c_1119620403.htm

Zhang, Z. (2020). Nanhai wenti shang de huayu boyi yu zhongguo guoji huayu quan [Discourse Game and China's International Discourse Power on the South China Sea Issue]. *Exploration and Free Views, 1*(7), 126–134.

Zhang, C., & Huang, D. (2015). Zhongguo beiji quanyi de weihu lujing yu celue xuanze [The Maintenance Route and Strategy Choice of China's Arctic rights and Interests]. Huadong ligong daxue xuebao (shehui kexue ban). *Journal of East China University of Science and Technology (Social Science Edition), 30*(6), 73–84.

Zhao, K. (2016). China's Rise and its Discursive Power Strategy. *Chinese Political Science Review, 1*(3), 539–564. https://doi.org/10.1007/s41111-016-0037-8.

**Mariia Kobzeva** is a postdoctoral fellow in Global Arctic studies at the Department of Social Sciences, UiT the Arctic University of Norway.

# Chapter 17
# Sustainable Development of the Arctic?

**Matthias Finger**

## Introduction

The now-global Arctic is an example of development that is unsustainable, both for the Arctic and for the planet. The unsustainable development of the Arctic is the result of unsustainable "Development" (capital "D") at the planetary scale, especially in the past 50–70 years, which is also called the period of the "Great Acceleration". This unsustainable Development worldwide has led, among other things, to global warming and more specifically to the melting of the Arctic sea ice. The ensuing accessibility of Arctic resources is likely to lead to a new "gold rush", made possible because of Development's hunger for energy and minerals resources. The exploitation of the Arctic's resources will only accelerate unsustainable development and leave the Arctic overall worse off. All in all, this is a truly vicious circle.

The chapter is a reaction to the numerous publications over the past 20 years that have called for, actively promoted, and probably believed that "sustainable development of the Arctic" is desirable and possible. Undoubtedly the most prominent and most comprehensive of these publications is the Arctic Human Development Report (2004), published by the Stefansson Arctic Institute, under the auspices of the Icelandic Chairmanship of the Arctic Council (2002–2004). It has served as an input to the Arctic Council's Sustainable Development program ever since, which, in turn, was intended to somewhat counterbalance the Council's foundational Arctic Environmental Protection Strategy in 1991. More generally, it places the Arctic within the broader UN discourse on sustainable development.

---

M. Finger (✉)
Northern Arctic Federal University (NArFU), Arkhangelsk, Russia
e-mail: matthias.finger@epfl.ch

The purpose of this chapter is not to discuss the concept of sustainable development nor to analyze its translation and application to the Arctic. Suffice to say that "sustainable development" is simply the latest version of a broader state-centric Development discourse and corresponding policies and practices, actively promoted since the end of the decolonization era by the United Nations system, development agencies, economists, corporations, and politicians of all colors. It is only logical that the Arctic Council and most other stakeholders in the Arctic have come to embrace sustainable development as the new "imaginary" of the Arctic (Kristoffersen and Langhelle, 2017). Controversies about the concept of sustainable development, and Development for that matter, abound, and the same controversies will without doubt also take place when it comes to the "sustainable development" of the Arctic. All these controversies pertain to the inequitable distribution of the benefits (or negative effects) of development, while leaving Development's fundamental assumptions, its feasibility, and its desirability untouched.

In this chapter I will place Development and the "sustainable development of the Arctic" within the broader context of Earth System dynamics. A study of the "development" of the Arctic, whether sustainable or not, cannot avoid the more fundamental context and question of Earth System dynamics, of which the Arctic is an integral part and in which it plays a central role. At least, this is the claim of this chapter. More precisely, I want to show that, within such an Earth Systems dynamics framework, there can be no such thing as a sustainable development of the Arctic. Worse, the unsustainable development of the Arctic will exacerbate the unsustainable development path that industrial civilization has been engaging in for quite some time now.

The remainder of chapter is structured into the following four sections, which together form a circle. In the first section, I summarize the key elements of the Earth System's dynamics conceptual framework, especially during its most recent period called the "Great Acceleration", which is currently leading to the well-known Arctic tipping points. In the second section, I show how the Arctic is opening up for Development, as the combined result of a warming Arctic, the availability of resources, and the world's hunger for them. In the third section, I identify the main actors of Arctic Development and show how unsustainable development in the Arctic will inevitably unfold as a result of it. The fourth and concluding section explains why the concept of "sustainable development of the Arctic" has prevailed, despite its bio-physical improbability.

## From Earth System Dynamics to Arctic Tipping Points

The purpose of this first section is to outline the basic conceptual elements underlying this chapter, namely the Earth System dynamics framework on the one hand and Development on the other hand. The combination of Development with Earth System dynamics leads to Arctic (and other) tipping points.

## Earth System Dynamics

This section is not about the complex "functioning of the Earth System"; rather, it is about the conceptual framework looking at the Earth as a complex and dynamic bio-geo-physical system along geological time periods since its beginning 4.5 billion years ago. This approach has become prominent recently because of the inflationary use of concept of the "Anthropocene". The International Commission of Stratigraphers has proposed that the Anthropocene be officially recognized as the most recent geological epoch, during which "humans" are said to have become a significant, if not the most significant geological force (Thomas et al., 2020). There is still much debate about whether "human" impact on the Earth System justifies the creation of such a new epoch of geological proportions and, if so, when exactly this new epoch of the Anthropocene would have started. The main controversy in this regard is between the 1950s, which is the beginning of the so-called "Great Acceleration" (see below), and the beginning of the Agricultural Revolution, in which case the concept would not be justified, as the Anthropocene would be identical to the Holocene (11,650 years ago at the end of the last glacial period).

What is more relevant than this terminological debate among stratigraphers and geologists is the conceptual framework of Earth System dynamics, which is much broader, as it also encompasses the Biosphere (Grinevald, 2007). The Biosphere is basically what sustains life on planet Earth. With the exception of the sun, the Biosphere is basically a closed system. It can be decomposed, in a simplified manner, into the lithosphere (soil, Earth crust), the cryosphere (ice), the hydrosphere (water), and the atmosphere (air), as well as geographically more or less distinct ecosystems. All of these interact with one another in a systemic manner. Life emerged approximately 3.5 billion years ago from an-organic matter and the energy of the sun and evolved thereafter. A major milestone in the evolution of life is the so-called "Cambrian explosion" between 540 and 520 million years ago, during which most animal life, as well as the diversification of most plant organisms, occurred. Since then, scholars have counted five mass extinctions (of animal and plant species within a short period of time), the most radical one having taken place 250 million years ago and the most recent one 66 million years ago. It is debated whether the current biodiversity loss (due to "human" activities) constitutes the sixth such mass extinction (Kolbert, 2014).

Within the Earth System dynamics conceptual framework, the geological (Geosphere) and the biospherical (Biosphere) dimensions come together. Organizationally, they come together within the International Geosphere Biosphere Program (IGBP). The IGBP was launched in 1987 by ICSU, the International Council of Scientific Unions (and closed down in 2015). Its role was to coordinate *"international research on global-scale and regional-scale interactions between Earth's biological, chemical and physical processes and their interactions with human systems. IGBP views the Earth system as the Earth's natural physical, chemical and biological cycles and processes and the social and economic dimensions.*"[1] The IGBP was the first – and so far the only – systematic effort to understand the interaction between "humans" and the Earth system. The IGBP was itself the synthesis of two different strands of research activities during the early 1980s. One strand was NASA (the US National Aeronautics and Space Administration) created in 1958 and its Apollo program, which led to the first human moon landing in 1969. Seeing the Earth from the moon led to an epistemological change and especially to the new concept of "Global Habitability" and related research and exploration efforts by NASA.[2] In a second strand, ICSU integrated NASA's Earth Systems view under its own concept of "Global Change". During the 1984 landmark Symposium on Global Change, ICSU subsequently floated the idea of the IGBP, arguing that *"human impacts (on the Earth System) have grown to approximately those of the natural processes that control the global life-support systems"* (Malone and Roederer, 1985: xiv). The underlying conceptualization of Earth System dynamics by the IGBP in 1986 is represented in Fig. 17.1 below.

What has changed since 1986, when this conceptualization was originally created, is the fact that "human activities" are no longer at the margin of the Earth System, affected by climate change and affecting in turn tropospheric chemistry.

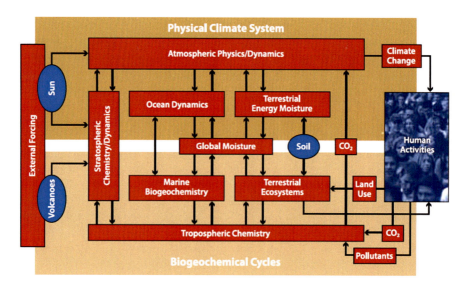

**Fig. 17.1** Conceptualization of the Earth System dynamics. (Source: *Global Change*, Issue 84. November 2015, page 10)

---

[1] http://www.igbp.net/about.4.6285fa5a12be4b403968000417.html, accessed on May 7th, 2021.

[2] *"A planet's habitability, or ability to harbor life, results from a complex network of interactions between the planet itself, the system it's a part of, and the star it orbits. The standard definition for a habitable planet is one that can sustain life for a significant period of time."* https://seec.gsfc.nasa.gov/what_makes_a_planet_habitable.html, accessed May 14th 2021.

Today, as the concept of the Anthropocene tries to capture, one would have to conceptualize "human activities" as affecting every single variable of the Earth System and perhaps even affecting external forcing (because melting of the Arctic ice is changing the tilting of the Earth). Also, one would have to conceptualize "human activities" in a much more sophisticated manner, which is what I do in the next section.

## "Development Was the Problem, Not the Solution"

The reason why "human activities" are conceptualized in a such generic way by Earth System experts is the result of the profound epistemological divide between the natural and the social sciences. This is less the fault of the Earth System experts than a result of the failure of the social sciences to epistemologically acknowledge the very bio-physical foundations (and limitations) of the economic and social activities they are studying. This failure has led the social scientists to believe that Development can be sustained indefinitely into the future (for example, "sustainable development"), a belief that is particularly widespread among economists (Georgescu-Roegen, 1971). As a result, the social sciences have become a tool of (human, social, economic, industrial) Development, rather than the critical analysis of it. Yet, from an Earth Systems dynamics point of view, it is fair to say that *"Development is the problem, not the solution"*.[3]

The most recent and the most obvious expression of the unsustainable nature of Development is the so-called "Great Acceleration" of the past 50–70 years (McNeil and Engelke, 2016). During this most recent epoch – which some equate to the Anthropocene, as shown above – all the main indicators reflecting threats to the habitability of planet Earth, such as greenhouse gas emissions, population growth, top-soil depletion, ocean acidification, biodiversity loss, global warming, and many others more have increased exponentially. Even more worrisome is that all these indicators correlate almost perfectly with indicators of Development, such as real GDP, foreign direct investment, primary energy use, telecommunications, transportation, and many others. Figure 17.2 graphically represents a few of the key (socio-economic) variables measuring Development during the past 250-plus years.

While the Great Acceleration is not limited to the above 12 variables, it can basically be extended to almost every single variable reflecting some aspect of industrial development, such as cell phone uptake, chicken production, cardio-vascular diseases, obesity, healthcare expenditures, McDonald's outlets, or global tourism for that matter. It is probably fair to say that, from such an Earth System Dynamics point of view, the Great Acceleration is most probably going to be very short-lived, as such a great acceleration cannot be sustained, even if it were to be stabilized. This is both because of input (resources) and output (global warming, pollution) of industrial development.

---

[3] This the provisional title of a book the author is working on, to be published by Routledge in 2022.

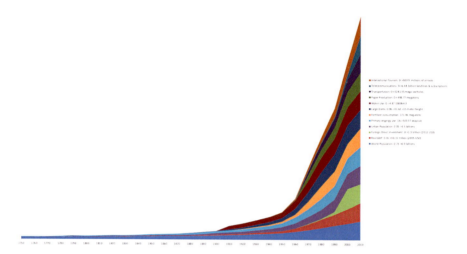

**Fig. 17.2** The Great Acceleration. (Source: https://en.wikipedia.org/wiki/Great_Acceleration, accessed 5.5.2021)

As a result, it is also no longer appropriate to summarize the effects of industrial development on the dynamics of the Earth System as stemming simply from "human activities". A much more sophisticated conceptualization of "human activities" is in order. Also, these "human activities" must be put into an institutional and historical perspective going beyond the past 50–70 years (Great Acceleration) or even 150 years (Industrial Revolution). This Great Acceleration is merely the last stage of a process I call "Development", which is the result of the co-evolution of institutions (rules) and technology (applied science).

The origins of Development can be traced back to the militant Christianization of Europe in the aftermath of the Fall of the Roman Empire (Buc, 2015; White 1967). Christianism laid the cultural foundation for Western technological development, which became secularized during the protestant Reformation and the Renaissance (White, 1968). Over the next 200 years, rationalism, the inseparable scientific/military Revolution and the rise of the modern nation-State on the model of the Vatican led to a mathematical understanding and from there on to the mastery of the world, also thanks to colonization powered by Western technology (Daly, 2014). Like manipulatory science came to replace religion, political theology came to be replaced by theological politics ("monarchies based on divine law"), ultimately confusing democracy and the army during the French Revolution (for example, "citizen-soldier") (Tilly, 1990). In the nation-State, militarism, science (the new religion), and engineering became inextricably interwoven to turn into the main drivers of conquest, colonialism, Europeanization or simply annihilation of entire civilizations paralleled by the spreading of the Christian and nation-State-led model of Development, an argument so aptly made by Ivan Illich (Hartch, 2015).

With the Industrial Revolution, Development became powered by non-renewable resources (coal and later oil and gas), along with minerals, and thus accelerated significantly. Simultaneously, Development became equated with economic growth. The nation-state, in addition to preparing for war, now also turned into a powerful "Development agency". Development as economic growth was promoted since the Second World War and for all UN Member States and citizens thereof, as *the* historical driving force towards human progress. During the Cold War and the period of decolonization, "fossil-fuel-powered development" (Auzanneau, 2015) rebranded itself as something new, even though developmentalism remains essentially rooted in the Western sense of time, history, progress, evolution and irreversibility. Consequently, the Western urban-industrialized society (and even more so the American consumer society) was, after the Second World War, considered to be the best and only model and the ultimate stage of Development. After the Second World War, the traumatic birth of the atomic age and the frightening long interlude of the Cold War, Development became the new name for "peace". However, after the unexpected auto-dissolution of the Soviet Empire, and with neoliberal hyper-industrial capitalism being declared the winner of the Cold War, the American Dream (happiness and opulence), or economic growth and technological progress for all and forever, became the new religion, a sort of "paradise" created by science, technology and the "market", yet a market whose lease on life is granted by nation-States.

## *Arctic Tipping Points*

The cumulative and recent effects of Development now significantly affect all dimensions of the Earth System, thus destabilizing the more stable Holocene, which had given rise to the Agricultural Revolution that allowed civilizations to emerge. $CO_2$ emissions, among other things, are altering the chemistry of the atmosphere and are leading to global warming. Deforestation, industrial agriculture, and urbanization, among other things, are altering the lithosphere and lead to biodiversity decline and loss of soil fertility. Water pollution and ocean warming are altering the chemistry of the hydrosphere and are leading to ocean acidification, biodiversity loss, and changing ocean currents, etc. Finally, global warming is altering the cryosphere, thus leading to loss of polar and mountainous ice cover.

The interactions among these dimensions, and especially global warming in the atmosphere, are triggering feedback loops, thus accelerating and exacerbating the destabilizing effects of Development on the Earth System. Some of the particularly powerful feedback loops trigger so-called "tipping points". Tipping points define thresholds that, if crossed, trigger runaway transformations of the Earth System, which may or may not lead to new stable states, although none of these were experienced by Homo sapiens during its 300,000 or so years of existence.

Among the main currently identified tipping points, the three most significant ones are in the Arctic (Lenton, 2012). All of them have global warming resulting from greenhouse gas emissions as their underlying cause:

- The loss of the Arctic sea ice alters the Earth's ice-albedo effect, thus further accelerating global warming; this tipping point already has the potential to trigger "run-away global warming".
- The melting of the Greenland ice sheet alters the Ocean currents, thus potentially disabling the ocean circulation system and in any case triggering unprecedented (in the Holocene) climatic effects worldwide.
- The warming of the Arctic oceans and landmass negatively affect permafrost, thus triggering methane release, an even more powerful greenhouse gas than $CO_2$, and another potential trigger of runaway global warming.

In this first section I have shown how Development has unfolded over the past almost 2000 years to end up in the Great Acceleration, which is now altering Earth System dynamics and potentially triggering tipping points. Today's most imminent such tipping points are in the Arctic. Instead of taking this as a warning, however, the warming Arctic is seen as an opportunity for more Development, both of the Arctic and worldwide. Re-branding Development as being sustainable does not change its destabilizing effects on the Earth System's dynamics. Even if it did, the cumulative negative effects of Development until today are unprecedented and have already destabilized the relatively stable Earth System of the Holocene and will continue to do so. The most significant indicator in this regard is $CO_2$ emissions and corresponding $CO_2$ concentration in the atmosphere, which correlates with global average temperature. Current $CO_2$ levels in the atmosphere are approximately 30 percent higher than at than at any point during the past 800,000 years. Temperature is likely to follow, albeit with a certain time lag, owing to Earth Systems dynamics.

## The "Arctic Paradox"

The cryosphere – that is, both poles along with the world's mountainous glaciers – appears to be particularly sensitive and vulnerable to global warming. The poles and the mountainous regions, but especially the Arctic, are warming much faster than the other ecosystems of the planet and temperatures have never been as high in the past 44,000 or perhaps 120,000 years.[4] In spring of 2020 the Siberian Arctic experienced a heatwave of 100 degrees Fahrenheit, probably the first in the region since Homo sapiens sapiens first saw the light of the day (approximately 300,000 years ago). However, in between now and the time the Arctic will become uninhabitable or inaccessible, there may still be some time for (sustainable) development. Indeed, the melting of the Arctic sea ice and the warming of the Arctic appears to be an

---

[4] https://www.livescience.com/40676-arctic-temperatures-record-high.html, accessed May 16th, 2021.

unprecedented opportunity for it. More precisely, it is a unique opportunity to access the Arctic's resources. In this section, I will first recall the world's hunger for resources in general and then show what such resources a more accessible Arctic has to offer (to the world).

## *The World's Hunger for Resources*

Development requires, among other things, non-renewable resources, notably energy (oil, gas, coal), as well as minerals. Corresponding resources extraction and subsequent consumption is closely correlated with GDP and population growth and most other indicators that measure one aspect of Development or another. Not surprisingly, then, *"global resource extraction grew more or less steadily over the past 25 years, from 40 billion tons in 1980 to 58 billion tons in 2005, representing an aggregated growth rate of 45%"*,[5] and *"is set to double again between 2015 and 2030"*,[6] paralleling the Great Acceleration. In other words, the Great Acceleration basically reflects accelerated fossil fuel use. This can be extended to other non-renewable resources such as minerals, quality agricultural land, water, fisheries, etc.

Over time, there is significant technological potential for resources and especially energy efficiency gains. However, for organizational reasons, efficiency gains are never exhausted to the point that the entropy law would allow. Also, with population growth and the Development of the newly emerging economies and their raising middle-class, there will be rebound effects. Thus, over the past 25 years *"the average resource extraction per capita remained almost stable, today amounting for nearly nine tons"*.[7] Even though this does not really matter from an Earth Systems dynamics perspective, it is worth mentioning that resource extraction, resource consumption, and resource efficiency gains are unevenly distributed among countries and social groups. Therefore, there is potential for conflict and perhaps even war around particularly essential resources.

Considering that non-renewable resources are finite and that even renewable resources have fixed regeneration rates, there will be resources limitations to Development (Meadows et al., 2004). In this regard, it is relevant to mention the concept of "peak oil", but the same concept can be applied to every other non-renewable resource. It is also not so important to precisely determine when peaks will occur or have already occurred. The fact is that all non-renewable resources will eventually peak. In other words, in light of intensifying global demand, and even with current demand levels, the pressure for resources, especially some non-renewable resources will only increase, making the warming Arctic attractive for their potential extraction.

---

[5] https://www.eea.europa.eu/soer/2015/global/competition, accessed May 16th 2021.
[6] ibid, accessed May 16th, 2021.
[7] https://www.wrforum.org/publications-2/publications/, accessed May 16th, 2021.

## What Does the Warming Arctic Have in Store?

In 2019, Nordregio, the Nordic Center for Spatial Development, published its most recent map indicating the main sites of existing, prospective, and potential non-renewable resources. This map is a compilation of most other available information at that time and can be considered as the most comprehensive representation of the current resource situation in the Arctic. (Fig. 17.3)

The map reflects both energy and minerals. In 2002 the Arctic accounted for approximately 10 percent of global oil and 25 percent of global gas production. These figures have remained steady since. The Arctic is said to hold similar percentages of the world's oil and gas resources. The fact that they have not yet been extracted is basically due to high costs, as there are oil and gas reserves available elsewhere in the world that are easier to extract. Minerals have a more than 100-year-long history of extraction in the Arctic (see Chap. 3 in this book). Some of these mineral resources are important for Development, such as phosphates (fertilizers), bauxite (aluminum), iron ore, copper, zinc (alloys and batteries), nickel (batteries), palladium (catalytic converters), and others. Greenland is said to hold rare earth metals, some of which – such as zinc, nickel, and palladium – are important for the so-called energy transition. There is also gold, silver and diamonds in the Arctic, even though these may be less strategic for Development. Finally, fisheries, fresh water (from the Greenland ice sheet), and hydropower are potentially exploitable Arctic resources.

The "Arctic paradox" (Finger, 2015; Finger & Heininen, 2018) defines the fact that Arctic resources that can and probably will be used for unsustainable Development worldwide become available precisely because of the world's unsustainable Development to begin with. While I argue that these resources should remain unexploited in light of the unsustainable Development path that the world is already engaged in, others will argue differently. At least some of the resources – especially the rare-earth-metals – are central to the energy transition (decarbonization) and should be mined precisely for this reason. Others will argue that resources extraction more generally will finally allow the Arctic to be become developed, something Arctic peoples are also entitled to. Therefore, it is unlikely that the Arctic resources will remain in the ground, and it is even more unlikely that the Arctic will remain off-limits for development.

## Sustainable Development of the Arctic?

In this section, I examine how the Development of the Arctic will most probably look like. We have already seen that the Development of the Arctic will be mainly about the exploitation of resources, particularly oil, gas, and minerals. There may also be Arctic shipping and fisheries exploitation for the (short) time that there will be fish. All of these activities will require investments into infrastructure for such

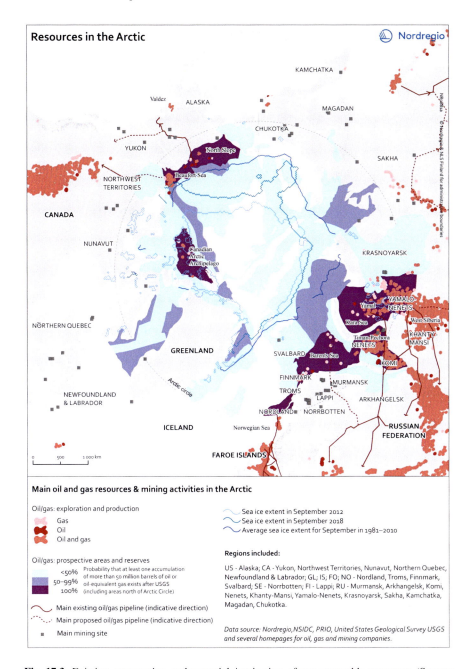

**Fig. 17.3** Existing, prospective, and potential Arctic sites of non-renewable resources. (Source: https://nordregio.org/maps/resources-in-the-arctic-2019/)

exploitation, as well as access and transport infrastructures. This may be accompanied by broader construction projects for housing. However, all of these infrastructures will mostly remain supportive of resources exploitation, even though they may generate important economic activities in themselves. Thus, I will only focus on resources exploitation here. In order to do that, I will proceed in two steps. I will first identify the main actors of such resources exploitation in the Arctic, as well as reflect on their motives (States) and their incentives (companies). In a second step, I will discuss what the exploitation of such resources will look like from the perspective of the peoples who are supposed to be the beneficiaries of "sustainable development".

## *Actors of Arctic Development*

Krivorotov and Finger (2018) identified the main actors of Arctic resource exploitation and concluded that they are mostly state-owned enterprises (SOEs), either owned be the Arctic States themselves or by governments of other countries interested in the Arctic (such as China and France). More private companies operate in the US Arctic. However, their relationship with government is similar to that of SOEs (Finger, 2013), a conclusion also supported by Kaznatcheev and Balazeva (2016). Furthermore, Krivorotov and Finger (2018) analyzed the strategies of the two most important SOEs operating in the Arctic – Russian Gazprom and Norwegian Statoil, now rebranded Equinor – and drew some interesting conclusions, that can be generalized to other development actors in the Arctic, beyond oil and gas and beyond Russia and Norway.

- Overall, exploiting resources in the Arctic has been and remains a risky business, which companies can rarely shoulder by themselves. Whatever the companies, they will have to need government support in various forms, be it in terms of financing supporting infrastructures (access roads, ports, icebreakers), risk guarantees, preferential concessions, preferential loans, tax brakes, operational subsidies, and others. Such government support is ever more likely, the better the relationship of the company with the government, something that favors SOEs.
- However, this supportive relationship is reciprocal: government support is more likely to be given to companies that a government trusts or more easily controls than to private and foreign operators. This again leads to SOEs, such as Gazprom and Equinor. But there are many others, such as Rosneft and Novatek in Russia, as well as Petoro in Norway. In the US Arctic, concessions are typically given to US companies; even though they are not state-owned, they build on long-term trusted relationships with government. There is naturally a need for particular technical expertise, given the harsh Arctic environment. Such expertise is often lacking among the Russian companies, so they team up (joint ventures) with semi-state-owned companies (such as ENI, Total).

- This reliance on SOEs is even more important for countries for which oil and gas is of strategic economic importance. To recall, *"more than 80% of Russia's exports are oil, natural gas, metals and timber, approximately half of which stem from the Arctic region. Similarly, Norway's export of oil and gas constitute 45% of its total exports and more than 20% of its GDP"* (Krivorotov and Finger, 2018: 58). Quite logically, government will be more likely to entrust their own with the exploitation of such strategic resources than be reliant on companies that are more subject to market dynamics with the risk of them eventually leaving less profitable ventures.
- Again, this relationship is reciprocal: foreign private companies that would like to become active in the Arctic minimize their investment and political risks by teaming up with SOEs or directly with governments.

These same considerations can also be extended to mineral resources, even though there are more mining than oil and gas companies active in Arctic resource extraction. However, this is probably due more to the fact that there are very different types of minerals available in the Arctic and many companies tend to be specialized (such as nickel, cobalt, copper, zinc, lead, tungsten, titanium, zirconium, gold, silver, platinum, palladium, diamonds, and uranium). Fishing and tourism companies are mostly privately owned.

Despite the discourse on sustainable development, this is what development in the Arctic is and will essentially be about. This is also the kind of development that governments prefer, as such development is centrally controllable. With the exception of Greenland and Iceland, we do not really have Arctic States. Furthermore, Iceland has no indigenous peoples and Greenland, despite being mainly indigenous, is not totally independent from Denmark. The Arctic portion of all other states is under the control of a center, located far more south. Developing their periphery means devolving power, sharing revenues and ultimately empowering local peoples, many of them indigenous, which is an additional reason not to empower them. Natural resources exploitation in the Arctic is certainly also fully in line with what their respective nation-states prefer. Since the Industrial Revolution, renewable and especially non-renewable resources have been essential for Development to proceed and there will therefore inevitably be a market. Of course, there are companies, sometimes even private ones, that do the job of extraction these resources. Mostly, however, states will prefer their own SOEs, even though foreign capital and expertise is always welcome. So, what can the Arctic peoples expect from such development?

## Human Development of the Arctic?

Today, less than 10 percent of the Arctic region's 4 million population is indigenous. Some of these peoples still make their living (at least partly) from renewable resources, namely fish. The warming of the Arctic already impacts these peoples to a significant degree (Dudley et al., 2015). Consequently, their traditionally sustainable lifestyles as well as their livelihoods are now being threatened, supposedly to be replaced by a more encompassing sustainable development of the Arctic. Other indigenous communities, especially nomadic groups, are directly affected by mining activities. Many researchers have highlighted the fact that overall mining, and resources extraction more generally, does not lead to long-lasting development at the local level or in the Arctic overall (Tolvanen et al., 2019). There can always be some showcase examples of successful local development thanks to mining projects. They also serve as proof that "sustainable mining" should be possible.

There are other potential industries in the Arctic with less damageable impacts on its ecosystems and more beneficial socio-economic consequences for the Arctic peoples. This is the case with fisheries, to a certain extent, but with the opening up of the Arctic and the over-fishing that will inevitably follow, the fishing industry will also be threatened (see Chap. 14 in this book). Tourism is an industry that may be developed in a more sustainable manner, but the appreciation of tourism as a new sustainable industry remains somewhat mixed, including in the Arctic (Chap. 7 in this book).

Can extractive industries be sustainable, in the Arctic or elsewhere? At least for the local and thus mainly indigenous peoples, Arctic (onshore) oil and gas extraction, along with mining, will more closely resemble a "resource curse", so well-known already by many developing countries and especially by their indigenous peoples (Murshed, 2018). Resources extraction is overall predatory in nature; that is, it is not sustainable in itself and it does not sustain local communities and their economies. This will not be different in the Arctic. Oil and gas are particularly unsustainable and should, in any case, be phased out. If decarbonization worldwide continues, oil and gas extraction, especially in harsh territories will not be a sustainable business model either. Some minerals in the Arctic will arguably be needed for the global energy transition, but they will not be needed for the energy transition of the Arctic nor will they benefit local communities much. Overall, therefore, the very nature of the extractive Arctic industries will not allow for the sustainable development of the Arctic and its communities. In other words, the sustainable development of the Arctic thanks to resources extraction is illusionary.

It is even more illusionary if one considers the Earth Systems dynamic's perspective. The Arctic peoples, especially the local and indigenous communities, will not even be able to enjoy the "resource curse". Global warming in the Arctic will affect the livelihoods of the Arctic peoples to the extent that their lifestyles will not be sustainable for long. The melting of the Arctic ice and permafrost, along with substantial transformations of the Arctic ecosystems, will make even the living conditions of the non-urbanized Arctic peoples impossible. Nevertheless, oil, gas, and minerals will continue to be extracted and fish will continue to be harvested. Planetary Development will demand it, in order to be able to continue (for a while) on its profoundly unsustainable path, which will continue to negatively affect the Arctic, both directly via resources extraction or indirectly via Earth System dynamics.

## Conclusion: Why "Sustainable Development of the Arctic" Still Matters

Why do we continue to use the term "sustainable development" when referring to the Arctic? More precisely, why do we continue to believe, and make believe, that the sustainable development of the Arctic is desirable, let alone possible? In order to answer these questions, one must revert to the psychoanalytical concept of "denial". However, such denial is not just individual and collective; it is actually cultural and even political: the prospect of sustainable development must be kept alive, especially in the Arctic where it is most improbable.

One must revert to the discipline of ethno-psychanalysis in order to explain why the term "sustainable development" is being embraced. Ethno-psychanalysis looks at the combined psychanalytical (and even psychiatric) and anthropological dimensions of shared human belief systems and corresponding narratives to explain institutions and ensuing social and economic behavior (Devereux, 1983). During the Cold War and the corresponding threat of nuclear annihilation of humanity, authors referred to ethno-psychanalysis in order to understand the lack of reaction and fatalism of peoples and nation-states vis-à-vis such threats (Richter, 2000). The continued quasi-religious belief in Development, with science and technology at its core (Noble, 1999), is of such ethno-psychoanalytical nature.

Within this broad belief, I argue that the concept of a "sustainable development of the Arctic" serves at least three functions, which are as many reasons for the exploitation of Arctic resources to continue and even to the accelerate.

- Firstly, the exploitation of Arctic oil, coal, gas, and minerals is needed to sustain Development on a global scale. While the world may be able to do without Arctic oil and coal, gas from the Arctic constitutes a significant share of global production today and has the potential to constitute an equally important share in the future. Even if decarbonization should, at some point in time, stabilize and hopefully even reduce the need for fossil fuels to power Development, the exploitation of certain minerals from the Arctic that will be necessary for the world's energy transition will become even more urgently needed. The exploitation of such resources will be more acceptable if it is labeled "sustainable development", and especially will be more accepted by the local peoples, which from the outset do not spontaneously and necessarily see it that way. This is at least what governments and investors hope for. However, such sustainable development will just be a fig leaf for tradition, industrial development, which is why the concept of "sustainable development" was invented by the UN system during the 1980s to begin with (Chatterjee and Finger, 1994).
- Secondly, more precisely, the global investor community will need the label "sustainable development" so it can profitably invest its funds. For example, the World Economic Forum's "global agenda council" has developed an Arctic Investment Protocol, which *"puts forward sustainability principles similar to*

*initiatives developed for mature economies in recent years. The focus is long-term: tap the expertise of indigenous communities and treat them as commercial partners, protect eco-systems (even as rising temperatures change them before our eyes) ...* ".[8] The two most valued types of such infrastructure projects in 2016 were energy (47 projects) and mining (119) projects. Interestingly, the above-cited article features one of the major global investment funds, Guggenheim, one of the financial pillars of the Arctic Circle Conference, whose arguments are particularly cynical in this context: *"Guggenheim has long provided infrastructure finance, because it is an area that can offer long-terms stable returns. That is partly why the firm was drawn to the Arctic. ... The financial measure of opportunities available there are difficult to estimate, but $1 trillion may be a solid first pass. That is the figure Guggenheim says may be needed to get the Arctic up and running in a manner that won't deplete it in the long run."*[9] It is unsurprising, then, that the annual Arctic Circle Conference has become the place where all stakeholders, especially governments, investors, companies and indigenous peoples, gather to celebrate their common belief in the sustainable development of the Arctic.

- Thirdly, investors will not just need the label in order to justify the channeling of pension and other funds into Arctic infrastructure projects. Companies will not just need to be seen and perhaps even to be certified as being active contributors to the Arctic's sustainable development. The world will also need to believe that energy exploitation and mining in the Arctic equates to "sustainable development of the Arctic" and that the sustainable development of the Arctic contributes to the sustainable development at the planetary scale. This is basically because the belief, or myth of Development forever and for all will have to be maintained at almost any price. Nation-states will need it in order to remain legitimate in the eyes of their citizens, but also in order to generate the financially resources to sustain their "lifestyle". They will also need the myth in order to avoid war with other nation-states, as sustainable Development has the ability to maintain the illusion of a win-win situation. Investors will also need the myth, but even more so they will need the nation-states and the international economic system to perpetuate this myth in order to justify their behavior in front of the people who entrust them with their money. Firms will need the myth, but even more so the nation-states to offer them the legal framework conditions and often the active financial support to continue to do business. Finally, for their sanity, people will need the myth of state-led and science- and technology-fueled Development for ethno-psychoanalytical reasons.

Overall, the "sustainable development of the Arctic" is just a myth. Believing in it is treacherous and promoting it is hypocritical and cynical. What will it take for the myth of "sustainable Arctic development" to become recognized precisely for what it is: a myth, a scam, a fraud? In this sense, the Arctic is not just a "laboratory

---

[8] The World Has Discovered a $1 Trillion Ocean. *Bloomberg Business News*, January 21, 2016.
[9] idem.

of the Anthropocene" from a bio-physical and Earth System dynamics point of view. It is also a laboratory from an ethno-psychoanalytical point of view: what does it take and especially when will the "Development emperor" be exposed as being naked? This may never happen, and if it does, it will be too late.

## References

Arctic Human Development Report. (2004). Akureyri: Stefansson Arctic Institute.
Auzanneau, M. (2015). *Oil, power and war. A dark history*. River Junction, VT: Chelsea Green Publishing.
Buc, P. (2015). *Holy war, martyrdom, and terror. Christianity, violence and the West*. Philadelphia: University of Pennsylvania Press.
Chatterjee, P., & Finger, M. (1994). *The Earth Brokers. Power, politics and world development*. London: Routledge.
Daly, J. (2014). *The Rise of Western Power: A Comparative History of Western Civilization*. London: Bloomsbury.
Devereux, G. (1983). *Essais d'ethnopsychiatrie générale*. Paris: Gallimard.
Dudley, J. P., Hoberg, E. P., Jenkins, E. J., & Parkinson, A. J. (2015). Climate change in the North American Arctic: A One Health Perspective. *EcoHealth, 12*, 713–725.
Finger, M. (2013). What does the Arctic teach us. An epistemological essay on business-government relations. In H. Exner-Pirot (Ed.), *The 2013 Arctic Yearbook*. Rovaniemi: University of the Arctic. Web publication.
Finger, M. (2015). The Arctic, laboratory of the Anthropocene (2015). In L. Heininen (Ed.), *Future security of the global Arctic* (pp. 121–137). New York: Palgrave McMillan.
Finger, M., & Heininen, L. (2018). The "Global Arctic" as a new geopolitical context and method (with Lassi Heininen). *Journal of Borderland Studies, 33*(2), 199–202.
Georgescu-Roegen, N. (1971). *The entropy law and the economic process*. Cambridge MA: Harvard University Press.
Grinevald, J. (2007). *La Biosphère de l'Anthropocène. Climat et pétrole, la double menace. Repères transdisciplinaires (1824–2007)*. Genève: Georg.
Hartch, T. (2015). *The Prophet of Guernavaca, Ivan Illich and the Crisis of the West*. Oxford: Oxford University Press.
Kaznatcheev, P., & Balazeva, R. (2016). A comparison of the role of privately and state-owned oil companies in developing the Arctic shelf. *Energy Exploration & Exploitation, 34*(1), 99–112.
Kolbert, E. (2014). *The Sixth extinction. An unnatural history*. London: Bloomsbury.
Kristoffersen, B., & Langhelle, O. (2017). Sustainable Development as a Global Arctic Matter: Imaginaries and Controversies. In K. Keil & S. Knecht (Eds.), *Governing Arctic Change* (pp. 21–41). Berlin: Springer.
Krivorotov, A., & Finger, M. (2018). State-owned enterprises in the Arctic. In M. Finger & L. Heininen (Eds.), *The Global Arctic Handbook* (pp. 45–62). Berlin: Springer.
Lenton, T. (2012). Arctic Climate Tipping Points. *Ambio, 41*, 10–22.
Malone, T. F., & Roederer, J. G. (Eds.). (1985). *Global Change. The proceedings of a symposium sponsored by the International Council of Scientifc Unions (ICSU) during its 20th General Assembly in Ottawa, Canada on September 25, 1984*. Cambridge: Cambridge University Press.
McNeill, J. R., & Engelke, P. (2016). *The Great Acceleration: an environmental history of the Anthropocene since 1945*. Cambridge MA: Harvard University Press.
Meadows, D., Randers, J., & Meadows, D. (2004). *Limits to Growth: the 30-year update*. White River Junction VT: Chelsea Green Publishers.
Murshed, S. M. (2018). *The resource curse*. Newcastle: Agenda Publishing.
Noble, D. (1999). *The religion of technology. The divinity of man and the spirit of invention*. London: Penguin.

Richter, H.-E. (2000). *Umgang mit der Angst*. Frankfurt: Econ Verlag.
Thomas, J. A., Williams, M., & Zalasiewicz, J. (2020). *The Anthropocene. A multidisciplinary approach*. London: Polity Press.
Tilly, C. (1990). *Coercion, Capital, and European States, A.D. 990–1990*. Oxford: Oxford University Press.
Tolvanen, A., et al. (2019). Mining in the Arctic environment – A review from ecological, socio-economic and legal perspectives. *Journal of Environmental Management, 233*, 832–844.
White, L. (1967). The historical roots of our ecological crisis. *Science, 155*, 1203–1207.
White, L. (1968). *Machina ex Deo: Essays in the dynamism of Western Culture*. Cambridge MA: MIT Press.

**Matthias Finger** is a Professor Emeritus from Ecole Polytechnique Fédérale in Lausanne, Switzerland (EPFL), a part-time professor at the European University Institute (EUI) in Florence, Italy, a full professor at Istanbul Technical University (ITÜ) and a research professor at Northern Arctic Federal University (NArFU), Arkhangelsk, Russia.

# Part IV
# Governance

# Chapter 18
# Between Resource Frontier and Self-Determination: Colonial and Postcolonial Developments in the Arctic

Peter Schweitzer

## Introduction

When did the colonization of the Arctic begin? An obvious starting date could be when the first people moved into the lands north of the Arctic Circle after the end of the last Ice Age, some 8000 years ago, as Yvon Csonka told us in his chapter. While the first of these settlement waves encountered "empty lands", to misuse an unfounded justification by European colonists much later, some of these migrations obviously resulted in pushing out or replacing earlier populations. The most prominent Arctic example in that regard are the so-called Thule migrations, which formed the basis of today's Inuit groups and largely replaced earlier Paleo-Inuit populations (Friesen, 2016). The first Indo-European forays into the High North were tied to the Vikings, who started pushing past the Arctic Circle along the Norwegian Coast – that is, into Sámi territory – during the first millennium AD. Even more spectacular were the northern maritime endeavors of the Vikings, at the end of the first millennium and the beginning of the second millennium AD, which led to a sustained settlement of the Faroe Islands and of Iceland, a temporary collocation with Inuit groups in southern Greenland, and brief forays into Labrador (Fitzhugh & Ward, 2000).

This article deals with a narrower understanding of "colonial" by limiting it to the large-scale European expansion into the rest of the World from the sixteenth century onwards. Such a "post-Columbus" perspective not only omits the

P. Schweitzer (✉)
Department of Social and Cultural Anthropology, University of Vienna, Vienna, Austria
e-mail: peter.schweitzer@univie.ac.at

above-mentioned Viking explorations and settlements, but leads to a focus on the non-European Arctic. As the settlement of the Faroe Islands and of Iceland occurred by Vikings more than 1000 years ago, and did not replace or marginalize an earlier indigenous population, their histories will be largely omitted here. The same is true for the first (Viking) episode of European settlement in Greenland and for the above-mentioned gradual northward movement of speakers of Indo-European languages into Sámi lands. Given this limitation, we start this chapter east of the European Arctic, in Siberia, and progress eastward to Alaska, Canada, and Greenland. We only touch briefly on Northern Fennoscandia toward the end of the chapter. As a kind of conclusion, we end with an attempt to summarize the similarities and differences of these colonial (and post-colonial) encounters.

Given the logic of this chapter, most parts of it will be structured along the administrative divisions of today's Arctic states. While this can be justified by the availability and accessibility of sources, it prevents our understanding of how this "national" borders in the North came into being in the first place. To put things into perspective, we need to go back to the days of Christopher Columbus. After his 1491 "discovery" of the Americas, the treaty of Tordesillas (in 1494) cemented Spain's and Portugal's control over the southern sea routes to the Orient (Vaughan, 2007: 55). The two other rising western European sea powers of the time, England and Holland, saw the possibility of a Northeast (or Northwest) Passage as a way out, which promised to be shorter than the conventional routes to the Orient. England took the initiative first. Back in 1527, Robert Thorne had already suggested to King Henry VIII that he finance an expedition from England via the North Pole to East India; the fate of this proposal remains unknown. The famous English politician Sebastian Cabot was convinced that the Northeast sea route to India was possible. His project was eventually taken up by London merchants, who founded the Muscovy Company dedicated to the search of the Northeast Passage (Vaughan, 2007: 56–60). In 1565, a Dutch trading post was founded on the Kola Peninsula. Brussels-born Olivier Brunel was particularly active in the Russian North. In 1584, he organized an expedition headed for China, which had to turn back just south of Novaya Zemlya. However, the most famous Arctic Dutch seafarer of the time was Willem Barentsz, who attempted to cross the Northeast Passage three times. During the final voyage, which began in 1596, they had to winter on Novaya Zemlya before Barentsz succumbed to the strains of the venture in June of 1597 (Veer, 1876).

These endeavors and parallel attempts by Martin Frobisher and others to find a Northwest Passage are the initial steps of the European overseas colonization of the Arctic. One is tempted to see them also as early phases in the ongoing process of globalizing the Arctic. In doing so, however, we should not forget that the indigenous peoples of the Arctic had been agents of world history in Eurasia and the Americas long before Europe "discovered" them.

## The Arctic by Region[1]

The following overview by region is largely organized along national boundaries, as the policies of (southern) nation states have become dominant for the lives of Arctic residents. The only exception is the final section about the land of the Sámi, where similarities across national borders are more widespread than elsewhere.

### *Siberia*

The beginning of the Russian colonization of Siberia is usually tied to "Yermak's Campaign" in the 1580s (Armstrong, 1975). At that time, the up-and-coming Principality of Moscow tried to extend its influence to the east and allowed the Stroganov merchant family to exploit the Perm region economically. In 1581, a Cossack detachment of 840 men, led by Yermak Timofeevich, marched against the West Siberian Tatars and conquered their capital. While Yermak himself fell in battle in 1585, it took years before Moscow sent its own troops to establish state power in Siberia. Thus, what had started as a private affair of the Stroganovs gradually became a matter of state (Dahlmann, 2009: 48–55, 60–75). In the following years there was a rapid consolidation and expansion of the Russian presence in Siberia. This development, which took place at an exceptionally fast pace (there are little more than 60 years between Yermak's campaign and the first Russian settlements on the Pacific), was driven by the search for furs (especially for sable) and, in terms of transport, was driven by the large rivers (flowing from south to north) of Siberia (Armstrong, 1965: 32–38, 46–57). Of course, the Russian colonization of Siberia had multiple effects on the local population. The collection of *yasak*, a tribute that was mostly paid in furs, changed the subsistence activities of most of the inhabitants of the Siberian Subarctic dramatically by making fur hunting an economic necessity. Therefore, it is surprising that resistance was limited. Nevertheless, there were repeated uprisings by the local population. For example, at the beginning of the seventeenth century, the Nenets tried to prevent the establishment respectively the consolidation of the Mangazeya settlement (Forsyth, 1992: 45).

The abolition of serfdom in Russia in 1861 was partly responsible for the fact that a powerful stream of migrants moved to Siberia in the decades before the First World War (Treadgold, 1957). In 1911 the Russians already made up the vast majority (86 percent) of the inhabitants of Siberia. While indigenous societies did not have any formal rights, the contact between the local population and Russians was often minimal, especially in the northern parts of Siberia. In the Russian-dominated south of Siberia, the so-called "Siberian regionalists" pushed for the establishment of a Siberian regional parliament. Their demands also included forms of limited

---

[1] The following section is partially based on a German-language article of mine exploring colonial and post-colonial developments in the Arctic (Schweitzer, 2016a).

indigenous self-government. The decisive upheaval in the first half of the twentieth century was certainly a result of the Russian Revolution. Immediately after the October Revolution of 1917, the "Declaration of the Rights of the Peoples of Russia" was announced, which – at least on paper – ushered in a new phase of minority rights. It spoke of the equality and sovereignty of all nationalities, of the abolition of national and religious discrimination, and of the right to self-determination, including the separation from Russia. The *yasak* had already been abolished with the February Revolution of 1917. However, it was not until the mid-1920s that all of what was then Siberia was under Bolshevik control. In the wake of the civil war there has been a bewildering variety of revolutions and counter-revolutions, from short-lived governments and local uprisings. In many areas of Siberia, the traditional economic and infrastructural connections were interrupted by revolution and civil war, which in some places led to famine and other dramatic situations for the local population (Slezkine, 1994: 131–141).

Since 1917 the "People's Commissariat for National Affairs", a kind of ministry, was responsible for ethnic affairs in the whole of Russia. In northern Siberia, the commissariat only had nominal power for many years, as there were neither organs nor means to influence local conditions. By the mid-1920s, however, "village soviets" were set up in most of the northern areas to take on this task. In 1924 a special "Committee for the Aid of the Peoples of the Far North" was founded, which is better known as the "Committee of the North" and was characterized by competing visions. While the "conservatives" advocated the slow and gradual incorporation of indigenous societies, the "progressives" primarily had the industrial development of the north in mind. It soon became apparent, however, that the opposition to industrial interests was losing out (Slezkine, 1994: 219–246).

With the entry of the Soviet Union into World War II, Russian chauvinism, which had been frowned upon – at least officially – after the revolution, had become socially acceptable again. Above all, Stalin needed the support of the Russian majority people and therefore cared less and less about specific concerns of the non-Russian population of the USSR. Still, there were a number of volunteers from among the indigenous peoples of Siberia who fought in the Second World War (Forsyth, 1992: 347–349). However, it was the second half of the 1950s, the post-Stalin era, which was a sigh of relief after the Great Terror in most areas of the Soviet Union, that was responsible for the most radical and dramatic changes in many areas of the north. The project of the industrial "mastering of the North" (*osvoenie Severa*) now also encompassed the most remote areas of the Soviet Union (Forsyth, 1992: 403–409).

When the article "Big Problems of the Small-Numbered Peoples" (Pika & Prokhorov, 1988, English translation 1989) was published in 1988, the first signs of *perestroika* had become noticeable in Siberia. In the years that followed, *perestroika* resulted in an avalanche of critical media reports, newly founded political organizations and international contacts among the local population. The Siberian Yupik of the extreme northeast of Siberia founded the "Regional Society of the Eskimos of Chukotka" and took part in meetings of the "Inuit Circumpolar Council" for the first time. The Sámi of the Kola Peninsula formed their own association and made contact with the Sámi of Fennoscandia. The Nenets organized the association "Yamal for our descendants" and mobilized against the ecological destruction of the Yamal peninsula. The indigenous peoples of the Khanty-Mansi region decided at their first conference to establish nature reserves that give priority to indigenous resource use. These and many other events culminated in March 1990 with the holding of the first congress of the "Association of the Peoples of the North" in Moscow. In those years the local population was also favored by legislation. It was decided to resume teaching in a number of indigenous languages, and laws were passed on local self-government and the establishment of indigenous territorial units.

Unfortunately, the optimism and zest for action of the early 1990s is already history. In the late 1990s, the social and economic situation had deteriorated significantly in most areas of Siberia. The decline in industrial production and the emigration of many Russian workers – despite initially being viewed as positive – have led to a dramatic deterioration in infrastructure in the areas of education, health, and social services. The villages in the North, which were highly subsidized during the Soviet era, found themselves in hopeless situations: food and other everyday goods were either unavailable or prohibitively expensive; the supply of heating materials was not guaranteed in many areas; and unemployment and alcoholism rates rose dramatically.

The Putin years have had ambiguous impacts in the Siberian North. On one hand, the resumption of state provisioning led to a certain degree of social and economic consolidation in remote villages. On the other hand, neo-paternalism and the ever-increasing centralism have severely restricted local and regional political options for indigenous and other rural residents. Currently, it is an open question as to whether the future of Siberia will be reminiscent of Soviet models or whether alternative scenarios will be possible. So far, all signs point towards a revitalized strong state that, to some extent, "cares" for its (indigenous) citizens but does not tolerate local initiative or critical questions.

## *Russian America/Alaska*

The initial colonization of Alaska, while it was still known as Russian America, was the immediate result and continuation of the Russian conquest of Siberia. The two Bering or Kamchatka Expeditions initiated by Tsar Peter I and his successors played a major role in this. The First Kamchatka Expedition (1725–1730) had been ordered by Peter I himself, but could not answer the central question of whether Asia and North America are connected or not. Although the much larger Second Kamchatka Expedition (1733–1743) also failed to answer this scientifically and economically important question, this endeavor led to the Russian "discovery" of Alaska in 1741. During the second half of the eighteenth century, a series of privately financed trips targeted Alaska's rich fur resources, primarily the black fur of the sea otter, an economic incentive similarly strong as the sable furs of Siberia. The fact that Alaska was an overseas colony (Vinkovetsky, 2011), and that a state-controlled trading company – the Russian-American Company (RAC) – directed its economic and administrative fortunes, created different conditions than in Siberia.

Initially, the sale of Alaska to the United States of America in 1867 was barely noticeable to the local population. The purchase agreement only mentions that indigenous people, like the Russians, are not entitled to US citizenship (Shiels, 1967). The "First Organic Act" of 1884 marked the beginning of civil government and a judicial system in Alaska (Naske & Slotnick, 1987: 72–74). The "gold rushes" that hit Alaska and neighboring areas in Canada in the 1890s (from the Klondyke to the coasts of Nome) made a decisive contribution to a change in the relationship between the native and settler populations (Naske & Slotnick, 1987: 77–87). On one hand, within just a few years the demographic preponderance of the indigenous population that had hitherto existed through all phases of colonial history was lifted. By the early twentieth century, Alaska's non-indigenous population had become as numerous as the indigenous one. On the other hand, the sudden onslaught of thousands of "whites" forced the establishment of legal and military institutions. As a result, Alaska became increasingly an integral part of the USA, which found its political expression, among other things, in the creation of a legislative assembly for the territory in 1912 (through the "Second Organic Act").

In 1915 there was a meeting between delegate James Wickersham and "Indian" (Athapaskan) leaders in Fairbanks. This gathering, which went down in history as the Tanana Valley Chiefs Conference, was made necessary by rumors of waves of settlers forecast in the wake of the impending construction of the Seward to Fairbanks railway line (Schneider, 2018). Although these "chiefs" were political representatives of their tribal groups (they represented around 1500 "Indians" from the interior of the country), this meeting did not lead to a permanent and powerful political organization of the local population. Above all, the tribal leaders wanted assurance that their traditional ways of life would not be compromised and did not aim to increase their political influence in society as a whole. After the completed railway line did not lead to the expected wave of settlements, nothing more was heard from the "Tanana Valley Chiefs Conference" for almost 50 years.

World War II drastically changed the role of Alaska within the United States. With the Japanese attacks on Hawaii and the Aleutian Islands, Alaska suddenly became an important military area. The construction of roads, airports and military bases set in motion dramatic demographic and infrastructural developments. For some of the Aleuts, the war years brought not only a Japanese invasion, but also a painful resettlement campaign by their own government. Underground atomic bomb tests carried out on the Aleut island of Amchitka during the 1960s and 1970s were the long-term consequences of the militarization of the region (Kohlhoff, 1995). For the white population of Alaska, the most important political event of the post-war period was that Alaska finally became a US state in 1959. However, little has changed for the indigenous population. The constitution of the new federal state also contained hardly any specific statements on the subject. The most important paragraph states, in connection with the transfer of land in federal property to the State of Alaska, that any indigenous property rights do not expire. Although these property rights were not explained further, this sentence formed an important point of argument for future disputes.

As the state of Alaska took possession of more and more land, subsistence farming was also increasingly impaired. At the same time, however, the political awareness of the indigenous population of Alaska grew. While the American civil rights movement certainly played a part in these developments, a number of local political developments were also decisive. For example, the "Project Chariot" – the crazy idea of creating an artificial harbor on the northwest coast of Alaska using hydrogen bombs – led to the mobilization of Inuit and Yupik representatives from the entire north and west coast of Alaska (O'Neill, 1994). Finally, in 1967 the Alaska Federation of Natives (AFN) was founded, the first real all-Alaskan association of indigenous activists.

The discovery of oil deposits on the north coast (Prudhoe Bay) in 1968 turned out to be the decisive leverage for the clarification of the long-fallow subject of indigenous land rights. After several Indian villages in Central Alaska had halted construction of the planned oil pipeline across Alaska (because of indigenous property rights), it became clear to both the interested oil companies and the federal government that a regulation was necessary. In 1971 the US Congress passed the so-called "Alaska Native Claims Settlement Act" (ANCSA), which for the time, and at first glance, looked like a triumphant result in favor of the indigenous population. Twelve for-profit regional corporations were formed in Alaska and received generous compensation and title (including mining rights) over a little over 10 percent of the total area of Alaska. In addition, village corporations were set up within the regional societies (Mitchell, 2001). Despite all of these positive achievements, it must be said that the rationale behind ANCSA was the ultimate assimilation of the local population. The concept of turning indigenous property rights into business affairs has nothing to do with indigenous values. It is not surprising that many critics of ANCSA from among the ranks of the indigenous population find this regulation unacceptable to this day.

The buzzword "subsistence" is a political issue for both indigenous and non-indigenous residents of Alaska. Typically, a majority of the non-indigenous population are opposed to any federal interference in hunting and fishing regulations in Alaska; however, the indigenous population often see the administration of the state of Alaska as more threatening. The state of Alaska has traditionally opposed the federal government in Washington, DC. This tendency to question the primacy of federal politics, which is widespread throughout the western United States, plays a central role in the arsenal of Alaska's non-indigenous political representatives (Haycox, 2002: 313–318).

The issue of sovereignty appears again and again among the demands of indigenous organizations. In 1993, the federal government recognized over 200 villages in Alaska as "tribes". The question remains whether Alaskan tribes are entitled to similar rights of self-government as recognized American Indian tribes in the rest of the USA. Conflicts over the exploitation of (non-renewable) resources are typical in that respect, such as the dispute over the "Arctic National Wildlife Refuge (ANWR)". Controversies over the oil resources of ANWR have been raging for decades and cannot be reduced to a dichotomy of indigenous vs. non-indigenous interests, but have developed along the fault lines of environmental protection vs. resource development, a controversy that transcends ethnic and social boundaries in Alaska (Schweitzer, 2016b). During Donald Trump's presidency, the pendulum clearly swung towards resource development, while incoming President Joe Biden has sought to restore the environmental protection that Trump had removed from ANWR.

## *Canada*

As in Siberia and Russian America, the fur trade was the driving force behind the colonization of northern Canada, although here the focus was on beaver pelts (Moore, 2000: 107). Starting with Jacques Cartier, who explored the St. Lawrence River area for France in the first half of the sixteenth century, and the English discovery of Baffin Island in the second half of the sixteenth century, the colonial history of what is now Canada is characterized by the competition of these two colonial powers (Moore, 2000). In the second half of the seventeenth century, the Hudson's Bay Company was founded, which, like the RAC in Alaska and the Royal Greenlandic Trading Company (KGH), symbolized the close connection between monopolistic capitalism and state administration. The main colonial actors in the fur trade were the so-called *voyageurs*, fur hunters and traders of European origin, some of whom settled in the North and started families with indigenous women, which led to the formation of a mixed group, the Métis (Podruchny, 2006). In contrast to the only other named mixed group in the circumpolar North, the Alaskan Creoles, the creation and naming of the Métis cannot be traced back to colonial interventions, but had to be fought for by the groups themselves.

Canada became independent in 1867 as the "Dominion of Canada," a self-governing colony of the British Empire. The decisive piece of legislation that regulates the relationship between Canada and its First Nations is the "Indian Act," passed in 1876. This law, which originally only referred to "Indians" and today also includes Inuit and Métis, defined a special legal status that differs from that of other Canadians. There are over 2000 reserves in Canada. The first were set up by the Catholic Church in "New France" and later reservations were established through treaties with England and Canada. The land belongs to the state and a certain indigenous group (or "band") has the right to use it. Eleven contracts were signed between the local population or the respective state power. The legal basis of the first contracts was the English "Royal Proclamation" of 1763, which recognized the property rights of "Indians". The last treaty was signed between Canada and the Dene north of the Great Slave Lake in 1921.

The Second World War marked a turning point in Canada as well. This was the first time that military investments were made in the North, which served as a transport and transshipment point for military equipment destined for the Pacific and Europe. The construction of the Alaska Highway changed the situation in the North, as did the construction of air strips between Edmonton and Fairbanks. In Norman

Wells, oil production was increased and a pipeline to Whitehorse was installed. In the east of the Canadian Arctic, landing pads were built for planes en route to Greenland and England. In the 1950s, three additional radar systems were developed and installed (Coates et al., 2008: 53–79).

After the end of the Second World War, a general social sensitivity developed for the economic and social well-being of the inhabitants of the north of Canada. However, this was mainly shaped by paternalism that saw the well-being of indigenous northerners endangered by the advance of "southern civilization". During the 1950s there were a number of tragic resettlements of Inuit groups to unpopulated areas of the far north. In 1969, the government under Prime Minister Pierre Trudeau submitted a radical draft concerning "Indian" administration, the so-called "White Paper on Indian Policy". The proposal provided for the revocation of the "Indian Act" and with it any special legal status for the indigenous population. Although the government hailed the draft as "repealing apartheid policy," the proposal sparked an avalanche of extremely negative criticism from indigenous representatives. The "white paper" was never put into practice; however, the political mobilization of the indigenous population did not stop.

In the years and decades since then, a lot of energy has been invested in negotiating land claims. So-called "specific land claims" are mostly complaints about compliance with contracts or specific complaints from individual groups aimed at compensation. The "comprehensive land claims" are much more far-reaching; these can be negotiated by "status Indians" (including Inuit and Métis) whose ancestors did not sign any treaties. At least six such comprehensive agreements have been reached so far and several others are still in the negotiation stage. The first of those agreements historically was the James Bay and Northern Quebec Agreement, which was signed in 1975 by the Cree and Inuit of the area. This agreement, made necessary by the construction of huge hydropower plants, involved the exchange of indigenous property rights for money, minimal property rights, somewhat more generous rights of use and various social programs. This contract is very similar to the Alaska Native Claims Settlement Act and is now largely viewed as a sell-out of indigenous rights.

The conclusion of several other agreements in the Northwest Territories is closely linked to the plans developed in the early 1970s for an oil pipeline from Alaska through the Mackenzie Valley to the central parts of the United States. COPE ("Committee for Original People's Entitlement") was founded as early as 1969, an organization that today mainly represents the Inuit of the Mackenzie region. The Dene or Athapaskans of the Mackenzie Valley also organized and created the "Indian Brotherhood". In 1984, the "Western Arctic (Inuvialuit) Settlement Region", which mainly affects the Inuit areas in the Mackenzie Delta, was contractually secured. Similar to the case of the James Bay Agreement, indigenous land rights were exchanged for limited land ownership, mining rights, hunting and fishing rights and financial compensation. Warned by the negative consequences of the James Bay Treaty, the Inuvialuit settlement was much better. In recent years agreements have also been signed with the Athapaskans and Métis of the area.

The latest and most spectacular agreement concerns the "Nunavut Settlement Area," which was negotiated under the leadership of the "Tungavik Federation of Natives" (TFN), an Inuit organization, and approved by the Canadian Parliament in 1993. The prehistory of this treaty dates back to the early 1960s, when first proposals to split the Northwest Territories into two parts, along the tree line, were made by the Dene (Athapaskans). Similar proposals were made by the Inuit in the late 1970s (Purich, 1992). Instead of a hasty solution, however, the affected groups took a lot of time; it was not until 1992 that the finalized proposal was accepted by 69 percent of those entitled to vote (who were mainly Inuit). On April 1, 1999, the new "Nunavut Territory" (carved out from the Northwest Territories) became a reality. This created the first and only Canadian territory in which the Inuit constitute the vast majority of the population. The treaty guarantees ownership and mining rights over vast stretches of land, generous financial compensation, a percentage of federal income from Nunavut's mineral resources, and increased efforts to employ Inuit in the federal service.

More recently, a number of disputes over the Canadian Constitution have addressed the issue of indigenous self-government. In the debates surrounding Canada's new constitution, adopted in 1982, it has long looked as though there would be no mention of indigenous rights. Ultimately, massive protests led to at least a guarantee of existing rights. However, many of the constitutional debates focused on the autonomy rights of Quebec's French-speaking society. Today, hopes for constitutional reforms seem to have largely been buried; instead, most representatives of First Nations rely on bilateral agreements with the federal government, such as the aforementioned land claims. In 1992, the "Royal Commission on Aboriginal Peoples" met for the first time. This commission, made up of prominent representatives of First Nations, submitted its final five-volume report in 1996, dealing with a wide range of problems, ranging from social and economic issues to education and health to indigenous justice, land rights, and the constitution. In 2008, the Truth and Reconciliation Commission (TRC) was set up, which, in 2015, submitted a report on the effects of residential schools on indigenous peoples (TRC, 2015).

## Greenland

In 1721, the Lutheran priest Hans Egede from Norway, which was part of Denmark at the time, started the second Scandinavian colonial incursion into Greenland. He was looking for remnants of the old Viking colony, established a mission station near contemporary Nuuk, and thereby commenced the Danish colonial rule over Greenland. Egede and his followers and successors learned Inuit and translated the Bible, provided a first comprehensive description of southwest Greenland, and published the first Inuit dictionary, as well as other books and articles on the history and ethnography of Greenland. In 1733, the Moravians established a mission in south Greenland, which operated until 1900. In 1776, the Royal Greenlandic Company was founded, which held a monopoly of trade with Greenland until the end of the Second World War. In 1782, this company adopted new administrative rules, which remained in force until 1862. They contained rules for mixed marriages, which stated that children from such unions should be brought up as "competent Greenlanders"; they outlawed the sale of alcohol, and stated that hunters should not be lured away from their profession. Traders were instructed not to buy so much from locals that the latter would have too few supplies for the winter. While these rules are expressions of a paternalistic attitude, they nevertheless constituted attempts to preserve Greenlandic culture and life-ways (Gad, 1973, 1982).

While Danish rule of Greenland could be seen as "soft colonization" in a circumpolar comparative perspective, the fact remains that the Greenland Inuit have not been able to determine their own fate for centuries. It should not be forgotten that most of what appears in historical light as a "positive" development were concepts invented by Danes. This includes the "renaissance" of Greenlandic culture directed by Heinrich Rink in the nineteenth century, which resulted in the establishment of the world's first Inuit newspaper, *Atuagagdliutit*, in 1861. Another result of this movement was the establishment of local councils in which Greenlanders were also represented. Ultimately, however, it was the Inuit themselves who used Denmark's paternalistic policy for their own purposes and thus laid the foundations for the sovereignty movement of the twentieth century.

The isolationism and paternalism cultivated by Denmark vis-à-vis Greenland came to an abrupt end with the Second World War. In 1941, a treaty between the USA and Denmark was ratified, which - with the recognition of Danish supremacy - made American military bases in southwest Greenland possible. Due to the events of the war, Greenland was cut off from Denmark for five years. At the same time there was an advancing commercialization of fishing, which since the beginning of the twentieth century - due to global warming - has increasingly replaced the seal catch. This forced end to Danish attempts to isolate Greenland from the outside world led to the breakdown of traditional values and social norms of the Inuit. It became apparent that Greenland at the time was characterized primarily by poverty and inadequate and unsanitary living conditions. In the years and decades after the end of World War II, Denmark's Greenland policy was therefore primarily geared towards "modernization".

A constitutional amendment in 1953 changed Greenland from a colony to a "normal" part of Denmark. This marked the beginning of a phase of intensive social planning and welfare policy, which was expressed primarily in two large development projects (G-50 and G-60) of the post-war period. There was generous investment in health and education, as well as in housing. The fishing industry in south-west Greenland was treated as a priority and the KGH's trade monopoly, which had lasted for almost 200 years, was lifted. Although these measures led to a quantitative increase in the standard of living, at the same time the culture of the Greenlanders was suddenly much more threatened than in the centuries before. The modernization policy encouraged the emigration from smaller settlements and led to a concentration of the population in the cities. Some villages that had been declared "non-profitable" were forced to give up under "gentle pressure" (e.g. by closing schools and shops).

During the 1970s, the political awareness of the Greenlanders grew. The resulting anti-colonial and nationalist movement was largely sustained by a Danish-speaking Inuit elite trained in Denmark in the 1950s and 1960s. The founding of the Social-Democratic *Siumut* Party drove the political process further in the direction of autonomy. This led to the establishment of an autonomy commission at government level in 1975, which passed the Greenlandic Autonomy Act in 1978. The Autonomy Act came into force through a referendum the following year, in which 79 percent of the Greenlanders voted for autonomy. This included the establishment of a Greenland parliament (based in Nuuk) in which Inuit and resident Danes have the right to vote and stand for election. This means that Greenlandic autonomy is based on territorial and not on ethnic categories (Braukmüller, 1990).

Greenland is a member of NATO (through Denmark). The US military presence, which began in World War II, continues to this day through the military and radar base in Thule. In addition, Danish troops are stationed in South Greenland. The American base in Thule, in the far northwest of the country, has a controversial past. On the one hand, the construction of the military facility led to a painful resettlement of the local Inuit. On the other hand, there have been accidents and inadequate education of the population, such as the crash of a B-52 aircraft loaded with hydrogen bombs, where it has not yet been fully clarified whether all the bombs were recovered (Lynge, 2002).

Greenland left the European Economic Community (EEC) in 1985 after a referendum in 1982 resulted in a narrow majority against remaining in the EEC. Today, Greenland is linked to the European Union by an Association Treaty. Fishing is the country's most important industry and accounts for 80 percent of export income. However, the Greenland government is increasingly forced to force the exploitation of mineral resources, although this does not correspond to its original goals of "indigenous" development (for example, the ban on uranium mining was lifted in 2013). Tourism is another new branch of the economy that is supposed to bring much needed foreign currency to the country. Government-owned Royal Greenland is a multinational (and Greenland's largest) company primarily engaged in fishing, fish processing and distribution. Subsistence sea animal hunting is increasingly being affected by international developments. For example, the European Union's ban on the import of seal skins, the catch quotas of the International Whaling Commission and international protests by animal rights groups have threatened the continued existence of indigenous subsistence practices.

The autonomous government then gradually took over the administration of education, taxation, economy, transport and social affairs. Another important step towards independence was taken on June 12, 2009 (on the 30th anniversary of the Autonomy Act), namely the signing of the Greenland Self-Government Act, which meant extended autonomy by transferring the areas of external affairs and mining management from Denmark to Greenland. There is local self-government within Greenland, which is financed through local income taxes and transfer payments from the government. However, Greenland is still dependent on Denmark: be it financially (almost half of Greenland's budget is financed by Denmark) or in terms of administrative staff, where the shortage of local skilled workers continues to result in a high Danish share. Still, Greenland is certainly the best candidate for full state independence for a former colony in the Arctic region.

## Sápmi – The Land of the Sámi (or, Northern Fennoscandia and Northwest Russia)

As mentioned above, the colonial history of Sápmi is much older than in the Americas or Siberia. I will, however, not review this lengthy process (see Hansen & Olsen, 2014) but jump right into the early twentieth century. The reason for that is that Norway became a sovereign country in 1905, that is the union with Sweden was dissolved, which led to a strengthening of Norwegian nationalism and to laws that discriminated against the Sámi. For example, the Norwegian parliament had passed a land purchase law for Finnmark, i.e. for the Sámi-inhabited areas of northern Norway. This law made the government the owner of all land that was not privately held, with the right to sell or rent it; proof of knowledge of Norwegian was required to purchase the land. These developments led to a first wave of Sámi activism that remained largely ineffective politically.

In 1931, the "Society for the Promotion of Sámi Culture" was founded in Finland. The members of this society were mostly academics and prominent representatives of public life, but there were only few Sámi among them. Similar organizations were soon established in neighboring countries: in 1945 in Sweden and in 1948 in Norway. These activities culminated in the organization of the first Sámi conference in Jokkmokk, Sweden in 1953. This organizational network enabled the foundation of the "Nordic Sámi Council", which was brought into being in 1956 during the second Sámi conference in Karasjok, Norway. The "Nordic Council", founded in 1952, a cooperation body between the parliaments and governments of the Nordic countries served as a model and role model. Especially in the early days there were close contacts between the Sámi Council and the Nordic Council.

The structure of the national development associations behind the Nordic Sámi Council enabled good contacts to government agencies and academic centers (the reports of the first four Sámi conferences were published in English). On the other hand, there was very little support from local Sámi communities. Although most of the delegates at those conferences were Sámi, they were not representatives of grassroots organizations. Rather, they belonged to a Sámi elite that had been developing since the 1950s and whose political activities had been characterized by the catchphrase "Sámi Movement" since the 1960s. At the end of the 1960s, the creation of pure Sámi organizations on a national level became a priority. One of these newly created organizations was the "National Sámi Association of Norway", which began its work in 1969. Before that there were mainly Sámi reindeer herder associations, such as B. the "Confederation of Norwegian Reindeer Owners" or the "National Swedish Sámi Union (1950), which mainly represented reindeer keepers.

In Finland there are only regional and no national Sámi organizations to this day, which was partially offset by the establishment of the first Sámi Parliament to be discussed below. Eventually the "Confederation of the Swedish Sámi" was created, which is a counterbalance to the strong advocacy of reindeer herders. The culmination of the early activities of the Nordic Sámi Council was the establishment in 1973 of the Nordic Sámi Institute in Kautokeino, Norway.

One of the most important moments in the recent history of the Norwegian Sámi was certainly the so-called Alta conflict. In 1970, the first plans for the construction of the Alta Kautokeino hydropower plant were published, which also included the flooding of the Sámi village of Masi. Although an official report published in 1974 warned of the catastrophic consequences for reindeer herding, the Norwegian Parliament decided to start construction in 1979. This decision led to the creation of the "Sámi Action Group" (SAG) in the same year, after a "People's Action Group" (PAG) had been created in 1978, in which Norwegians and foreigners protested and practiced civil disobedience. The SAG organized a hunger strike of seven Sámi youths in front of the Norwegian Parliament in Oslo, where they erected a *lavvo* (a traditional Sámi conical tent). The ethnologist Robert Paine called this action a successful "ethnodrama" because – in contrast to the actions of the PAG – it succeeded in gaining the attention and sympathy of a broad public (Paine, 1985). The Alta conflict attracted international attention and brought the Sámi media coverage from around the world, but it split the Sámi movement into a radical and a moderate wing. Still, although the power plant was eventually built, the Alta conflict marked a turning point in the political history of the Sámi.

The most spectacular recent developments certainly concern the establishment of so-called Sámi parliaments in all countries in the region, with the exception of Russia. The first of these parliaments was set up in Finland in 1973, where it has had advisory functions ever since. Since 1991 the Finnish Sámi, or the Sámi Parliament, must be consulted on all decisions that affect them. In 1989, there was the first election to the Norwegian Sámi Parliament (*Sámediggi* in Sámi), which also has a predominantly advisory role; the areas in which the Sámi Parliament will have legislative functions remain to be clarified. In 1993, a Sámi parliament was finally set up in Sweden.

In 1987, the so-called Sámi Act was passed, which provided the legal basis for the Norwegian Sámi Parliament, together with Article 108 of Norwegian Constitution, which recognizes the Sámi as an indigenous people and gives the state the responsibility to guarantee that the Sámi can maintain and develop their society and culture. This article, which can be seen as a long-term consequence of the disputes over the Alta-Kautokeino hydropower plant, enables a new language law that practically equated Sámi and Norwegian as official languages. Finland has also allowed Sámi to be used in courts and authorities since 1991. Sámi language teaching has only been intensified since the 1960s and there have been a number of university institutes for this purpose since the 1970s.

In the last few years the concept of *Sápmi* ("Land of the Sámi") has become more and more important. It is an expression of a new, pan-Sámi identity and implies claims to an area that transcends national borders. At the same time, the problem of Sámi land rights has not yet been resolved. A recent decision in a Swedish court case that has been going on for over 20 years recognizes the existence of Sámi grazing rights and characterizes them as a special form of property rights, but all governments in the region have so far refused to recognize these property rights politically. These rights are of course particularly important in reindeer herding, which is increasingly threatened by external interests (through forestry, mining, environmental protection, tourism, etc.). Reindeer herding is the Sámi ethnic symbol par excellence, although today only a small percentage of the total population - a maximum of 15 percent - is engaged in this country. In Norway and Sweden, only Sámi are allowed to own or graze reindeer. Neither Finland nor Russia limit reindeer herding to the Sámi, but in Finland it is necessary to be located in one of the reindeer-herding districts, which are located in the north of the country.

It has been over thirty years since Russian and Fennoscandian Sámi groups have been able to re-establish formal contacts across borders. Today, the distinct political worlds in which these Sámi groups live are becoming increasingly evident. Although a Kola Sámi assembly (*Kuelnegk Soamet Sobbar*) was founded in Murmansk in 2010, it is still not yet recognized by the Russian government. While the political, social and cultural situation of the Sámi in Finland, Sweden and Norway is better than in Russia, this does mean that the Sámi in Fennoscandia live in ideal postcolonial conditions. There are also important differences between Fennoscandian countries, which is reflected in the fact that neither Finland nor Sweden (but Norway) have ratified the ILO Convention 169 (Indigenous and Tribal Peoples Convention, 1989).

## Concluding Remarks

As the pages above have demonstrated, the Arctic – with the exception of northern Fennoscandia and northwestern Russia – has come under "southern control" relatively late. While early forays (western Siberia, eastern Canada) date back to the sixteenth century, other parts of the North (Alaska, Greenland) were colonized during the eighteenth century, while parts of arctic Canada and the northeastern part of Siberia were hardly under state control before the twentieth century. During the heydays of arctic colonialism – the eighteenth and nineteenth centuries –, state-controlled companies with trade monopolies dominated the economic and political affairs of large parts of the Arctic. As the colonizers of the Faroe Island and of Iceland did not meet any aboriginal population on the islands, their political (colonial) histories differ from the other parts of the Arctic. These other parts are classical settler colonies, where non-indigenous settlers often dramatically outweigh the aboriginal residents.

The Arctic (and Subarctic) has attracted outside economic interests for centuries. Unlike more moderate zones of globe appropriate for agriculture, the economic attraction of North has always been its resources and not its lands. From the sable and beaver pelts of the early colonial periods to oil, gas and other mineral resources of today, the Arctic has been a resource frontier for quite some time. What is different today is that the resources of the High North have become more accessible.

Partly due to climate change, partly due to resource depletion elsewhere, and because of technological advances, the possibility of non-renewable resource exploitation is becoming more and more realistic in many areas of the North. It remains to be seen whether this form of economic development will prove to be beneficial for local communities. In many parts of the Arctic, social, economic and cultural marginalization of indigenous groups continues up to the present. At the same time, various attempts at cultural revitalization seem to be successful, even in the Russian Arctic where art and culture seem to offer politically acceptable venues of self-determination.

In the end, one is tempted to ask whether it is really possible to speak of decolonization in the Arctic. One dimension of that question is whether there is public acknowledgment of the colonial past (and present). Obviously, there are significant differences in how individual arctic countries deal (or not deal) with their problematic heritage. While some countries (e.g., Canada) have made quite some progress recently in publicly discussing the damage done by colonial policies (such as boarding schools), others (e.g., Russia) seem to be unwilling to apply the colonial label to their own history. Overall, there seems to be a tendency among arctic states to recreate certain Cold War divisions, which group the seven "western" states on one side, and Russia on the other. While this reflects global geopolitical constellations, it also speaks to rather distinct political relations between arctic states and arctic/indigenous communities. While the devolution of political decisions and the empowerment has been a general characteristic of the Arctic in recent decades, the political development of the Russian Federation has gone into the opposite direction of centralization. Still, notwithstanding these differences, all regions of the Arctic still display traces of "internal colonialism" (Hechter, 1975; Etkind, 2011). While "sovereignty" is on the agenda of many arctic indigenous groups, full-fledged political independence can only be expected for Greenland in the foreseeable future. What the costs and benefits of state sovereignty will be remains to be seen.

## References

Armstrong, T. (1965). *Russian settlement in the north*. Cambridge University Press.
Armstrong, T. (Ed.). (1975). *Yermak's campaign in Siberia*. The Hakluyt Society.
Braukmüller, H. (1990). *Grönland – gestern und heute. Grönlands Weg der Dekolonisation*. Lit Verlag.
Coates, K. S., Lackenbauer, P. W., Morrison, W. P., & Poelzer, G. (2008). *Arctic front: Defending Canada in the far north*. Thomas Allen Publishers.
Dahlmann, D. (2009). *Sibirien. Vom 16. Jahrhundert bis zur Gegenwart*. Ferdinand Schöningh.
Etkind, A. (2011). *Internal colonization: Russia's imperial experience*. Polity Press.
Fitzhugh, W. W., & Ward, E. I. (Eds.). (2000). *Vikings: The North Atlantic saga*. Smithsonian Institution Press.
Forsyth, J. (1992). *A history of the peoples of Siberia: Russia's north Asian colony 1581–1990*. Cambridge University Press.
Friesen, T. M. (2016). Pan-arctic population movements: The early paleo-Inuit and Thule Inuit migrations. In T. M. Friesen & O. K. Mason (Eds.), *The Oxford handbook of the prehistoric Arctic* (pp. 673–691). Oxford University Press.

Gad, F. (1973). *The history of Greenland II: 1700–1782*. McGill-Queen's University Press.
Gad, F. (1982). *The history of Greenland III: 1782–1808*. McGill-Queen's University Press.
Hansen, L. I., & Olsen, B. (2014). *Hunters in transition: An outline of early Sámi history*. Brill.
Haycox, S. (2002). *Alaska: An American colony*. University of Washington Press.
Hechter, M. (1975). *Internal colonialism: The celtic fringe in British national development, 1536–1966*. University of California Press.
Kohlhoff, D. (1995). *When the wind was a river*. University of Washington Press.
Lynge, A. (2002). *The right to return: Fifty years of struggle by relocated Inughuit in Greenland*. Forlaget Atuagkat.
Mitchell, D. C. (2001). *Take my land, take my life: The story of Congress's historic settlement of Alaska native land claims, 1960–1971*. University of Alaska Press.
Moore, C. (2000). Colonization and conflict: New France and its rivals 1600—1760. In C. Brown (Ed.), *The illustrated history of Canada* (pp. 105–188). Key Porter Books.
Naske, C.-M., & Slotnick, H. E. (1987). *Alaska: A history of the 49th state*. University of Oklahoma Press.
O'Neill, D. (1994). *The firecracker boys*. St. Martin's Griffin.
Paine, R. (1985). Ethnodrama and the 'fourth world': The Saami action group in Norway, 1979–1981. In N. Dyck (Ed.), *Indigenous peoples and the nation state: "Fourth world" politics in Canada, Australia and Norway* (pp. 190–235). Institute of Social and Economic Research, Memorial University of Newfoundland.
Pika, A., & Prokhorov, B. (1988). Bol'shie problemy malykh narodov. *Kommunist, 16*, 76–83.
Pika, A., & Prokhorov, B. (1989). Soviet Union: The big problems of small ethnic groups. *IWGIA Newsletter, 57*, 123–135.
Podruchny, C. (2006). *Making the voyageur world: Travelers and traders in the North American fur trade*. University of Toronto Press.
Purich, D. (1992). *The Inuit and their land: The story of Nunavut*. James Lorimer.
Schneider, W. (2018). *Tanana chiefs: Native rights and western law*. University of Alaska Press.
Schweitzer, P. (2016a). Der hohe Norden als Teil des globalen Südens? Koloniale und postkoloniale Entwicklungen in der Arktis und Subarktis vom 19. bis zum 21. Jahrhundert. In G. Saxinger, P. Schweitzer, & S. Donecker (Eds.), *Arktis und Subarktis. Geschichte, Kultur, Gesellschaft* (pp. 46–66). New Academic Press.
Schweitzer, P. (2016b). Fallstudie: Wer bestimmt über die Ressourcen der Arktis? Der Konflikt um das Arctic National Wildlife Refuge. In K. Fischer, G. Hauck, & M. Boatcă (Eds.), *Handbuch Entwicklungsforschung* (pp. 351–354). Springer VS.
Shiels, A. W. (1967). *The purchase of Alaska*. University of Alaska Press.
Slezkine, Y. (1994). *Arctic mirrors: Russia and the small peoples of the north*. Cornell University Press.
Treadgold, D. W. (1957). *The great Siberian migration: Government and peasant in resettlement from emancipation to the First World War*. Princeton University Press.
Truth and Reconciliation Commission of Canada (TRC). (2015). *Honouring the truth, reconciling for the future: Summary of the final report of the Truth and Reconciliation Commission of Canada*. Truth and Reconciliation Commission of Canada.
Vaughan, R. (2007). *The Arctic: A history. Revised edition*. Sutton Publishing.
Veer, G. de (1876). *Three voyages of William Barentz to the arctic regions (1594, 1595, and 1596)*. The Hakluyt Society.
Vinkovetsky, I. (2011). *Russian America: An overseas colony of a continental empire, 1804–1867*. Oxford University Press.

**Peter Schweitzer** is Professor at the Department of Social and Cultural Anthropology at the University of Vienna and Professor Emeritus at the University of Alaska Fairbanks. He is past president of the International Arctic Social Sciences Association (IASSA) and served as director of the Austrian Polar Research Institute (APRI) from 2016 to 2020.

# Chapter 19
# Understanding Cold War Trust-Building Between Norway and the Soviet Union. Norwegian-Russian Relations – A History of Peaceful Coexistence

**Gunnar Rekvig**

> *In a moment of private candor at Camp David in 1959, President Eisenhower confessed to Nikita Khrushchev that every time he tried to cut defense funds he would end up by backing down before aggressive military advisors who warned him that the Russians were developing new weapons systems which would reduce the United States to a second-rate power. Khrushchev, according to his memoirs, replied, "It's just the same. Some people from our military department come and say, Comrade Khrushchev, look at this. The Americans are developing such and such a system. And we take the steps which our military people have recommended."*
>
> —Senator J. Williams Fulbright (November 1974 [Willens, 1984, p. 65])

## Introduction

This chapter examines the trust-building activities that took place during the Cold War between the Kingdom of Norway and the Union of Soviet Socialist Republics (USSR).[1] To do this, one must first understand the historical background of the relations between Norway and Russia in its different iterations – czarist, communist, and finally as federation – without which, the trust-building activities would likely not have been possible. The Cold War itself was essentially a bipolar geopolitical and ideological confrontation that carried a potential for war between the United States and the Soviet Union and their respective allies. For the Barents Region in

---

[1] Hereafter the Soviet Union.

G. Rekvig (✉)
UiT The Arctic University of Norway, Tromsø, Norway
e-mail: gunnar.rekvig@uit.no

northern Europe, the regional division had, accordingly, Norway, a founding member of the NATO, in the Western Bloc, and the Soviet Union and the Warsaw Treaty Organization (WTO) in the Eastern Bloc (Johnson, 2020).[2] Lastly, Sweden and Finland, the neutral and non-aligned powers, existed between the blocs to varying degrees (Elgström & Jerneck, 1997; Singleton, 1978).[3]

Against this divisive backdrop, deterrence versus reassurance strategies were played out. Deterrence was represented by the military strategies of the two superpowers, which were driven by a nuclear weapons arms race and, perhaps more importantly, one of conventional arms, especially in Europe (Jervis, 2001). Norway was the only NATO member to share a border with the Soviet Union when the organization was founded. As such, the northern part of the country was developed as a first line of defense against the Soviet Union in case war broke out (Kernan, 1989). The deterrence strategies of Norway occurred in parallel to reassurance policies that were implemented to deescalate tensions and to function as confidence-building measures in the north, the sole purpose of which was to build trust with the Soviet Union. As these activities were initiated and instituted during a period that carried the risk of not only war, but nuclear war, what was accomplished was highly significant: the establishment of a zone of low tensions in the Barents Region of the Nordic Arctic between two countries in opposing camps during the Cold War (Larson, 1991; Lund, 1990). This also served to repudiate the *realist* thinking that dominated the field of international relations (IR) at the time.

## Current Outlook

The trust-building initiatives that took place between Norway and the Soviet Union created a momentum of their own that pushed tension reduction in the Barents Region and, consequently, also for the whole of northern Europe (Bitzinger, 1989). Thus, when the Cold War came to an end with the dissolution of the Soviet Union in 1991, the stage was already set for regional collaboration. The Barents Euro-Arctic Council (BAEC) was launched with ease only two years later in 1993. The council opened the door for close cooperation across the Barents Region, a cooperation that has been lauded as highly successful and that transitioned the Barents Region towards a zone of peace (Archer & Joenniemi, 2003; Kacowicz, 1995); *The success of the Barents Cooperation* n.d.). "In the post-Cold War period, the Nordic governments have aimed at preventing aspects of the new uncertainty in European security from spreading to their part of the world. In particular, the establishment of the Barents Euro-Arctic Council … have demonstrated the determination of the five

---

[2] The Warsaw Pact, officially the Treaty of Friendship, Cooperation and Mutual Assistance, was established in 1955 as a counterpart to NATO.

[3] For Sweden, neutrality has historically been something akin to sacred since its inception in 1812. For Finland to retain its neutrality, it engaged in so-called "Finlandization" as a means of keeping its sovereignty, which was the realpolitikal cost of being the neighbor of the Soviet Union.

countries to decrease the expectations of conflict growing in their proximity" (Archer, 1996).

Today, however, northern Europe in general, and the Barents Region in particular, is moving away from peaceful coexistence that has been a mainstay since the fall of the Soviet Union. Instead, we are witnessing the regional security situation again deteriorate quickly. A schism that is threatening the regional peace is developing in the Arctic. The West, represented by NATO, and Russia are again returning to the Cold War lines of the past. Compounding this divide is the arrival of new actors, especially China, which is aligning itself with Russia (Havnes & Seland, 2019). In 2020, the Norwegian Intelligence Service identified China and Russia as the largest threats to Norway, stating that "these two countries consider themselves to be in a persistent conflict with the United States and parts of the West" (*Focus* 2020 *The Norwegian Intelligence Service's assessment of current security challenges*, 2020). While there are several factors behind this degradation of Arctic security, of special significance is global warming, which is the main cause of the development in making the Arctic global; as the polar ice melts, activity in the Arctic will only increase. China, which in 2018 defined itself as a "Near-Arctic State" in its white paper *China's Arctic Policy,* has clear ambitions of becoming a key regional actor (*China's Arctic Policy*, 2018; Havnes, 2020). To achieve this, China is building its capacity for developing the Arctic. The Northern Sea Route (NSR) is fast becoming a reality as the sea ice melts in the Arctic (Tianming & Erokhin, 2021).

Consequently, with the Arctic returning to the global arena it was during the Cold War, the impact of rivalries, disputes, and conflicts elsewhere are and will be driving confrontations in the Arctic. For example, the Russo-Georgian War of 2008 and the Ukrainian Crisis of 2014 have seen Russia sanctioned and increasingly isolated by the West. China, however, following the Ukraine crisis, "responded with a willingness to expand economic and strategic cooperation with Russia" (Korolev, 2016). China, currently the world's second-largest economy, is projected to overtake the United States' economy in 2028 (Kawate, 2021).[4] Accordingly, China is rising to challenge the hegemony that the United States has enjoyed since the end of the Cold War (Layne, 2008). This will see bipolarity return to the global order and, consequently, for the Arctic as well. Since 2014, as relations have deteriorated, the military capacity of regional actors in the Barents Region has been increasing.

Therefore, as relations are worsening and a new Cold War is in the making, it is vital that the trust-building activities that lowered tensions and opened the door for nonconfrontational engagement between Norway and the Soviet Union are revisited and understood. To do this, it is important to examine the relations Norway has had with the Soviet Union – or Russia, in all its iterations – historically, and especially in the north.

---

[4] This timeline has shifted forward by five years due to how China suppressed the coronavirus early, which led to Chinese industrial output recovering quickly.

## Historical Background of Norway–Russia Relations

The relations between Norway and Russia have a long history, especially for northwest Russia and northern Norway in what today is the Barents Euro-Arctic Region. All the way back to the Viking Sagas, one can trace relations of travels to Bjarmeland, the area that today makeup the Kola Peninsula and the White Sea in northwest Russia. The earliest written work on this subject, which dates from the ninth century, describes the travels of a chieftain named Ohthere, from Hålogaland in northern Norway (Blom, 1984). From the Viking age and over the centuries, Norwegians moved further and further north in Norway and interactions with the Russians increased. Historical trade ties between Norwegians, the Sami (the indigenous peoples of the Nordic Region), and the Russians developed, first informally but becoming increasingly organized over time. During the sixteenth and seventeenth centuries, the trade in the north became "increasingly subjected to decrees and political intervention by the states that wanted to channel the trade to permanent markets under their control" (Hansen & Olsen, 2014, p. 230).

### *Pomor Trade*

During the eighteenth century, the so-called Pomor trade began. This trade was a continuation of the trade in the north, where tradesmen from the White Sea area initiated trade relations throughout northern Norway.[5] The trade began in Varanger, in the northeastern part of Norway, and then spreads west and southwards (Niemi, n.d.). The Pomor trade was a barter system, where the Norwegians traded especially fish for grain, flour, and tools from the Pomors. The trade was very important for people in northern Norway. "The social and economic conditions of people in Norway in the eighteenth century were different in the South from the North … Norwegian fishermen [in the north] were enslaved by monopolies from South Norway, and no representatives from any other country whatsoever were allowed to enter Norwegian waters" (Shrader, 2017, p. 111). Shrader went on to describe a monopoly system where trade-houses from Bergen, Trondheim, and Denmark had absolute control over provisions to northern Norway. Thus, the contact and trade with the Pomors started out as illegal, but over time became legal and regulated. It provided northern Norway with much-needed provisions during times of hardship, e.g., during poor harvests in the south, the Pomor trade brought grain and flour to northern Norway (Shrader, 2017, p. 112). Hardship was compounded in times of war. During the Napoleonic Wars, Norway, which was then part of the kingdom of Denmark–Norway, had sided with France and Napoleon (Planert, 2016, p. 199). This led to a British blockade that effectively stopped the trade between north and south in Norway. Therefore, the Pomor trade became a lifeline to northern Norway

---

[5] The word Pomor refers to a people by the sea.

as the Russians were able to avoid the blockade and maintain the trade for parts of the coast (Niemi, n.d.). The Pomor trade developed the bond between Norway and Russia and created a strong relationship in the north. This relationship was further cemented as the Napoleonic Wars ended.

While Napoleon was not defeated until 1815 in Waterloo, Belgium, 1814 marked the year in which the Napoleonic Wars ended for the Nordic Region. In the same year, Sweden, part of the sixth coalition against Napoleon, was victorious over the kingdom of Denmark–Norway, an ally of France. Sweden then proceeded to break up the kingdom in order to annex Norway (Riley, 2002; *Treaty of Concert and Subsidy Between His Britannic Majesty and the King of Sweden; Signed at Stockholm the 3d of March 1813*, 1813).[6] Sweden forced Norway to enter into a personal union in which the king resided in Sweden (Lindgren, 1959, pp. 26–28).[7] Sweden underwent a further great shift from 1814, especially in terms of how Sweden engaged in foreign policy. Its newly elected heir-presumptive, Jean Baptiste Bernadotte, a former marshal of Napoleon who went on to become Charles XIV & III John, instituted Swedish neutrality for the first time.[8] Sweden, which until then had been a great power in northern Europe, realigned itself to become a middle power and, with its newfound neutrality, turned away from matters of war. In an address to the Swedish Riksdag [Parliament], Bernadotte outlined the future for Swedish and Norwegian foreign policy as follows[9]:

> *Separated as we are from the rest of Europe, our policy and our interests will always lead us to refrain from involving ourselves in any dispute which does not concern the two Scandinavian peoples [Swedes and Norwegians]. At the same time, in obedience to the dictates both of our national duty and of our national honour, we shall not permit any other power to intervene in our internal affairs.* (Barton, 1930)

For the Scandinavian countries, this change to neutrality was significant regarding their place in Northern Europe, as a period of Nordic regional peace was about to begin. The year 1814 marked the onset of the Nordic Peace, a peace in which no Nordic country has gone to war with another Nordic country since 1814, and for

---

[6] The makeup of the Sixth Coalition that would defeat Napoleon was Russia, Great Britain, Portugal, Spain, Austria, Prussia, Sweden, and several German states. The Swedish annexation Norway was seen as revenge for the loss of Finland to Russia in 1809. For more on this see: Barton, H. A. (1972). Russia and the Problem of Sweden-Finland, 1721–1809. *East European Quarterly*, 5(4), p. 431., Kuldkepp, M. (2019). National Revanchism at a Critical Juncture: Sweden's Near-Involvement in the Crimean War as a Study in Swedish Nationalism. *Scandinavica*, 58(2), 115–133.

[7] A personal union, as opposed to a real union, denotes two separate independent kingdoms under one monarch, unlike the Kingdom of Denmark–Norway, which signified one kingdom – an absolute monarchy – where the monarch held absolute autocratic power. The implication for Norway was that Norway was independent in all matters except foreign affairs.

[8] Bernadotte became Charles XIV John in Sweden and Charles III John in Norway.

[9] The Riksdag is the Swedish parliament.

which all intra-Nordic conflicts have been resolved peacefully (Archer & Joenniemi, 2003).[10]

After 1814, the rapprochement grew for Sweden and the Russian Empire, not only because of the victory over Napoleon, of which Sweden was part as an ally of Russia, but also because of the relationship the Pomor trade had created in the north between Russians and Norwegians (Niemi, n.d.). Niemi referred to the relations in the north that had been created between Norwegians and Russians as historically persisting characterized by peace and *soft power* (See: Nye Jr., 2004, for Soft Power), which goes hand in hand with the Swedish power realignment. Thus, the relationship between Norway and Russia took a new direction with implicit trust as a foundation. Building on this, the next step was to establish the land-border between Norway and Russia. This happened with the border convention between the Russian Empire and the Kingdom of Norway of May 14, 1826 (Roginsky, 2005, p. 62). This border treaty was unique in that it was not the result of conflict and war as was the norm, but that the two states had reached a point of maturity for a formal border agreement that was settled peacefully.

The Pomor trade developed, became institutionalized, and flourished. Trade relations peaked in the nineteenth century. "These relations helped decidedly to bring the Russian and the Norwegian nations closer together" (Schrader, 1988). The beginning of the twentieth century marked the end of the Pomor trade. The Russian Revolution and its outcome was the cause for this and from 1919, the trade came to an end, never to return (Balsvik, 1994: Cited in Holtsmark & Mankova, 2015, p.37; Friis, 2020; Niemi, 1992, pp. 82–83 & 127). The border between Norway and Russia closed, for all intents and purposes, and did not reopen until after the fall of the Soviet Union in 1991.

## *The Second World War*

The Pomor trade came to an end in the aftermath of the Bolshevik takeover in the October Revolution of 1917. Subsequently, the Norwegian-Russian border closed in 1919 and relations cooled. Tensions rose further in the early twentieth century as there is a scramble for the Arctic, with several territories treated as *terra nullius;* that is, territories that are legally deemed unoccupied and not belonging to any country. Of these, Svalbard and Bear Island were significant for Norway and the Soviet Union and were behind the rise in tensions. Despite this, the territorial disputes were resolved peacefully.[11]

---

[10] All wars that the Nordic countries have been part of since 1814 have been started by a country outside the Nordic region.

[11] Svalbard has historically been a latent conflict between Norway and Russia that continues today. The treaty that recognized the sovereignty of Norway over Svalbard was signed in 1920 as part of the settlements of the First World War that were negotiated in Paris. The Soviet Union did not sign the treaty until 1924. For more on the Svalbard question, see: Grydehøj, A. (2020). Svalbard:

Peace lasted until the start of the Second World War in 1939 with the German invasion of Poland. By then, Germany and the Soviet Union had signed the Molotov-Ribbentrop Pact, a non-aggression pact. Notwithstanding Norwegian endeavors for neutrality, Norway and Denmark were invaded and occupied in 1940 by Germany as part of Operation Weserübung (Lunde, 2009; Nilesh, 2012). The Soviet Union did not object to this, and instead blamed Great Britain and France for forcing Germany's hand (Corum, 2004; Holtsmark & Mankova, 2015, p. 235).[12] Germany did not respect Norwegian wishes for neutrality, nor would it respect the non-aggression pact it had with the Soviet Union. Germany launched Operation Barbarossa in 1941, which was the invasion of the Soviet Union. The main front was in central Europe, but Germany would also invade from Norway and Finland in the north. Kirkenes in Finnmark County would be used as a staging area for the invasion as part of Operation Silver Fox (Carruthers, 2013; Jacobsen, 2017, pp. 8–9; Sallinen, 2020).[13] The Germans made it to the River Litza, where the Red Army stopped their advance, only 60 km from the port city of Murmansk, the only all-year ice-free port of the Soviet Union in the northwest. The frontlines in the Arctic remained stable until the Soviet offensive, Operation Petsamo-Kirkenes, that liberated Norway in 1944.

The Petsamo-Kirkenes operation was the first military operation to liberate a Norwegian territory from the Germans in the Second World War. Germany, which was hard-pressed in the Arctic, lost Finland as the Finns negotiated an armistice with the Soviet Union in August, 1944 (Miloiu, 2008). Germany, through Operation Birke, had ambitions of holding Petsamo [today Pechenga] because of the nickel mining and smelting industry that was vital to the war effort: "*Ohne Kolosjoki [today Nikel] kein Nickel, ohne Nickel kein Nickelstahl, ohne Nickelstahl keine Panzer, ohne Panzer keinen Sieg, daher: Haltet Kolosjoki!*" [Translation to English: Without Kolosjoki [today Nikel] no nickel, without nickel no nickel steel, without nickel steel no tanks, without tanks no victory, therefore hold Kolosjoki!" (Gorter et al., 2005, p. 21). The Red Army attacked Norway in October 1944 with a three-to-one troop superiority, and in late October the Germans started the forced evacuation of the Varanger Peninsula. The German generals realized that they did not have the manpower to face the Soviet offensive, so they halted Operation Birke and instituted Operation Nordlicht under the command of Lothar Rendulic. The order

---

International relations in an exceptionally international territory. In *The Palgrave handbook of Arctic policy and politics* (pp. 267–282). Springer., Holtsmark, S. G. (1994). Soviet strategic interests in Norway: Svalbard and the northern borderlands. *Norwegian Institute for Defence Studies*, 12–16., Holtsmark, S. G., & Mankova, P. (2015). *Naboer i frykt og forventning: Norge og Russland 1917–2014*. Pax.

[12] Operation Weserübung was a preventative occupation as Germany expected a French-British occupation of Norway under the so-called "Plan R 4."

[13] Operation Silver Fox was launched to coincide with Operation Barbarossa. The goal of the operation was to capture Murmansk and secure the nickel mines around Petsamo. While Petsamo was captured, Murmansk would not fall to the Germans. For more on Germany's operations in the north, see: Mann, C., & Jörgensen, C. (2003). *Hitler's Arctic War: The German Campaigns in Norway, Finland, and the USSR 1940–1945*. Macmillan.

was to withdraw the German troops to Lyngen in Troms County (Gebhardt, 1989). Hitler, fearing that the Soviets or the Norwegians would secure a foothold in the north, gave the order for the entire civilian population of more than 70,000 people, including children, the elderly, and the sick, to be forcibly evacuated, and he instituted the scorched earth policy, burning down and destroying everything in Finnmark and North-Troms counties (Norsk Folkemuseum, n.d.).[14] The order from Hitler ended with the words: "Compassion for the civilian population is uncalled for" (*Hostage Case: Lothar Rendulic Testimony*, 1947).[15] While the majority of the population was forcibly evacuated, many, especially in the east, fled the Germans into the mountains or even mine shafts (Eriksen, 2014).

The order itself stated that the scorched earth and forced evacuation of Finnmark was to protect the Norwegians from Bolshevism, but this was nothing more than propaganda. In reality, the destruction and forced evacuation of Finnmark and North-Troms was, next to the persecution and deportation of the Jews and other victims to the Nazi extermination camps, the greatest war crime in Norwegian history (Abrahamsen, 1983; *Killing Centers: An Overview*, 2021; Olsen, 2019, pp. 243–302).[16] The reason behind the scorched earth order was twofold: (1) to deny the Soviet troops any facility or aid from the local population in the area, and (2) so that neither the Soviets nor the Norwegian government in exile should be allowed to gain a foothold in Norway (Ziemke, 1960, p. 308).

As the Red Army entered Finnmark and pushed the Germans west and south, the people who had fled the forced evacuation came out of hiding and were met by their liberators, Russian soldiers. For these people, the occupation had ended, but there was no time for celebrations. While they were happy to be free of the Germans and to be liberated, everything was gone. And, unlike the celebrations that came in the south when the war ended on May 8, 1945, in the north they started by cleaning up, searching for anything useful in the absolute destruction (Eriksen, 2014). The rebuilding of Finnmark and North-Troms Counties took years (Dancke, 1986; Lund, 1947). The Red Army went as far as the Tana River Valley and remained in northern

---

[14] "12,000 residential buildings for 50,000 people. 4700 barns and outhouses. 150 schools. 27 churches with 15 vicarages. 21 hospitals and nursing homes. 500 industrial firms of all sizes. 200 fish processing centres. 53 hotels and inns. 350 bridges. 180 lighthouse stations. 22,000 telegraph poles. 430,000 metres of overhead lines and cables. 1066 electric motors. 118 power plants. 12 telephone exchanges. 11 telegraph stations. 350 motor boats. Several thousand rowing boats, and quays and harbours. All domestic animals were to be slaughtered. Finnmark was mined, with close to 100,000 laid in the sea and onshore." Norsk Folkemuseum. (n.d.). *SCORCHED EARTH*. Retrieved 10 April 2021 from https://norskfolkemuseum.no/en/scorched-earth-

[15] Finnmark county is, at 48,618 km$^2$, larger than the Kingdom of Denmark, which is 42,933 km$^2$.

[16] In his book, Olsen further described how the Norwegian Attorney General and his staff did not prioritize collecting evidence of the war crimes in northern Norway; northern Norway and its experiences in the war were treated as inconsequential in Southern Norway. Consequently, Lothar Rendulic was found not guilty of the war crimes he committed in Northern Norway. Olsen, P. K. (2019). *Jevnet med jorden: brenningen av Finnmark og Nord-Troms 1944, Part III: Rettsoppgjøret som forsvant [Tr. Razed to the Ground: the Burning of Finnmark and North-Troms 1944, Part III: The Justice that Disappeared]*, pp. 243–302. Aschehoug.

Norway until the fall of 1945. The Norwegian government in exile sent a small number of Norwegian troops to Finnmark county so that freed Norway, from Finnmark to the Lyngen Fjord in North-Troms, would be governed by a Norwegian (military) mission (Suprun, 2020; Udgaard, 2015). Norwegian–Russian relations saw the rapprochement grow with the subsequent unilateral Soviet withdrawal in the fall of 1945 from North Norway. By then, the Soviet Union had occupied the countries that it entered in Eastern Europe during the Second World War. Norway is the only country from which the Soviet Union withdrew unilaterally.

This unilateral withdrawal was tremendously important for relations. The Soviet Union liberated Norway in the North and was viewed as liberator and the Red Army as war heroes. Norway's ruling labor party considered cooperation with the Soviet Union, which it emphasized as a partner for Norway.

The reason for the Soviet withdrawal from Norway was that the Soviet leadership viewed Norway as part of Great Britain's sphere of interest in Europe. In return, the Soviet Union demanded that the United States and Great Britain recognize a Soviet "sphere" in the east (Udgaard, 2015). On September 25, 1945, the last Soviet troops left Norwegian soil. Less than a year later in the spring of 1946, Winston Churchill used the term "Iron Curtain" to describe the Soviet Union's sphere of influence and from there, antagonism grew between East and West. The world shifted to bipolarity, represented by two spheres that competed for dominance. As the Second World War ended, the Cold War began. In the aftermath of the Second World War Norway turned to bridge-building between the emerging Eastern and Western blocs. The Norwegian government had been in exile in Great Britain. Having been liberated by the Soviet Union, and being its neighbor while in the sphere of interest of the Western powers, Norway pursued policies that would not antagonize either bloc. In 1946, the Soviet Union recognized that Norway's bridge-building efforts were genuine and that this presented an opportunity to balance Norway's close relations to Great Britain and the US (Holtsmark & Mankova, 2015, p. 302). However, Norway ended up antagonizing both blocs and subsequently turned to the West. Part of the reason for this was that the Marshall Plan to rebuild Europe mandated a choice, but also that the Soviet Union was installing Communist Party regimes in eastern Europe and striking down dissent (Pharo, 1976). The Western powers moreover wanted Norway to enter into the Atlantic partnership. In 1948, the Norwegian Foreign Minister, Halvard Lange, stated Norway's position that a military treaty with the Soviet Union is unthinkable. Sweden proposed a Scandinavian Defense Union, which failed due to Swedish doggedness on non-alignment and neutrality, where Norway insisted on linking to a broader collective security system (German, 1982; Petersson, 2012). In 1949, Norway became a founding member of NATO.

## Cold War Trust-Building

Norway was different from the other member states in NATO. The bridge-building policy of the post-World War II years had never truly disappeared, and throughout the Cold War Norway instituted policies that lowered tensions in Northern Europe and the Barents Region where, in 1949, NATO and the Soviet Union shared its only border. Even before signing the treaty itself, Norway set itself apart by stating that "the Norwegian government declared that it would not allow foreign military bases on Norwegian territory in peacetime as long as Norway was not attacked or threatened by attack" (Petersson & Lunde Saxi, 2013). Thus, upon entering NATO, Norway would have a base policy that reassured the Soviet Union and that showed that Norway respected and understood Soviet security concerns. This policy had a knock-on effect, as Denmark instituted a similar policy in 1953, with an exception for Greenland. As the Cold War progressed, deterrence was overt and visible in both blocs. However, to balance this in the border area of NATO and the Soviet Union, Norway started several cooperative activities that ended up building and then maintaining trust.

### *Culture Agreement*

At the height of the Cold War in 1955, Norway entered talks with the Soviet Union to start cultural cooperation. The Soviet Union in the post-Stalin period was opening up under the leadership of Nikita Khrushchev, who denounced Stalin and renewed the focus on "peaceful coexistence" with the West (Lerner, 1964; Rettie, 2006). On October 12, 1956, Norway and the Soviet Union signed a culture agreement, the USSR's first with a Western country (*Kulturavtale mellom Norge og Sovjet-Samveldet [Tr. Culture Agreement between Norway and the Soviet Union]*, 1956). The agreement encompassed people in culture and science. In the border area of Finnmark County in northern Norway and the Soviet Union, where in 1944, the Red Army liberated Norway, this agreement would also include sports. This meant that the people living there could compete with their neighbors in Murmansk across the border. On the sports collaboration with the Soviet Union, Arne Ulvang from Kirkenes who participated recalled: "it started as an opening in the iron curtain. And I believe that it was significant for what would later happen here" (Jentoft, 2014). Ulvang went on to describe the sport exchanges as a revival of the Pomor trade as they exchanged more modern equipment with Russian vodka and fur hats. The contact over the border was to have a focus on sports rather than politics: despite the crackdown in Czechoslovakia in 1968, the collaboration was only temporarily stopped. It resumed after a year. Einar Niemi, professor emeritus in history at UiT the Arctic University of Norway, grew up in the border region and also participated in the sport collaboration. He said of the collaboration that they were always greeted with kindness and interest from the Russians (Berg, 2014). Two years after the

Norwegian-Soviet Culture Agreement, the United States followed with the Lacy-Zaroubin Agreement (Department of State Bulletin, 1958).

After the culture agreement, Norway and the Soviet Union entered several other agreements in the 1950s, and many of these were in the north in the border area. The sea boundary in the Varanger fjord was decided in 1957 (Vassdal & Espelund, 2012). On the use of hydropower resources on the river Paz (Norwegian: Pasvikelva; Finnish: Paatsjoki), an agreement was also reached in 1957 (*Energy of Cooperation. Borisoglebskaya HPP – 55 Years of Operation*, n.d.). In 1956, Norway and the Soviet Union signed an agreement on search-and-rescue cooperation in the Barents Sea (Eie, 2011).

## *Science Cooperation and Fisheries Management*

Since the mid-1950s, scientific cooperation between Norway and the Soviet Union has existed to manage the fish stocks in the Barents Sea. The fish stocks that are shared have been properly managed and have been sustainably exploited. The science cooperation stemmed from concern over fluctuations in the catch of fish stocks. A need to regulate the fisheries brought about the first joint research session in "1957 in Murmansk, involving researchers from the Institute of Marine Research in Norway … and the Knipovich Polar Research Institute of Marine Fisheries and Oceanography (PINRO, Murmansk) … the researchers agreed that it would be expedient to compile a joint program for surveillance of the condition of fish stocks and the environment in the Barents Sea" (*The Fisheries Commission - History*, n.d.). In 1975 the Join Norwegian-Russian Fisheries Commission was established to set the Total Allowable Catch (TAC) as a means of maintaining sustainable fisheries in the Barents Sea (Hammer & Hoel, 2012).

## *Barents Sea Boundary Delimitation and a Moratorium on Hydrocarbon Extraction*

While the land border was established in 1826, and the sea boundary in the Varanger Fjord in 1957, the sea boundary in the Barents Sea was not settled until 2010. The negotiations to settle this boundary started, however, in the 1970s. The area in the Barents Sea was determined by two principles on how to divide a disputed territory: (1) the median line principle in the east (Norway); and (2) the sector line principle in the west (Soviet) (Henriksen & Ulfstein, 2011). The area between these two principles is 175,000 km$^2$. In 1977 both Norway and the Soviet Union established Exclusive Economic Zones with overlapping claims. This area would be called the Grey Zone, and in 1978, the Grey Zone Agreement was signed, which governed how fishing rights were managed in this disputed area (Stabrun, 2008). The area in the Barents Sea is also rich in oil and gas and a moratorium on the exploration and exploitation of these resources was in place while the negotiations were ongoing.

The solution to delimiting the boundary in the Barents Sea was a compromise that modified the claims of Norway and the Soviet Union and later Russia. "Under Article 1 of the Treaty, the former disputed area in the Barents Sea is now divided into two areas of almost equal size. The agreed demarcation line lies, apart from certain deviations, practically half-way between Norway's old median line claim and Russia's sector line claim" (Jensen, 2011). The delimitation agreement was signed on 15 September 2010 and could be seen as "reflecting a broader cooperative trend in Russian foreign policy under President Medvedev" (Moe et al., 2011). On 8 June 2011, the day after the delimitation agreement entered into force, seismic testing for hydrocarbon deposits began in the previously disputed territory (*Oljedirektoratet tror på enorme mengder olje i Barentshavet nordøst [Tr. The Norwegian Petroleum Directorate believes in enormous amounts of oil in the Barents Sea in the northeast]*, 2017).

## *Self-Imposed Military Restrictions*

The above agreements and accomplishments underline the high level of activity between Norway and the Soviet Union during the Cold War, despite being in opposing blocs. These activities lowered tensions and built trust between the two countries. While deterrence is what the Cold War is generally known for, the above activities fall under the reassurance side of relations. The most reassuring policies that Norway instituted unilaterally toward the Soviet Union were the self-imposed military restrictions that started with the Norwegian base policy of 1949 that Norway proclaimed prior to entering NATO as a founding member. The base policy restricted foreign military bases on Norwegian soil in peacetime. It did not, for example, restrict foreign military bases in times of war, or when Norway was threatened with attack, or participating in military exercises (Lodgaard & Gleditsch, 1977). Lodgaard also highlighted Norway's nuclear policy of 1957, which stated that Norway would not permit nuclear weapons on Norwegian territory in peacetime. Lodgaard furthermore states that as nuclear weapons can enter Norway during times of war, Norway participates in NATO's Nuclear Planning Group (NPG). The nuclear policy was, however, to be further cemented in 1988 when the Minister of Defense stated Norway's nuclear policy as follows:

> *In accordance with international agreements, Norway will not test, produce, or in any other way attain, nuclear weapons; nuclear weapons will not be stored in or deployed to Norway; Norwegian armed forces will not be trained in the use of nuclear weapons; Norway will not enter into any cooperation agreement with an aim to transfer nuclear weapons or information about nuclear weapons to Norway; special storage sites for nuclear weapons will not be established in Norway; Norwegian weapon systems will not be certified for use of nuclear munitions.* (Børresen et al., 2004, pp. 108–109)

The last of the key restrictions are related to the north and state that there shall be no military operations by allied forces "in the county of Finnmark, next to the Soviet border, or from Norwegian airfields or harbors or at sea or in the air east of 24 degrees east longitude" (Børresen, 2011).

The Labor Party was behind many of these policies as it was in power in the postwar years in the Cold War. The Labor Party instituted the trust-building policies mentioned here after the failed bridge-building policy between East and West. The party furthermore harbored the pre-World War II ideas of neutrality which is reflected in these policies and initiatives. These policies were thus a continuation of the bridge-building that Norway attempted before entering NATO and they ultimately created a safer northern Europe.

## Trust in Norway-Soviet Union Relations

The Cold War was at the core an ideological confrontation. Therefore, competition takes place in relation to the political systems of the countries the two blocs competed for. It was the life and death struggle of two incompatible socioeconomic systems. The overt nature of the conflict was the nuclear (and conventional) arms race that balanced power under the mutual assured destruction (MAD) doctrine. Kenneth Waltz, one of the founders of neorealism, stated on the effectiveness of this doctrine that "nuclear forces are useful, and their usefulness is reinforced by the extent to which their use is forestalled" (Waltz, 1979, p. 187). Retaliatory or second-strike capability, the ability to respond to a nuclear attack in equal or greater force, were thus maintained as deterrent by both sides. For the Soviet Union, this became the bastion strategy. The Barents Sea was militarized as a bastion where there is an area of inner defense, or control, around the Kola Peninsula, and an area of outer defense for ambition of denial, stretching toward the Greenland, Iceland, United Kingdom (GIUK) gap in the North Atlantic (Boulègue, 2019). The area of control is to ensure the survivability of second-strike capabilities. The Norwegian north is inside the area of inner defense of the bastion, while the outer area follows the coast of Norway southward.

Outside the framework of deterrence, Norway instituted the trust-building policies to placate the Soviet Union, the most overt of which were the self-imposed military restrictions. Norway chose to not threaten the Soviet Union in the north by denying any allied military activity in Finnmark County, from where the Soviet Union had been invaded in Operation Barbarossa in the Second World War. By implementing this and other restrictions, Norway showed it recognized the security concerns of the Soviet Union and balanced deterrence with reassurance.

## Trust-Building and Costly Signaling

Andrew Kydd, the political scientist behind the theory *costly signaling,* argues that to decisively signal trustworthiness, an actor must send a "costly signal." A country that wants to be trusted should be willing to take some risk (Kydd, 2007, p. 197). "Costly signals, in this context, are signals designed to persuade the other side that one is trustworthy by virtue of the fact that they are so costly that one would hesitate to send them if one were untrustworthy" (Kydd, 2000). In relation to this, Norway signals trustworthiness by entering NATO after having stated its base policy that prohibits foreign military bases in peacetime on Norwegian soil. This is then strengthened with the policy prohibiting nuclear weapons in Norway, and the policies restricting military activity in Finnmark County and near the Soviet border. Norway, in NATO, signals trustworthiness to the Soviet Union. The trust then becomes rational in the sense that by Norway showing it understands Soviet security concerns, the threat aspect is lessened in the north and tensions are reduced. Consequently, the Soviet Union could trust Norway. The rational aspect of the trust is connected to a competency that had been built on regional knowledge wherein the dynamics of Norway-Soviet Union relations were understood. In the north this was further compounded by the shared historical narrative: the Soviet Union had liberated the north and then withdrawn and thus showed that it saw Norway in the western sphere. Johan Jørgen Holst, Norway's former defense minister stated that Norway explicitly instituted the trust-building policies vis-à-vis the Soviet Union to "communicate peaceful intentions and avoid challenging vital Soviet security interests during peacetime" (Holst, 1985).

## A Departure from Cold War Trust

Today, when the Cold War is a distant memory and we live in a time of peace, tensions are rising, and Norway is departing the self-imposed military restrictions. Norway has left its base policy and is allowing allied troops on Norwegian soil in peacetime, is holding international military exercises in Finnmark County, and is providing a port for American nuclear-powered attack submarines in Tromsø, the largest city in Troms & Finnmark County (Balsvik, 2021; *Forsvarsministeren besøkte øvelse Joint Viking 2017 [Tr. The Minister of Defense visited Exercise Joint Viking 2017]*, 2017; *Joint Viking*, 2021; Sagflaat, 2021-05-06). Norway is doing this in addition to and alongside the West denouncing and sanctioning Russia because of the Ukraine crisis of 2014. This crisis is used to exemplify a more dangerous Russia and a Russia that could have ambitions of invading Norway (Wormdal, 2018). Russia has, however, never threatened Norway. It has invaded Norway once and that was to liberate it from a shared enemy. Relations have historically been good in the Barents Region, even during the Cold War when comparing to other regions where the Eastern and Western blocs met. And while there were episodes that heightened tensions, the reassurance policies continued irrespective of the crises that developed.[17]

The trust that was built up during the Cold War is diminishing today. As it was built on experiences based on cooperation and costly signals, regaining it is hard once it is gone. The relationship between Norway, as a part of NATO, and the Soviet Union (and Russia today) is and has been asymmetric. Thus, the balancing that took place in the Cold War was based on a reciprocal trust that the security concerns of the other was respected. The Soviet Union did not threaten Norway unnecessarily, and Norway understood and respected Soviet security concerns. The asymmetry is today, as then, the foundation of the relationship. The Soviet Union did not need to fear Norway as long as Norway was a rational actor that maintained the trust by upholding the military restrictions. Conversely, Norway did not need to fear the Soviet Union rationally by respecting the Soviet Union's security concerns. Therefore, Norway needed to trust that the Soviet Union trusted Norway, and vice versa, the Soviet Union needed to trust that Norway trusted the Soviet Union.

Trust is underpinned by knowledge and competency that comes from understanding an adversary's concerns and needs. This important understanding of the adversary is created from interactions over time. As the understanding of the adversary grows, one can better accommodate their concerns and thus start to build peaceful relations. This understanding is what Norway had at the onset of the Cold War. On the basis of this understanding, a competency was built that was able to reduce tensions in Northern Europe during the Cold War by the trust-building policies.

## Concluding Remarks

When the Cold War ended in 1991 with the dissolution of the Soviet Union, Norway and its Nordic neighbors were in a unique position to start a new era of cooperation with the Russians. On January 11, 1993, the Barents Euro-Arctic Cooperation was launched and formalized with the signing of the Kirkenes Declaration (*The Kirkenes Declaration from the Conference of Foreign Ministers on Co-operation in the Barents Euro-Arctic Region*, 1993). The regional cooperation comprises the 13 territories in the northern parts of Norway, Sweden, Finland, and Northwest Russia that makeup the Barents Region (Fig. 19.1). Iceland, Denmark, and the European Commission are also signatories to the declaration and, as such, are integral parts of the regional cooperation. A regional council, the Barents Euro-Arctic Council (BAEC), has been established between the partners. And while the Barents

---

[17] For example, in 1968, as Norway hosted the military exercise "Polar Express" in Troms County close to Finnmark County, the Soviet Union mobilized on the border and opened fire with blanks on the Norwegian border posts as a demonstration of power and to signal discontent with the military exercise. Kjøllberg, E., Kårstad, G., & Hansen, S. (2017). Tårnvakten [Tr. The Tower Guard]. *NRK*. https://www.nrk.no/dokumentar/xl/hemmelige-rom-ii_-episode-5-1.12945607

**Fig. 19.1** Map of Barents Region. (Source: Arctic Center, University of Lapland)

Cooperation exists on both state and local levels, it is within the Barents Region that most activity occurs, as it is here that the people-to-people cooperation takes place. The cooperation has since the end of the Cold War, been lauded as a success in terms of peaceful transformation and cooperation. The future of this cooperation hinges on having good relations between the countries that makeup the region. And as relations sour and the divisions of old are reappearing, it is time to recollect the competencies that created a peaceful Northern Europe during the Cold War. And instead of moving forward in peaceful coexistence as we could, we are moving toward deterrence without reassurance.

> *The Pentagon defends MIRV [Multiple Independently targeted Re-entry Vehicle] as necessary to insure penetration of a heavy ABM [Anti-Ballistic Missile] defense, which the Soviets might build. Pentagon Research Chief John Foster has testified there is no evidence the Soviets have started such a system and that if they do, it will take five years to build it. In other words, the Pentagon is deploying this weapon at least four years in advance of the Soviet deployment it reportedly is a reaction to. If that sounds as fishy to Soviet diplomats as it does to us … their generals would inevitably want to press harder with their own multiple warhead testing.*
>
> —*The Wall Street Journal (1969 editorial)*
>
> *In 1969 and 1970 I wish we had thought through the implications of a MIRV'ed world.*
>
> —*Henry Kissinger (December 1974 [Willens, 1984, p. 29])*

# References

Abrahamsen, S. (1983). The Holocaust in Norway. In *Contemporary views on the Holocaust* (pp. 109–142). Springer.

Archer, C. (1996). The Nordic area as a 'Zone of Peace'. *Journal of Peace Research, 33*(4), 451–467. https://doi.org/10.2307/424569.

Archer, C., & Joenniemi, P. (2003). *The Nordic peace*. Ashgate.

Balsvik, R. R. (1994). Pomorbyen Vardø og Russland. *Den menneskelige dimensjon i nordområdene*, 151–166.

Balsvik, R. R. (2021). Kunsten å berolige folket [Tr. The art of reassuring the people]. *Nordlys*.

Barton, D. P. (1930). *The amazing career of Bernadotte* (pp. 1763–1844). Kessinger Publishing, LLC.

Barton, H. A. (1972). Russia and the problem of Sweden-Finland, 1721–1809. *East European Quarterly, 5*(4), 431.

Berg, S. H. (2014). Idrettssamarbeid i aust opna jernteppet. Sport cooperation in the east opened the iron curtain. Retrieved 2020-12-17 from https://forskning.no/sport-historie-krig-og-fred/idrettssamarbeid-i-aust-opna-jernteppet/578692

Bitzinger, R. A. (1989). *Denmark, Norway, and NATO: Constraints and challenges*. Santa Monica: RAND.

Blom, G. A. (1984). The participation of the kings in the early Norwegian sailing to Bjarmeland (Kola Peninsula and Russian Waters), and the development of a royal policy concerning the Northern Waters in the Middle Ages. *Arctic, 37*(4), 385–388.

Børresen, J. C. (2011). Alliance naval strategies and Norway in the final years of the Cold War. *Naval war college review, 64*(2), 97.

Børresen, J., Gjeseth, G., & Tamnes, R. (2004). *Norsk forsvarshistorie: 1970–2000*. Allianseforsvar i endring. Eide.

Boulègue, M. (2019). *Russia's military posture in the Arctic*. Chatham House. Retrieved 2020-05-12 from https://www.chathamhouse.org/2019/06/russias-military-posture-arctic/2-perimeter-control-around-bastion.

Carruthers, B. (2013). *Hitler's forgotten armies: Combat in Norway and Finland*. Coda Books Ltd..

*China's Arctic Policy*. (2018). The State Council, The People's Republic of China. http://english.www.gov.cn/archive/white_paper/2018/01/26/content_281476026660336.htm

Corum, J. S. (2004). Uncharted waters: Information in the First Modern Joint Campaign–Norway 1940. *Journal of Strategic Studies, 27*(2), 345–369.

Dancke, T. M. (1986). *Opp av ruinene: gjenreisningen av Finnmark* 1945–1960 [Tr. Up from the ruins: the reconstruction of Finnmark 1945–1960]. Gyldendal. https://www.nb.no/items/URN:NBN:no-nb_digibok_2009012804050?searchText=%2522f%C3%B8r%2520krigen%2522

Department of State Bulletin. (1958). *UNITED STATES AND USSR SIGN AGREEMENT ON EAST-WEST EXCHANGES@ Text of Communique*. Retrieved from https://heinonline.org/HOL/Page?collection=journals&handle=hein.journals/dsbul38&id=243&men_tab=srchresults

Eie, H. H. (2011). Sjøsikkerhet i det russiske nord [Tr. Security at Sea in the Russian North].

Elgström, O., & Jerneck, M. (1997). Activism and adaptation: Swedish security strategies, 1814–85. *Diplomacy & Statecraft, 8*(3), 210–236.

*Energy of Cooperation. Borisoglebskaya HPP – 55 Years of Operation.* (n.d.). TGC-1. https://www.tgc1.ru/en/press-center/special/2019/energy-cooperation/

Eriksen, I. (2014). *De gjemte seg i gruvegangene med fare for å bli tilintetgjort [Tr. They hid in the miningshafts under the danger of destruction]*. NRK. Retrieved 2020-10-15 from https://www.nrk.no/tromsogfinnmark/tusenvis-av-mennesker-gjemte-seg-i-tunnelen-1.12002423

*Focus 2020 The Norwegian Intelligence Service's assessment of current security challenges*. (2020). The Norwegian Intelligence Service. https://www.forsvaret.no/aktuelt-og-presse/publikasjoner/fokus/rapporter/Focus%202020%20english.pdf/_/attachment/inline/7bc5fcbd-e39c-4cb6-966a-0f7115205b44:d282e733ce4f5697c9e4a7afc5a63c16dab6c151/Focus%202020%20english.pdf

*Forsvarsministeren besøkte øvelse Joint Viking 2017 [Tr. The Minister of Defense visited Exercise Joint Viking 2017]*. (2017). Government of Norway. Retrieved 2021-03-08 from https://www.regjeringen.no/no/aktuelt/forsvarsministeren-besokte-ovelse-joint-viking-2017/id2542412/

Friis, K. (2020). Norway, NATO, and the Northern flank 1. In *NATO and transatlantic relations in the 21st century* (pp. 67–84). Routledge. https://www.taylorfrancis.com/chapters/edit/10.4324/9781003045434-5/norway-nato-northern-flank-1-karsten-friis.

Gebhardt, M. J. F. (1989). The Petsamo-Kirkenes operation: Soviet breakthrough and pursuit.

German, R. K. (1982). Norway and the bear: Soviet coercive diplomacy and Norwegian security policy. *International Security, 7*(2), 55–82.

Gorter, A. A., Gorter, W. T., & Suprun, M. N. (2005). *Frigjøringen av Øst-Finnmark 1944–1945 Osvobozhdenie vostochnogo Finnmarka, 1944–1945 [Tr. Liberation of East-Finnmark 1944–1945]*. Arkhangelsk-Vadsø: "Arkhangelsk Pomor".

Grydehøj, A. (2020). Svalbard: International relations in an exceptionally international territory. In *The Palgrave handbook of Arctic policy and politics* (pp. 267–282). Springer.

Hammer, M., & Hoel, A. H. k. (2012). The development of scientific cooperation under the Norway–Russia fisheries regime in the Barents Sea. *Arctic Review, 3*(2).

Hansen, L. I., & Olsen, B. (2014). 5 state integration and Sámi rights ca. 1550–1750. In *Hunters in transition* (pp. 229–311). Brill.

Havnes, H. (2020). The polar silk road and China's role in Arctic governance. *Journal of Infrastructure, Policy and Development, 4*(1), 121–138.

Havnes, H., & Seland, J. M. (2019). The increasing security focus in China's Arctic Policy. *The Arctic Institute. July*, 16.

Henriksen, T., & Ulfstein, G. (2011). Maritime delimitation in the Arctic: The Barents Sea treaty. *Ocean Development & International Law, 42*(1–2), 1–21.

Holst, J. J. (1985). The military build-up in the high north: Political implications for regional stability: A Norwegian perspective. *NUPI Notat No. 318B, Norwegian Institute for International Affairs*.

Holtsmark, S. G. (1994). Soviet strategic interests in Norway: Svalbard and the northern borderlands. *Norwegian Institute for Defence Studies*, 12–16.

Holtsmark, S. G., & Mankova, P. (2015). *Naboer i frykt og forventning: Norge og Russland 1917–2014*. Pax.

Hostage Case: Lothar Rendulic Testimony. (1947). Nuremberg Transcripts. https://commons.und.edu/cgi/viewcontent.cgi?article=1007&context=nuremburg-transcripts

Jacobsen, A. R. (2017). *Miracle at the Litza: Hitler's first defeat on the eastern front*. Casemate Publishers.

Jensen, O. (2011). The Barents Sea. *Int'l J. Marine & Coastal L., 26*, 151.

Jentoft, M. (2014). Brøt Jernteppet med skismurning og vodka [Tr. Broke the Iron Curtain with Ski Wax and Vodka]. *NRK (Urix)*. https://www.nrk.no/urix/brot-jernteppet-med-skilop-1.11478926.

Jervis, R. (2001). Was the cold war a security dilemma? *Journal of Cold War Studies, 3*(1), 36–60.

Johnson, A. R. (2020). The Warsaw pact reconsidered: International relations in Eastern Europe, 1955–1969 by Laurien Crump. *Journal of Cold War Studies, 22*(1), 249–252.

*Joint Viking*. (2021). Norwegian armed forces. Retrieved 2021-03-08 from https://www.forsvaret.no/om-forsvaret/operasjoner-og-ovelser/ovelser/joint-viking

Kacowicz, A. M. (1995). Explaining zones of peace: Democracies as satisfied powers? *Journal of Peace Research, 32*(3), 265–276. https://doi.org/10.2307/425664.

Kawate, I. (2021, 2021-03-01). China gains momentum in race to overtake US economy. *Nikkei Asia*. https://asia.nikkei.com/Economy/China-gains-momentum-in-race-to-overtake-US-economy

Kernan, R. F. (1989). *Norway and the northern front: Wartime prospects*. Defense Technical Information Center.

*Killing Centers: An Overview*. (2021). United States Holocaust Memorial Museum. Retrieved 2021-05-16 from https://encyclopedia.ushmm.org/content/en/article/killing-centers-an-overview

Kjøllberg, E., Kårstad, G., & Hansen, S. (2017). Tårnvakten [Tr. The Tower Guard]. *NRK*. https://www.nrk.no/dokumentar/xl/hemmelige-rom-ii_-episode-5-1.12945607.

Korolev, A. (2016). Systemic balancing and regional hedging: China–Russia relations. *The Chinese Journal of International Politics, 9*(4), 375–397.

Kuldkepp, M. (2019). National Revanchism at a critical juncture: Sweden's near-involvement in the Crimean war as a study in Swedish nationalism. *Scandinavica, 58*(2), 115–133.

*Kulturavtale mellom Norge og Sovjet-Samveldet [Tr. Culture Agreement between Norway and the Soviet Union]*. (1956). Norges Regjering Retrieved from https://www.stortinget.no/no/Saker-og-publikasjoner/Stortingsforhandlinger/Lesevisning/?p=1958&paid=2&wid=a&psid=DIVL889&pgid=a_0354; https://www.stortinget.no/no/Saker-og-publikasjoner/Stortingsforhandlinger/Saksside/?pid=1955-1961&mtid=61&vt=a&did=DIVL74828

Kydd, A. (2000). Trust, reassurance, and cooperation. *International Organization*, 325–357.
Kydd, A. H. (2007). *Trust and mistrust in international relations*. Princeton University Press.
Larson, D. A. (1991). *Reinforcing Norway in war: A dilemma in Norwegian National Security Policy*. Ft. Belvoir: Defense Technical Information Center.
Layne, C. (2008). China's challenge to US hegemony. *Current History-New York Then Philadelphia, 107*(705), 13.
Lerner, W. (1964). The historical origins of the Soviet Doctrine of Peaceful Coexistence. *Law & Contemp. Probs., 29*, 865.
Lindgren, R. (1959). *Norway-Sweden: Union, Disunion and the Scandinavian Integration*. Princeton N.
Lodgaard, S., & Gleditsch, N. P. (1977). Norway—The not so reluctant ally. *Cooperation and Conflict, 12*(4), 209–219.
Lund, D. H. (1947). The revival of northern Norway. *The Geographical Journal, 109*(4/6), 185–197.
Lund, J. (1990). *Scandinavian NATO policy: The next five years*. Santa Monica: CA. Rand Corporation.
Lunde, H. O. (2009). *Hitler's pre-emptive war: The Battle for Norway, 1940*. Casemate Publishers.
Mann, C., & Jörgensen, C. (2003). *Hitler's Arctic War: The German campaigns in Norway, Finland, and the USSR 1940–1945*. Macmillan.
Map of Barents Region (n.d.). Arctic center, University of Lapland.
Miloiu, S.-M. (2008). Diverging their destinies. Romania, Finland and the September 1944 armistices. *Valahian Journal of Historical Studies, 10*, 41–55.
Moe, A., Fjærtoft, D., & Øverland, I. (2011). Space and timing: Why was the Barents Sea delimitation dispute resolved in 2010? *Polar Geography, 34*(3), 145–162. https://doi.org/10.1080/1088937x.2011.597887.
Niemi, E. (1992). *Pomor: Nord-Norge og Nord-Russland gjennom tusen år*. Gyldendal norsk forlag. https://www.nb.no/nbsok/nb/ea321f16a77c546ca9dc8c603e3d0328?index=3#7.
Niemi, E. (n.d.). Grenseoverskridende transaksjon, soft diplomacy og mat på Nordkalotten i tidlig ny tid.
Nilesh, P. (2012). *Norway and World War II: Invasion, occupation, liberation*. Proceedings of the Indian History Congress.
Norsk Folkemuseum. (n.d.). *Scorched Earth*. Retrieved 10 April 2021 from https://norskfolkemuseum.no/en/scorched-earth
Nye, J. S., Jr. (2004). *Soft power: The means to success in world politics*. Public affairs.
*Oljedirektoratet tror på enorme mengder olje i Barentshavet nordøst [Tr. The Norwegian Petroleum Directorate believes in enormous amounts of oil in the Barents Sea in the northeast]*. (2017). Retrieved 2018-06-08 from https://www.nrk.no/tromsogfinnmark/oljedirektoratet-tror-pa-enorme-mengder-olje-i-barentshavet-nordost_-1.13519272
Olsen, P. K. (2019). *Jevnet med jorden: Brenningen av Finnmark og Nord-Troms 1944, Part III: Rettsoppgjøret som forsvant [Tr. Razed to the ground: The burning of Finnmark and North-Troms 1944, Part III: The justice that disappeared]* (pp. 243–302). Aschehoug.
Petersson, M. (2012). Sweden and the Scandinavian defence dilemma. *Scandinavian Journal of History, 37*(2), 221–229.
Petersson, M., & Lunde Saxi, H. (2013). Shifted roles: Explaining Danish and Norwegian alliance strategy 1949–2009. *Journal of Strategic Studies, 36*(6), 761–788.
Pharo, H. Ø. (1976). Bridgebuilding and reconstruction: Norway faces the Marshall Plan. *Scandinavian Journal of History, 1*(1–4), 125–153.
Planert, U. (2016). *Napoleon's empire: European politics in global perspective*. Springer.
Rettie, J. (2006). How Khrushchev leaked his secret speech to the world. *History Workshop Journal*.
Riley, B. J. (2002). The partnership of unequals: A short discussion on coalition war. *Defence Studies, 2*(3), 103–118.
Roginsky, V. (2005). The 1826 delimitation convention between Norway and Russia: A diplomatic challenge. *Russia—Norway. Physical and Symbolic Borders*, 162–168.

Sagflaat, E. (2021-05-06). Farvel til norsk basepolitikk [Tr. Goodbye to Norwegian Base Policy]. *Dagsavisen*.
Sallinen, M. (2020). De-escalation amid a total war?: An interpretivist-constructivist analysis of Finland's involvement (or lack thereof) in the Siege of Leningrad and Murmansk during the Continuation War 1941–1944. In.
Schrader, T. A. (1988). Pomor trade with Norway. *Acta Borealia, 5*(1–2), 111–118.
Shrader, T. (2017). Across the borders: The Pomor trade. *Russia and Norway. Physical and Symbolic Borders, 105*.
Singleton, F. (1978). Finland between east and west. *The World Today, 34*(8), 321–332.
Stabrun, K. (2008). The Grey zone agreement of 1978. *Institute for Defence Studies, 43*.
Suprun, M. (2020). The liberation of northern Norway in Stalin's post-war strategy. *The Journal of Slavic Military Studies, 33*(2), 277–291.
*The Fisheries Commission – History*. (n.d.). Joint Russian-Norwegian Fisheries Commission. Retrieved 2021-02-16 from https://www.jointfish.com/eng/THE-FISHERIES-COMMISSION/HISTORY.html
*The Kirkenes Declaration from the Conference of Foreign Ministers on Co-operation in the Barents Euro-Arctic Region*. (1993). Kirkenes.
*The success of the Barents Cooperation*. (n.d.). European Comission Retrieved from https://ec.europa.eu/regional_policy/archive/consultation/terco/pdf/8_external/4_3_nordland_troms_en.pdf
Tianming, G., & Erokhin, V. (2021). China-Russia collaboration in shipping and marine engineering as one of the key factors of secure navigation along the NSR. *China's Arctic Engagement, 9*, 234.
*Treaty of Concert and Subsidy Between His Britannic Majesty and the King of Sweden; Signed at Stockholm the 3d of March 1813*. (1813). Great Britain.
Udgaard, N. M. (2015). Norge "delt i tre" da frigjøringen endelig kom [Tr.: Norway "divided in three" when the liberation finally came]. *Aftenposten*. https://www.aftenposten.no/norge/i/dQ51/norge-delt-i-tre-da-frigjoeringen-endelig-kom.
Vassdal, T. O., author, & Espelund, T. E., revisor. (2012). *Historisk sammendrag vedrørende riksgrensen Norge - Russland [Tr. Historical summary Concerning the border of Norway – Russia]*. https://www.webcitation.org/6A3Ot59xt?url=http://www.statkart.no/filestore/Landdivisjonen_ny/Fagomrder/dGrenser/grensefiler/Grensenotat_0821012.pdf
Waltz, K. N. (1979). Theory of international politics/Kenneth N. Waltz. In: Reading, Mass.: Addison-Wesley Pub. Co.
Willens, H. (1984). *The trimtab factor: How business executives can help solve the nuclear weapons crisis*. William Morrow.
Wormdal, B. (2018-11-24). Slik kan Russland invadere Norge [Tr. How Russia might invade Norway]. *NRK*.
Ziemke, E. F. (1960). *The German northern theater of operations, 1940–1945*. US Government Printing Office.

**Gunnar Rekvig** is an Associate Professor at UiT The Arctic University of Norway (UiT), Norway, and the 2019–2021 Nansen Professor at the University of Akureyri (UNAK), Iceland. He was an Assistant Professor at Tokyo University of Foreign Studies (TUFS), Tokyo, Japan (2018–2019).

# Chapter 20
# Regional Governance: The Case of the Barents Region

**Florian Vidal**

## Introduction – Between the Arctic and Europe

During the 1990s, the Arctic took a new turn with the emergence of the Arctic Council, a key intergovernmental institution that empowered the circumpolar states to build a common policy agenda on the eve of the twenty-first century. Many scholars emphasized the novelty of the Arctic governance for enabling stakeholders to engage in critical issues, particularly those related to the environment (see Young, 2010, 2012; Schram Stokke, 2011). Nevertheless, this critical development significantly affected the Barents region, while it settled its own subregional institutions. Against the backdrop of the Arctic institutions, Barents governance is rather different, due to several factors.

To start with, this subarctic region has been tightly connected to the European continent since the Middle Ages (Hofstra & Samplonius, 1995; Zachrisson, 2008). Unlike the rest of the Arctic, this region has long established singular political dynamics, including a sophisticated trading mechanism. Since then, the relations between Norway and Russia have fueled the regional paradigm. In other words, both countries sustain a matrix for the Barents dynamics, which Finland and Sweden mostly rely on. When the Cold War era came to an end, Norway took the initiative to mobilize the Barents community and to generate new momentum. As Norwegian scholars and diplomats largely expressed (Schram Stokke & Tunander, 1994; Kvistad, 1995, Hønneland, 1998), this political initiative brought specific institutions and gave expression to the so-called "Barents spirit". In this chapter, we discuss the historical background of this subarctic region as well as its specificities in light of the current challenges. However, it is important to stress that the Barents is also a door for the European continent in the Arctic. On one hand, this region is

F. Vidal (✉)
French Institute of International Relations, Paris, France

interlocked to continental geopolitical trends, as seen in 2014 with the Ukrainian crisis. On the other hand, the region is at the forefront of the socio-ecological crisis unleashed by the European Industrial Revolution.

## A Long Journey Through the Barents Region Metabolism

To better understand the current geopolitical landscape of the Barents area, it is necessary to comprehend the long processes of its historic development. The interactions between people predate the contemporary era, with the Middle Age identified as the starting point to this long-term building. The social and political evolution of the Barents region over time can be explained using the metaphor of social metabolism. Despite obvious differences, there is a structural similarity between a human society and a living organism, as the laws of behavior and evolution ruling can be applied in the same way (González de Molina & Toledo, 2014). In the Arctic, this subregion is a singular case in which human interactions were forged in a true continuum.

### *From the Constitutive Mythology of the Viking Era to the Prosperous Fisheries*

The notion of the Barents region as a political entity in the European history dates back to the ninth century, when the first reference to the Viking voyages to the White Sea region was found in an Old English text (Hofstra & Samplonius, 1995). The voyage made by Ohthere, a Nordic chief of Hålogaland, changed the course of the region by reaching the Northern borders of the continent. Its inclusion in the continental dynamic is illustrated by the connection between the Norse people and the Sámi. From the Arctic shores to the Southern areas of the Scandinavian peninsula, the Sámi were well-known for performing "magic" that influenced Scandinavian society and its rulers (Zachrisson, 2008). For example, the first Christian Norwegian King Olaf Tryggvason is said to have visited a Sámi to listen to prophecies about his future. The practice of learning magic and asking for the prophecies of the Sámi was preserved until the thirteenth century (Kusmenko, 2009).

While the Northerners settled in the Barents region, these vast territories remained contested areas "where the Russian state of Novgorod, Denmark–Norway and Sweden taxed the Sami peoples in an area referred to as *fellesdistriktet*, the common district" (Rønning Balsvik, 2015, p. 87). To protect the Finnmark area, a Norwegian fortress was built in Vardø in the early fourteenth century as a means of stopping incursions led both by Russian and Karelian peoples, and regularly threatened settlement activities mainly dedicated to fisheries. Despite these skirmishes, the cultural and political entanglement paved the way for increasing maritime navigation through the Norwegian coastlines up to the Northern part and the Kola peninsula. The development of fisheries helped to build a regional trader's community overlapping thousands of kilometers of sea lanes.

True to this spirit, fisheries can be regarded as a structural factor due to trade exchanges, which flourished in the Northern part of the Fennoscandia. To understand its tremendous significance, the Sámi population shaped their economic organization around fishing, hunting, and trapping for centuries. During the Early Modern period, this model shifted, allowing fisheries to become a central component in their activities. They were also associated with the rising of trade exchanges between Russians and Norwegians (Hansen, 2006). These lasting communications can be regarded as the main historical factor of the Barents region metabolism. As it became stronger over centuries, trade exchanges reached a new high with the development of the Pomor trade. The latter has been tracked back to the sixteenth century, when the Pomors went fishing every year during springtime in the Murman peninsula. These activities led to the foundation of the town of Kola in 1550 as fisheries progressively expanded to the Finnmark area. To complete the picture, a trade exchange system enabled Russians to sail along the coast of Northern Norway during the summer period and bring with them several commodities (rye flour, wheat flour, groats, meat and butter). In return, the Russian traders secured the purchase of fish, including coalfish, cod, and halibut. The latter was particularly prized (Hansen, 2006; Rønning Balsvik, 2015). Founded in 1583, Arkhangelsk became the center of trade for fishing communities trading with Norwegian communities in Northern Russia. During the eighteenth century, cod fish was very popular in the diet of the urban centers of the Russian Empire, such as Moscow, St. Petersburg, and Nizhny Novgorod (Schrader, 2005).

Cereals from Russia were brought to Arkhangelsk and then exported to Norway. Access to these commodities was critical to Finnmark, particularly during the British blockade executed during the Napoleonic wars. Conducted as a form of barter, the Pomor trade run until 1919, before it ended due to a combination of factors: the private enterprise was not welcome by the emerging communist regime; Russia aimed to create a modern fishery; and the Norwegian authorities wished to enhance trade exchanges between North and South instead of East-West (Thuen, 1993). However, the Pomor idea survived this new era and flourished again in the 1990s once the Cold War ended.

## *The Memory of the Second World War and the Cold War Proof*

When the Second World War blew up in Europe, Northern Europe was one of the strategic targets for the German Nazis when they started their military offensive. The Arctic held a specific position throughout the conflict in terms of controlling lines of communication and key continental resources. More than it ever had before, the region became a war front on every dimension. Following technical and strategic developments during the First World War, the Barents region became vitally important for Russia. The port cities of Arkhangelsk and Murmansk offered a vital logistical window to supply the Empire at war. On the eve of the Second World War, the militarization of the Arctic region became a focal point through the conflict. As a result, the Arctic route was undoubtedly risky for navigation, while the German Kriegsmarine and Luftwaffe, based in occupied Norway, patrolled the Kara Sea and up to the Novaya Zemlya in search of Allied shipping (Luzin, 2007; Kikkert & Lackenbauer, 2019).

Most importantly, the Lend-Lease Act concluded between the USSR and the Allies during the war enhanced a strategic passage, securing a vital logistic route to allow the Arctic convoys to pass. By mid-1942, the Allies had delivered to Arkhangelsk and Murmansk ports 1903 aircrafts, 2314 tanks, and 1550 tankettes. In return, the United States acquired strategic commodities (chrome, manganese, platinum, gold, and wood) (Krasnozhenova et al., 2020). As Kikkert and Whitney Lackenbauer (2019) recalled, the region "became the scene of high-intensity conflict between 1939 and 1945" (p. 494). In addition to the Winter War that confronted the Soviet army, and the Finnish one in the Lapland region, the Petsamo-Kirkenes offensive in October 1944 has remained in the memory of the region until today. As the tide of the war ran against the Nazi regime, this offensive aimed to remove German troops from the Barents region and initiated the end of the northern front. Between October 7 and October 29, more than 6000 Soviet soldiers and over 8000 on the German side fell or were captured during the operation. The liberation of Finnmark by the Soviet troops was achieved under the "Soviet-Norwegian Agreement on Civil Administration and Jurisdiction on Norwegian Territory After its Liberation by the Allied Expedition Forces". This agreement, signed by both sides, gave formal permission to the Soviet army to intervene in the northern part of Norway. As part of the military operations during these conflict years, the Barents region finally settled in the security landscape, which impacted the second half of the twentieth century. This episode has significantly impacted bilateral relations from both national and local levels ever since. The memory of this conflict remains among Norwegian and Russian people in the Barents region.

As the Second World War ended, Norway seeks to preserve its "bridge-building policy" between the Soviet Union and the Western allies. However, the establishment of NATO in 1949 stopped this ambition, as Norway became a founding member of the Western military alliance (Suprun, 2020). During the Cold War, Northern Norway occupied a geostrategic position throughout this period. The militarization of the Arctic on the Soviet side intensified through these decades. Although the Kola Peninsula offered a strategic window for the Russian surface vessels and submarines with the free-ice naval bases, it became the most militarized area in the Arctic. Until the 1970s, tense military activities occurred on the GIUK (Greenland–Iceland–United Kingdom) Gap to curb Soviet activities in the Northern Atlantic region. The Soviets' successful development of longer-range submarine-launched ballistic missiles (SLBMs) reduced the need for Soviet nuclear-powered ballistic missile submarines (SSBNs) to transit the GIUK Gap ("The GIUK Gap", 2020).

Despite the constant climate of confrontation between the West and the East, the Barents region witnessed some positive outcomes under the auspices of the

Soviet–Norwegian relationships. Back to the regional metabolism, fisheries remain the best option to cooperate on main regional economic issues. In the same spirit, the establishment of the Norwegian–Russian Joint Fisheries Commission in 1976 proved the ability to spark scientific cooperation for economic purposes, which resulted in quotas being set for Barents Sea fishing. This included three of the most significant fish stocks: cod, haddock, and capelin. Since 1993, the bilateral commission's activities have overlapped an exchange of catch data, inspections, and other issues related to compliance control. The Commission subsequently made significant progress in managing the Barents Sea's fishery stocks and it became one of the best sustainable fishing areas on the planet. Despite the events during the Cold War years, this joint work and agreements made undoubtedly confirmed the long standing of the regional dynamic. In 1987, the Murmansk speech by Soviet leader Mikhail Gorbachev (1987) highlighted "the idea of cooperation between all people also from the standpoint of the situation in the northern part of this plane" (para. 20). The choice of Murmansk for this speech was symbolic, in several ways: (1) it offered prime access to the Russian Arctic waters; (2) it reflected the long history of the Barents spirit; and (3) it had been a main military theater over the previous 50 years. In this last respect, the Great Patriotic War became a political instrument because of its unifying role for the Soviet regime. Today, its remembrance intensified for domestic purposes and brought Murmansk – named "Hero cities" – in a new light (Aas, 2012; Malinova, 2017).

Gorbachev's speech concluded the era of global tension in which the Arctic region had engulfed for several decades. As we discussed, the Barents region metabolism proved to be resilient and the end of the Cold War made it possible to engineer a new phase in the regional process. As a part of this new governance, regional institutions welcome all stakeholders to take an active role.

## *The Promise of an Era of Peace and Cooperation*

As the Cold War faded away in the 1990s, the removal of the Iron Curtain led to a new regional agenda to support the establishment of new institutional bodies. It spawned a new set of networks and paved the way for reviving the regional cooperation. Accordingly, environmental protection and economic development were the critical points to this process, which initiated new political and institutional dynamics in the Barents region. In these circumstances, political changes in Moscow unveiled good intentions to restore Russia's close ties with its Nordic neighbors.

From that perspective, Norway pursued other diplomatic motives to promote the regional initiative. Norwegian diplomats and academics both advocated this Barents initiative. At this point, this opportunity could bolster connections not only "between Russia and the Nordic countries, but also between the Barents region and the continent of Europe" (Zimmerbauer, 2012, p. 93). Another conventional argument related to the historic regional background: the Norwegian negotiators often cited the Pomor trade to support their project. Aside from this romantic vision, Geir

Hønneland (1998) recalled that the first objective of the Norwegian side was "reducing the military tension, the environmental threat and the East-West gap in standards of living in the region" (p. 278).

To support the idea of a new regional governance, the Barents Euro-Arctic Council (BEAC) was created in 1993 under the guidance of the Kirkenes Declaration.[1] The signatories agreed to contribute substantially to both political stability and social and economic progress. This interregional cooperation forum was organized around several working groups, including environment, transport and health. The establishment of the BEAC is a continuum to shape the lifelong idea of Nordic cooperation, as previously promoted by the Nordic Council of Ministers in the 1960s. The newly regional architecture encompasses 13 subnational territories as members of the Barents Regional Council (BRC).[2] In a practical way, this institutional body encourages all types of cooperation at the local level. In the post-Cold War context, the Kirkenes Declaration is the founding act of this regional dynamic.

Trust between Russia and Nordic countries – founded as a common policy principle – is a cornerstone in this governance arrangement. The political normalization aimed first to improve economic networking and create a lasting business community in the Northernmost part of Europe. True to this spirit, this institutionalization process in this area is in line with the European construction. The purpose is to overcome the confrontation and division that tore apart the continent and to ensure stability and security in the region. In this regard, the Barents cooperation must be analyzed within the framework of the European cooperation and the regional integration process.

This subregional process went hand in hand with the establishment of the Arctic Council (AC) in 1996. The regional organization was engaged to support sustainable development in the Polar region, which covers economic and social development, in line with the Barents countries' agenda (AC, 1996). In some way, the regional governance body supplements the Barents institutions. For instance, the decisions adopted by the AC strengthen the feasibility of sustainable development in the Barents region through a specific environmental framework.

Regarding the diplomatic stance of the Russian Federation, it regularly expresses its support for the Arctic regional bodies to strengthen international cooperation and to resolve the many challenges while ensuring sustainable development of the region. Set up alongside the AC, the Barents regional architecture is built around the principle of consensus, which is both a strength and a fulcrum for the stabilization of Russian-Norwegian relations. This regional architecture was created under optimal conditions in the early 1990s. Its capacity and effectiveness are based on the desire of the states to propose a strategy and means to promote and support this

---

[1] The Kirkenes declaration was signed by the foreign ministers and other delegates from Denmark, Finland, Iceland, Norway, Russia, Sweden and the Commission of the European Communities.

[2] It includes the following Barents administrative units: Lapland, Oulu Region, Kainuu and North Karelia (Finland); Nordland and Troms og Finnmark (Norway); Arkhangelsk Region, Murmansk Region, Karelia, Komi and Nenets (Russia); Norrbotten and Västerbotten (Sweden).

cooperation. The synergy of these intergovernmental institutions within the Arctic area reinforces the interlocked and complex dynamics between Norway and Russia.

That being said, climate change effects in the region and the interrelated economic prospects are challenging the regional balance. The acceleration of the biophysical transformation is changing the balance of power that prevailed since the post-Cold War era. For instance, faster ice-melting is opening new lanes and offering unprecedented economic opportunities. Among the hypotheses, maritime transport, hydrocarbon exploitation, and the fishing industry are regularly mentioned.

In 2013, on the eve of the 20th anniversary of its establishment, the BEAC renewed its ambition to foster sub-regional cooperation. Nevertheless, geopolitical change in Europe interrupted this positive momentum in 2014. In fact, the outbreak of the armed conflict in Eastern Ukraine and the annexation of Crimea by the Russian Federation permanently broke the political trust between Moscow and its Nordic neighbors.

## 2014 and the Epoch of Uncertainties

Since 2014, Russia has become the main security threat for the Nordic countries, while their relations deteriorated sharply following the Ukrainian crisis. These new geopolitical circumstances undermined the Barents cooperation, while it certainly concluded the convergence phase that had been initiated in the early 1990s.

### *Crimea and the Spillover Effect: Military Posture and Diplomatic Chill*

After two decades of positive development in the Barents region, the geopolitical shift in Europe that started in 2014 led to serious challenges in relationships between Russia and the West. The political and military deterioration raised questions about "the nature and density of cooperation with the Russian political regime" and highlighted uncertainties regarding Nordic commitment "to maintain constant and fruitful relation with their Arctic neighbor" (Vidal, 2016, p. 273). The diplomatic chill that created the Ukrainian crisis between Western countries and Russia led to a global security reassessment of the European security architecture. As part of the NATO's doctrine, the Barents region has been particularly scrutinized as military activities amplified under the post-Crimean phase. There is room for sharp tension in the area, whereas military assets and training operations are reciprocally strengthening.

Inasmuch as the influence of the United States within the Barents region took a new turn, its activities with its regional allies increased significantly. For instance, Norway and the US invested in upgrading the GLOBUS radar system in light of increasing tensions with Russia. Giving Norway better eyes and ears for monitoring Russian military activities, the station has operated in Vardø since the 1950s (Nilsen, 2018). In accordance with Article 5 of the North Atlantic Treaty (1949), Norway is committed to the collective defense of its allies and vice-versa. This collective commitment ensures NATO's security architecture. However, the Svalbard archipelago, under Norwegian sovereignty, is excluded from that treaty. Under the legal jurisdiction of the 1920 Svalbard Treaty, it remains a security blackhole in case a major dispute should arise between Russia and Norway (Whither, 2018; Kelman et al., 2020). In recent years, we have witnessed diplomatic tensions around the Svalbard archipelago, but these have not yet led to a major incident. Although Russia does not dispute the legal regime in force, as defined by the Spitsbergen Treaty, it intends to defend its interests there.

More broadly, the geopolitical tensions between the United States and Russia shape and impact Northern Europe. Increasing military activities foster conditions of the Barents–Baltic continuum that knotted each other in the European security landscape. In 2009, the five Nordic countries (Denmark, Finland, Iceland, Norway, and Sweden) established the Nordic Defense Cooperation (NORDEFCO), which stands as a ministerial-level security forum. In 2019, Sweden chaired this security forum, which included several priority projects, such as strengthening Nordic and transatlantic relations, consolidating NORDEFCO's role as a crisis consultation mechanism, and continuing joint military exercises in the Arctic space (*Swedish chairmanship*, 2019). Both Finland and Sweden have become further tangled in NATO's activities, such as their participation in the large-scale NATO exercise, Trident Juncture, held in Norway in 2018 (Møller, 2019). In the meantime, this development reflects Swedish concerns about security and defense policy in the Baltic region, but also in the Arctic area. In recent years, Sweden encouraged cooperation between the Baltic and Nordic states within the framework of the N5 + 3 intergovernmental forum.

Similarly, Sweden signed a Memorandum of Understanding (MoU) with NATO relating to the host nation support (HNS) agreement in September 2014. The *Riksdag* (Swedish Parliament) validated this agreement in 2016 confirming the rapprochement with this intergovernmental military alliance. In those terms, Sweden can provide support to military units of any NATO member states on its territory. In light of the ever-growing tensions between Western countries and Russia, Sweden is strengthening the transatlantic link. In a reduced format, the countries of the Fenno-Scandinavian Peninsula concluded a trilateral agreement aimed at strengthening cooperation in defense matters, which allows them to coordinate their efforts on security threats. According to the trilateral declaration, the parties involved want to improve their interoperability and their ability to provide mutual assistance (*Statement of Intention*, 2020a). Notwithstanding, this operational enhancement does not entail a mutual defense obligation. Within this framework, it is stressed that Denmark and Iceland will be informed on a regular basis regarding the development of this cooperation.

Now that the security landscape in Europe has shifted dramatically, the Norwegian authorities have taken this new paradigm to reorganize the country's defense policy.

However, numerous publications on strategy and defense have defined new priorities and specifically highlighted Russia as the main threat to the Norwegian state. In line with this shift, *Fokus 2018* (2018) reported the rapid modernization of Russian military infrastructure and the significant deployment of armed forces in the Arctic region. In these circumstances, Russia's strategic development in the region has dramatically altered the security architecture, obliging Norway to take the necessary measures, including a new defense doctrine. In *Capable and Sustainable* (2016), the Norwegian Ministry of Defense warned of "Russia's growing military capability and its use of force" (p. 8). Due to the post-Crimean geopolitical conditions, the Kingdom of Norway reversed its defense doctrine to protect its territorial integrity, which stands as a top priority.

As a reminder, at the end of the Cold War, Norway undertook a profound transformation of its defense apparatus to adapt it to new challenges, requiring flexible armed forces in order to engage in remote theaters of operations. Norway's contribution to the NATO operation in Afghanistan from 2001 illustrates this change in the doctrine of the Norwegian army. This situation was thwarted in 2014 in light of the Ukrainian crisis. For the Norwegian armed forces, this moment ushered in a paradigm shift in line with this new security situation on the European continent.

Similarly, Russia has adopted a new military posture in the area, which yearns to enforce the country's sovereignty over its territory and borders (Staun, 2017). However, the modernization of the Russian military infrastructure is challenging the North Atlantic area in reviving the GIUK gap as a strategic maritime line for NATO's allies, which constitutes the key transatlantic routes for military support and procurement coming from North America (Boulègue, 2019; "The GIUK Gap", 2019). That said, Russian military activities in the region again underline the strategic dimension of the Arctic region. In this respect, the development of the Northern Sea Route (NSR) – a priority of Moscow – positions the Barents region as a strategic lock to access to Eastern Asia. This area continues to be a hotspot in Europe.

## *Local Cooperation as a Strong Driver to Dodge the Crisis*

Before delving into the Barents context, we must recall that the Ukrainian crisis has fashioned new dynamics in the West-Russia relationship. In 2014, the imposition of financial and economic sanctions by the European Union and the United States particularly targeted Russian energy projects in the Arctic region. In response, the Russian government imposed counter-sanctions with a food embargo. Since then, the mutual sanctions regime has undermined these relations and drastically lowered business exchange. These sanctions and counter-sanctions have negatively impacted the local economy in the region (Vidal, 2016). For instance, the embargo on food products hit the fishery industries in Norway extremely hard. As a result, the fishery industries turned to other international markets as the counter-sanctions became a structural business feature. These economic conditions curbed the scope for cooperation between the Nordics and the Russian Federation.

Having said that, local stakeholders depend on these geopolitical conditions for the continuation of their cooperation in the long term. In other words, their relations in various spots could be shaken wherever Russian and Western interests confront one another. Based on neighborly relations, however, the Barents governance mechanism demonstrated resistance since 2014, underlining the depth of the established contacts.

The Barents region is again at a crossroads: are these geopolitical conditions going to close the door of cooperation and shut the gateway that existed for three decades? As Hønneland (2013) argued, cooperation in the Barents region has never been easy since the 1990s, particularly in the economic domain. Instead, the regional project focused on people-to-people cooperation. Clearly, the very essence of the Barents identity is a bridge between the people, making it possible to survive cyclical crises. The 1993 Kirkenes Declaration has been the key to this success because it laid the foundations for a dynamic and sustainable regional governance structure and resulted in a sophisticated cooperation scheme between member countries. The BEAC has remained as an important landmark for all gathered communities of this region.

Nearly three decades after the Kirkenes Declaration, the Barents spirit remains and has resisted geopolitical whims. Both Russia and Norway have sustained their cooperation in several fields (science, education, culture) which has significantly impacted local development. Esperitu (2015) commented that "research and academic exchange, and public engagement are some of the most effective ways of continuing and strengthening the important cross-border relations in the Barents Region" (p. 8). After all, cultural and science diplomacy is an effective way to prevent decline in multilateral and bilateral cooperation.

While each stakeholder holds its own agenda, the regional governance is complexified and lies on multiple layers. In the light of the West-Russia rift, national governments are having a stronger influence in the region, whereas central authorities deal directly with security matters. Against the backdrop of this critical phase, this situation highlights the division between and the south and the north of each Barents country, while both sides stand with diverging interests. Rather than question the evolution of internal dynamics of the Barents region, this political spectrum affects the national narrative of these countries. This poses a particular issue on the idea of northernness, in which hegemonic elites of the south enforce a political agenda (Spracklen, 2016). It breaks the Barents symbiose, where local populations are at the center of this political and geographical semantic (Hønneland, 2013; Medby, 2014). Even though the Barents region is subject to major disruption, we observe the resilience of this regional governance. From a local perspective, these geopolitical tensions have not so far broken discussions and cooperation. If the level of trust between Nordic countries and Russia has been eroded since the outcome of the Ukrainian crisis, the spirit of this cooperation remains at the local level.

All in all, the Barents spirit can be defined as the ability to engage various stakeholders despite external and uncontrollable factors that may ignite tensions. At a time when challenges are increasing, the Barents region must ensure critical choices as climate change effects move the Arctic into a new dimension. On the eve of the climate crisis and the quest for socio-ecological transition, these societies are engaged in a journey of historic transformation. This raises questions about both the economic development model and the manner of organizing the democratic process and debate between the stakeholders.

## Climate Change and the Barents Governance: The Search for Socio-Ecological Balance

If the Arctic region and its imaginary attract a longstanding curiosity from scientists and artists, it also acts as a magnet for the business community. Large companies have demonstrated a particular interest in investing in several development projects in the Barents area (O&G, mining, tourism, ICT, etc.). Thus, the prospect of an alternative economic model in the coming decades is at stake, although it is still based upon the foundations of the thermo-industrial civilization. Nonetheless, this civilization finds itself trapped considering growing climate change consequences, and largely reflected in this polar region. For instance, the ecosystem of the Barents region is already under pressure as benthic species composition is changing (Meredith et al., 2019). These ecological changes are directly disturbing human communities and, therefore, questioning the human-environmental equilibrium.

### *Ecological Dynamic vs. Economic Rationale*

The case of fisheries highlights these increasing difficulties in coming decades. Accelerating global warming increases the pressure on Arctic waters. Via a cascade effect, the living marine resources are going to significantly transform the regional ecosystem (Lind et al., 2018; Koenigstein, 2020). Between Russia and Norway, this situation might challenge fisheries management cooperation in the medium and long term. The impending local ecosystem transformation threatens to jeopardize this established fisheries governance. This example underlines the need for further economic and political arrangements, whereas fish stocks could move into new marine areas and disrupt the ecological balance (Vidal, 2018). In other words, a new understanding is urgently needed to cope with the ecological transformation and maintain the social and economic stability.

Today, one of the main challenges is to ensure the decarbonization of society (see Gras, 2016). To this end, the Norwegian government's policy is a driving force to enable the rise of economic segments that lean towards a disconnected carbon system. At the end of the day, the objective of this transition is to move to end fossil fuel dependency and to eventually switch to a post-carbon society over the course of this century.

The Barents region offers valuable and diversified economic opportunities that local and national authorities encourage. Ironically, the climate change and the extraction of natural resources are interlocked, while the acceleration of ice-melting could enable further economic opportunities (such as maritime transportation, oil and gas exploitation, and fishing industry). The business community is involved in several investment projects, which the Barents states have promoted in recent years. As the driving force, these countries have conducted a strategic policy in which their northern part reflects their economic ambitions. In *Sveriges strategi för den arktiska regionen* (2020b), a strategic Arctic policy document, the Swedish government described the Arctic as an area for innovation and international exchange. For this purpose, Norbotten and Västerbotten relied on competitive and ground-breaking research institutes and universities, which are connected to companies and the business community. In addition to creating jobs for local population, such national strategies would aim to support innovation and create favorable condition for private actors in the Barents region. Given this ambitious prospect, extractive activities accentuate the complexity to manage economic drivers and ecological conditions.

The mining industry in the Barents region dates back to the nineteenth century; Luossavaara-Kiirunavaara Aktiebolag (LKAB), established in 1890, started operations in Kiruna in 1898. The Kiruna mine remains the largest underground iron ore mine in the world, but threatens to partly collapse the Swedish town. In response to this challenge, LKAB funded an innovative operation to relocate it few kilometers further east (Sjöholm, 2017). Likewise, in the 1930s the Soviet Union developed mining activities in the Kola Peninsula. Thereafter, the Russian company Norilsk Nickel continued the production and processing of nickel ore. The ageing infrastructure, still in operation, generates massive $SO_2$ emissions, which bring major environmental and health concerns. For several decades, the pollution coming from the border town Nikel to Norway alerted the public authorities (Nilsen, 2019; "Protest against heavy pollution", 2020). To reverse the carbon emissions model, raw materials are needed to develop carbon-free technologies. This, in turn, has attracted investors to develop extensive mining projects in the region (Sakatti copper and nickel mine in Sodankylä, Finland; Nussir copper mine in Finnmark, Norway, etc.). Apart from negative environmental and health effects, extractive activities contribute up to 50 percent of $CO_2$ emissions (IRP, 2019). To curb these emissions, LKAB aspires to implement a carbon-free strategy from extraction to ore processing, whereby Norbotten County would become the epicenter of "green industrial transformation" (LKAB, 2020). To sum up, the mining sector gathers all complexities that the Barents region must deal with, as a path must be found between economic opportunities and the need to take out fossil fuels from the equation.

Based on the Industrial Revolution, the development of these industrial activities is an integral part of the modern society. As Fressoz (2016) argued, the green policy is finally a means of avoiding financial loss, as the main interest of the industrial ecology is to relate ecology and benefit. Indeed, the development of the extractive activities is a direct response to the material needs for the shifting of energy model (wind power, solar power, transportation electrification, etc.). In this regard, the electrification of the vehicle fleet in Norway is an important step in the decarbonization process. While it encompasses the whole country, the Barents region is also included in this transition. Concretely, and symbolically, rapid battery chargers for electric vehicles were installed in Murmansk and along the road that links the Russian city to Kirkenes, Norway, and Inari, Finland (Nilsen, 2017). In creating such a transport corridor, the region is situating itself at the vanguard of the automotive market transformation towards a post-carbon model. Such initiatives improve the narrative about the shifting social and economic model. However, the political dimension must embed local communities and other stakeholders in that process.

## *The Need to Boost and (Re)shape Institutional Arrangements*

Considering the growing business opportunities promoted by both national and local authorities, Barents governance faces new challenges. Indeed, the various projects that have been discussed or are in progress have encountered local resistance. Local communities oppose projects that could directly jeopardize socio-ecological balances. Besides, the changes occurring in the region could hasten tensions, which will undermine political stability in the long run. Put simply, increasingly vulnerable social, economic, and environmental conditions threaten local and indigenous communities (Hossain et al., 2016). On one hand, new activities bring jobs and income for locals, but also modern infrastructures. On the other hand, they may generate negative outcomes with unplanned and detrimental impacts. This is the case with the proposed Arctic Railway project, which would connect Kirkenes, Norway, to Rovaniemi, Finland. This infrastructure project has not yet been decided upon, but it is very costly. Meanwhile, Finnish authorities promote this project as part of an integrated European transport route for natural resources (minerals, timbers, fish, etc.) and interlink it to the Northern Sea Route (*The Arctic Railway*, 2020).

However, this project threatens Sámi's activities, as the railway would further change and diminish the land area that they rely upon to maintain their traditional livelihoods. In other words, this intended project would undeniably deprive Sámi of access to land. Although the Arctic Railway remains unclear, the organization of discussions have denied Sámi the opportunity to properly participate in its decision-making process (Ott, 2020). This example shows the need to offer larger and more inclusive means of democratic debate. In the Barents region, the Sámi must play a part in the democratic institutions, which enable them to express their voice and actively participate in the democratic making-decision process. As Hossain et al. (2016) wrote, these major development projects are "one of the most significant sources of abuse of the rights of indigenous peoples" (p. 60). Toivanen (2019) rightly points out the significance of preserving the vitality of their cultures, which is connected to their livelihoods.

After all, the indigenous community in the Barents region has been an important eyewitness of the socio-ecological disequilibrium. They have progressively changed their way of life as technology has entered their daily lives (for example, the introduction of the snowmobile in the 1960s). If they maintain cultural practices and traditional knowledge, however, they will not be outsiders of the modern civilization. That being said, and for obvious ethical reasons, they should ensure their right to participate in the democratic debate within the Barents community.

Inasmuch as climate change requires to implement new practices, the Sámi community may influence the ongoing socio-economic transition. Based on traditional knowledge and know-how, these practices could inspire modern technologies. This approach would make it possible to rebalance according to the preservation of ecosystems and the traditional living of indigenous people. Altogether, the current democratic and ethical challenges about the Barents socio-economic model reflect broader issues. Meanwhile, they recall the complexity involved in shifting away from the foundations of the thermo-industrial civilization.

## Conclusion – In the Age of the Anthropocene

True to this spirit, the Polar region covers equivocal perspectives gathered around permanence and transformation, mythology and futurism, clarity and darkness. While the Arctic region is subject to contrast and diversity, the Barents area is interlocked with Europe on a deep historic continuum. Today, the Barents region attracts both geostrategic and economic attention, while dramatic changes are visible on the horizon. The Arctic is warming twice as fast as the planet overall, which causes irreversible changes. The reverse side of the coin is a global competition for accessing to natural resources and economic benefits. As Finger (2016) commented, the Arctic emerges as a laboratory, bringing together bio-geo-physical forces. However, the Barents governance is clearly under increasing pressure to cope with those multiple challenges. The first direct consequences are already altering the ecosystem such as the "Atlantification" of the Barents Sea, which results in changes to fishing stocks (Lind et al., 2018).

Local communities have their livelihoods directly threatened as extreme weather events will increase in the coming decades (wildfires, floods, higher prevalence of snow and storms, etc.). If these emerging climate conditions are here to stay, the Barents societies will have to adapt their infrastructures, and adjust their social and economic order. To succeed in this path, the Barents governance must bargain two opposing trends: the socio-ecological transition and the global market dynamic. While Russia is fully committed to expanding fossil fuel production and exports, and not bent on reducing $CO_2$ emissions anytime soon, the Nordic countries are engaged in a decarbonization strategy. In contrast with Russia, these countries are committing to reach carbon neutrality thanks to the development of renewables. Nevertheless, investors are flocking to the Barents region to further develop the natural resources. These activities, such as mining, have a high energy demand, and therefore increase the use of materials. From this point of view, Fressoz (2020) underlined that material accumulation has never stopped, and has even accelerated since the start of the twenty-first century. Finally, the Barents region is an integrated element of this historic development, whereas one of the outcomes of the Anthropocene is a geological transformation of the Arctic reduced to the human scale.

## References

Aas, S. (2012). Norwegian and Soviet/Russian World War II memory policy during the cold war and the post-Soviet years. *Acta Borealia: A Nordica Journal of Circumpolar Societies, 29*(2), 216–239.

Arctic Council. (1996). *Declaration on the establishment of the Arctic council 1996 – Ottawa declaration*. Ottawa: Arctic Council.

Boulègue, M. (2019). Russia's military posture in the Arctic. Managing hard power in a "low tension" environment. NDC Research Paper. Rome: NATO Defence College.

EJA (2020). Protest against heavy pollution from Russian nickel plant in Kirkenes, Norway. *Environmental Justice Atlas*. https://ejatlas.org/conflict/protest-against-heavy-pollution-from-russian-nickel-plant-kirkenes-norway. Accessed 5 December 2020.

Esperitu, A. (2015). Moving forward: Strengthening cooperation in today's Barents region. *Barents Studies: Peoples, Economies and Politics, 1*(3), 7–11.

Finger, M. (2016). The Arctic, laboratory of the Anthropocene. In L. Heininen (Ed.), *Future security of the Global Arctic: State policy, economic security and climate* (pp. 121–137). Basingstoke: Palgrave Macmillan.

Fressoz, J.-B. (2016). La main invisible a-t-elle le pouce vert? Les faux-semblants de "l'écologie industrielle" au XIX$^e$ siècle. *Techniques & Culture*, "Réparer le monde. Excès, reste et innovation" 65–66, 324–339.

Fressoz, J.-B. (2020). L'Anthropocène est un "accumulocène". *Regards croisés sur l'économie, 26*, 31–40.

González de Molina, M., & Toledo, V. M. (2014). *The social metabolism. A socio-ecological theory of historical change*. Cham: Springer. https://doi.org/10.1007/978-3-319-06358-4_3.

Gorbachev, M. (1987). Mikhail Gorbachev's speech in Murmansk at the ceremonial meeting on the occasion of the presentation of the order of Lenin and the gold star to the City of Murmansk. Murmansk. https://www.barentsinfo.fi/docs/Gorbachev_speech.pdf. Accessed 5 December 2020.

Gras, A. (2016). The deadlock of the thermo-industrial civilization: The (impossible?) energy transition in the Anthropocene. In E. Garcia, M. Martinez-Iglesias, & P. Kirby (Eds.), *Transitioning to a post-carbon society* (pp. 3–35). London: Palgrave Macmillan.

Hansen, L. I. (2006). Sami fisheries in the pre-modern era: Household sustenance and market relations. *Acta Borealia, 23*(1), 56–80.

Hofstra, T., & Samplonius, K. (1995). Viking expansion northwards: Mediæval sources. *Arctic, 48*(3), 235–247.

Hønneland, G. (1998). Identity formation in the Barents Euro-Arctic region. *Cooperation and Conflict, 33*(3), 277–297.

Hønneland, G. (2013). Borderland Russians. In *Identity, narrative and international relations*. London: Palgrave Macmillan.

Hossein, K., Zojer, G., Greaves, W., Roncero, M., & Sheehan, M. (2016). Constructing Arctic security: An inter-disciplinary approach to understanding security in the Barents region. *Polar Record, 53*(1), 52–66.

IRP. (2019). *Global resources outlook 2019: Natural resources for the future we want*. Nairobi: United Nations Environment Programme.

Kelman, I., Sydnes, A. K., Duda, P. I., Nikitina, E., & Webersik, C. (2020). Norway-Russia disaster diplomacy for Svalbard. *Safety Science, 130*. https://doi.org/10.1016/j.ssci.2020.104896.

Kikkert, P., & Lackenbauer, W. (2019). The militarization of the Arctic to 1990. In K. Coates & C. Holroyd (Eds.), *The Palgrave handbook of Arctic policy and politics* (pp. 487–505). London: Palgrave Macmillan.

Koenigstein, S. (2020). Arctic marine ecosystems, climate change impacts, and governance responses: An integrated perspectives from the Barents Sea. In E. Pongracz, V. Pavlov, & N. Hänninen (Eds.), *Arctic marine sustainability. Arctic maritime businesses and the resilience of the marine environment* (pp. 45–71). Cham: Springer.

Krasnozhenova, E., Kulik, S., & Lokhova, T. (2020). Arctic transportation system during World War II. *IOP Conf. Series: Earth and Environmental Science, 404*. https://doi.org/10.1088/1755-1315/434/1/012002.

Kusmenko, J. (2009). Sámi and Scandinavians in the Viking Age. *Scandinavistica Vilnensis, 2*, 65–94.

Kvistad, J. M. (1995). *The Barents Spirit. A bridge-building project in the wake of the cold war*. Oslo: Institutt for forsvarsstudier.

Lind, S., Ingvelsen, R. B., & Furevik, T. (2018). Arctic warming hotspot in the northern Barents Sea linked to declining sea-ice import. *Nature, 8*, 634–639.

LKAB. (2020). *Historic transformation plan for LKAB: "The biggest thing we in Sweden can do for the climate"*. Luleå: LKAB.

Luzin, D. (2007). The northern sea route during World War II, 1939–1945. *The Journal of Slavic Military Studies, 20*(3), 421–432.

Malinova, O. (2017). Political uses of the great patriotic war in post-Soviet Russia from Yeltsin to Putin. In J. Fedor, M. Kangaspuro, J. Lassila, & T. Zhurzhenko. *War and memory in Russia, Ukraine and Belarus* (pp. 43–70). Cham: Springer.

Medby, I. A. (2014). Arctic state, Arctic nation? Arctic national identity among the post-Cold War generation in Norway. *Polar Geography, 37*(3), 252–269.

Meredith, M., Sommerkorn, M., Cassotta, S., Derksen, C., Ekaykin, A., Hollowed, A., Kofinas, G., Mackintosh, A., Melbourne-Thomas, J., Muelbert, M. M. C., Ottersen, G., Pritchard, H., & Schuur, E. A. G. (2019). Polar Regions. In H.-O. Pörtner et al. (Eds.), *IPCC special report on the ocean and cryosphere in a changing climate* (pp. 203–320). Geneva: Intergovernmental Panel on Climate Change.

Møller, J. E. (2019). Trilateral defence cooperation in the North: An assessment of interoperability between Norway, Sweden and Finland. *Defence Studies*. https://doi.org/10.1080/14702436.2019.1634473.

Nilsen, T. (2017). Europe's northernmost cross-border highway to get charger network for electric cars. *The Barents Observer.*. https://thebarentsobserver.com/en/ecology/2017/11/europes-northernmost-cross-border-highway-get-charger-network-electric-cars. Accessed 5 December 2020.

Nilsen, T. (2018). US and Norway upgrade eye on border to northern Russia. *The Barents Observer*. https://thebarentsobserver.com/en/security/2018/11/us-and-norway-upgrades-eye-northern-russia. Accessed 5 December 2020.

Nilsen, T. (2019). Norway's enviro minister brings dispute on cross-border pollution to Moscow. *The Barents Observer*. https://thebarentsobserver.com/en/ecology/2019/01/norways-enviro-minister-brings-cross-border-pollution-moscow. Accessed 5 December 2020.

Norway. (2016). *Capable and sustainable. Long Term Defence Plan*. Oslo: Forsvarsdepartementet.

Norway. (2018). *Fokus 2018. Etterretningstenesta si vurdering av aktuelle tryggingsutfordringar*. Oslo: Etterretningstenesta.

Ott, A. (2020). Sámi's enactments of the Arctic railway: Sustainable development or environmental injustice?. *Versus*. https://www.versuslehti.fi/gradusta-asiaa/samis-enactments-of-the-arctic-railway-sustainable-development-or-environmental-injustice/. Accessed 5 December 2020.

Rønning Balsvik, R. (2015). Russia and Norway: Research collaboration and comparison of asymmetrical relations. *Acta Borealia*. https://doi.org/10.1080/08003831.2015.1029850.

Schram Stokke, O. (2011). Environmental security in the Arctic: The Case for Multilevel Governance. *International Journal*. https://doi.org/10.1177/002070201106600412.

Schram Stokke, O., & Tunander, O. (1994). *The Barents region. Cooperation in Arctic Europe*. London: SAGE Publications.

Shrader, T. (2005). Across the Borders: The Pomor trade. In T. Jackson & J. P. Nielsen (Eds.), *Russia-Norway: Physical and symbolic Borders* (pp. 105–115). Moscow: Languages of Slavonic Culture.

Sjöholm, J. (2017). Authenticity and relocation of built heritage: The urban transformation of Kiruna, Sweden. *Journal of Cultural Heritage Management and Sustainable Development, 7*(2), 110–128.

Spracklen, K. (2016). Theorising northernness and northern culture: the north of England, northern Englishness, and sympathetic magic. *Journal for Cultural Research, 20*(1), 4–16.

Staun, J. (2017). Russia's strategy in the Arctic: Cooperation, not confrontation. *Polar Record, 53*(3), 314–332.

Suprun, M. (2020). The liberation of northern Norway in Stalin's post-war strategy. *The Journal of Slavic Military Studies, 33*(2), 227–291.

Sweden. (2019). *Swedish chairmanship priorities for Nordefco 2019*. Stockholm: Försvarsdepartementet.

Sweden. (2020a). *Statement of intention enhanced operational cooperation among the Ministry of Defence of the Republic of Finland, the Ministry of Defence of the Kingdom of Norway and the Ministry of Defence of the Kingdom of Sweden*. Stockholm: Försvarsdepartementet.

Sweden. (2020b). *Sveriges strategi för den arktiska regionen*. Stockholm: Regeringskansliet.

The GIUK Gap. (2019). The GIUK Gap's strategic significance. *Strategic Comments, 25*(8), i–iii. https://doi.org/10.1080/13567888.2019.1684626.

The GIUK Gap. (2020). *The Arctic railway. Risks and Opportunities*. Rovaniemi: House of Lapland.

Thuen, T. (1993). Two Epochs of Norwegian-Russian trade relations: From symmetry to asymmetry. *Acta Borealia: A Nordic Journal of Circumpolar Societies, 10*(20), 3–18.

Toivanen, R. (2019). European fantasy of the Arctic region and the rise of indigenous Sámi voices in the global arena. In N. Selheim, Y. V. Zaika, & I. Kelman (Eds.), *Arctic triumph. Northern innovative and persistence* (pp. 23–40). Cham: Springer.

Vidal, F. (2016). Barents region: The Arctic council as a stabilizing magnet. In L. Heininen, H. Exner-Pirot, & J. Plouffe (Eds.), *Arctic yearbook 2016* (pp. 224–246). Akureyri, Iceland: Northern Research Forum.

Vidal, F. (2018). The Barents Sea: Environment cooperation in the Anthropocene era. *E-International Relations*. www.e-ir.info/2018/05/06/the-barents-sea-environment-cooperation-in-the-anthropocene-era. Accessed 5 December 2020.

Wither, J. K. (2018). Svalbard. *The RUSI Journal, 163*(5), 28–37.

Young, O. R. (2010). Arctic governance – Pathways to the future. *Arctic Review on Law and Politics, 1*(2), 164–185.

Young, O. R. (2012). Building an international regime complex for the Arctic: Current status and next steps. *The Polar Journal, 2*(2), 391–407.

Zachrisson, I. (2008). The Sámi and their interaction with the Nordic peoples. In S. Brink & N. Price (Eds.), *The Viking World* (pp. 32–39). London: Routledge.

Zimmerbauer, K. (2012). Unusual regionalism in Northern Europe: The Barents region in the making. *Regional Studies, 47*(1), 89–103.

**Florian Vidal** is a research fellow (since 2020) at the French Institute of International Relations in Paris, France (Ifri) and an associate research fellow (since 2019) at the Paris Interdisciplinary Energy Research Institute (PIERI), University of Paris, France.

# Chapter 21
# The Arctic Council at 25: Incremental Building of a More Ambitious Inter-governmental Forum

**Timo Koivurova and Malgorzata Smieszek**

## Introduction

The 25th anniversary of the Arctic Council (AC) marks a suitable point for reflection on the Council's performance thus far and the challenges that lay ahead. Established at a time when the world's interest in the Arctic was historically low, the institution has managed to adapt to a rapidly changing biophysical and geopolitical environment over the past 25 years and has established itself as a primary inter-governmental body for co-operation on matters pertaining to the region. During the same period, the Arctic has transformed profoundly due to the combined pressure of climate change and globalization. Named 'the bellwether of global climate change', the Arctic has warmed at more than twice the rate as the rest of the planet, resulting in decreasing Arctic sea ice, accelerated melting of the Greenland Ice Sheet, thawing permafrost, and the acidification of the Arctic Ocean. Simultaneously, economic prospects related to opening the Arctic Ocean, including new transport routes and offshore energy resources, have piqued the interest of the international community and non-Arctic actors that have begun to articulate their stakes in the region.[1]

---

Arctic Council Strategic Plan 2021 to 2030, at https://oaarchive.arctic-council.org/bitstream/handle/11374/2601/MMIS12_2021_REYKJAVIK_Strategic-Plan_2021-2030.pdf?sequence=1&isAllowed=y.

---

[1] According to Oran Young, in the last three decades the Arctic has experienced two fundamental state changes, each of which has major consequences for Arctic policymaking and governance in

---

T. Koivurova (✉)
University of Lapland, Rovaniemi, Finland
e-mail: timo.koivurova@ulapland.fi

M. Smieszek
UiT The Arctic University of Norway, Tromsø, Norway
e-mail: malgorzata.smieszek@uit.no

© The Author(s), under exclusive license to Springer Nature Switzerland AG 2022
M. Finger, G. Rekvig (eds.), *Global Arctic*,
https://doi.org/10.1007/978-3-030-81253-9_21

The worldwide media have speculated about a race to tap the melting Arctic's riches and have constructed narratives of alleged geopolitical conflicts and a rising new Cold War. Put together, these developments have presented a challenge to the AC, both from the outside and from within. Additionally, all these developments have placed the AC in the spotlight, where it has stayed ever since. This global attention manifested dramatically in one of the most recent AC meetings of foreign ministers in Rovaniemi, Finland, in May 2019, where, because of the US administration's rejection of language referring to climate change, it was, for the first time, impossible for Arctic states to agree on a ministerial declaration at the meeting (Koivurova, 2019).

How has the AC reacted to this interest and these new developments? How has it coped with the regional challenges? What do past experiences say about the Council's ability to address the challenges to come? At the heart of this enquiry is an examination of how the Arctic Council has changed over the course of these last 25 years. Our claim is that the Council has retained its core functions whilst also changing other functions, some quite dramatically, in response to the needs witnessed presently.

To examine these issues, we claim that it is important to go back in time and understand the foundations, laid in the late 1980s and early 1990s, upon which the Arctic Council was built, as most of them are still relevant and demonstrate the continuity in the AC's operations. Concurrently, the general provisions of the Council's founding documents left enough space for interpretation, which has enabled the body to make incremental changes and develop. Among others, over the course of the last decade, they have allowed for a negotiating space for legally binding agreements and have catalyzed the formation of external bodies to address specific topics of increasing importance, such as the Arctic Economic Council, the Arctic Coast Guard Forum, and the Arctic Offshore Regulators' Forum. Hence, we observe both continuity and change in the AC. Arguably, whereas the former may be perceived as a limitation, the latter represents the possibilities that the Council has at hand regarding emerging issues. We argue that there is a clear path-dependence in terms of how the Council has evolved, even if it has, to some degree, been able to evolve in order to address the endogenous and exogenous challenges.

The remainder of this chapter is structured as follows. First, it covers the Arctic Environmental Protection Strategy (AEPS), signed in 1991, and the transition period during which the body was subsumed into the Arctic Council, established in

---

broader terms. The first change, "a delinking or decoupling shift", took place in the late 1980s and early 1990s and was closely linked to the waning of the Cold War and the collapse of the Soviet Union. It resulted in the launch of numerous formalized structures of collaboration, was marked by a strong focus on Arctic-specific matters, and allowed for the gradual development of "the idea of the Arctic as [a] distinctive region with a policy agenda of its own". This process also caused a disconnect between Arctic governance and the governance unfolding on a global scale. The second state change, "a linking change", began in the Arctic in the early 2000s and continues today. To a great extent, the change has been driven by the processes of global environmental change and globalization; in other words, a mix of forces of environmental and socioeconomic character. See Oran R. Young, "The Arctic in Play: Governance in a Time of Rapid Change," *The International Journal of Marine and Coastal Law* 24 (2009): 423–442 (Young, 2009a).

1996. Second, it provides an overview of the early days of the AC until 2007 when the Arctic, following a series of hyped-up events, became a focus of the global community. This part also includes initiatives that the Council employed to cope with this unprecedented attention and its related challenges. This same part then moves to examine the more recent changes to the Council. The chapter concludes with a discussion of the record of the Arctic Council's performance and its ways of moving forward and addressing future challenges. First and foremost, we ask how much of the Council's past has dominated its evolution, given that it inherited rather weak foundations from the AEPS. How has the Arctic Council been able to respond to external and internal challenges over all these years without any overall strategy to guide its work? How is it currently positioned to tackle new challenges?

## From the AEPS to the Arctic Council

The origins of today's circumpolar co-operation date back to the late 1980s and the Finnish initiative, which followed the historical speech of Mikhail Gorbachev in Murmansk in 1987. In his talk, the then-general secretary of the Communist Party of the Soviet Union called for the Arctic to become a zone of peace and fruitful co-operation via coordinated scientific research in the North and co-operation in protecting the Arctic's natural environment, supported by a circumpolar environmental monitoring program. In this international arena deeply divided by the Cold War, science and environment were perceived as relatively neutral grounds to help further alleviate political tensions (Nilsson, 2007). Hence, the latter idea induced Finland to convene a meeting of all eight Arctic states in Rovaniemi, Finland, in 1989. At that gathering, eight Arctic states discussed the prospect of such a collaboration, and this was the initiative that gave birth to the *Declaration on the Protection of Arctic Environment* and the *Arctic Environmental Protection Strategy* (AEPS), signed in 1991.

The AEPS aimed to deepen scientific understanding of the causes, pathways and effects of pollution in the Arctic, as well as to continuously assess threats to the Arctic environment. Following consultations and previous work, it was decided that the AEPS would focus primarily on particular pollution issues, including oil, acidification, persistent organic contaminants, radioactivity, noise, and heavy metals (Arctic Environmental Protection Strategy, 1991). To achieve its objectives, the strategy called for the establishment of four working groups: the Arctic Monitoring and Assessment Programme (AMAP); Conservation of Arctic Flora and Fauna (CAFF); Emergency, Preparedness, Prevention and Response (EPPR); and the Protection of Arctic Marine Environment (PAME). These groups were to carry out the programmatic activities of the AEPS in accordance with their respective mandates (Young, 1998). As Håken R. Nilson (1997) reported, the initial stages of the AEPS were to set those programs in motion and develop arrangements to enable the specialist work to deliver results of some substance. Consequently, the process was largely driven by environmental experts and was of a 'bottom-up' nature, where working groups were given substantial autonomy in forming and developing their projects. The major deliverable of the AEPS came in 1997/1998 from the AMAP in the form of a thick report presenting the state of the Arctic environment (AMAP, 1998). Next to critical information on heavy metals, radioactivity, and persistent organic pollutants (POPs), the report also provided substantial texts on polar ecology and the peoples of the North, and it established a lasting precedent for conducting assessments, which became the hallmark of the AEPS/Arctic Council work, recognized as its most effective products (Kankaanpää & Young, 2012; Stone, 2016).

In the meantime, while the AEPS was carrying out its projects and working on its preliminary assessments, the negotiations continued to expand Arctic co-operation from focusing primarily on environmental issues to encompassing more human dimensions and matters of sustainable development. The idea of a significantly broader circumpolar collaboration was not new *per se*. In fact, it had been considered much earlier by the Canadians, who had pondered over the conceptions of the Arctic Basin Treaty as early as the 1970s (Cohen and Pharand as quoted in Keskitalo, footnote 192 in Nilsson, 2007). The idea gained little support at that time, and it found equally meagre support in the early 1990s due to resistance by the United States. Consequently, the bargaining over the establishment and form of the Arctic Council was highly protracted as well. The US insisted on setting up the Council as a purely consultative forum with few mutual obligations (Scrivener, 1999: 55) and ultimately agreed to the formation of a body without legal personality (Bloom, 1999) – a high-level forum designed to promote co-operation "among the Arctic states, with the involvement of the Arctic indigenous communities and other Arctic inhabitants on common Arctic issues" (*Declaration on the Establishment of the Arctic Council*, 1996; hereinafter, Ottawa Declaration). As a result of the position held by the United States during the negotiations, the AC emerged as a fraction of what had been earlier envisaged (Scrivener 1996, as in Scrivener, 1999; see more English, 2013) and a small-scale institution without a permanent chair, secretariat, or budget.[2] The newly established AC subsumed four working groups of the AEPS,[3] and while the Ottawa Declaration marked a shift in focus of Arctic co-operation from environmental protection alone toward a broader concept of sustainable development, the AC began its operations from work on the rules of procedure and drafting terms of reference of a sustainable development program.[4] However, there was considerable discord among Arctic states over the meaning and definition of the

---

[2] It was assumed that rotating chairmanship would serve as a cost-sharing measure, where the temporary host country would provide much of the operational and support resources needed to run the Council in the given two-year period.

[3] Technically speaking, the formal transition of these programs to the Arctic Council, and the termination of the AEPS, occurred only at the AEPS Ministerial meeting held in Alta, Norway in June, 1997 (Fenge & Funston, 2015).

[4] Whereas the AEPS had focused primarily on environmental protection through its working groups, its institutional structure was extended somewhat in 1993 when the Task Force on Sustainable Development (TFSDU) was established in an attempt to broaden the scope of the strategy's activities.

concept of sustainable development (for more, see: Keskitalo, 2004), so instead of a comprehensive program, it was decided at the Ministerial meeting in Iqaluit that the Sustainable Development Program would comprise a series of specific projects (Bloom, 1999); this practice still largely prevails today, even if the SDWG did adopt a strategic framework in 2017 (SDWG Strategic Framework, 2017).

In addition to the category of Members of the Arctic Council, reserved for eight Arctic states (Canada, the Kingdom of Denmark, Finland, Iceland, Norway, the Russian Federation, Sweden, and the United States), the Ottawa Declaration provided for the categories of permanent participants and observers. The former has been an innovative and largely unprecedented arrangement under which a number of organizations of indigenous peoples have their representatives sit alongside ministers and Senior Arctic Officials (SAOs)[5] and have a strong voice in the Council's activities (Arctic Governance Project, 2010; Bloom, 1999; Fenge & Funston, 2015).[6] The latter category encompasses non-Arctic states, global and regional inter-governmental and inter-parliamentary organizations, and non-governmental organizations "that the Council determines can contribute to its work" (*Declaration on the Establishment of the Arctic Council*, 1996). Originally, 14 observers were present at the ceremonial signing of the declaration in Ottawa in 1996; today there are 38, including actors such as China, India and Japan, plus the European Union (EU) is recognized as a *de facto* observer of the AC. As the Council's rules of procedure stipulate, the primary role of observers is to observe the work of the Arctic Council and they are expected to contribute and engage predominantly at the level of AC working groups (Arctic Council, 2013a). All decisions of the AC and its subsidiary bodies are made by consensus of all eight Arctic states (*Declaration on the Establishment of the Arctic Council*, 1996).

## Emerging Challenges and Institutional Response

During much of its first decade in operation, the Council resembled more of a scientific forum than a policy forum, as the Arctic remained largely on the sidelines of the mainstream foreign and national politics in most Arctic countries. The practices

---

[5] Senior Arctic officials (formerly senior Arctic Affairs Officials, SAAOs) are the high-ranking officials (usually at the ambassador level), designated by each Arctic state, who meet at least twice a year. Their main task is to oversee the work of the AC's working groups and its other subsidiary bodies in order to ensure the implementation of the mandates issued by Arctic ministers at ministerial meetings. Additionally, over the last decade SAOs have been given some flexibility to review and adjust the work plans and mandates of the working groups while ensuring they function in accordance with the overarching guidance provided in the biennial ministerial declarations.

[6] As Fenge and Funston noted in the overview of AC practices, in most respects permanent participants take part in discussions in the same manner as states and even though only Arctic states are (technically) considered capable of determining whether there is a consensus for decision, indigenous representatives have been occasionally able to influence the course of the resolutions due to their moral authority (Fenge & Funston, 2015).

coined under the AEPS largely prevailed and the Council's work was carried out predominantly by the working groups. Their priorities and work plans were identified and elaborated on by scientists and officials in each working group and were usually approved by SAOs and Ministers without many modifications and following only limited debate (Fenge & Funston, 2015). Whereas the first Canadian chairmanship of the Council (1996–1998) was mostly dedicated to developing rules of procedure and other operational measures, the United States stood at the helm of the institution during the first active phase of the Council's existence (1998–2000). It was during that period that the Council initiated work on its most seminal product to date, the Arctic Climate Impact Assessment (ACIA), which was the first such regional assessment of climate change impacts (Duyck, 2015; Nilsson, 2007).[7] The ACIA was approved as a project by the Arctic Council in 2000 and it was a joint effort of the AMAP, CAFF, and the International Arctic Science Committee (IASC). In 2004 the assessment team delivered to the AC the report "Impacts of a Warming Arctic", which was intended for policy-makers, and in 2005 a full scientific report of more than one thousand pages was released. The results of ACIA contributed greatly to understanding the implications of climate change in the Arctic, and many of the report's findings were subsequently incorporated into work and assessments of the Intergovernmental Panel on Climate Change (IPCC). Moreover, ACIA attracted world-wide media attention and pointed the spotlight at the circumpolar region, resulting in rising speculation on new economic opportunities and challenges as well as the security risks related to opening the formerly frozen and largely inaccessible Arctic Ocean (Koivurova et al., 2015). This discourse was further escalated by a number of events in 2007–2008, including the planting of a Russian flag on the seabed underneath the North Pole in August 2007; the collapse of Arctic sea ice in September 2007; and a publication in May 2008 by the United States Geological Survey (USGS) of the estimates of Arctic oil and gas reserves, according to which around 13 percent of the world's undiscovered oil and 30 percent of the undiscovered natural gas could be located in the region (U.S. Geological Survey, 2008).[8] The alleged scramble for Arctic territory and resources (Borgerson, 2008 TIME, 1 October 2007, cover, "Who Owns the Arctic?"), along with surging public and international interest in the Arctic, led five littoral Arctic states to reassert their rights in the region in the Ilulissat Declaration issued in May 2008. In that Declaration, Canada, Denmark (Greenland), Norway, Russia, and the United States proclaimed that "[b]y virtue of their sovereignty, sovereign rights and jurisdiction in

---

[7] Climate change was identified as one of the two most significant threats to the Arctic environment in preliminary studies leading to the AEPS. At that stage, however, it was decided that the Strategy should focus its work on pollutants while the responsibility for strengthening the knowledge of causes and effects of climate change would lie with other international groupings and fora (Arctic Environmental Protection Strategy, 1991; Nilsson, 2012).

[8] The International Polar Year (IPY) (2007–2009) was another major event that brought together the global scientific community and focused its research efforts on the Arctic. The results of those extensive international scientific collaborations advanced the understanding of the importance of the Arctic as an indicator of global changes and the interconnectedness of the Arctic and global systems.

large areas of the Arctic Ocean the five coastal states are in a unique position to address […] possibilities and challenges" in the region and that they have a stewardship role in protecting the ecosystem of the Arctic Ocean (Ilulissat Declaration, 2008). Whereas some of the states present at the meeting justified the Arctic Five format as necessary at a time of heightened interest in the Arctic (Pedersen, 2012), Iceland, Finland, Sweden, and the permanent participants expressed their deep discontent with the newly emerged assembly undermining already existing patterns and rules of collaboration in the circumpolar north. It also raised questions about the Arctic Council as the preeminent region-specific forum – a debate that stalled in 2010, when the then United States Secretary of State Hilary Clinton signaled a significant shift in US Arctic policies and practically dismissed the A5 gathering, making the Arctic Council the only relevant forum for general discussions on Arctic matters at that time (Pedersen, 2012).

The rapidly changing climate and ice conditions in the Arctic Ocean firmly placed the Arctic on the international agenda and translated into various non-Arctic actors expressing interest in the regional matters and work of the Arctic Council (Koivurova et al., 2015; Pedersen, 2012). The number of applications to become observers to the AC has risen dramatically since 2007, and existing state observers began to push the motion to enhance their role in the Council (Graczyk, 2011; Koivurova, 2009; Young, 2009b). However, significant concerns were expressed among the main AC actors, both states and permanent participants, regarding the increased role and number of observers. Following discussions and initiatives undertaken by the AC Danish Chairmanship (2009–2011), at the Ministerial meeting in Nuuk in 2011 the Council decided to establish a Task Force for Institutional Issues (TFFI) to "implement the decisions to strengthen the Arctic Council" (Arctic Council, 2011), which would serve to address the issues and to elaborate on the principles and rules concerning the admission, function, and position of the AC observers. Whereas the Ottawa Declaration and AC Rules of Procedure from 1998 only stipulated that the applicants and holders of observer status should contribute to the Council's work, which was to be determined by the AC (Arctic Council, 1998; *Declaration on the Establishment of the Arctic Council*, 1996), the Task Force developed a more specific set of criteria to be adhered to by new and ongoing observers. The set was first presented in the annex of the SAO report to ministers in 2011 and was eventually incorporated into the revised rules of procedure adopted by the Council in 2013 (Arctic Council, 2013a).[9]

The Nuuk Ministerial meeting in 2011 was also significant in the Council's history in two other respects. First, at that meeting, foreign ministers of Arctic states made the decision to establish a permanent Arctic Council secretariat to be located in Tromsø, Norway, and to be operational no later than the beginning of 2013 (Arctic Council, 2011). The decision to establish a standing secretariat was

---

[9] Following the adoption, an applicant for observer status is now to, inter alia, recognize Arctic states' sovereignty, sovereign rights, and jurisdiction in the Arctic; respect the values, interests, culture and traditions of Arctic indigenous peoples; and demonstrate a concrete interest and ability to support the work of the Arctic Council (Arctic Council, 2013a).

a major step in the AC's effort to strengthen the capacity of the Council to respond to emerging challenges and opportunities and provide it with the institutional memory it had not had with a secretariat rotating every 2 years with the chairmanships. It also constituted a principal difference in the practice of the AEPS and the Council's first 15 years in operation. Second, at Nuuk ministers signed the first legally binding agreement negotiated under the auspices of the Arctic Council. It was the Agreement on Cooperation on Aeronautical and Maritime Search and Rescue in the Arctic (SAR Agreement), followed later by the Agreement on Cooperation on Marine Oil Pollution, Preparedness and Response in the Arctic (Oil Spills Agreement) signed at the Ministerial in Kiruna, Sweden in May 2013.[10] Because the Arctic Council is founded on a declaration rather than an international treaty, it cannot itself enact any legally binding decisions. Therefore, both agreements are between the eight Arctic states, rather than being "Arctic Council agreements". Nonetheless, even if the Council served primarily as the catalyst for their negotiations and signature, the agreements marked an evolution of the AC from a body "set up to discuss, inform and potentially shape decisions by national governments" (Fenge & Funston, 2015, p. 10) toward a more decision-making one.

The Ministerial meeting in Kiruna drew global media attention for another reason. As questions relating to the appropriate role for observers were a key part of the negotiations that led to the creation of the Council in 1996, the decision on granting a status to new actors and non-Arctic states was a major point for discussion in 2013. In parallel with including the agreed set of criteria for observers into the revised AC rules of procedure, the Council decided to admit six new countries as observers: China, India, Italy, Japan, Singapore, and South Korea, while receiving the application of the European Union 'affirmatively', with a final decision awaiting 'implementation' (Arctic Council, 2013b).[11] In addition to more precisely defining the criteria for observers, in Kiruna the Council also adopted the AC observers Manual, which is to guide the Council's subsidiary bodies in matters of meeting logistics and the role played therein by observers.[12]

The example of the aforementioned legally binding agreements points to another development in the institutional architecture of the Arctic Council, namely the

---

[10] Agreement on Cooperation on Aeronautical and Maritime Search and Rescue in the Arctic (signed in Nuuk on 12 May 2011, entered into force 19 January 2013) 50 ILM 1119 (2011) (SAR Agreement); Agreement on Cooperation on Marine Oil Pollution, Preparedness and Response in the Arctic (signed in Kiruna on 15 May 2013) <www.arctic-council.org/eppr> accessed 15 January 2017 (Oil Spills Agreement).

[11] In practice, the EU participates in meetings of the Arctic Council and its subsidiary bodies as any other observers and is recognized as a *de facto* observers of the AC. Even though Canada has lifted its objections to granting the EU an observer status since 2013, developments in relations with Russia and sanctions that the EU imposed on it following Russia's annexation of Crimea in 2014 have once more put the implementation of the Kiruna decision on hold.

[12] Since its adoption in 2013, the Manual has been updated twice: first at the SAO meeting in October 2015 and second at the SAO meeting in October 2016.

increasing use of task forces as vehicles for targeting specific matters within a given time frame. According to the AC Rules of Procedure, the Council may establish working groups, task forces or other subsidiary bodies to carry out programs and projects under the guidance and direction of Senior Arctic Officials, with their composition and mandates agreed by the Arctic states in a Ministerial meeting (Arctic Council, 2013a). The AC began the practice of establishing task forces in 2009,[13] and since then three of them have paved the way for three legally binding agreements,[14] while the Task Force to Facilitate the Circumpolar Business Forum laid the grounds for establishing the Arctic Economic Council (AEC), the first in a series of bodies aimed to address specific matters related to developments in the region, in co-operation with, but independently from the Arctic Council. Whereas some close observers of the AC have seen the use of task forces as an illustration of the Council's "increasing commitment to translate its science and assessment work into policy and action" (Fenge & Funston, 2015: 11), there have been also concerns about how the relationship between newly established bodies and working groups may lead to competition over limited human and financial resources (Supreme Audit Institutions of Denmark, Norway, The Russian Federation, 2015) and duplication of efforts. Another group of questions has related to the development of relationships between the AC and the bodies it led or facilitated the creation: the Arctic Economic Council (AEC), the Arctic Offshore Regulators' Forum,[15] and the Arctic Coast Guard Forum that were formed in 2014 and 2015. While all of them operate independently from the Council, their composition largely resembles that of the AC (for example, in the case of the AEC, each Arctic state and permanent participant organization of the AC can name up to three of its business representatives) and their chairmanships work in tandem with the rotation cycle of the Arctic Council.[16] They all also intend to provide information to the AC, serve as the synthesis of

---

[13] Examples include the Task Force on Short-Lived Climate Forcers (2009); Task Force on Search and Rescue (2009; Task Force on Arctic Marine Oil Pollution Preparedness and Response (2011); Task Force for Institutional Issues (2011); Task Force to Facilitate the Circumpolar Business Forum (2013); Task Force on Black Carbon and Methane (2013); and Scientific Cooperation Task Force (2013). There are currently no task forces.

[14] Next to the SAR and Oil Spills Agreement, the Scientific Cooperation Task Force agreed (before the ministerial meeting in Alaska) *ad referendum* on the text of a third legally binding agreement negotiated under the auspices of the AC, the Agreement on Enhancing International Arctic Scientific Cooperation. It was signed by representatives of Arctic states at the Ministerial meeting in Fairbanks, Alaska in May 2017.

[15] The Arctic Offshore Regulators' Forum had its roots in the work of the Task Force on Arctic Marine Oil Pollution Prevention (TFOPP) and the Framework Plan for Cooperation on Prevention of Oil Pollution from Petroleum and Maritime Activities in the Marine Areas of the Arctic that it produced. Among other actions, the Framework Plan enabled a strengthening of the co-operation of national regulators that resulted in the creation of the Forum.

[16] Arguably, the relationship between those bodies and the Council also depends on the position of the AC Chair on the given issue. For example, while Canada had a positive view of the Arctic Economic Council, for which it provided initial impetus, and remained in favour of forging close bonds with it, the United States held a much more reserved position toward engagement with the business community.

Arctic business perspectives for consideration by the Arctic Council (Arctic Economic Council), tap into the work of the EPPR working group of the AC (Arctic Coast Guard Forum), and complement the AC work in the field of offshore petroleum safety (Arctic Offshore Regulators' Forum).

In more recent years, the Council has had to deal with various challenges, especially related to the US Trump administration and its actions, both in the Arctic in general and in the Arctic Council in particular. The US chairmanship of the Council (2015–2017) that began during the second Obama administration was highly ambitious and backed with an abundance of resources. Among others, during the US term at the helm of the AC, in order to enable the Council to make decisions regarding marine issues, the idea of a marine commission was supported through the work of the Task Force on Arctic Marine Cooperation (TFAMC). In addition, the work commenced to develop a long-term strategy for the whole Arctic Council to guide the activities of all its component parts. So far, the Council has grown without any overall strategy to guide its activities, although individual working groups have adopted their respective strategies and frameworks. Following the second US chairmanship, Finland, as the AC chair 2017–2019, was tasked with developing the ideas for the marine cooperation subsidiary body and a long-term strategy for the Council, but this proved to be difficult during Finland's chairmanship. Gradually, the functioning of the Council was affected by the Trump Administration's priorities, which differed significantly from the Obama-era US chairmanship priorities. The US was one player in the group that was now against establishing the marine cooperation subsidiary body, and the idea was effectively rejected in the middle of Finland's chairmanship. Similarly, even though Finland tried to advance the adoption of the first ever long-term strategy for the Council, this too proved to be impossible because of the US stance against any reference to climate change. Given those circumstances, it did not make sense to attempt to adopt such a strategy at the time when the US – conversely to its own earlier position – was not willing to even acknowledge the phenomenon of climate change, the biggest driver of changes in the Arctic (Koivurova, 2019).

Finland's tumultuous chairmanship culminated with the Rovaniemi ministerial meeting in May 2019, which was preceded by a presentation from the US Secretary of State Mike Pompeo, just a day before the actual ministerial gathering. In his speech, Secretary Pompeo voiced the Trump administration's concerns against China and Russia in terms of their Arctic policy and presence in a manner that starkly contrasted with the casual tone of meetings of the AC and Arctic states. Similarly, in the ministerial meeting, the US adamantly resisted any reference to climate change in the ministerial declaration, resulting in an unprecedented non-adoption of a ministerial declaration. In fact, there were indications that the US stance toward the ministerial declaration caused problems that could have been sufficient to contest the functioning of the Arctic Council. In the end, however, the Council, Iceland as the next AC chair, and the Council's working groups received their mandates through the adoption of a joint statement and the adoption of the SAO report. Nonetheless, as Koivurova has argued elsewhere, the US position towards the Arctic Council and Arctic governance in general became increasingly controversial during the Trump era, and this stance posed a real challenge to not only the Arctic Council but also to international co-operation in the region in general (Koivurova, 2019).

In their joint statement, Arctic ministers, among others, instructed the Senior Arctic Officials to continue strategic planning of the AC and to review the roles of the ministerial meetings, the SAOs, and the permanent participants by the ministerial meeting scheduled for May 2021. The strategic planning has been envisaged as a two-part process to be started during Iceland's AC Chairmanship and continued during the consecutive Russian one (2021–2023). From mid-March 2020 onwards, however, the attention of the Arctic Council, as of all states and other international bodies has been nearly singularly drawn to the outbreak and challenges of the COVID-19 pandemic in the region, including the AC's own mode of operation and functioning. The pandemic has challenged international co-operation and diplomacy in general and Arctic Council co-operation in particular. As in many other international fora, where much of the diplomacy takes place outside of formal meeting rooms, the online form of organizing Arctic Council meetings has proven to be occasionally challenging, but the work has continued unabated nonetheless. Among others, the Council was able to act swiftly on the Covid-19 pandemic, and per the request of the SAOs, it produced a report on the pandemic situation in the regions of the circumpolar North and specified the actions taken against it (Covid report, 2020). Following the release of the study, the Council has pondered the follow-up to this report, including a likely increase in its own capacity to battle the pandemic situations in some way in the future.

Taken jointly, the present collection of various bodies and entities formed both within and outside the Council paints a far more complex picture than at the birth of the Council in the mid-1990s (Smieszek, 2020). At its 25th anniversary in 2021, the Arctic Council represents both continuity and change in the formalized Arctic co-operation of the last two decades. On one hand, the body has inherited many practices formed at the time of the Arctic Environmental Protection Strategy and in the early days of its own existence. These practices include a non-legal foundation for its work, a relatively strong position of the working groups that perform the majority of the AC's scientific and technical work, but also a precarious funding situation, until now based on grants and voluntary and in-kind contributions from some of the participating member states. On the other hand, the Council is currently considered the preeminent forum for political discussions on region-specific matters, firmly placed in the institutional landscape of Arctic governance. It has a permanent Secretariat, it catalyzes legally binding circumpolar agreements, and its meetings draw ever-increasing attention from much higher political levels. Concurrently, throughout its existence the Council has been subjected to critique and proposals for reform – both from within and externally.

## *Critique and Proposals for Reform*

By the second Ministerial meeting in 2000 in Barrow, Alaska, "much of the original excitement and enthusiasm associated with the Ottawa Declaration had been dampened" (Nord, 2016: 44) and soon after its establishment, the Council became a subject of criticism and a long series of reviews and reform proposals, both from actors engaged in the process and from outside observers (Fenge & Funston, 2015). First among the proposals was a request from Arctic Ministers to the SAOs at the meeting in Barrow to consider and recommend appropriate ways to improve the structure and functioning of the Arctic Council. In response to this request, Finland, as the consecutive chair of the Council (2000–2002), commissioned Pekka Haavisto, a former Finnish minister of environment, to prepare a study on the topic. The report, delivered in 2002 at the ministerial meeting in Inari, Finland, identified several limitations and challenges in the operations of the AC, including missing co-ordination of actions between the AC and other Arctic actors, deficient communication among the AC working groups and between them and the senior Arctic officials, competition for resources among the WGs, deficient outreach, and an imprecise definition of the role of observers (Haavisto, 2001).

In a similar vein, when Norway took over the AC chairmanship at the Council's tenth anniversary in 2006, it included among its priorities the review of the AC's structures to provide for regular evaluation of the institution and to consider ways of improving its efficiency and effectiveness (Norway's Ministry of Foreign Affairs, 2006). The proposal was also included in the joint program that Norway, Denmark, and Sweden announced in 2007 for their consecutive Arctic Council chairmanships, where among other common objectives for 2006–2012 they listed issues related to management of the Council (Norwegian, Danish, Swedish common objectives for their Arctic Council chairmanships 2006–2012). The efforts to strengthen the AC during that period corresponded directly with the international community's rapidly growing interest in both the Arctic and the Arctic Council's work, and these efforts involved, *inter alia*, a review of observers carried out by the Danish Chairmanship (2009–2011), as well as the aforementioned undertakings and deliverables of the Task Force on Institutional Issues (TFII, 2011–2013).

In addition to actions taken by the Council itself, non-governmental organizations and academics involved and interested in Arctic affairs have considered AC's shortfalls and presented proposals for addressing them (Graczyk, 2012; Smieszek, 2019). The concerns they raised included a soft-law profile of the Council and its basis in the declaration, not a treaty, which precludes imposition of legally binding obligations on its participants (Koivurova, 2009; Koivurova & Molenaar, 2010). The AC was also criticized for a lack of systematic evaluation and the absence of follow-up of guidelines produced within its framework, missing long-term and strategic policy of the institution, as well as exclusion of certain issues from its deliberations (such as fisheries or military security). The list encompassed difficulties in securing regular funding for the Council's projects and debated the place of the AC within the broader structures of Arctic governance and its exclusive character, leaving out voices of northern regions from the discussions.

Whereas some of the identified hindrances have been addressed over time, others have continued to impede the work of the Council and have been noted again by Danish, Norwegian, Russian, Swedish, and US institutions that collaborated on a multilateral audit of their national participation in the Arctic Council, the results of which were published in May 2015. The audit in the United States came ahead of the second US chairmanship, during which the US Government Accountability Office (GAO) examined matters related to US Council participation and the Council's organization in order to "position the United States for a successful Arctic Council chairmanship" (United States Government Accountability Office, 2014, p. 42). The audit was one of a series of measures taken by the US government to prepare for the Council's chairmanship, which was much altered in comparison to country's first chairmanship 1998–2000, both domestically and internationally. As the challenges ahead of the Arctic mount and the AC continues to operate in the increasingly dynamic international setting, calls for its reform and to enhance its efficiency and effectiveness are likely to continue (Balton & Ulmer, 2019).

## The Arctic Council on Its 25th Anniversary – Can the Structure of the Arctic Council Meet the Challenges of Tomorrow?

The COVID-19 pandemic is still with us as we celebrate the 25th anniversary of the Arctic Council in 2021. Russia is to assume the chairmanship of the Council from Iceland in May 2021and as it will lead the AC into the second quarter of the century of AC's existence, it is important to reflect on how much the Arctic Council has changed from the days of the AEPS and how much it is still captive to the foundations of the AEPS. It is also of significant interest to ponder how much the current structure of the Arctic Council can meet the challenges of tomorrow. The questions are: what has stayed the same in the Council and what has changed, and is this change enough to meet the challenges to come?

To begin with, there is a clear path dependency in terms of how Arctic intergovernmental co-operation has progressed from the adoption of the AEPS in 1991 and the Arctic Council in 1996 to the present. There are still essentially the same actor categories: member states, permanent participants,[17] and observers (Graczyk & Koivurova, 2014). Decisions are still made by the consensus of its members, after consulting with the permanent participants. There is still no legal foundation for the Council, as both the AEPS and the Arctic Council co-operation were commenced with the adoption of a declaration rather than a treaty, and the co-operation overall does not have any stable, continuous funding mechanism. The institutional forms

---

[17] Even though the status of permanent participants was formalized with the founding of the Arctic Council, at the 1996 Inuvik ministerial meeting of the AEPS they were already referred to as the AEPS permanent participants, before the adoption of the Ottawa declaration.

are still quite similar to those during the AEPS and the beginning stages of the Arctic Council. Chairmanship still plays a big role in the Arctic Council, even if the permanent secretariat's significance has changed it somewhat by securing institutional memory of the AC beyond ministries of foreign affairs of Arctic countries. The main political decision-making mechanisms, ministerial meetings, and SAO meetings are still the same, even though there have been changes in how they function. While working groups remain the main workhorses of the Council, there are now also other subsidiary bodies like the expert groups and temporary task forces. The Council still works for environmental protection and sustainable development in the region, and it has not taken security issues onboard, even though there have been proposals as such (Conley & Melino, 2016).

There have not been many actual structural changes to the Council either, so the clearest changes are not structurally related; rather, they are most evident in the Council's practical functioning. There are two more working groups compared to the AEPS, and, perhaps most importantly, the permanent secretariat was established, effective from 2013 (although there was already a semi-permanent secretariat in Tromsø during the three consecutive Nordic chairmanships). The Council has also seen the development of many new procedural guidelines and operational manuals, all of which were designed to streamline work of the Council, not only internally but also in its relations with the outside world (particularly related to the working groups) (Smieszek, 2020). Now there are also tested ways to establish new subsidiary bodies, expert groups and task forces, which have proven both complementary and, in some cases, overlapping with the work of the six working groups. Much discussion has centered on the outcome of the task forces, legally binding agreements, and how they have changed the Arctic Council from a decision-shaping to a decision-making body, even if this appears to be a rather simplistic view of those developments. Even if it is indeed correct that these legally binding agreements have been negotiated in these task forces, they are still independent from the Council, they have their own meetings of the relevant parties, etc. It is also noteworthy that the content of the first two agreements was mostly legally binding on the Arctic states, given that they were already parties to global legally binding agreements on search and rescue and oil spill preparedness (Byers, 2013, pp. 212–213). The science co-operation agreement did not lay out many concrete obligations and is considered to be fairly soft in character. It is also important to observe that despite the hype about task forces and their providing legally binding agreements to govern various aspects of Arctic affairs, at the point of writing this chapter, during the Icelandic chairmanship (2019–2021) the Council has not had any task forces, nor it is considering conceiving any new agreements.

One clear difference that can be noted is that there is arguably more interest now in making an impact on Arctic policy, and this has evidently manifested in the work of the Council. There are increasing numbers of observers, including non-Arctic states, and also an increase in the overall interest in participating in the work of the Council. This change has also been reflected in the ministerial meetings, as these are no longer left to the civil servants as in the past; now Arctic foreign ministers participate in AC ministerial meetings. In addition, the SAO's role vis-à-vis working groups has become more relevant as a guiding body compared to its former role.

It is clear from the above that Arctic inter-governmental co-operation has developed in an incremental manner, with each change built on the foundations that prevailed during the AEPS and the early days of the Arctic Council. Therefore, it is relevant to ask whether this type of council can meet future challenges, given that the region is changing with increasing rapidity.

Even though the Council's turning from a decision-shaping to decision-making body has been widely lauded, it does seem difficult (and for many it is even undesirable) for the Council to become a major regulator in the Arctic, given that its foundational elements do not allow it to take legal action and many of the region's major problems have their sources in the southern latitudes and beyond the circle of exclusively Arctic nations. With this structure, the Council can indeed catalyze legally binding agreements and organizations, and in particular it can advance influential scientific assessments of Arctic environment and societies, but it will be challenging to engage with more ambitious Arctic governance. The Council has had some impact on the workings of the global climate change regime via the ACIA (Duyck, 2015), but its overall contribution in this respect has been minimal. In a similar vein, the Council has conducted some soft work on the extractive industries, but these are regulated and governed on a local and sub-regional scale and are considered domestic in most Arctic states (Arctic Council, 2016).

Also, the Council does not appear to be a major actor nor even the principal venue for addressing matters related to Arctic governance in areas such as shipping, fisheries, climate change, or biodiversity (Young, 2016b). The AC refrains from identifying and communicating the related concerns and position of Arctic constituencies at, for instance, the United Nations Framework Convention on Climate Change (UNFCCC) conferences of the parties. Moreover, Arctic states themselves seem to prefer other fora for taking action concerning governance, as exemplified by the case of fisheries in the central Arctic Ocean (Young, 2016a), where discussions took place within the extended Arctic Ocean coastal states (A5) format, including China, Iceland, the European Union, Japan and South Korea (A5 + 5), and not in the AC Task Force on Arctic Marine Cooperation (TFAMC).

While there are many issues that fill up the Arctic agenda, one of the major future challenges for the Arctic Council is how it will respond to calls to strengthen its marine governance capacity. The Agreement to Prevent Unregulated High Seas Fisheries in the Central Arctic Ocean will enter into force soon,[18] and the recent negotiations to develop an implementing agreement under the United Nations Convention on the Law of the Sea (UNCLOS) on conservation and the sustainable use of marine biodiversity of areas beyond national jurisdiction (BBNJ), if successfully completed, would be of major relevance to the Arctic Ocean. In particular, the possible BBNJ convention could be viewed as posing a direct challenge to the AC, which already does a significant amount of work on marine biodiversity, also in areas beyond national jurisdiction. At the same time, it is clear that the Council cannot effectively manage and govern biodiversity in Arctic waters. Given that the current draft of the BBNJ seems to rely on regional bodies for its implementation, the

---

[18] As of August 2020, nine of the 10 signatories have completed the ratification process for the agreement, which will enter into force 30 days after receipt of the tenth and final instrument of ratification. The Agreement is to remain in effect for 16 years. It will be automatically be extended for additional five-year periods if the parties agree.

Council is presently exploring how it could make a positive contribution to governing biodiversity in areas beyond national jurisdiction (SAO marine mechanism presentations, 2020). It is clear that the Council has increased its capacity to tackle questions of Arctic marine governance, as exemplified in the establishment of the SAO marine mechanism during the Iceland's Chairmanship 2019–2021, but it remains to be seen whether this is enough for it to have a role in implementing the BBNJ convention, if/when it becomes a reality.

The Council has traveled in time with largely the same foundations as during the early days of the AEPS. It has clearly been able to incrementally adopt new ways to function within its limited foundations, but one may well question whether it can face the major challenges the region will confront in the years to come. Governing biodiversity in areas beyond national jurisdiction in the Central Arctic Ocean will be a litmus test for the Council as to whether it can find a way to tackle these challenges without revising its foundations. As for any inter-governmental forum or organization, the Council needs to consciously think about its place in Arctic governance, which is dominated by nation-state systems (and their sub-systems) and global governance forms. It is a hopeful sign that the Council has now adopted a long-term strategy, pointing to a conscious process whereby the Arctic states and permanent participants can seriously consider the best possible niche for the Council in the multilevel governance reality of the Arctic.

## Bibliography

AMAP. (1998). *AMAP assessment report: Arctic pollution issues*. Arctic Monitoring and Assessment Programme, https://www.amap.no/documents/doc/amap-assessment-report-arctic-pollution-issues/68

Arctic Council (1996, September 19). *Declaration on the establishment of the Arctic Council*. (1996). http://hdl.handle.net/11374/85

Arctic Council. (1998, September 17–18). *Arctic Council rules of procedure (as adopted by the Arctic Council at the first Arctic Council Ministerial meeting, Iqaluit, Canada)*, pp. 1–11.

Arctic Council. (2011). *Nuuk Declaration on the occasion of the Seventh Ministerial Meeting of the Arctic Council. 12 May 2011*, Nuuk, Greenland. https://oaarchive.arctic-council.org/handle/11374/92

Arctic Council. (2013a). Arctic Council Rules of procedure. Revised by the Arctic Council at the Eight Arctic Council Ministerial Meeting, Kiruna, Sweden, May 15, 2013 *the Eighth Arctic Council Ministerial Meeting, Kiruna, Sweden, May 15, 2013*, https://oaarchive.arctic-council.org/handle/11374/940

Arctic Council. (2013b). *Kiruna Declaration*, https://oaarchive.arctic-council.org/handle/11374/93

Arctic Council. (2016). Arctic Resilience Report. In M. Carson & G. Peterson (Eds.), *Arctic herald*. Stockholm Environment Institute and Stockholm Resilience Centre. http://hdl.handle.net/11374/1838

Arctic Environmental Protection Strategy. Declaration on the protection of Arctic environment. (1991). Rovaniemi, Finland, http://library.arcticportal.org/1542/1/artic_environment.pdf

Arctic Governance Project. (Ed.) (2010). *Arctic Governance in an Era of Transformative Change: Critical Questions, Governance Principles, Ways Forward*.

Balton, D. & Ulmer, F. (2019). *A strategic plan for the arctic council: Recommendations for moving forward*. Harvard Kennedy School Belfer Center for Science and International Affairs/Wilson Center: Cambridge, MA/Washington, DC.

Bloom, E. (1999). Establishment of the Arctic Council. *American Journal of International Law, 93*(3), 712–722. https://doi.org/10.2307/2555272

Borgerson, S. G. (2008, March/April). Arctic meltdown. The economic and security implications of global warming. Foreign Affairs, 87(2), 63–77.

Byers, M. (2013). *International law and the Arctic*. Cambridge University Press.

Conley, H. A., & Melino, M. (2016). *An Arctic redesign. Recommendations to Rejuvenate the Arctic Council*. Center for Strategic & International Studies. https://www.csis.org/analysis/arctic-redesign

Covid report. (2020). *Covid-19 in the Arctic: Briefing document for Senior Arctic Officials.* Arctic Council, 2020, June 24–25. https://oaarchive.arctic-council.org/handle/11374/2473.

Duyck, S. (2015). What role for the Arctic in the UN Paris Climate Conference (COP-21)?. In L. Heininen, H. Exner-Pirot, & J. Plouffe (Eds.), *Arctic yearbook*. https://arcticyearbook.com/arctic-yearbook/2015/2015-briefing-notes/157-what-role-for-the-arctic-in-the-un-paris-climate-conference-cop-21

English, J. (2013). *Ice and water. Politics, peoples, and the Arctic Council*. Penguin Group.

Fenge, T., & Funston, B. (2015). *The practice and promise of the Arctic Council. Independent report commissioned by Greenpeace.*

Graczyk, P. (2011). Observers in the Arctic Council – Evolution and prospects. In G. Alfredsson & T. Koivurova (Eds.), *Yearbook of polar law, 3* (pp. 575–633). Martinus Nijhoff Publishers.

Graczyk, P. (2012). The Arctic Council inclusive on non-Arctic perspectives: Seeking a new balance. In T. S. Axworthy, T. Koivurova, & W. Hasanat (Eds.), *The Arctic council: Its place in the future of Arctic governance* (pp. 261–305). The Gordon Foundation.

Graczyk, P., & Koivurova, T. (2014). A new era in the Arctic Council's external relations? Broader consequences of the Nuuk observer rules for Arctic governance. *Polar Record, 50*(3), 225–236. https://doi.org/10.1017/S0032247412000824

Haavisto, P. (2001). *Review of the Arctic Council structures. Consultant's study.* http://hdl.handle.net/11374/449

Ilulissat Declaration. (2008, May 27–29). *Arctic ocean conference*, Ilulissat, Greenland, https://cil.nus.edu.sg/wp-content/uploads/2017/07/2008-Ilulissat-Declaration.pdf

Kankaanpää, P., & Young, O. R. (2012). The effectiveness of the Arctic Council. *Polar Research, 31*, 1–14. https://doi.org/10.3402/polar.v31i0.17176

Keskitalo, E. C. H. (2004). *Negotiating the Arctic. The Construction of an international region.* Routledge. <–Tästä löytyi myös 2003-vuoden kirja?

Koivurova, T. (2009). Limits and possibilities of the Arctic Council in a rapidly changing scene of Arctic governance. *Polar Record, 46*(2), 146–156. https://doi.org/10.1017/S0032247409008365

Koivurova, T. (2019, December 11). Is this the End of the Arctic Council and Arctic Governance as we Know it? *Polar Connection*. https://polarconnection.org/arctic-council-governance-timo-koivurova/

Koivurova, T., Kankaanpaa, P., & Stepien, A. (2015). Innovative environmental protection: Lessons from the Arctic. Journal of Environmental Law, 27(2) (2013), 1–27. doi:https://doi.org/10.1093/jel/equ037.

Koivurova, T., & Molenaar, E. J. (2010). *International governance and regulation of the marine Arctic: Overview and gap analysis.* WWF International Arctic Programme. https://doi.org/10.4337/9781781009413.00012

Nilson, H. R. (1997). *Arctic Environmental Protection Strategy (AEPS): Process and organization, 1991–97. An assessment.* (Rapportserie no.103). http://hdl.handle.net/11250/173498

Nilsson, A. E. (2007). *A changing arctic climate. Science and policy in the Arctic Climate Impact Assessment* [Doctoral thesis, Linköping University]. Linköping University Electronic Press. http://urn.kb.se/resolve?urn=urn:nbn:se:liu:diva-8517

Nilsson, A. E. (2012). Knowing the Arctic: The Arctic Council as a cognitive forerunner. In T. S. Axworthy, T. Koivurova, & W. Hasanat (Eds.), *The Arctic Council: Its place in the future of Arctic governance* (pp. 190–224). The Gordon Foundation.

Nord, D. C. (2016). *The changing Arctic. Creating a framework for consensus building and governance within the Arctic Council*. Palgrave Macmillan.

Norway's Ministry of Foreign Affairs. (2006). *The Norwegian Chairmanship of the Arctic Council 2006–2008*. Retrieved September 4, 2015, from https://www.regjeringen.no/en/aktuelt/the-norwegian-chairmanship-of-the-arctic/id436983/

Norwegian, Danish, Swedish common objectives for their Arctic Council chairmanships 2006–2012 November 2007. https://oaarchive.arctic-council.org/handle/11374/2103

Pedersen, T. (2012). Debates over the role of the Arctic Council. *Ocean development & International Law, 43*(2), 146–156. https://doi.org/10.1080/00908320.2012.672289

SAO marine mechanism presentations. (2020). Retrieved at https://oaarchive.arctic-council.org/handle/11374/2537

Scrivener, D. (1999). Arctic environmental cooperation in transition. *Polar Record, 35*(192), 51–58. https://doi.org/10.1017/S0032247400026334

SDWG Strategic Framework (2017), *The human face of the Arctic: Strategic framework 2017*. Arctic Council. http://hdl.handle.net/11374/1940, https://oaarchive.arctic-council.org/handle/11374/1940.

Smieszek, M. (2019). Do the cures match the problem? Reforming the Arctic Council. *Polar Record, 55*(3), 121–131. https://doi.org/10.1017/S0032247419000263

Smieszek, M. (2020), "Steady as She Goes? Structure, Change Agents and the Evolution of the Arctic Council", The Yearbook of Polar Law, Volume 11, pp. 39–80.

Stone, D. P. (2016, November 29). *A short history of AMAP – 25 Years of connecting science to policy* [presentation]. AMAP 25th Anniversary Seminar, Helsinki, Finland.

Supreme Audit Institutions of Denmark, Norway, The Russian Federation, S. and the U.S. of A. (2015). *The Arctic Council: Perspectives on a Changing Arctic, The Council's Work, and Key Challenges. A Joint Memorandum of a Multilateral Audit on the Arctic States' national authorities' work with the Arctic Council*. Retrieved from https://www.riksrevisjonen.no/globalassets/reports/en-2014-2015/arcticcouncil.pdf

U.S. Geological Survey, U. (2008). *Circum-Arctic Resource Appraisal: Estimates of Undiscovered Oil and Gas North of the Arctic Circle* (Vol. 2000).

United States Government Accountability Office. (2014). *GAO-14-435 Report to Congressional Requesters. Better direction and management of voluntary recommendations could enhance U.S. Arctic Council participation*. https://www.gao.gov/products/GAO-14-435

Young, O. R. (1998). *Creating regimes: Arctic accords and international governance*. Cornell University Press.

Young, O. R. (2009a). The Arctic in play: Governance in a time of rapid change. *The International Journal of Marine and Coastal Law, 24*(2), 423–442. https://doi.org/10.1163/157180809X421833

Young, O. R. (2009b). Whither the Arctic? Conflict or cooperation in the circumpolar north. *Polar Record, 45*(1), 73–82. https://doi.org/10.1017/S0032247408007791

Young, O. R. (2016a). Governing the Arctic Ocean. *Marine Policy, 72*, 271–277. https://doi.org/10.1016/j.marpol.2016.04.038

Young, O. R. (2016b). The shifting landscape of Arctic politics: Implications for international cooperation. *The Polar Journal, 6*(2), 1–15. https://doi.org/10.1080/2154896X.2016.1253823

**Timo Koivurova** is a research professor at the Arctic Centre, University of Lapland, Finland

**Malgorzata (Gosia) Smieszek**, PhD, is a researcher and project coordinator at UiT The Arctic University of Norway in Tromsø and an Adjunct Fellow at the East-West Center in Honolulu, Hawaii. Her research focuses on international environmental and Arctic governance, science-policy interface, Arctic scientific cooperation, and gender-environment nexus. She is a co-founder of a non-profit "Women of the Arcticry".

# Chapter 22
# The European Union and Arctic Security Governance

Andreas Raspotnik and Andreas Østhagen

## Introduction

Northern military activity is at its highest level since the end of the Cold War. The idea of the Arctic as a 'zone of peace' has been replaced with a narrative about the top of the world as an emerging arena for great power rivalry. As a consequence, in 2020 the US Sixth Fleet conducted three maritime security operations in the Barents Sea, just off the Arctic coast of Norway and Russia. This followed warnings by then US Secretary of the Navy, Kenneth J. Braithwaite, about increasing hostility in the Arctic. Braithwaite noted: "The Chinese and the Russians are everywhere, especially the Chinese" (Humpert, 2020). This explicit name-calling in Arctic politics is a shift from more sober approaches to circumpolar geopolitics, where resource development, mitigating climate change, and ensuring harmonious regional relations were pronounced priorities. The Arctic is in a state of flux, literally due to global climate change, but also figuratively due to increased global awareness. As former Norwegian Foreign Minister Jonas Gahr Støre once famously put it: "Geography is changing – even though we cannot change geography" (Støre, 2012). The deterioration in the Arctic is not so much a response to events in the Arctic as a consequence of the challenges and trends emerging in global politics over the last decade. In the middle of this stands the so-called "great powers": China, Russia, the United States, and – increasingly – the European Union (EU).

A. Raspotnik (✉)
Nord University, Bodø, Norway
e-mail: araspotnik@fni.no

A. Østhagen
Fridtjof Nansen Institute, Lysaker, Norway
e-mail: ao@fni.no

Also, the EU has experienced substantial change over the past few years. Brexit might be the most tangible one of these, with yet-to-be-defined implications for the EU's identity and global status in general, and its Arctic presence in particular. However, the EU has also felt the need to gradually adapt its posture on the increasingly conflicted world stage, whether because of emerging great power rivalry, a changing transatlantic relationship, a more assertive China, or its continuous clashes with Russia (Blockmans, 2020; Østhagen, 2019). EU Member States have become more skeptical about China's global intentions, leading to a new convergence of EU-rope's assessment of the challenges China poses to the Union (Oertel, 2020). In a post-Crimea world, EU–Russia relations have shifted from fostering interdependence to managing vulnerabilities (Raik and Rácz, 2019). The Ukraine crisis has affected the EU's understanding of its role in international relations with diplomats in Brussels and EU-ropean capitals having started to embrace the idea that the EU must have a more strategic and geopolitical approach in its foreign policy (Nitoiu and Sus, 2019). Ursula van der Leyen spoke of a geopolitical Commission, and French President Emanuel Macron argued for a more decisive European Union with geopolitical awareness.

With these trends, shifts, and developments in mind, the present chapter aims to examine the EU's role in Arctic security governance. We first discuss the Arctic's security environment and three levels of Arctic geopolitics. In a second step, we place these developments in an EU–Arctic context and ponder on how a Union of 27 states – only three of them Arctic, but most of them non-Arctic – will continue to engage with the Arctic region. How does the EU's Arctic endeavor fit with the emerging security concerns – related to great-power relations – in the Arctic? What role is there, if any, for the EU in such matters?

We rely on three levels of political dynamics in the Arctic to better understand security developments in the Arctic and the EU's role herein. We distinguish between the international (system level), the regional (Arctic) level, and sub-regional (national) level. Such an approach helps bring out the different dynamics that exist in the Arctic; it explains why ideas of conflict persist and why this is not necessarily at odds with ideas of regional cooperation and stability; and it allows us to examine what role the EU actually has, and can play, at these different levels of security dynamics in the North.

## Setting the Stage

### *Regional Relations in the Arctic*

Let us start with the regional relations among the eight Arctic states: Canada, Denmark (via Greenland), Finland, Iceland, Norway, Russia, Sweden, and the United States. Generally, interactions between geographically proximate states, whether positive or negative, will be increasingly intense, compounding over time

(Hoogensen, 2005). This can translate into regional security dilemmas informed by shared histories (Buzan & Wæver, 2003, p. 46).

Regionally, the Arctic states have recognized the value of creating a political environment that is favorable to investments and economic development. As the Cold War's systemic overlay faded, regional interaction and cooperation in the North started to flourish. Several organizations, such as the Arctic Council (AC), the Barents Euro-Arctic Council (BEAC), and the Northern Forum, emerged in the 1990s to tackle issues such as environmental degradation, regional and local development, and cultural and economic cross-border cooperation.

The decision to deliberately exclude military security issues led to the emergence of a plethora of cooperative arrangements between the Arctic countries in different constellations without getting bogged down in the security concerns at the time. Whereas interaction among Arctic states increased during this period, and also included Arctic indigenous peoples (as they gained more political visibility and an official voice), geopolitically the region seemed to disappear from the radar of global power politics. This allowed the circumpolar countries to recognize the value of creating a political environment that was favorable to investments and economic development, giving rise to the idea of the Arctic's political dynamics as exceptional (Exner-Pirot and Murray, 2018; Wilson Rowe, 2020).

The region was thrown back onto the international agenda in the early 2000s due to the increasingly apparent effects of climate change and the economic and political changes that followed. Arctic ice sheets were disappearing at an accelerated pace, which coincided with new prospects for offshore oil and gas exploration, as well as the opening of shipping lanes through sensitive areas such as the Northwest Passage and Northern Sea Route.

In the wake of this, environmental organizations and politicians outside the region led an outcry about the 'lack of governance' in the Arctic (Hoel, 2009).[1] In response, top-level political representatives of the five Arctic coastal states met in Ilulissat, Greenland, in 2008, where they publicly declared the Arctic to be a "region of cooperation" (Exner-Pirot, 2012). They also affirmed their intention to work within established international arrangements and agreements, in particular the United Nations Convention on the Law of the Sea (UNCLOS), an international agreement that binds states in shared pursuit of order, cooperation, and stability at sea.

Since then, the Arctic states have repeated the mantra of cooperation, articulating the same sentiment in relatively streamlined Arctic policy and strategy documents. The deterioration in relations between Russia and its Arctic neighbors since 2014 – a result of Russian actions in eastern Ukraine and the Crimean Peninsula – did not change this (Byers, 2017), although security and military concerns now occupy more space in Arctic discussions than ever. Indeed, the foreign ministries of all AC members, including Russia, keep pro-actively emphasizing the peaceful and

---

[1] *See* Greenpeace's Save the Arctic Campaign: https://www.peoplevsoil.org/en/savethearctic/

cooperative nature of regional politics (as long as this does not include security concerns) (Heininen et al., 2020; Wilson Rowe, 2020).

Some have also argued that low-level forms of regional interaction help ensure low tension in the North, despite not dealing with security matters (Keil and Knecht, 2016). The emergence of the AC as the primary forum for regional affairs in the Arctic plays into this setting (Rottem, 2017). The AC serves as a platform from which its member states can portray themselves as working harmoniously toward common goals (Exner-Pirot, 2015). Adding to its legitimacy, an increasing number of states have applied for and gained observer status on the AC: initially France, Germany, the Netherlands, Poland, Spain, and the United Kingdom; and, more recently, China, India, Italy, Japan, Singapore, South Korea, and Switzerland.

## *Global Power Politics Through the Arctic*

What happens *in* the Arctic is not the same as international global politics *through* the Arctic. During the Cold War, the Arctic held a prominent place in the political and military standoffs between the two superpowers. It was important not because of interactions in the Arctic itself (although cat-and-mouse submarine games took place there), but because of its wider strategic role in the systemic competition between the United States and the USSR. The Arctic formed the buffer zone between these two superpowers, its airspace comprising the shortest distance for long-range bombers to reach one another's shores.

From the mid-2000s onwards, the Arctic regained strategic geopolitical importance. A repeat of Cold War dynamics saw Russia strengthen its military (and nuclear) prowess in order to reassert Russia's position at the top table of world politics. Given the country's geography and recent history, its obvious focus would be its Arctic lands and seas. In this terrain, Russia could pursue its policy of rebuilding its forces and expanding its defense and deterrence capabilities in an unobstructed manner (Hilde, 2014, pp. 153–5).

This build-up did not occur primarily because of changing political circumstances in the Arctic, but because of Russia's naturally (that is, geographically) dominant position in the North and its long history of a strong naval presence, the Northern Fleet, on the Kola Peninsula. This is where Russia's strategic submarines are based, which are essential to the country's status as a major global nuclear power (Sergunin and Konyshev, 2014). Melting of the sea ice and increased resource extraction on the coast along the Northern Sea Route are only some of the elements that have spurred Russia's military emphasis in its Arctic development efforts: Russia's north matters for the Kremlin's more general strategic plans and ambitions in world politics. However, as we explain further in the next section, for Canada, the effects of these efforts are limited. It is the Nordic states that stand to bear the brunt of the emerging challenges.

Within these shifting geo-economic and geo-strategic dynamics, China has also emerged as a new Arctic actor, proclaiming itself as a "near-Arctic state"

(Kopra, 2013). With Beijing's continuous efforts to assert influence, the Arctic has emerged as the latest arena where China's presence and interaction are components of an expansion of power, whether through scientific research or investments in Russia's fossil fuel industries (Bennett, 2017; Koivurova & Kopra, 2020; Sun, 2014). This has led to the Arctic becoming relevant in a global power competition between China and the United States. In 2019, US Secretary of State Mike Pompeo warned that Beijing's Arctic activity risks creating a "new South China Sea" (US Department of State, 2019).

The sudden realization by the White House that Greenland occupies a strategically significant position, and that the United States has a military base there (Thule) links to strategic concerns and fears over Chinese investments in Greenland (Lulu, 2018). Although these concerns have failed to materialize on any great scale, a 2019 tweet by former US President Donald Trump about buying Greenland was not a coincidence. The US Government's reopening of its consulate in Nuuk, Greenland's capital, demonstrates how the US position on China as a strategic rival in the Arctic does indeed have an impact, and is yet another arena where systemic competition between the two countries is increasing.

While Arctic states continue to highlight cooperative traits in the region and positive regional affairs, politics between the great powers of China, Russia, the United States and (to some extent) the EU, increasingly impact on Arctic affairs. On one hand, tensions between NATO and Russia have an Arctic/North Atlantic component, as seen with an increasing number of military exercises in the area since 2014. On the other hand, the Trump administration's decision to challenge China globally has also led to a tougher stance against China in the Arctic, at least rhetorically. This suggests the need to distinguish between intra-regional dynamics in the Arctic, and the spill-over effect of events and power struggles elsewhere on Arctic issues.

## *Sub-regional Concerns Within the Arctic*

There is one further political dynamic that requires examination: interactions between Arctic states in close proximity to each other. If we truly want to understand the foreign- and security policy interests and motivations of specific states, it is insufficient to only examine the systemic (overarching) international level, or even the specific 'regional security complex' that each state belongs to. We must also look at national – sub-regional – interest formation and specific bilateral relations, where states interact on a daily/regular basis.

Central here is the role the Arctic plays in considerations of national security and defense. This varies greatly amid the Arctic Eight, because each country prioritizes and deals with its northern areas differently. For Russia, with its vast Eurasian domain, the Arctic is integral to broader national defense considerations (Sergunin, 2014). Even though these considerations are also linked to developments elsewhere, investments in military infrastructure in the Arctic have a direct regional impact, particularly for the much smaller countries in its western neighborhood: Finland, Norway, and Sweden.

For these three Nordic countries, the Arctic is fundamental to national defense policy, precisely because this is where Russia invests considerably in its military capacity (Jensen, 2017; Saxi, 2011). Especially Norway, a founding member of NATO located on the alliance's 'northern flank', is increasingly concerned with the expansive behavior of the Russian military in the North Atlantic and Barents Sea (Hilde, 2019).

Arguably, the Arctic does not play the same pivotal role in national security considerations in North America as it does in northern Europe. Even while pitted against the Soviet Union across the Arctic Ocean and Bering Sea during the Cold War, Alaska and northern Canada were primarily locations for missile defense capabilities, surveillance infrastructure, and a limited number of strategic forces (Østhagen et al., 2018).

The geographical dividing line falls between the European Arctic and the North American Arctic, in tandem with variations in climatic conditions. The north Norwegian and the northwest Russian coastlines are ice-free during winter. However, even though the ice is receding, it remains a constant factor in the Alaskan, Canadian, and Greenlandic Arctic. Due to the sheer size and inaccessibility of the region, the impact of security issues on either side of the dividing line is relatively low.

In conclusion, security and essentially defense dynamics in the Arctic remain anchored at the subregional and bilateral levels. Of these arrangements, the Barents Sea/European Arctic stand out. Here, bilateral relations between Russia and Norway are especially challenging in terms of security interactions and concerns. Despite rhetoric to the contrary, Russian investments in Arctic troops and infrastructure have had little impact on the North American security outlook. Approaches by Russian bombers and fighter planes may cause alarm, but the direct threat to North American states in the Arctic compared to that of its Nordic allies is limited. This is also why Canadian troops have been exercising in the Norwegian Arctic in recent years, and not vice versa.

However, bilateral dynamics like in the case of Norway and Russia are multifaceted, as the two states also engage in various types of cooperation, ranging from co-management of fish stocks to search-and-rescue operations and a border crossing regime. In 2010, Norway and Russia were able to resolve a longstanding maritime boundary dispute in the Barents Sea, partly in order to initiate joint petroleum ventures in the disputed area (Moe et al., 2011). These cooperative arrangements and agreements have not been revoked following the events of 2014 (Østhagen, 2016; Rowe, 2018), a clear indication of the complexity bilateral relations in the Arctic.

Clearly, the Arctic is not simply *one* or *the* Arctic but essentially defined by many *Arctics* and several different levels of Arctic relations. Where, then, does the European Union fit in this three-level separation of northern security dynamics and related governance? To start unpacking that, we must begin with the EU's quest for an Arctic policy.

## The European Union and Its Decade of Arcticness

Connoisseurs of the EU's past Arctic endeavors are well aware of the region's marginal importance in day-to-day EU-ropean political life. Although a dedicated set of Arctic-related documents has been developed by the EU's main institutions since 2007–2008, the region has not yet gained a prominent place in the hallways of Brussels. Ever since 2008, the EU and its various institutional actors have slowly but steadily developed a dedicated EU Arctic policy, setting common positions, stressing the Union's Arctic credentials and prominently expressing its very own 'Arcticness' (Raspotnik, 2018).

Essentially, these documents identify the EU as part of and linked to the Arctic, affecting and affected by the Arctic (Stępień & Raspotnik, 2019, p. 1). To date, the list of EU Arctic policy documents includes 13 policy documents, *see* Table 22.1. Additionally, the Arctic region has also been cross-referenced in, *inter alia*, the Integrated Maritime Policy of 2007, the Maritime Security Strategy of 2014 and, most recently, in the 2016 Global Strategy on Foreign and Security Policy for the European Union (Commission of the European Communities, 2007; Council of the European Union, 2014; High Representative, 2016).

At the time of writing, the most recent policy statement from the European Commission and the High Representative (HR) was the Joint Communication on an integrated European Union policy for the Arctic, published in 2016. The document was an attempt to emphasize the most important areas of EU engagement around three broad themes: (1) climate change and safeguarding the environment; (2) sustainable development in the (European) Arctic, and (3) international cooperation on Arctic issues (European Commission & High Representative, 2016). The 2016 Joint

**Table 22.1** The European Union's Arctic Policy Milestones (2008–2021)

| Year | Document |
|---|---|
| 2008 | **EP Resolution** on *Arctic governance* |
|  | **Commission Communication** on *The European Union and the Arctic region* |
| 2009 | **Council Conclusions** on *Arctic issues* |
| 2011 | **EP Resolution** on *A sustainable EU policy for the High North* |
| 2012 | **Commission and High Representative Joint Communication** on *Developing a European Union Policy towards the Arctic Region: progress since 2008 and next steps* |
| 2014 | **EP Resolution** on the *EU strategy for the Arctic* |
|  | **Council Conclusions** on *Developing a European Union Policy towards the Arctic Region* |
| 2016 | **Commission and High Representative Joint Communication** on *an integrated European Union policy for the Arctic* |
|  | **Council Conclusions** on *the Arctic* |
| 2017 | **EP Resolution** on *an integrated EU policy for the Arctic* |
| 2019 | **Council Conclusion** on the *EU Arctic policy* |
| 2021 | **Commission and High Representative Joint Communication** *on a stronger EU engagement for a peaceful sustainable and prosperous Arctic* |
|  | **EP Resolution** on the Arctic: opportunities, concerns and security challenges |

Source: Own compilation based on Raspotnik (2018, p. 93)

Communication received strong political support as it was launched by HR Federica Mogherini and the Commissioner for Maritime Affairs and Fisheries, Karmenu Vella. Nonetheless, the Communication remained primarily an overview of existing policies and actions, with only a few aspects being future-oriented (Stępień and Raspotnik, 2019).

For instance, the EU proposed that the European northernmost regions work on key investment and research priorities for the Arctic, as well as launch a new EU-Arctic meeting place, the annual Arctic Stakeholder Conference. However, the lack of a long-term vision, a limited number of future-oriented actions, and changing geopolitical circumstances led to calls to adopt a more ambitious Arctic policy framework, urging the Commission and the HR to propose a new policy (European Political Strategy Centre, 2019). The first signals towards another recalibration of the EU's Arctic policy are currently set with a new Joint Communication aimed to be issued in autumn 2021. It can be assumed that an updated EU Arctic policy will remind an international audience of the Union's Arctic objectives and competences, and will be built around the three familiar themes of climate change, sustainable development, and international cooperation.

Generally, the EU's decade-long involvement in the Arctic can be characterized by ambivalence, with 'the Arctic' essentially residing within the realm of 'soft (security) policy' – not written into the Treaties, with no distinct budget line and no set rule book on how (or what) to protect (in) the Arctic (Raspotnik, 2020). Clearly, the EU is an Arctic actor and no stranger to its northern backyard. It has an obvious presence in the north in terms of geography, legal competence, market access, or its environmental footprint and contribution to Arctic science. Its Member States Denmark (on behalf of Greenland), Finland, and Sweden are located in the region and the EU has close relationships with the other five Arctic states. Moreover, the EU's economy and population also affects the region via an environmental and climate footprint, as well as its market influence, essentially contributing to the demand for Arctic resources. Additionally, EU policies – such as climate change mitigation efforts, clean air policy, or raw materials strategies – have an impact on these environmental and economic footprints (Raspotnik and Stępień, 2020, p. 132).

However, three factors have made the EU's efforts to become constructively involved in the Arctic both controversial and complex. The first is its lack of direct access to the Arctic Ocean, which seems to be key of conventional 'Arcticness' (Dodds, 2012). The second is its slightly paternalistic Arctic policy statements, which portray the EU as part of the 'solution' to the region's real or perceived challenges without sufficiently taking into considerations Arctic sensitivities, often dominated by special interest groups concerned with climate change, animal protection, or economic ventures. The third factor is the sustained difficulty of finding a convincing Arctic narrative that would attract broader attention throughout the Member States (Østhagen, 2019; Raspotnik, 2018; Stępień and Raspotnik, 2019).

In geographical and legal terms, the EU is, by virtue of its Member States and its relations via the European Economic Area (EEA), represented at the AC, either via the AC's member states (Denmark, Finland, Iceland, Norway, and Sweden) or its observers (France, Germany, Italy, Poland, the Netherlands, and Spain) (Stępień and Koivurova, 2016).[2] However, international emphasis concerning an active and politically participative role of the EU in the Arctic has been predominantly placed on the geographical fact that the EU does not have an actual European shoreline on the Arctic Ocean (Koivurova et al., 2012), which has been the case since Greenland withdrew from the European Economic Community in 1985.

Thus, the EU's externality regarding the majority of Arctic states represents a major constraint on the EU's 'Arcticness'. The EU has no coastline to the Arctic Ocean, and EU law applies in the Arctic directly only to Finland and Sweden, and via the EEA Agreement, to Iceland and Norway (excluding the Archipelago of Svalbard). Hence, foreign policy plays an essential role in respect to EU Arctic activities. This includes, for instance, the EU's cooperative efforts with Russia in the European Arctic, and its engagement within the AC (Raspotnik and Stępień, 2020).

In 'solution'-oriented terms, it was particularly the EU's ban on seal products that led to controversial political and legal debate in Arctic international circles, initially affecting the EU's applications for AC observer status (Raspotnik, 2018). In regional engagement terms, the EU's regional commitment has fluctuated over the last decade as more pressing issues have arisen on EU-rope's agenda (Raspotnik and Stępień, 2020). The fact that the Arctic is largely only of peripheral concern for EU policymakers leaves the Arctic as a niche policy domain, dominated by special interests ranging from environmental protection to fisheries to regional development in Northern Fennoscandia. In a way, it is the democratic deficit turned on its head. Few pay attention to one-off controversial statements by national politicians in Arctic states, especially when these are non-official or issued in Russian or a Scandinavian language. Yet, as soon as Members of the European Parliament state that a moratorium on Arctic oil and gas should be implemented, strong northern reactions are guaranteed.

Other strong reactions to the EU Arctic policy relate to the position of the EU as a market for Arctic resources. It may be about the – implemented – ban on seal products or about the remote possibility of setting special standards for Arctic resources imported into the single market. In all such cases, the concern of Arctic actors is that a solely environmentally focused EU Arctic policy would result in limitations of placing Arctic resources on the EU market. At the same time, the produce of other regions – unaffected by the environmental symbolism associated with the Arctic – would be free from such limitations and take place of current and future Arctic resources in the EU market (Raspotnik and Stępień, 2020).

---

[2] On November 9, 2020, Estonia submitted an application for AC observer status (Estonian Ministry of Foreign Affairs, 2021). Ireland also submitted an application in January 2021. Both applications will be considered at the next Arctic Council Ministerial meeting in May 2021.

## Analysis: Three Arctic Levels and the European Union

Returning to the three levels of Arctic geopolitics, how can we make sense of the EU's decade-long Arctic engagement and its involvement in Arctic security governance?

Starting with the *international (system) level*, this is perhaps where the EU has the most obvious role to play. Recently, the tensest relations with an Arctic impact have concerned the growing hostility between what some refer to as the new 'two poles': the United States and its perceived challenger China (Tunsjø, 2018). Some scholars have stressed the anarchic state of the international system, where relative power considerations and struggles determine the path taken by states and thus inexorably lead to conflict (Tripp, 2013; Tunsjø, 2018; Waltz, 1959). However, such analyses focused on relative power do not have to become self-fulfilling prophecies. After all, measures can be taken to alleviate concerns and possible rivalry at the international level, by cooperation, by putting in place agreements, or by developing joint institutions, thereby fostering greater trust.

China's increasing global engagement and influence has, thus far, been rather subdued in the North. For all its rhetoric about its in interests in a 'Polar Silk Route', Beijing has used all the correct Arctic buzzwords about cooperation and restraint in tune with the preferences of the Arctic states (State Council of the People's Republic of China, 2018). However, there are legitimate fears that this may be just be a mollifying tactic – merely the beginning of a more assertive Chinese presence where geo-economic actions; that is, financial investments with motivated by geopolitical goals (Sparke, 1998) – are part of a more ambitious political strategy aimed at challenging the hegemony of the 'West' and also the balance of power in the North (Brutschin and Schubert, 2016; Lanteigne, 2015). The Arctic speech by Secretary of State Pompeo in 2019 fed directly into this narrative (US Department of State, 2019). The question is whether Chinese actions in the region are meant to challenge the United States by an engagement that appears to assume predominantly softpower characteristics (Lulu, 2018). At the same time, shifting power balances and greater regional interest from Beijing need not lead to tension and conflict; on the contrary, they might spur efforts to find ways of including China in regional forums, alleviating the (geo-economic) concerns of Arctic states (Sverdrup-Thygeson and Mathy, 2020).

The other Great Power with global (international) status as much as Arctic influence is of course Russia, which is, by nature, an Arctic state, unlike China. As the largest country of the circumpolar region (by far) and the most ambitious in terms of military investments and activity, Russia sets the parameters for much of the Arctic security trajectory. This is not likely to change, although exactly what the future Arctic security environment will look like depends on the West's response to Russian actions predominantly taking place in other regions around the world.

Here, the EU has a role to play. In an Arctic context, considerations on matters of security have a long history for the EU (Airoldi, 2020). Both the establishment of the BEAC in 1993 and the introduction of the Northern Dimension (ND) aimed to foster relations with Russia in order to mutually tackle a broad range of security challenges in the European Arctic. Over the last few years, however, hard security issues have only been mentioned in a general, implicit way: the strengthening of low-level regional and multilateral cooperation, the allegiance to international legal order, and the vision of a cooperative Arctic that is not affected by any spill-over effects.

The Global Strategy of 2016 took the same line, highlighting the Arctic as a potential venue of selectively engaging with Russia (High Representative, 2016). However, the EU is increasingly aware of the Arctic's changing geo-political dynamics and the need to address those in light of regional and global security considerations (Coninsx, 2019). Also, both Germany's updated Arctic policy of August 2019 (The Federal Government of Germany, 2019) and France's Defense Policy for the Arctic of October 2019 (Ministry for the Armed Forces, 2019) specifically responded to the changing security aspects of the Arctic. In its new Arctic Strategy from October 2020 (Ministry for Foreign Affairs of Sweden, 2020), Sweden urged the EU to identify its strategic interest in the Arctic given a changing security policy environment. Similarly, Poland will highlight the rise of political tensions in the Arctic as having consequences for international cooperation and dialogue in its upcoming Polar Policy.

Taking up von der Leyen's challenge for the EU to become a more 'geopolitical' actor, the Arctic of the twenty-first century could be an area where the EU could play to its strength. Geopolitics is actually nothing new to the European Union. Over the last years, the EU has steadily developed a tacit geopolitical discourse, exhibiting international ambitions alongside its own conceptualization of world order, core values, rule of law, and good governance (Raspotnik, 2018). From civilian to regulatory or market power, the labels of such geopoliticized European Union are plentiful.

Also, the Global Strategy changed the EU's own perception to that of a power broker, keen on defending its own interest, insisting on principled pragmatism in foreign policy and strengthening third countries' resilience. As such, the European Arctic could be an area where the EU could seek talks with Russia based on its continuing northern cross-border bond and a potential willingness on the part of Russia. The peripheral Arctic might be the area where the EU and Russia find common ground again, not only improving their relationship but also promoting Arctic stability. Using the Arctic as arena for renewed relations, or at least talks thereof, might have positive spill-over effects on other areas of dispute.

In other words, an EU Arctic policy approach that separates regional from systemic components, and sustainable development and environmental protection from questions of hard security, offers the opportunity to delineate clear and ambitious goals for the EU's Arctic involvement. This would start with specific discussions on Chinese localization tactics in Greenland (Lulu, 2018) to the possibility of the Arctic being an important regional prologue to the currently developed Strategic Compass, an initiative to strengthen the EU's Common Security and Defense Policy (CSDP) and inject coherence into European security and defense (Raspotnik, 2020; Scazzieri, 2020).

Turning specifically to the *regional security dynamics* and the EU's role, the central question in the Arctic, in terms of national security concerns, is the extent to which developments occurring at a regional level can be insulated from events and relations elsewhere. If the goal is to keep the Arctic as a separate, exceptional region of cooperation, the Arctic states have managed to do a relatively good job, despite setbacks due to the Russian annexation of Crimea in 2014.

The most pressing regional challenge, however, is how to deal with, and talk about, Arctic-specific security concerns, which are often excluded from cooperative fora and venues. The debate about what mechanisms are best suited for further expanding security cooperation has been ongoing for a decade (Conley et al., 2012). Debates have concerned whether the AC could acquire a security component (Graczyk & Rottem, 2020), whereas others look to the Arctic Coast Guard Forum or other more ad hoc venues (Wilson Rowe et al., 2020b). The Northern Chiefs of Defense Conference and the Arctic Security Forces Roundtable were initiatives that were established to this end in 2011/2012 (Depledge et al., 2019), but fell apart after 2014.

Any Arctic security dialogue is fragile, and risks being overshadowed by the increasingly tense NATO–Russia relationship in Europe at large (ref. spill-over effect from the international 'level' to the regional). Precisely what such an arena for dialogue is intended to achieve (that is, preventing the spill over of tensions from other parts of the world into the Arctic) is the very reason why progress here is so difficult. For the same reason, it is difficult to imagine a more pan-Arctic political role for NATO.

The EU's role in these endeavors is limited. Although brimming with relevant competence and experience for an expansion in such efforts, the Arctic states – Russia in particular, but also Canada, Norway and the United States – are sensitive to EU interference in what is seen as internal/regional issues. Even Denmark – the only EU member with a direct link to the Arctic Ocean – has chosen to opt out of the EU's CSDP.

Still, the EU's role as a global maritime security provider, promoting maritime multilateralism and the rule of law at sea, while also acting as dominant economic maritime actor, could lead to a stronger EU voice in Arctic affairs as well. This could particularly be the case if there is further emphasis on the Union's space 'actorness' and its related centerpieces: the Galileo satellite navigation system and the Copernicus earth observation satellite system; two infrastructure assets that bring significant degrees of strategic autonomy. Space governance and Arctic governance are closely related. Satellite navigation and earth observations are essential for operating in the Arctic, from civilian traffic to military operations, and for grasping the climatic changes that are transforming the Arctic (Wilson Rowe et al., 2020a).

The 2016 Joint Communication essentially refers to the measures for effective stewardship of the Arctic Ocean that were also advocated in the Maritime Strategy (Airoldi, 2020, p. 343). As such, the Commission was part of the 2018 agreement creating a moratorium on commercial fisheries in the Central Arctic Ocean. Here, the EU is a signatory alongside Canada, Denmark, Norway, Russia, United States, China, Iceland, Japan, and South Korea. Moreover, the EU has extensive regulations that are relevant to soft security in the maritime Arctic. These cover pollution from ships, rules for ship inspection, port state control, support for the implementation and improvement of the Polar Code, or contribution activities on marine oil pollution, black-carbon reduction, and search and rescue (including cooperation between respective coast guard institutions) (Airoldi, 2020, p. 343).

In many ways, the *regional level* is the one where the EU has had the most extensive Arctic engagement, albeit not with a strictly security dimension. As previously mentioned, the EU has invested heavily, *inter alia*, in Arctic research and regional development programs. The regional and sub-regional level of the European Arctic policy space is characterized by direct application of EU laws and policies or the operation of EU cross-border and intra-regional programs. The geographical definition of the European Arctic – defined as a region stretching from Greenland to northwest Russia – is fluid from the EU's perspective.

Especially central here, as per the levels outlined previously, is Russian military engagement and the effect of this in the various sub-regions of the Arctic. Even if old Russian bases are revived and new ones are built along its Northern shoreline and islands, its emphasis is concentrated in the North Atlantic/Barents Sea portions of the wider circumpolar area. This is where the *sub-regional level* comes into play. Geographic proximity does play a role. After all, neighbors are forced to interact regardless of the positive or negative character of their relations. In turn, centuries of interaction compound and form historic patterns that influence relations beyond the immediate effects of other crisis and developments, on regional or global levels.

For the EU, the most severe security challenge in its northern near-abroad is the Barents region and interactions between NATO-member Norway and Russia. On one hand, the two countries collaborate on everything from dealing with environmental concerns to cultural exchange and border crossings, independent of events elsewhere. Moreover, the regional upsurge in Arctic attention around 2007/2008 had a positive impact on bilateral relations. In 2010, a new 'era' of Russo-Norwegian relations was announced (Ims, 2013; Lavrov and Støre, 2010), after various forms of bilateral cooperation had been established as the Cold War receded.

On the other hand, these bilateral relations are also at the behest of power asymmetry, rivalry, and the tendency of states to revert to power balancing (for example, via alliance systems). Moreover, they are influenced by international events. When events in Ukraine brought a deterioration in NATO–Russia relations, Norway–Russia relations were negatively affected (Friis, 2019; Østhagen et al., 2018). Since then, mistrust and accusations of aggressive behavior have returned, reminiscent of Cold War dynamics.

In this relationship, the EU's role is limited, even unwanted. When looking at specific security issues, the EU's own set-up and lack of a clear defense and security policy has hindered any clear regional security role of the EU. However, the Union's limited *regional* role might be misleading, as the EU actually has a broad toolbox of regional competences, expertise, and initiatives at its disposal. This 'EU Arctic spectrum of capabilities' could serve as a framework for a currently developed and to-be-updated policy and – if properly implemented – act as a trigger to a more confident and trustworthy relationship with Russia.

A framework starts with concepts on small but nevertheless important confidence-building measures, such as search and rescue efforts and cross-border environmental cooperation; possibly extending to tougher cooperation nuts to crack, such as the salvage of nuclear submarines. This 'spectrum structure' would offer the possibility for the EU to be the region's honest broker and to act in the Arctic without artificially fueling conflict narratives or being perceived as an Arctic security actor (Raspotnik, 2020).

## Conclusion

In this chapter, we have employed a stylized separation of Arctic security involving three different levels – the regional, the international, and the sub-regional – to tease out the different national security dynamics in the Arctic, and the European Union's role therein. Such an approach helps to bring out the different dynamics that exist in the Arctic and to not fall into an either-or trap; either the Arctic is a region of multilateral cooperation or it is a region of (emerging) conflict. This approach explains why ideas of conflict persist and why this is not necessarily at odds with ideas of regional cooperation and stability. Further, it allows us to examine what role the EU actually has, and can play, at these different levels of security dynamics in the North.

Clearly, the EU is an Arctic actor and multidimensional regional stakeholder, part of and linked to the Arctic, affecting and affected by the Arctic region. However, hard security has not yet been the EU's Arctic pet issue, and has mainly been tackled as constant repetitions of allegiance to the international/Arctic legal order or the articulated vision of keeping the region a low-tension area. However, if our three-level analysis is applied, it becomes evident that the EU's Arctic security role might not be less limited as often analyzed and perceived.

On the *international (system) level*, the civilian, regulatory and market power that is the European Union has the opportunity to set some of the agenda in global politics and help shape politics concerning the emerging China–US rift and the ongoing NATO–Russia tension. Such efforts can also have an Arctic component and impact Arctic relations. On the *Arctic regional level*, limitations on regional influence are indeed given, yet the EU's global maritime role might offer the potential for further involvement, especially in combination with the Union's space capacities. The EU as a space actor, owner, and operator of significant infrastructure can make important contributions to Arctic communities relating to communications, data-sharing, and creating global attention to the findings of earth observation.

On the *sub-regional (national) level*, the EU's role is perhaps the most limited, albeit with its Arctic member states Denmark, Finland, and Sweden increasingly engulfed by NATO–Russia tension in the Barents Sea-region. It might be worthwhile for policymakers in Brussels to start thinking of how the Union could contribute to reducing tension in its near-abroad. Such efforts are not likely to be welcomed by either Russia or Norway and/or NATO, but that does not mean they are not in the interests of the EU.

In this chapter, we have also raised more broad questions concerning (security) governance related both to whether the Arctic states will continue to attempt to insulate the region from great power politics elsewhere, and how to improve intra-regional cooperation on security matters. Regarding the former, it is clear that the Arctic will not become any less important on the strategic level simply because the US and Russia are already *in* the region, and China is increasingly demonstrating its (strategic) northern interests. If global relations continue to deteriorate among these actors (through increasingly bellicose statements, military posturing and exercises, and sanction regimes) greater tensions in the Arctic may well result. The Arctic is then, to some extent, used as an arena for symbolic gestures and power projection, which has little to do with resources or territory in the Arctic specifically. Nevertheless, the Arctic states and the EU can *choose* to keep the region 'separate' in their statements and regional interactions, even if this is predominantly a rhetorical instrument in order to reduce northern tensions.

Regarding the latter question, it is clear that increasing attention has been paid for some time now to northern security challenges by Arctic actors (including Russia, the US, and the EU) and those with a growing interest in the Arctic, like China. Which forum or institution that might be an appropriate venue for remains open to debate. The most purposeful arrangement is likely to be an 'Arctic security council', separate from existing structures and involving officials from the military as well as politicians and the small community of Arctic security scholars. Here, the EU has a specific role to play. This could, in turn, help ensure that Arctic relations remain relatively peaceful, even as the region is becoming a focal point in global politics.

## References

Airoldi, A. (2020). Security aspects in EU Arctic policy. *Routledge Handbook of Arctic Security*, 337–347.

Bennett, M. M. (2017). Arctic law and governance: The role of China and Finland (2017). *Jindal Global Law Review, 8*(1), 111–116. https://doi.org/10.1007/s41020-017-0038-y.

Blockmans, S. (2020, September 15). *Why the EU needs a geopolitical commission.* Centre for European Poliy Studies. https://www.ceps.eu/why-the-eu-needs-a-geopolitical-commission/

Brutschin, E., & Schubert, S. R. (2016). Icy waters, hot tempers, and high stakes: Geopolitics and Geoeconomics of the Arctic. *Energy Research and Social Science, 16*, 147–159.

Buzan, B., & Wæver, O. (2003). *Regions and powers: The structure of international security.* Cambridge University Press.

Byers, M. (2017). Crises and international cooperation: An Arctic case study. *International Relations*, 1–28.

Commission of the European Communities. (2007). An integrated maritime policy for the European Union (COM(2007) 575 final). *Brussels, 10*(10), 2007.

Coninsx, M.-A. (2019). The European Union: A key and reliable partner in the Arctic and beyond. *European Foreign Affairs Review, 24*(3), 237–242.

Conley, H. A., Toland, T., Kraut, J., & Østhagen, A. (2012). A new security architecture for the Arctic: An American perspective. In *Naval war college review.* Center for Strategic & International Studies (CSIS). http://csis.org/files/publication/120117_Conley_ArcticSecurity_Web.pdf.

Council of the European Union. (2014). *European Union maritime security strategy* (11205/14), Brussels*, 24 June 2014*.

Depledge, D., Boulègue, M., Foxall, A., & Tulupov, D. (2019). Why we need to talk about military activity in the Arctic: Towards an Arctic Military Code of Conduct. *Arctic Yearbook, 2019*. https://arcticyearbook.com/arctic-yearbook/2019.

Dodds, K. (2012). Anticipating the Arctic and the Arctic council: Pre-emption, precaution and preparedness. In T. S. Axworthy, T. Koivurova, & W. Hasanat (Eds.), *The Arctic council: Its place in the future of Arctic governance* (pp. 2–28). Munk-Gordon Arctic Security Program & University of Lapland.

Estonian Ministry of Foreign Affairs. (2021, January 28). *Estonia as an aspiring Arctic Council observer state: the Arctic's inventive neighbour*. https://vm.ee/en/estonia-aspiring-arctic-council-observer-state-arctics-inventive-neighbour

European Commission, & High Representative. (2016). An integrated European Union policy for the Arctic (JOIN(2016) 21 final). *Brussels, 27*(4), 2016.

European Political Strategy Centre. (2019). *Walking on thin ice: A balanced Arctic strategy for the EU* (EPSC strategic notes; issue 31). European Commission. https://ec.europa.eu/epsc/publications/strategic-notes/walking-thin-ice-balanced-arctic-strategy-eu_en

Exner-Pirot, H. (2012). New directions for governance in the Arctic region. *Arctic Yearbook*.

Exner-Pirot, H. (2015). Arctic council: The evolving role of regions in Arctic governance. *Alaska Dispatch*.

Exner-Pirot, H., & Murray, R. W. (2018). Regional order in the Arctic: Negotiated exceptionalism. *Politik, 20*(3), 47–64.

Friis, K. (2019). Norway: NATO in the north? In N. Vanaga & T. Rostoks (Eds.), *Deterring Russia in Europe: Defence strategies for neighbouring states* (1st ed., pp. 128–145). Routledge.

Graczyk, P., & Rottem, S. V. (2020). The Arctic council: Soft actions, hard effects? In G. H. Gjørv, M. Lanteigne, & H. Sam-Aggrey (Eds.), *Routledge handbook of Arctic security* (pp. 221–234). Routledge.

Heininen, L., Everett, K., Padrtová, B., & Reissell, A. (2020). Arctic policies and strategies – Analysis, synthesis, and trends.

High Representative. (2016). *Shared Vision, Common Action: A Stronger Europe: A Global Strategy for the European Union's Foreign and Security Policy*. http://europa.eu/globalstrategy/sites/globalstrategy/files/pages/files/eugs_review_web_13.pdf

Hilde, P. S. (2014). Armed forces and security challenges in the Arctic. In R. Tamnes & K. Offerdal (Eds.), *Geopolitics and security in the Arctic: Regional dynamics in a global world* (pp. 147–165). Routledge.

Hilde, P. S. (2019). Forsvar vår dyd, men kom oss ikke for nær. Norge og det militære samarbeidet i NATO [Defend our virtue, but do not get too close. Norway and the military cooperation in NATO]. *Internasjonal Politikk, 77*(1), 60–70. https://doi.org/10.23865/intpol.v77.1626.

Hoel, A. H. (2009). Do we need a new legal regime for the Arctic Ocean? *The International Journal of Marine and Coastal Law, 24*(2), 443–456.

Hoogensen, G. (2005). Bottoms up! A toast to regional security? *International Studies Review, 7*(2), 269–274.

Humpert, M. (2020, May 25). U.S. warns of Russian Arctic military buildup: "Who puts missiles on icebreakers?". *High North News*. https://www.highnorthnews.com/en/us-warns-russian-arctic-military-buildup-who-puts-missiles-icebreakers.

Ims, M. (2013). *Russiske oppfatninger om delelinjeavtalen i Barentshavet (Russian perceptions concerning the maritime boundary agreement in the Barents Sea)*. University of Tromsø.

Jensen, L. C. (2017). An Arctic 'marriage of inconvenience': Norway and the othering of Russia. *Polar Geography, 40*(2), 121–143. https://doi.org/10.1080/1088937X.2017.1308975.

Keil, K., & Knecht, S. (2016). Governing Arctic change: Global perspectives. In *Governing Arctic Change: Global Perspectives*. Palgave Macmillan. https://doi.org/10.1057/978-1-137-50884-3.

Koivurova, T., & Kopra, S. (Eds.). (2020). Chinese policy and presence in the Arctic. Brill Nijhoff.

Koivurova, T., Kokko, K., Duyck, S., Sellheim, N., & Stępień, A. (2012). The present and future competence of the European Union in the Arctic. *Polar Record, 48*(4), 361–371.

Kopra, S. (2013). China's Arctic interests. *Arctic Yearbook, 2013*, 1–16.

Lanteigne, M. (2015). The role of China in emerging Arctic security discourses. *S+F Security and Peace, 33*(3), 150–155.

Lavrov, S., & Støre, J. G. (2010, September). Canada, take note: Here's how to resolve maritime disputes. *The Globe and Mail*.

Lulu, J. (2018, October 22). *Confined discourse management and the PRC's localised interactions in the Nordics*. https://sinopsis.cz/en/confined-discourse-management-and-the-prcs-localised-interactions-in-the-nordics/

Ministry for Foreign Affairs of Sweden. (2020). *Sweden's strategy for the Arctic region*. https://www.regeringen.se/land%2D%2Doch-regionsstrategier/2011/05/sveriges-strategi-for-den-arktiska-regionen/

Ministry for the Armed Forces. (2019). *France and the new Strategic Challenges in the Arctic*. https://www.defense.gouv.fr/english/dgris/international-action/regional-issues/l-arctique

Moe, A., Fjærtoft, D., & Øverland, I. (2011). Space and timing: Why was the Barents Sea delimitation dispute resolved in 2010? *Polar Geography, 34*(3), 145–162.

Nitoiu, C., & Sus, M. (2019). Introduction: The rise of geopolitics in the EU's approach in its eastern neighbourhood. *Geopolitics, 24*(1), 1–19.

Oertel, J. (2020, September 7). *The new China consensus: How Europe is growing wary of Beijing*. European Council on Foreign Relations. https://ecfr.eu/publication/the_new_china_consensus_how_europe_is_growing_wary_of_beijing/

Østhagen, A. (2016). High north, low politics maritime cooperation with Russia in the Arctic. *Arctic Review on Law and Politics, 7*(1), 83–100.

Østhagen, A. (2019). *The new geopolitics of the Arctic: Russia, China and the EU* (policy brief April 2019). Wilfried Martens Centre for European Studies. https://martenscentre.eu/publications/new-geopolitics-arctic-russia-china-and-eu

Østhagen, A., Sharp, G. L., & Hilde, P. S. (2018). At opposite poles: Canada's and Norway's approaches to security in the Arctic. *Polar Journal, 8*(1), 163–181.

Raik, K., & Rácz, A. (Eds.). (2019). *Post-Crimea shift in EU-Russia relations: From fostering interdependence to managing vulnerabilities*. International Centre for Defence and Security. https://icds.ee/post-crimea-shift-in-eu-russia-relations-from-fostering-interdependence-to-managing-vulnerabilities/

Raspotnik, A. (2018). *The European Union and the geopolitics of the Arctic*. Edward Elgar.

Raspotnik, A. (2020, December 17). *A quantum of possibilities: The strategic spectrum of the EU's Arctic policy*. Centre for European Poliy Studies. https://www.ceps.eu/a-quantum-of-possibilities/

Raspotnik, A., & Stępień, A. (2020). The European Union and the Arctic: A decade into finding its Arcticness. In J. Weber (Ed.), *Handbook on geopolitics and security in the Arctic: The high North between cooperation and confrontation* (pp. 131–146). Springer.

Rottem, S. V. (2017). The Arctic council: Challenges and recommendations. In S. V. Rottem & I. F. Soltvedt (Eds.), *Arctic governance: Law and politics. Volume 1* (pp. 231–251) I. B. Tauris.

Rowe, L. (2018). Fornuft og følelser: Norge og Russland etter Krim (Sense and Sensibility: Norway and Russia after Crimea). *Nordisk Østforum, 32*, 1–20.

Saxi, H. L. (2011). Nordic defence Cooperation after the Cold War. In *Oslo Files: Vol. March 2011*. Norwegian Institute for Defence Studies.

Scazzieri, L. (2020). *Can the EU's strategic compass steer European defence?* (Issue 134). Centre for European Reform. https://www.cer.eu/publications/archive/bulletin-article/2020/can-eus-strategic-compass-steer-european-defence

Sergunin, A. (2014, November). Four dangerous myths about Russia's plans for the Arctic. *Russia Direct*.

Sergunin, A., & Konyshev, V. (2014). Russia in search of its Arctic strategy: Between hard and soft power? *The Polar Journal, 4*(1), 69–87.

Sparke, M. (1998). From geopolitics to geoeconomics: Transnational state effects in the borderlands. *Geopolitics, 3*(2), 62–98.

State Council of the People's Republic of China (2018). *China's Arctic policy*. Chinese Government.

Stępień, A., & Koivurova, T. (2016). The making of a coherent Arctic policy for the European Union: Anxieties, contradictions and possible future pathways. In A. Stępień, T. Koivurova, & P. Kankaanpää (Eds.), *The changing Arctic and the European Union: A book based on the report "strategic assessment of development of the Arctic: Assessment conducted for the European Union"* (pp. 20–56). Brill.

Stępień, A., & Raspotnik, A. (2019). *The EU's Arctic policy: Between vision and reality* (CEPOB #5.19). College of Europe. https://www.coleurope.eu/system/files_force/research-paper/stepien_raspotnik_cepob_5-19.pdf?download=1

Støre, J. G. (2012, August 27). *The Norwegian perspective on Arctic resource development and management (Speech held at ONS 2012 Summit – The Geopolitics of Energy, Sola, 27. August 2012)*. Norwegian Government. https://www.regjeringen.no/no/aktuelt/arktis_ressurser/id698106/

Sun, K. (2014). Beyond the dragon and the panda: Understanding China's engagement in the Arctic. *Asia Policy, 18*(1), 46–51. https://doi.org/10.1353/asp.2014.0023.

Sverdrup-Thygeson, B., & Mathy, E. (2020). Norges debatt om kinesiske investeringer: Fra velvillig til varsom (Norway's debate about Chinese investments: From willing to cautious). *Internasjonal Politikk, 78*(1), 79–92.

The Federal Government of Germany. (2019). *Germany's Arctic policy guidelines: Assuming responsibility, creating trust, shaping the future*. Federal Foreign Office. https://www.auswaertiges-amt.de/en/aussenpolitik/themen/internatrecht/einzelfragen/arctic-guidelines/2240000

Tripp, E. (2013). Realism: The domination of security studies. *E-International Relations, June 14*.

Tunsjø, Ø. (2018). *The return of bipolarity in world politics: China, the United States, and Geostructural realism*. Columbia University Press.

US Department of State. (2019). *Looking North: sharpening America's Arctic focus*. Remarks.

Waltz, K. N. (1959). *Man, the state, and war*. Columbia University Press.

Wilson Rowe, E. (2020). Analyzing frenemies: An Arctic repertoire of cooperation and rivalry. *Political Geography, 76*.

Wilson Rowe, E., Bertelsen, R. G., Kobzeva, M., & Raspotnik, A. (2020a, November 26). Unexplored resources for EU Arctic policy: Energy, oceans and space. *High North News*. https://www.highnorthnews.com/en/unexplored-resources-eu-arctic-policy-energy-oceans-and-space

Wilson Rowe, E., Sverdrup, U., Friis, K., Hønneland, G., & Sfraga, M. (2020b). *A Governance and Risk Inventory for a Changing Arctic: Background Paper for the Arctic Security Roundtable at the Munich Security Conference 2020*. https://www.wilsoncenter.org/publication/governance-and-risk-inventory-changing-arctic

**Andreas Raspotnik** is a Senior Researcher Fellow at the Fridtjof Nansen Institute in Lysaker, Norway (since 2018), a Senior Fellow at The Arctic Institute – Center for Circumpolar Security Studies in Washington, DC (since 2011), and a part-time Senior Reseacher at the High North Center for Business and Governance, Nord University in Bodø, Norway (since 2017), and, since 2019, a Senior Fellow at the Institute of European Studies, University of California, Berkeley.

**Andreas Østhagen** is a Senior Research Fellow at the Fridtjof Nansen Institute in Lysaker, Norway (since 2017), a Senior Fellow at The Arctic Institute – Center for Circumpolar Security Studies in Washington, DC (since 2011), a part-time Advisor at the High North Center for Business and Governance, Nord University in Bodø, Norway (since 2014), and, since 2018 an Associate Professor at Bjørknes University College in Oslo.

## Chapter 23
# Global Conventions and Regional Cooperation: The Multifaceted Dynamics of Arctic Governance

**Cécile Pelaudeix and Christoph Humrich**

## Introduction: Analyzing Institutional Dynamics in the "Globally Embedded" Arctic

The Arctic "is a 'globally embedded space', interlinked politically, economically, environmentally and socially with global structures and processes" (Keil & Knecht, 2017, p. 4). As such, there are "global impacts within" and "worldwide implications of the Arctic" (Finger & Heininen, 2019, p. 2). At least since the seventeenth century, multifaceted dynamics in the Arctic result from the interplay of global and regional level (cf. Sale, 2009). The incremental expansion of regional cooperation and governance institutions that started after the Cold War with the Arctic Environmental Strategy (AEPS, 1991), and in 1996 saw the founding of the Arctic Council (AC), the "pre-eminent regional forum" (Arctic Council, 2013a, b), is no exception to this observation. Rather, the significance of the global-regional interplay for Arctic governance only seems to have grown since the end of the first decade of the 2000s. In a short period of time, a series of events put the Arctic into the global spotlight: the planting of the Russian flag on the seafloor at the geographic North Pole in 2007, the US Geological Service's publication in 2008 of estimates of abundant undiscovered oil and gas reserves in the region's subsoil and seabed, and record low ice-covers that made visible the dramatic consequences of climate change in the region in the same years. These events created global attention as well as interest and initiatives from non-Arctic actors. Providing powerful imaginaries of geopolitical grandstanding, of an Arctic gold rush, and of the opening of a

C. Pelaudeix (✉)
Sciences Po Grenoble, Grenoble, France
e-mail: cecile.pelaudeix@umrpacte.fr

C. Humrich
University of Groningen, Groningen, The Netherlands
e-mail: c.humrich@rug.nl

© The Author(s), under exclusive license to Springer Nature Switzerland AG 2022
M. Finger, G. Rekvig (eds.), *Global Arctic*,
https://doi.org/10.1007/978-3-030-81253-9_23

new ocean, the above-mentioned events certainly worked as drivers of change in Arctic governance. However, these events do not determine how the respective institutional dynamics unfold. We want to explore these dynamics by focusing on how the nexus between regional cooperation and global conventions developed; that is, on the institutional interplay that makes the Arctic a "globally embedded space" in governance terms.

Such interplay has been the object of research for quite some time already (e.g., Oberthür & Stokke, 2011), and has been analyzed both regarding issue-specific governance complexes *in* the Arctic or *the* Arctic governance complex with the Arctic Council at the center (e.g., Stokke & Hønneland, 2007; Young, 1996). However, we contend that, so far, this literature has mainly looked for functionally effective interplay in which regional cooperation adds value to global governance. In such a view, regional cooperation is, for instance, found to implement global conventions or complement them with regional governance output. New institutions are shown to select or adapt to functional niches or fill gaps in the governance complexes. While this kind of interplay can certainly be observed in the Arctic, it is not the only one. More recently, international governance research has focused on less benign forms of interplay, as captured, for instance, in the notion of conflictive fragmentation (Biermann et al., 2009) or regional challenges to global governance (e.g., Kahler, 2017). Distinguishing different types of links between the regional and the global level, this chapter assesses the extent to which regional cooperation in the Arctic established a nexus to global conventions in either harmonious, cooperative, conflictive or indifferent ways. Thus, we hope to reveal both the impacts of global governance on regional cooperation as well as the global implications of Arctic governance, and to capture dynamics of governance in the globally embedded Arctic.

The following section briefly presents our analytical framework. This framework is then applied to four issue areas with global conventions at the center (the governance of marine and maritime spaces, indigenous peoples' affairs, climate governance, environmental protection and conservation), and as a fifth area to the legally binding agreements concluded on the regional level.[1] For each of these areas, we seek to determine the governance dynamic as it unfolded from the early days of the AEPS to the years in which the Arctic entered the global spotlight. While there is variation between the five areas under focus, we found that the links between regional and global governance have generally been explicitly acknowledged in the beginning of regional cooperation, but did not emerge as a harmonious nexus, and became even less cooperative and more conflictive the more the Arctic was drawn into the maelstrom of globalization. Based on this assessment, in the conclusion we argue that, overall, and against prevalent praise of the achievements of Arctic regional cooperation, our findings cast some doubts on the current capacity of Arctic governance to effectively engage in the solving of regional or global problems.

---

[1] For a similar overview endeavor in which we build, see also Stokke (2009).

## The Nexus Between Regional and Global Governance: An Analytical Framework

In order to capture the regional-global nexus of Arctic governance we build on and slightly amend for our purposes an analytical framework proposed by Arie Kacowicz (2018). This framework looks at the nexus as established by or emerging from rhetoric, decisions, and behavior of respective states or their governments on the regional level in view of global governance. States in a region can act on and react to global governance in four different ways: indifferently, conflictively, cooperatively, or harmoniously. Kathrin Keil and Sebastian Knecht (2017, p. 12) argued that one can take either an inside-out or an outside-in perspective on the Arctic as a globally embedded space. We take this distinction as characterizing perspectives that regional states can assume. In the outside-in perspective, states or governments react to the global governance framework, which they accept as given and not amenable to change from the regional level. States that take the inside-out perspective, by contrast, aim to act on the regional level for global governance. Combining these two perspectives and Kacowicz's four ways of establishing the regional-global nexus, we get a taxonomy of eight different types of linkages that can be applied to the relationship between regional cooperation and global conventions (Table 23.1).

In a harmonious link, the regional cooperation is regarded as "a building bloc, and as a desirable and necessary part of any GG architecture" (Kacowicz, 2018, p. 70). This first type is representing one of the most common views of a desirable regional-global nexus. Considering regions as "important elements of the architecture of world politics", Hurrell, for instance, saw the UN system's central role as setting the standards, "with regional bodies entering the story principally in terms of more detailed specification of rights and implementation" (2007, p. 142). Insofar as the regions become active towards global governance in a harmonious relationship, our second type, they might act as innovators and entrepreneurs to facilitate the setting of standards on the global level. Such a role for regions has, for instance, been advocated in the context of ocean governance in which regional institutions were envisioned to take initiatives for environmental governance of the high seas, which

**Table 23.1** The nexus between regional cooperation and global conventions

|  | Value-added | | Value subtracted | Value neutral |
|---|---|---|---|---|
|  | Harmonious | Cooperative | Conflictive | Indifferent |
| Outside (-> In) | Regional implementation of global conventions | Seeking regional niches in global governance | Defensive regionalization | Global conventions do not affect regional cooperation |
| Inside (-> Out) | Regions as legal innovators and entrepreneurs in global governance | Filling global regulatory gaps regionally | Challenging global norms on the regional level | Regional cooperation does not affect global conventions |

Source: Own compilation, categories cf. Kacowicz (2018), and Keil and Knecht (2017)

would then feed into the process of devising new global standards (e.g., Mahon & Fanning, 2019; Rochette et al., 2014).

When we speak of a cooperative link between regional cooperation and global conventions, regions are seen less as an integral part of global governance, but co-existing in a way that still provides some added value. Taking an outside-in perspective, regional cooperation seeks to find what Olav Stokke has called its own governance niche within the framework established by the global convention (2007b, p. 17). States taking the inside-out perspective seek to contribute to a global framework regionally by complementing it, for instance in matters that can be dealt with better on the regional level or where gaps in the global framework exist, because agreement on a global convention is (still) out of reach. For instance, the 2010 move of state parties to the OSPAR Convention for the Protection of the Marine Environment of the North-East Atlantic to establish marine protected areas in areas beyond national jurisdiction was interpreted in this way (e.g., Molenaar & Oude Elferink, 2009; Rochette et al., 2014, p. 111).

In our fifth type, regional cooperation does not any more add value to global conventions; rather, the former seeks to resist or alleviate compliance pressure from the latter. It is defensive in the sense of cooperating in order to deal with negative regional externalities from global governance. Kacowicz mentions examples of regional cooperation in South America to better cope with the impact of what is perceived to be a hegemonic global order (2018, p. 68). Our sixth type refers to challenges to global norms from the regional level in the sense that regional cooperation aims to revise global standards. For instance, Randall Henning has identified respective tendencies in global financial governance (2017).

In our final two types, the regional cooperation and global conventions are (made) independent from each other. In both the outside-in and the inside-out perspective, this means that norms and obligations emerging on either level are not regarded as relevant for the cooperation on the other. Note that when states decide that one or the other level should be the sole arena for certain issue-specific rules, this can either be because they accept that the respective level is the best functionally regarding scale or scope, or because one or the other level simply better matches with their particular interests.

## The Regional-Global Nexus in Arctic Governance

The aim of this section is now to apply the framework developed above in an overview-like fashion to see what kind of nexus emerges between global conventions and regional cooperation in the case of the Arctic. We start by looking at two constitutional issues: the assignment of respective marine and maritime spaces (including shipping and marine pollution governance) according to United Nations Convention of the Law of the Sea (UNCLOS), and the question of Indigenous Peoples rights regarding participation in Arctic governance. Next, we focus on two core functional issue areas of regional cooperation in the AC: climate governance as concerned with the core challenge the region is facing, and pollution and conservation of the Arctic environment as the original focus of regional cooperation when it emerged in the early 1990s. The picture that develops from the application of our analytical framework is that the nexus between regional and global governance is adding much less value than would be expected or is usually assumed. We bolster this impression by further looking at the binding regional agreements the Arctic states concluded.

## *Marine and Maritime Space Delimitation, Shipping, and Marine Pollution*

The Arctic is an ocean surrounded by three continents. Thus, one would expect UNCLOS to play a major and explicit role in the region's governance. As the so-called constitution for the oceans, UNCLOS not only is the central legal reference for the delimitation of marine and maritime spaces and the accompanying rights of coastal states and other users of the sea; it also is a governing framework for shipping, and emphasizes the significance of and explicitly endorses regional activities to protect the marine environment. Yet, when attention on the Arctic grew, some even opined that "there are currently no overarching political or legal structures that can provide for the orderly development of the region or mediate political disagreements over Arctic resources or sea-lanes", because "UNCLOS cannot be seamlessly applied to the Arctic" (Borgerson, 2008, pp. 71–72). It seems many agreed at the time, and thus called for a respective complementary legal document for the region.

Such a perception might be understandable given that, until that time, UNCLOS had not played any significant role in the regional cooperation, despite being mentioned as constitutive framework in the AEPS. There might be two reasons for that. First, when regional cooperation started in the early 1990s, UNCLOS was not even in force, even though it had been signed in 1982. When it entered into force at the end of 1994, only Iceland had ratified the convention. In the three years after the entry into force, ratifications from Finland, Norway, Sweden, and Russia followed. Canada and Denmark ratified in 2003 and 2004, respectively. Until today, the US senate has held its almost forty-year-old position of not acceding to UNCLOS.[2] Second, however, just as the region was emerging from the Cold War, it proved that marine environmental protection was still too "politically sensitive" (Young, 1998, p. 137), as it unavoidably touched on highly politicized or even securitized matters such as marine and maritime jurisdiction, resource and fishing rights, or nuclear submarine conduct. With the respective working group on the protection of the Arctic marine environment (PAME) being "relatively quiescent" during these early

---

[2] Despite administrations from at least Bill Clinton via George Bush to Barack Obama calling for ratification.

years of cooperation (Young, 1998, p. 137), regional cooperation and global ocean governance remained in a state of politically endorsed indifference to each other.

Even more surprising is that this indifference continued into the first decade of the new millennium. The AEPS had noted that the A8 "agree to apply the principles concerning the protection and preservation of the Marine Environment as reflected in [UNCLOS]" (AEPS, 1991, p. 33), and thus suggested a harmonious nexus of implementation. A legally binding regional seas arrangement, as envisioned in the respective United Nations Environment Program (UNEP) policy initiative and legally underpinned by UNCLOS's part XII, section 2, seemed to be a natural course of action. Its advantages had been emphasized by PAME's very first report (PAME, 1996). Yet, despite significantly expanding their marine environmental activities from the end of the 1990s, the AC members did not take up negotiations for a legally binding regional seas arrangement until 2015, when the US chairmanship brought it into the discussion. This led nowhere, however. The respective task force could not agree on anything significantly beyond the current institutional structure of the AC (cf. Humrich, 2018, pp. 226–230).

However, regional governance offered an opportunity to establish a cooperative nexus with UNCLOS by complementing the global rules for shipping – as laid down in the International Maritime Organization's (IMO) conventions. In combination with the 2004-published Arctic Climate Impact Assessment (ACIA), PAME was tasked to work on an Arctic Marine Shipping Assessment (AMSA), which was presented to the AC in 2009. It recommended the negotiation of updated and mandatory guidelines for Arctic shipping. While these developments certainly contributed to getting negotiations going on the mandatory Polar Code at the IMO, it was not Arctic regional cooperation that left its mark on the result. The negotiations of the Polar Code have not witnessed coalition of interests amongst Arctic states. For instance, comparing the approaches of Russia and Canada in the negotiations Dorottya Bognar-Lahr shows different emphases stemming from the two states' political and economic realities and capacities (Bognar-Lahr, 2020). Their one common position was to support the special rights they have under Article 234 of UNCLOS and ensuring that the Polar Code buttresses the international legal basis for these.[3]

Ironically, at the time when observers perceived the alleged lack of an overarching legal framework for Arctic governance, UNCLOS's significance for driving the dynamics in the region governance was probably greatest. When, according to Scott Borgerson (2008, p. 64) the "Arctic powers" were "racing to carve up the region", that was neither a race nor due to alleged resource riches or new shipping highways. Of the five Arctic Ocean rim states (A5),[4] the four UNCLOS parties were simply submitting proposals for the limits of their continental shelves or preparing to submit these to the respective UN commission within the legally prescribed timeframe

---

[3] Article 234 is the so-called ice-paragraph, which allows a coastal state to set stricter rules for shipping in its Exclusive Economic Zone in ice-infested waters.

[4] Norway, Denmark/Greenland, Canada, US, Russia.

(10 years after ratification). This initially proceeded without regional cooperation. However, so concerned were non-Arctic states about the future of the region or so interested in its potential, and so persistent were the calls for an international Arctic treaty, that the A5 felt compelled to meet and initiate some sort of defensive regional cooperation. The resulting 2008 Ilulissat Declaration was "explicitly designed to challenge the notion that the Arctic should be internationalized" (Dodds, 2013, p. 49).[5] On first view it endorses UNCLOS as part of an existing "extensive international legal framework" (A5, 2008). On second view, however, this only relates to those parts that concern the rights of the coastal states. Not only does the declaration marginalize the AC, and thus the remaining three Arctic states, it also leaves out matters of the High Seas and the Area, in which also other nations would have rights.[6] Instead, it claims a special "stewardship role" for the A5 and seems to reserve the right to involve other parties in the management of the Arctic Ocean as they see fit (cf. Humrich, 2018, pp. 231–232). The emerging governance of fisheries in the Central Arctic Ocean could serve as an example of subsequent institutional dynamics. With the Oslo Declaration of 2015, the A5 set the agenda and a political precedent first, even though they cannot claim any exclusive jurisdiction (Kuersten, 2016, p. 437).[7] Only then did the A5 invite other interested parties to join the club and to negotiate a more comprehensive and legally binding agreement, which was signed in 2018.

## *Indigenous Peoples' Rights and Participation*

If the AC has been praised for one thing in particular (e.g., Heininen, 2018), it has been its inclusion of indigenous peoples' organizations (IPOs) as permanent participants with full consultative rights in all deliberations of the AC, supported by an indigenous peoples' secretariat administratively, and since 2017 by a fund that also supports the participation economically. However, the truly innovative character of IPO inclusion might make one forget that a substantive link between global conventions and regional cooperation also exists here, and state practice establishes a certain regional-global nexus. Moreover, the character of this nexus was a matter of long negotiations and deep disagreements, of which the respective documents reveal only a small part (cf. English, 2013).

The AEPS emphasized the important role of indigenous peoples and their special vulnerability to pollution and relevance for conservation, and the states committed to facilitate the participation of IPOs in their efforts under the strategy. What is notably absent in the AEPS is the 1989 International Labor Organization's (ILO) convention 169, the Indigenous and Tribal Peoples Convention, which at the time

---

[5] For a respective history of the Ilulissat Declaration, see Rahbek-Clemmensen and Thomasen (2018).
[6] As asserted, for instance, by China in its Arctic policy white paper (2018).
[7] See also Zou (2016, p. 460).

was arguably the most important legally binding instrument dealing with indigenous peoples' rights and participation at the global level. When the AEPS was negotiated, all Arctic states were members of the ILO, but only Norway had ratified the global Convention 169, in 1996 followed by Denmark.[8] In the negotiations for the set-up of the Arctic Council, the indigenous peoples' participation emerged as a major point of contention. In the Ottawa Declaration (AC, 1996), the permanent participant status remained reserved for certain IPOs only (representing either one people in several countries or several people in one country) and limited in number to seven (at any time less than the member states). Moreover, after lengthy discussions over the "s" in peoples, a footnote stated that "the use of the term 'peoples' in this Declaration shall not be construed as having any implications as regard the rights which may attach to the term under international law" (AC, 1996, p. 2). This footnote epitomizes what must be regarded as a nexus of indifference between the regional cooperation at that time and the existing global standards.

The major global convention (ILO, 169) did not have an acknowledged influence on regional cooperation, nor did the unique institutional innovation of the AC – the permanent participant status – seem to have significant impact on the AC's actions regarding the global level of indigenous peoples' governance. In 2007 the United Nations Declaration on the Rights of Indigenous Peoples (UNDRIP), even though not legally binding, arguably replaced ILO 169 as the most important global governance instrument. Over two decades in the making, the core of UNDRIP is the principle of the right of indigenous peoples to self-determination and governance over their natural resources. Arctic IPOs, notably the Inuit Circumpolar Council (ICC), have been active and instrumental in the drafting process for the Declaration, yet there is no record of the Arctic Council being involved or a regional effort being made. While most AEPS and AC declarations usually acknowledged new and relevant global governance instruments, the UNDRIP is, again, notably absent in the years after its adoption. Moreover, in the voting on UNDRIP, the US and Canada voted against and Russia abstained. Russia upheld its abstention and the US endorsed UNDRIP fully in 2010. Canada endorsed it with certain reservations in the same year. Only at the end of 2020 was a bill brought before the parliament by the Canadian government, which would fully implement UNDRIP in Canadian legislation.

In this context, the discussion on the Nordic Sami Convention is also noteworthy. Proposed in 2005, after being negotiated since 2011, and signed by the three involved governments of Norway, Sweden and Finland January 2017, it is still under legislative review in the respective countries, also considering changes proposed by the participating Sami institutions. The proposed convention was criticized by indigenous peoples' representatives in Norway for going below some of the standards already in force since Norway's ratification of ILO convention 169. While the Norwegian government rightly points to a clause in the proposed convention preventing the undermining of further going commitments by the convention, this is not the case for Finland and Sweden, which did not ratify the ILO document.

---

[8] Until the time of writing this chapter the other Arctic states have not ratified ILO 169.

Investigating why the two latter states did not commit to ILO169 while Norway did, Semb pointed to the competing interests of forestry and mining industries with reindeer husbandry in Sapmi (2012). At least for these latter, the draft Nordic Sami Convention might, in our terms, be regarded as constituting a conflictive nexus directed against more demanding global standards (defensive).[9]

## *Climate Change*

As climate change brings more challenges to the Arctic and is one of the most daunting issues of our present time, the question of whether the existing regional cooperation in the Arctic is adequate to confront it is of paramount importance. In 2018, when several scholars nominated the AC for the Nobel Peace prize, they clearly seemed to think it was adequate. One of the mentioned reasons in support of the nomination was the forum's contribution to climate governance. The AC was credited as being "a leader in climate science and advocacy" and "the first organization to take climate-specific action at the regional level" (Heininen, 2018).

Two particular achievements stand out. First, the already mentioned ACIA was commonly regarded as a landmark document that made a global contribution and managed to put the Arctic into the spotlight, not only nationally, particularly in the US, but also to include the Arctic in the Intergovernmental Panel on Climate Change's knowledge production on the global level. Alf Håkon Hoel convincingly described this achievement as the result of a process on the regional level to find a respective governance niche in global climate governance (2007). However, this process was not necessarily inevitable. Contrary to some beliefs, neither the AEPS nor the Arctic Council were primarily set up to deal with climate change in the region. Rather, the AEPS explicitly stated that this topic was thought of being dealt with already in other fora – that is, the global level – upon which the United Nations Framework Convention on Climate Change (UNFCCC) was negotiated at the time (AEPS, 1991, p. 12). This desired state of indifference to existing global climate governance efforts gave way to initial efforts to establish a cooperative nexus during the 1993 AEPS Ministerial Meeting in Nuuk, which took place in the wake and under the impression of the 1992 United Nations Conference on Environment and Development (UNCED) and concomitant signing of the UNFCCC. "Noting the existing global cooperation", the Arctic Monitoring and Assessment Working Group (AMAP) was tasked "to identif[y] gaps […] with a view to ensuring that specific issues related to the Arctic region are placed on the agenda of the appropriate international bodies" (AEPS, 1993). However, in the run up to and negotiations for the AC, climate change did not matter much (e.g., English, 2013, pp. 228–251). Specific activities were not considered before 1998 when preparatory steps were

---

[9] For how ILO 169 mattered in initial deliberations on the Nordic Saami Convention see Åhrén (2007).

taken to assess consequences of climate change in the Arctic, and then led to the launch of the ACIA process in 2000.

The second achievement is the AC's work on short-lived climate forcers (SLCFs), particularly black carbon and methane. At its 2009 Ministerial Meeting, the Arctic states decided to establish a Task Force on SLCFs, which produced two reports in 2011 and 2013. These led to a new task force mandated at the 2013 Ministerial Meeting "to develop arrangements on actions to achieve enhanced black carbon and methane emission reductions in the Arctic" (AC, 2013a, b, p. 4). This task force, in turn, produced a Framework for Action by 2015, which among other things set an aspirational goal for emission reductions. By adopting this aspirational goal, "the Arctic states seek to fill the gap in the legal and governance landscape on black carbon" (Yamiveva & Kulovesi, 2018, p. 214).

Looking through various AC declarations confirms the nexus between Arctic regional and global climate governance as mostly cooperative in both variants. The respective formulations and actions envisioned suggest the niche-seeking and gap-filling approach, while refraining from establishing a more harmonious link. Declarations mostly encourage the Arctic states to take effective measures to meet their global obligations individually, as well as remind states to do their bit to reach "adequate agreed outcome[s]" on the global level (AC, 2009, p. 2). However, a direct role for the regional level in norm-setting within the global regime or in its implementation is not spelled out. As Sebastian Duyck (2012) found, the legal documents of global climate governance did not pay special attention to the Arctic either. Even in the 2015 negotiations for the Paris agreement, the Arctic states did not develop a particular regional contribution nor did they individually bring a specific Arctic aspect to the global norm-setting process: they "have seldom referred to the region in their negotiating positions and, when they did, these references related primarily to the need for further scientific research on regional climate processes" (Duyck, 2015, 416). Moreover, the Arctic states did not manage to influence global norm-setting by repeatedly failing to form a respective coalition and leaving their imprint on the negotiations (Selin, 2017). Given the acknowledged urgency of climate change – in the Arctic Council's own words, "one of the greatest challenges facing the Arctic" (AC, 2009, p. 1) – and having bolstered this evaluation with its own scientific assessments, this is certainly not much to speak of. It was mostly the US that prevented more here. During the administrations of the two George Bushes and Donald Trump, advances in regional climate policy were blocked, but also with Bill Clinton and Barak Obama, progress was small and slow and mostly limited to furthering scientific assessments.

## *Environmental Protection and Conservation*

At the time the AEPS was negotiated, Arctic states deemed pollution and "accidental discharges and uncontrolled releases of pollutants" into the vulnerable Arctic environment the major problem to tackle on the regional level (AEPS, 1991: 12).

Thus, over half of the document is devoted to outlining six respective pollution problems, reviewing existing international governance instruments (including global conventions), dealing with these, and then outlining actions to be taken. These actions are often framed as resulting from a gap-analysis or niche-seeking, but also as implementation and joint actions in relevant international fora to strengthen the global side of Arctic governance (AEPS, 1991). Arguably one of the most important founding documents of Arctic regional cooperation displays a vision of a cooperative or even harmonious nexus with global conventions regarding pollution. Ever since then, the explicitly expressed awareness of and link to existing conventions has been a feature of the regional environmental cooperation and also shows in subsequent reports and assessments. However, the question is to what degree this rhetoric is also reflected in actions.

In general, it can be stated that the AEPS or AC does not seem to have influenced the ratification patterns of identified relevant global conventions significantly. However, shifting away from an overly legalistic reading, the analysis needs to look into the policy processes and examine the ways in which Arctic states have collectively shaped global conventions of special importance for the Arctic or implemented these on the regional level. In the field of pollution and conservation, this can only be done in an exemplary way. To determine the actual merits of regional cooperation, it makes sense to look at cases of global conventions, which are usually considered successes of regional cooperation. First among these would certainly be the 2001 Stockholm Convention on Persistent Organic Pollutants, which aims to reduce emissions of these substances at the global level (cf. Selin, 2017). POPs are carried into the Arctic from different parts of the world via long-range atmospheric transport as well as by ocean currents and rivers, and were mentioned the first of the pollution problems in the AEPS. The AEPS already envisioned joint action on the international level, particularly through AMAP. The role of Arctic science and work done under the AEPS and AC in addressing POPs has been highlighted as a significant contribution to the global norm-setting process (e.g., Koivurova et al., 2015; Selin, 2017). However, Stokke suggested that the impacts of AC activities on the adoption or contents of the Århus Protocol, upon which the Stockholm convention is built, "should not be over-estimated." The credit should go to "research and monitoring activities by certain Arctic states – especially Canada and Sweden – rather than AMAP or other Arctic Council activities" (2007a, p. 406). Stokke also highlighted the active role played by the ICC. Moreover, it remains unclear whether the engagement of some AC member states to influence global levels of governance, in coordination with UNEP, was a consensus-based strategy from the AC or more a result of the AMAP secretariat's rather independent, proactive role (Platjouw et al., 2018). In any case, a study of the negotiation process for the convention does not show any evidence of specific Arctic collaboration (Karlaganis et al., 2001). Rather, the particular role of individual states stands out; for instance Norway, which has acted as a green ambassador in the negotiation process (Rottem, 2017). That the convention might be less of a collective achievement than often assumed is also supported by the fact that the convention has not been ratified by the US yet, and that it does not apply to Greenland. Denmark lifted the reservation for the Faroes in 2012, but not for Greenland.

A second prominent case often presented as a positive example of global norm-setting in which Arctic states participated collectively, which suggests a harmonious nexus between regional cooperation and global conventions, is the 2013 Minamata Convention on mercury. While mercury is of global concern, its levels in the Arctic can be five to 50 times higher than those usually measured in Europe and North America (Platjouw et al., 2018). While mercury does not figure as prominently in the AEPS as POPs, it rose on the agenda through later work of AMAP. The respective insights were then also instrumental in paving the way for the convention: "To a large extent the pathway towards the Minamata Convention has been shaped through important science-policy interfaces, where the Arctic Council and its working group AMAP have been heavily involved" (Platjouw et al. 2018, 229). However, on closer look there are again some doubts about the degree to which this is a collective achievement of Arctic regional cooperation. On one hand, there is the lack of an initial Arctic regional coalition. When Norwegian government officials proposed a legally binding approach to the regulation of mercury in 2003, they did not go through the Arctic Council, but actually preferred to avoid the risk of a slow-down by those Arctic countries favoring a voluntary instead of legally binding approach to deal with the problem (Canada, US, and Russia). Instead, Norwegian government officials chose to go directly to UNEP and thus not to rely on the regional level. On the other hand, one could also again question the extent to which AMAP's crucial role in the process is an achievement of regional cooperation. It seemed to have been more the executive staff and researchers associated with AMAP directly who left a mark on the global process than the group of governments representing the AC on the basis of AMAP's findings. Moreover, lacking a funding mechanism at that time, it was not Arctic regional cooperation that raised the money for many of the related AMAP activities, but the Nordic Council of Ministers.[10]

Looking at two cases of 'inside-out' environmental governance, we found a potentially harmonious nexus between global conventions and regional cooperation. What about the outside-in perspective? As mentioned above, the AEPS and later AC reports listed relevant global conventions. As already indicated in the case of the UNFCCC, however, regional relevance did not necessarily translate into regional implementation. We have again selected two cases to show that what might appear as a harmonious nexus first turns out to be much less ambitious or probably even conflictive at second view.

Our first example is the 1991 Espoo Convention, which made it mandatory for its parties to take necessary legal, administrative, or other measures to ensure proper environmental impact assessments (EIA) in transboundary cases. The convention had already been identified as clearly relevant for the Arctic in the Nuuk Declaration (AEPS, 1993, §9) and explicitly endorsed again at the following Ministerial

---

[10] There are also doubts about the effectiveness of the convention for meeting the problem (VanderZwaag, 2015).

Meeting, where the Arctic states called for it to be ratified and implemented. Two years later, the regional cooperation under the AEPS had brought about the "Guidelines for Environmental Impact Assessment in the Arctic". These were received "with appreciation" and states agreed "that these Guidelines be applied" (AEPS, 1997, §3). Real added value was seen in specifying standards for its implementation under Arctic conditions. Yet, as Timo Koivurova and David VanderZwaag pointed out 10 years later, the Guidelines "did not appear to have influenced any environmental impact assessment processes in the Arctic" (2007, 158). Here, as in other cases, there simply was no collective regional follow-up, as states were left alone to implement the outcomes of their cooperation individually. Until today, the Espoo convention has been poorly ratified among the Arctic states. Despite having signed it, Iceland, Russia, and the US are still not parties.

As with the Espoo Convention, the "early ratification" of the UN's Convention on Biological Diversity 1992 (CBD) was endorsed by the Arctic states in the Nuuk Declaration – stating in the same paragraph "the need for effective application of existing legal instruments" (AEPS, 1993, §10).[11] For instance, the Nuuk meeting tasked the newly constituted working group on the Conservation of Arctic Flora and Fauna (CAFF) with developing a Circumpolar Protected Area Network Plan (CPAN), which was adopted in 1996. However, while some efforts were made to understand and synchronize representation of protective status according to the World Conservation Union's (IUCN) standardized criteria, little in the plan was really a collective regional effort to substantiate globally set ambitions. Instead, the states retained full control over the respective policies and did not commit to significant collective goals. Thus, while the plan involved some reporting and stock-taking, as well as knowledge production over the years, not much else happened. After 2005, the status of the respective sub-working group even became 'dormant' (cf. Koivurova, 2009), despite the 2002 World Summit on Sustainable Development (WSSD) explicitly calling for the establishment of representative protected area networks by 2012, and other regional cooperation arrangements ramping up their respective activities (for example, OSPAR from 2003). In the Arctic, new activities only emerged in 2013 when the processes started to develop a Framework for a Marine Protected Area Network, adopted in 2015. That was rather late to meet the so-called 2010 Aichi targets of the CBD, to which the Arctic states had globally committed. In these, a role for regions was envisioned, especially in the creation of protected area networks, and states were called on to put a respective percentage of land and marine areas under protection by 2020. However, the Framework explicitly leaves the implementation to each state individually. In the context of the long break in activities and the obvious disconnection with the global norm-setting processes in the CBD, it appears that rather than implementing CBD goals, the new activities come as defensive action against two other processes: the progress in the pre-negotiation phase for an international legally binding instrument for the protection of biodiversity beyond national jurisdiction, and the OSPAR activities for marine

---

[11] The US is still not a party to the CBD.

protected areas in the convention's area 1: the Arctic North-East Atlantic (cf. Rochette et al., 2014).

In the former, the question of the relation between global and regional arrangements was on the table early, with emerging agreement that existing regional arrangements should not be affected by the new global norms. To protect potential interests against intervention from the global level, it made sense – as in the fisheries case – to establish a regional precedent. In the OSPAR case, its Arctic members, Iceland and Norway, dodged the attempt to include marine spaces in area 1 in OSPAR's network strategy by invoking the AC's role as pre-eminent and thus competent forum for the region. The framework then undergirded that claim.

## *Arctic Legal Instruments*

In the case of pollution, regional governance activities of the Arctic states have exerted some modest influence on the negotiation of legally binding global conventions. Within the Arctic, however, we so far have dealt with soft-law instruments only. This corresponded to the regional cooperation's underlying political and functional rationale since its inception, as well as to the later-expressed conviction of the A5 that there was no need for a comprehensive legally binding instrument regulating Arctic affairs (cf. A5, 2008). Yet, as a follow-up to AMSA, the Arctic states negotiated two legally binding issue-specific agreements. The first agreement signed under the auspices of the Arctic Council was the 2011 Agreement on cooperation on aeronautical and maritime search and rescue in the Arctic (SAR). Only two years later, the Agreement on Cooperation on Marine Oil Pollution Preparedness and Response in the Arctic (MOSPA) was signed. In 2017 the Arctic states added the Agreement on Enhancing International Arctic Scientific Cooperation. Thus, the question is whether these added value to the global governance of the Arctic.

All three agreements have a substantive link to global conventions. The SAR agreement covers commitments the Arctic States already have under the IMO SAR Convention and the Convention on International Civil Aviation (the Chicago Convention) (Kao et al., 2012). The 1990 International Convention on Oil Pollution Preparedness, Response, and Co-operation to which MOSPA is related in substance seems to be the only convention of major significance for the Arctic that is signed by all Arctic states. However, it lacks a specific Arctic focus and, arguably, measures attuned to the special circumstances of the vulnerable Arctic environment. While the Science agreement is not limited to the Arctic Ocean, in this area it clearly relates to the respective provisions for marine scientific research in UNCLOS' part XIII (Shibata, 2019).

For both the SAR and MOSPA it has been argued that the obligations of the respective agreements do not add anything to existing commitments of the Arctic states from either the mentioned global conventions or other bilateral and multilateral agreements (for example, with respective neighboring states).[12] Svein Rottem stated that the texts "do not indicate a desire [...] to press ahead regarding binding international agreements" (2015, p. 56). This impression is further confirmed when considering that the A8 also deliberated on an agreement about the prevention (and not only preparedness and response) of oil spills in the Arctic. As Cécile Pelaudeix (2018) showed, differences in regulation among Norway, Denmark/Greenland, and Canada were tightly related to the respective energy policies. Thus, agreement would have required "advances to be made across the very delicate sphere of industrial standards and corporate secrets" (Vasiliev, 2015, p. 147). Because variance in national regulation was too broad to be harmonized, cooperation on the topic of oil spill prevention remained on the soft law level. If the agreements do not add anything, what do they do instead?

We feel that these agreements are best understood as establishing a conflictive nexus and thus defensive link between the global conventions and regional governance. By agreeing on the SAR and MOSPA, the competence to discuss related matters was moved to the regional level at a time where they became more relevant. That they do not go beyond existing obligations suggests that they might have a symbolic function primarily. They signal who is in charge (Rottem, 2015). The same seems to apply to the Science agreement, for which Akiho Shibata has pointed out that, against the backdrop of UNCLOS provisions, it raises the question of the "balance between universalism and regionalism" (Shibata, 2019, p. 214). While Shibata concluded that there is no necessary legal conflict in the provisions of the agreement with UNCLOS, he also pointed out that the agreement signals a shift to a regional "paradigm" for scientific activities. By limiting the opportunities for other polar research nations from actively participating in the agreement (such as meeting of the parties), here as in the other two cases, Arctic states assumed at least political prerogatives on the regional level against the rights and interests of non-Arctic states.

## Conclusion: How to Make Sense of the Regional-Global Nexus in the Arctic

Our brief analysis of the links between regional cooperation and global conventions in the Arctic has revealed that, in most of the issue areas under scrutiny here, the nexus rather falls short of adding value for Arctic governance or the governance of the Arctic, and where it does not, the contribution is modest at best.

Regarding UNCLOS, we argued that regional cooperation initially featured indifference due to political sensitivities. Even when cooperation intensified, the Arctic states failed to establish a more cooperative nexus. Rather, the A5 have taken defensive action to limit other states' claims in the Arctic Ocean. Regarding

---

[12] For MOSPA see Byers and Stoller (2013); for SAR see Kao et al. (2012), and Rottem (2015).

indigenous peoples' rights and participation, the Arctic Council is certainly special in terms of how it has institutionally involved IPOs. However, that has not translated into a harmonious or cooperative role of the AC in the global governance of indigenous affairs. Even on the problem that transforms the entire region and arguably is the most serious environmental threat today – climate change – regional cooperation has, again after starting with stated indifference of regional cooperation as to global governance in the AEPS, established a limited cooperative nexus to global governance at best. A harmonious nexus regarding inside-out activities can be observed in the often-quoted cases of the Stockholm and the Minamata Conventions. However, even these cases need to be qualified to the effect that the contribution of regional cooperation to global governance is modest at best. While some conventions were tackled on the regional level, the Arctic cooperation did not significantly influence ratification patterns and where the regional contribution to implementation produced regional recommendations (such as for EIA in the Arctic), there was little implementation review or follow-up. Therefore, it is difficult to see that the Arctic Council has added more than just minor attempts to make the regional level relevant for the implementation of global conventions. Finally, the merits of the regional agreements reached under the auspices of the Arctic Council make the most sense when understood as a signal to non-Arctic actors that the Arctic states are and want to remain in control of governance in the region.

Regional cooperation and the AC's institutional set-up, output, and outcomes do not show significant signs of shared sovereignty or of delegation of authority or effective embedding in a global governance architecture for the region. In a sense, Arctic governance remains strictly "Westphalian" (cf. Pelaudeix, 2014). Regional cooperation is determined by differing and narrowly interpreted national interests, with respective pivotal states setting the lowest common denominator for agreement, and by governments seeking to preserve their prerogatives. Moreover, it becomes more regional and less global the more it came into the global spotlight and non-Arctic states developed interest in the region. It seems that Arctic states can agree to keep commitment low and non-Arctic actors out (cf. Humrich, 2017). This casts some doubts on the capacity of Arctic regional cooperation to successfully manage the problems of global impacts *within* and world-wide implications *of* the Arctic. Therefore, making global conventions count for regional cooperation will require the engagement of civil society. It needs to pressure the governments nationally, regionally, and globally to live up to the magnitude of the challenges brought about in the region by globalization, climate change, and pollution. The fact that the Arctic is in the global spotlight might help in that, at least.

# References

A5. (2008). *The Ilulissat Declaration*. Ilulissat.
AEPS. (1991). *Arctic environmental protection strategy*. Rovaniemi.
AEPS. (1993). *The Nuuk declaration on environment and development in the Arctic*. Nuuk.

AEPS. (1997). *The Alta declaration on the Arctic environmental protection strategy*. Alta.
Åhrén, M. (2007). The Saami convention. *Gáldu Čála, 3*, 8–39.
Arctic Council. (1996). *Arctic Council declaration*. Ottawa.
Arctic Council. (2009). *Tromsø declaration*. Tromsø.
Arctic Council. (2013a). *Kiruna declaration.*. Kiruna.
Arctic Council. (2013b). *Kiruna vision for the Arctic*. Kiruna.
Biermann, F., Pattberg, P., van Asselt, H., & Zelli, F. (2009). The fragmentation of global governance architectures: A framework for analysis. *Global Environmental Politics, 9*(4), 14–40.
Bognar-Lahr, D. (2020). In the same boat? A comparative analysis of the approaches of Russia and Canada in the negotiation of the IMO's mandatory Polar Code. *Ocean Development & International Law, 51*(2), 143–161.
Borgerson, S. G. (2008). Arctic meltdown. The economic and security implications of global warming. *Foreign Affairs, 87*(2), 63–77.
Byers, M., & Stoller, M. (2013). What small teeth you have: Arctic oil spill response agreement weakened by conflicting interests. EU Arctic Information Centre. www.arcticinfo.eu/en/features-what-small-teeth-you-have.
China. (2018). *China's Arctic policy* (White Paper). State Council Information Office of the People's Republic of China.
Dodds, K. (2013). The Ilulissat declaration (2008): The Arctic states, 'law of the sea', and Arctic Ocean. *SAIS Review of International Affairs, 33*(2), 45–55.
Duyck, S. (2012). Which canary in the coalmine? The Arctic in the international climate change regime. *The Yearbook of Polar Law Online, 4*(1), 583–617.
Duyck, S. (2015). What role for the Arctic in the UN Paris climate conference (COP-21)? In L. Heininen, H. Exner-Pirot, & J. Plouffe (Eds.), *Arctic yearbook 2015. Arctic governance and governing* (pp. 413–421). Northern Research Forum.
English, J. (2013). Ice and water. In *Politics, people, and the Arctic Council*. Allen Lane.
Finger, M., & Heininen, L. (2019). Introduction. In M. Finger & L. Heininen (Eds.), *The global Arctic handbook* (pp. 1–8). Springer.
Heininen, L. (2018). *Arctic Council nomination letter for the Nobel Peace Prize* (15 January 2018). University of the Arctic Thematic Network on Geopolitics and Security. Retrieved from https://de.scribd.com/document/369274017/Arctic-Council-Nomination-Letter-for-the-Nobel-Peace-Prize
Henning, C. R. (2017). Avoiding fragmentation of global financial governance. *Global Policy, 8*(1), 101–106.
Hoel, A. H. (2007). Climate change. In O. S. Stokke & G. Hønneland (Eds.), *International cooperation and Arctic governance. Regime effectiveness and northern region building* (pp. 112–137). Routledge.
Humrich, C. (2017). The Arctic Council at Twenty: Cooperation between governments in the global Arctic. In E. Conde & S. I. Sánchez (Eds.), *Global challenges in the Arctic region. Sovereignty, environment and geopolitical balance* (pp. 149–169). Routledge.
Humrich, C. (2018). Souveränitätsdenken und Seerecht. Regionalisierung von Meerespolitik in der Arktis als neue Staatsräson. In M. Albert, N. Deitelhoff, & G. Hellmann (Eds.), *Ordnung und Regieren in der Weltgesellschaft* (pp. 211–241). Springer.
Hurrell, A. (2007). One world? Many worlds? The place of regions in the study of international society. *International Affairs, 83*(1), 127–146.
Kacowicz, A. M. (2018). Regional governance and global governance: Links and explanations. *Global Governance, 24*(1), 61–80.
Kahler, M. (2017). Regional challenges to global governance. *Global Policy, 8*(1), 97–100.
Kao, S.-M., Pearre, N. S., & Firestone, J. (2012). Adoption of the arctic search and rescue agreement: A shift of the arctic regime toward a hard law basis? *Marine Policy, 36*(3), 832–838.
Karlaganis, G., Marioni, R., Sieber, I., & Weber, A. (2001). The elaboration of the 'Stockholm Convention' on Persistent Organic Pollutants (POPs): A negotiation process fraught with obstacles and opportunities. *Environmental Science and Pollution Research, 8*(3), 216–221.

Keil, K., & Knecht, S. (2017). Introduction: The Arctic as a globally embedded space. In K. Keil & S. Knecht (Eds.), *Governing Arctic change. Global perspectives* (pp. 1–18). Palgrave Macmillan.

Koivurova, T. (2009). Governance of protected areas in the Arctic. *Utrecht Law Review, 5*(1), 44–60.

Koivurova, T., & VanderZwaag, D. L. (2007). The Arctic Council at 10 years: Retrospect and prospects. *UBC Law Review, 40*(1), 121–194.

Koivurova, T., Kankaanpää, P., & Stępień, A. (2015). Innovative environmental protection: Lessons from the Arctic. *Journal of Environmental Law, 27*(2), 285–311.

Kuersten, A. (2016). The Arctic Five vs the Arctic Council. In L. Heininen, H. Exner-Pirot, & J. Plouffe (Eds.), *Arctic yearbook 2016. The Arctic council: 20 years of regional cooperation and policy shaping* (pp. 434–440). Retrieved from www.arcticyearbook.com

Mahon, R., & Fanning, L. (2019). Regional ocean governance: Polycentric arrangements and their role in global ocean governance. *Marine Policy, 107*, 103590.

Molenaar, E. J., & Elferink, A. G. O. (2009). Marine protected areas in areas beyond national jurisdiction. The pioneering efforts under the OSPAR convention. *Utrecht Law Review, 5*(1), 5–20.

Oberthür, S., & Stokke, O. S. (Eds.). (2011). *Managing institutional complexity. Regime interplay and global environmental change*. MIT Press.

PAME. (1996). Working Group on Protection of the Arctic Marine Environment (Report to the 3rd Ministerial Conference on the Protection of the Arctic Environment, Inuvik, Canada, 20–22.03.1996). Miljøverndepartementet, Oslo.

Pelaudeix, C. (2014). What is "Arctic governance"? A critical assessment of the diverse meanings of "Arctic governance". *Yearbook of Polar Law, 6*(1), 398–426.

Pelaudeix, C. (2018). Governance of offshore hydrocarbon activities in the Arctic and energy policies: A comparative approach between Norway, Canada and Greenland/Denmark. In C. Pelaudeix & E. M. Basse (Eds.), *Governance of Arctic offshore oil and gas* (pp. 108–126). Taylor & Francis Group.

Platjouw, F. M., Steindal, E. H., & Borch, T. (2018). From Arctic science to international law: The road towards the Minamata convention and the role of the Arctic Council. *Arctic Review on Law and Politics, 9*, 226–243.

Rahbek-Clemmensen, J., & Thomasen, G. (2018). *Learning from the Ilulissat initiative. Report*. Centre for Military Studies, University of Copenhagen.

Rochette, J., Unger, S., Herr, D., Johnson, D., Nakamura, T., Packeiser, T., … Cebrian, D. (2014). The regional approach to conservation and sustainable use of marine biodiversity in areas beyond national jurisdiction. *Marine Policy, 49*, 109–117.

Rottem, S. V. (2015). A note on the Arctic Council agreements. *Ocean Development & International Law, 46*(1), 50–59.

Rottem, S. V. (2017). The use of Arctic science: POPs, the Stockholm Convention and Norway. *Arctic Review on Law and Politics, 8*, 246–269.

Sale, R. (2009). The scramble for the Arctic. In *Ownership, exploitation and conflict in the far North*. Frances Lincoln.

Selin, H. (2017). Global environmental governance and treaty-making: The Arctic's fragmented voice. In K. Keil & S. Knecht (Eds.), *Governing Arctic change. Global perspectives* (pp. 101–120). Palgrave Macmillan.

Semb, A. J. (2012). Why (not) commit? – Norway, Sweden and Finland and the ILO Convention 169. *Nordic Journal of Human Rights, 30*(2), 122–147.

Shibata, A. (2019). The Arctic science cooperation agreement: A perspective from non-Arctic actors. In A. Shibata, N. Sellheim, M. Scopelliti, & L. Zou (Eds.), *Emerging legal orders in the Arctic: The role of non-Arctic actors* (pp. 207–225). Routledge.

Stokke, O. S. (2007a). A legal regime for the Arctic? *Marine Policy, 31*(4), 402–408.

Stokke, O. S. (2007b). Examining the consequences of Arctic institutions. In O. S. Stokke & G. Hønneland (Eds.), *International cooperation and Arctic governance. Regime effectiveness and northern region building* (pp. 13–26). Routledge.

Stokke, O. S. (2009). Protecting the Arctic environment: The interplay of global and regional regimes. *The Yearbook of Polar Law Online, 1*, 349–369.

Stokke, O. S., & Hønneland, G. (Eds.). (2007). *International cooperation and Arctic governance. Regime effectiveness and northern region building.* Routledge.

VanderZwaag, D. (2015). The 2013 Minamata convention and protection of the Arctic environment: Mercurial promises and challenges. *China Oceans Law Review, 11*(2), 224–243.

Vasiliev, A. (2015). Agreement on cooperation on Arctic marine oil pollution preparedness and response. In N. Loukacheva (Ed.), *Polar law textbook* (pp. 145–154). Nordic Council of Ministers.

Yamiveva, Y., & Kulovesi, K. (2018). Keeping the Arctic white: The legal and governance landscape for reducing short-lived climate pollutants in the Arctic region. *Transnational Environmental Law, 7*(2), 201–220.

Young, O. R. (1996). Institutional linkages in international society: Polar perspectives. *Global Governance, 2*(1), 1–23.

Young, O. R. (1998). *Creating regimes. Arctic accords and international governance.* Cornell University Press.

Zou, L. (2016). Stirred waters under the ice cap: An analysis on A5's stewardship in the Central Arctic Ocean fisheries. In L. Heininen, H. Exner-Pirot, & J. Plouffe (Eds.), *Arctic yearbook 2016. The Arctic council: 20 years of regional cooperation and policy shaping* (pp. 457–469). Retrieved from www.arcticyearbook.com

**Cécile Pelaudeix** is a Research Fellow at PACTE Sciences Po Grenoble since 2020, and external lecturer in International Relations at Sciences Po Grenoble since 2013. She was Associate Professor at Aarhus University at the Law Department, and Associated Researcher at the Arctic Research Center (2014–2018).

**Christoph Humrich** is an Assistant Professor for International Relations and Security Studies at the Centre for International Relations Research, and Associated Researcher at the Arctic Centre, University of Groningen (RUG), The Netherlands.

# Chapter 24
# Arctic Order(s) Under Sino-American Bipolarity

**Rasmus Gjedssø Bertelsen**

## Introduction: International Order and Arctic Order

For centuries, the Arctic has been part of international political, economic, security, and technology systems (Heininen & Southcott, 2010).[1] The current Arctic order reflects the current international system, and the Arctic order of the future will reflect future international system and order. The international order and the Arctic order reflects the international system with its distribution of power between the most powerful state actors. This chapter analyzes and discusses Arctic order in light of international order inspired by John Mearsheimer (Mearsheimer, 2019).

International order is how superpowers and great powers organize their coexistence in the international system concerning security and economy and other cross-border issues, such as climate change today. The strongest power(s) in unipolar, bipolar, or multipolar systems have created orders and institutions to advance their interests. Therefore, superpowers or great powers usually follow the institutions of their order, but superpowers or great powers will violate their own order and institutions, if necessary. The US has and continues to support autocratic regimes. These strongest powers will coerce weaker states to follow the order and the institutions. It is questionable to what extent institutions have a life of their own independent of the powers behind them. Arctic orders also have superpower-sponsored or -tolerated institutions to ensure cooperation or advancement of interests such as the Arctic Council, Arctic Economic Council, or the Polar Code today.

---

[1] I thank postdoc Mariia Kobzeva, UiT The Arctic University of Norway, for comments and exchanges preparing this chapter.

R. G. Bertelsen (✉)
UiT the Arctic University of Norway, Tromsø, Norway
e-mail: rasmus.bertelsen@uit.no

The relationship among the distribution of power in the international system, the international order, and the Arctic order was clear during the Cold War and the post-Cold War era. This chapter looks at international order and Arctic order during the Cold War, post-Cold War, and emerging Sino-American bipolarity. The close connection between Cold War and post-Cold War order and the Arctic order makes it credible to discuss the likely emerging Arctic order under foreseeable international order.

The international systemic developments shaping the future Arctic order are the decline of the post-Cold War liberal international order (J. J. Mearsheimer, 2019), democratic and socioeconomic crisis in the West (Case & Deaton, 2020), the return of Russia as a great power (J. J. Mearsheimer, 2014), and power transition from West to East with the return of Asia to its historical relative position in the world economy, especially due to four decades of historically unprecedented growth in China (Asian Development Bank, 2011). It would also be interesting to look at the Arctic order in the multipolar order before World War II and the Cold War, but that is beyond the scope of this paper.

International order can be categorized along several parameters, which will be used in this chapter to analyze and discuss the relationship between international order and Arctic order. The premise of international order is the anarchic nature of the international system forcing the strongest states into self-help and weaker states in reliance on strong states for survival (Herz, 1950; Waltz, 1959).

## *Unipolar, Bipolar, and Multipolar Systems*

The international system's first characteristic is the distribution of power between the strongest states. The fundamental measure of power is the absolute size and sophistication of the national economy, which is the basis of military, science, technology, etc. Economic power itself is dependent on other factors, such as demography, natural resources, and political system, which are assumed here. An international system with one overwhelming state is unipolar, which was the case with the US after the Cold War. When two states stand out from other strong states, the system is bipolar, as was the case with the US and the USSR during the Cold War, and with the US and China increasingly today. When three or more strong states stand out, a system becomes multipolar, as was the case with the European great powers (including Russia), the US, and Japan before World War II. Any international order other than unipolarity force superpowers and great powers into security competition for the anarchic logic of the international system. A bipolar world order forces the two competing superpowers into security competition between them at a global scale (Mearsheimer, 2019; Waltz, 1979).

To describe states relatively to each other, I will use the term superpower for the largest national economies and usually with global economic, military, political, scientific, and technological reach. The US is, and historically the USSR was, a superpower. Today, the People's Republic of China today is the other (by far) largest national economy, together with the US, and has global economic engagements, but does not yet have global military reach. The old great powers – Britain, France, Germany, and Russia – are relatively great powers for their diplomatic status (United Nations Security Council Permanent Members, Britain, France, and Russia), nuclear weapons (Russia in particular, but also Britain and France), global military reach (Britain and France and somewhat Russia), or size and sophistication of economy (Germany). Japan is an economic and science and technology great power that has a very large and sophisticated navy for the Western Pacific, and could most probably develop nuclear weapons very quickly, if it decided to do so. India is a great power for its demography, absolute size of its economy, nuclear weapons, and position in the Indo-Pacific region.

The eight Arctic Council member states and the Arctic Council observer states represent a range of states with their respective structural position in the international system, international order, and Arctic order (Bertelsen et al., 2016). The US is a superpower and thinks and acts like one, Russia is a great power, Canada is a middle-power (another similar middle-power would be Australia), and the five Nordic states are small states. Among the Arctic Council observer states, China is an emerging superpower, while Britain, France, Germany, India, and Japan are great powers in different diplomatic, military, economic, or science and technology aspects.

## *Realist, Agnostic, or Liberal International Order and Arctic Order*

The international system can be realist, agnostic, or ideological depending on the distribution of power among superpowers and great powers and their domestic political systems and ideologies (Mearsheimer, 2019). The ideological nature of the international system and its leading powers have deeply affected Arctic order and will continue to do so. Any bipolar or multipolar system is, by necessity, realist, because the security competition will force the leading powers to ignore their ideological preferences for competing with the other superpower. This dynamic would also be clear between competing liberal powers, as it was clear before World War I. It was clear during the Cold War that the US supported autocracies violating human rights to compete with the USSR regionally and continue to support and tolerate regimes as the Kingdom of Saudi Arabia or Egypt.

If the system is unipolar, the ideology of the leading power can have widespread effects, which is clear from the post-Cold War liberal international order. The US, unchecked by a peer competitor, could engage in spreading liberal democratic practices and institutions throughout the world. The USSR also sought to spread its domestic ideology and would have spread its brand of communism had it won the Cold War. An agnostic leading power could certainly be ideological at home and would act forcefully abroad to protect domestic political stability, but would not engage in regime change abroad. Current-day China is such an agnostic power, concerned with domestic political stability and highly attentive to outside threats to domestic regime stability (as any recognition of the Dalai Lama), but has no interest in exporting its domestic ideology of communist party-state legitimized by economic growth and nationalism. In contrast, Maoist China was exporting its brand of communist revolution to developing countries and Maoist political groups in the West. Unlike the USSR, Russia today has no ideology to export to other societies.

## *International Order and Bounded Order, Also in the Arctic*

There can be order at two levels. International order involves all or most of the superpowers and great powers and is global in nature. The Arctic reflects the international order. Bounded regional orders are organized by superpowers to control and harness their smaller allies and clients for security competition with competing superpowers or great powers (Mearsheimer, 2019). The Arctic was divided into competing Cold War bounded orders and is being divided again.

Super powers and great powers design the international order to take care of the security and some economic questions, as well as possible cross-cutting questions. During the Cold War, the international order by the US and the USSR focused on stable mutual deterrence, arms control, and non-proliferation, where the two superpowers cooperated to keep their nuclear monopoly. In contrast, there was very little economic exchange and integration between the Eastern and Western blocs, so there was very little need for international economic order and coordination between the US and the USSR. It was clear that there was little Cold War international order in the Arctic outside the strategic nuclear domain.

If the international system is bipolar or multipolar, superpowers and great powers will be forced into security competition by the logic of anarchy. This security competition will most likely lead to each superpower establishing its regional bounded orders. The US built a Western bounded order with NATO, OECD, GATT, and Bretton Woods Institutions, supporting the European Communities. These institutions of the US bounded order all had the clear aim of strengthening the West in the security and economic competition with the USSR and its bounded order organized in the Warsaw Pact, Comecon, etc. The US and Soviet bounded orders clearly divided the Cold War Arctic.

## *Thick or Thin Order*

Orders can differ in the breath and depth of their influence. How wide are the topic areas they cover? How deep is their influence on state behavior? The Cold War Arctic order only reflected the security competition and mutual deterrence of the superpowers, but it deeply affected Canadian and Nordic behavior, where NATO allies followed American policy closely. The post-Cold War Arctic order is much broader in scope with sustainable development, environmental protection, indigenous peoples' rights, but excluding military security.

## Cold War International Order and Arctic Bounded Orders

The Arctic order closely reflected the Cold War international order driven by US–USSR bipolarity. The Cold War international order was firm (rather than the emerging loose Sino-American bipolarity). The US and USSR were by far the relatively largest national economies, militaries, technology and science systems, etc., to emerge from the carnage of World War II, which destroyed Germany and Japan and seriously weakened Britain and France as great powers. These two superpowers directly confronted each other in Europe along the Iron Curtain, which was an exceptionally militarized zone for what was nominally peacetime, with millions of heavily armed troops facing off from the Barents Sea to the Mediterranean. Due to nuclear science and technology developments, this superpower competition froze into nuclear mutual deterrence with countless proxy wars in developing countries (Westad, 2005). The destructive potential of nuclear weapons made it imperative for the US and the USSR to manage a stable mutual deterrence or strategic stability with elaborate doctrine on both sides and extensive arms control diplomacy.

The Cold War Arctic closely reflected the bipolar Cold War international order. The Cold War Arctic was first and foremost a geostrategic theatre for strategic nuclear weapons systems and distant early warning because of the shorter flight-paths for bombers and missiles across the North Pole between Eurasia and North America. At the US and USSR superpower level, the Arctic was directly integrated into the two superpowers' institutions in mutual deterrence and arms control to manage their nuclear competition (Lindsey, 1989). This nuclear international order was imposed on the Arctic and its local and indigenous communities with no say, as was clear, for instance, in the forced relocation of the Inughuit community at Dundas/Uummannaq in Northwest Greenland in 1953 to make space for the Thule US Air Force base. Small US allies like the Kingdom of Denmark, Norway, and Iceland had no say in these Arctic mutual deterrence questions, although they did supply the territory in Greenland (Christensen & Nehring, 1997), the Keflavik Base (Ingimundarson, 2011), and deep Norwegian-US signal and electronic intelligence cooperation in Northern Norway (Wormdal, 2015). The Arctic as such clearly illustrated the thin Cold War international order between the US and the USSR with hardly any other topics of governance.

The Arctic was clearly separated into the two bounded orders of the US and the USSR. There was a strict separation between the Soviet Arctic and the Nordic and North American Arctic, with the Iron Curtain separating the Soviet and Finnish and Norwegian Arctic (Holtsmark, 2015), and the Ice Curtain through the Bering Strait separating traditionally linked indigenous peoples in the Soviet Far East and Alaska (Ramseur, 2017). In the overall Cold War order, there was very little economic or technological integration between the Eastern and Western blocs, unlike today. Rare and very small exceptions included the relatively large Icelandic economic relationship with the East Block in fish, oil and shipbuilding (Petkov, 1995), and the Faroese–Soviet fisheries relationship, which caused great concern among Danish officials (Jensen, 2003). There was extremely limited Circumpolar Arctic cooperation and exchange, with the 1973 Agreement on the Conservation of Polar Bears between Canada, the Kingdom of Denmark, Norway, the US and the USSR, or the 1976 Norwegian-Soviet joint fisheries commission for the Barents Sea as notable exceptions.

The nuclear international order and the two bounded orders dividing the Cold War Arctic were materially clear from the extensive strategic infrastructure built by the two superpowers in each their Arctics. The USSR built an extensive nuclear weapons, distant early warning, and air defense infrastructure across its Arctic from the Kola Peninsula to the Far East. The US built a parallel nuclear weapons, distant early warning, and air defense infrastructure from the Aleutians, Alaska, via the Canadian High North, Greenland, Keflavik, Northern Norway, and Britain (Lindsey, 1989). Much of this strategic infrastructure lives on and has been continuously developed, especially concerning missile defense and space security, which is touched upon below.

## Post-Cold War International Order – Post-Cold War Arctic Order

The end of the Cold War and the dissolution of the USSR was brought on by the internal socio-economic and political dissolution of the East Block (Paine, 2021). This collapse had severe social, health, and human security consequences within post-Soviet society. It ended the bipolar world order bringing on a unipolar US world order, where the US and its West-European allies in the European Union could pursue their liberal projects unrestrained. Expanding the EU and NATO in Central and Eastern Europe went well, until Russia could block this expansion in Georgia in 2008 and Ukraine 2014. Attempts at regime change and nation building in Afghanistan, Iraq, Libya, and Syria went disastrously wrong (Mearsheimer, 2014; Mearsheimer, 2019).

In the Arctic, the end of the Cold War and the post-Cold War, the international order was clearly reflected in small state and middle-power liberal initiatives. Finland sensed the end of the Cold War and initiated the Rovaniemi Process in 1989, leading to the Arctic Environmental Protection Strategy in 1991 with all eight Arctic states: Canada, the Kingdom of Denmark, Finland, Iceland, Norway, Sweden, the US and the USSR. For Finland, as a small state bordering the USSR/Russia, it was an adept use of environmental policy as soft foreign and security policy (Heininen, 2013). Norway followed in 1993 initiating the Barents Cooperation between Norway, Sweden, Finland, and Russia, and with the European Commission, the Kingdom of Denmark and Iceland as additional member states. The US, Canada, and a number of European countries are observers (Holtsmark, 2015). The middle-power Canada continued the Finnish Arctic Environmental Protection Strategy by establishing the Arctic Council in 1996 (English, 2013). The US tolerated these small state and middle-power ally liberal initiatives, but vetoed including military security in the Arctic Council, which would have threatened superpower security interests. Arctic military security is driven from the nuclear strategic level, and the Arctic Council membership is not relevant for this topic.

The relaxation of the dangerous international systemic and order constraints allowed these small and middle-power states to pursue such soft foreign and security policy. The dangerous Soviet opponent of these countries had been greatly weakened and subsequently dissolved, and the American protector of Canada and Norway was the undisputed liberal hegemon of the international system. These initiatives were liberal in nature, focusing on individual human rights, human security through environmental protection, sustainable development, and indigenous peoples' rights at local and regional level. These initiatives were not realist because they did not focus on state military security.

The dissolution of the USSR brought enormous socioeconomic and comprehensive security disruption to the former East Block. It led to severe social and health repercussions throughout post-Soviet society and open conflict in, for instance, the Caucasus, or further afield, the Balkans. The post-Soviet Arctic faced the abandonment by and great retraction of the state removing all kinds of basis of livelihoods.

At the circumpolar level, the end of Cold War bipolarity allowed for liberal Circumpolar Arctic cooperation in areas as environmental protection, sustainable development, indigenous peoples' rights because of the end of US–USSR global security competition (where the Arctic had played a central geostrategic role for mutual deterrence), US unipolarity and dissolution of the USSR with deep socioeconomic crisis in Russian society. This liberal cooperation and liberal order was clearly reflected in the Rovaniemi Process with the Arctic Environmental Protection Strategy, the Barents Cooperation, the Arctic Council and cross-Bering Strait exchanges and cooperation.

This liberal Circumpolar Arctic cooperation has been highly beneficial for the old Western bloc Arctic in the Nordic countries and North America. I am not able to estimate the net effect on the Russian Arctic, taking the welfare cost of the dissolution of the USSR into account. The benefits to the Nordic and North American Arctic fit into a general Western narrative of the benefits of the end of the Cold War and the dissolution of the USSR. The West benefitted greatly from these international systemic developments, which characterized the 1990s with the peace dividend, the possibility to reunify Germany and Europe, great technological and economic development, and growth with the internet. It was a golden age for the West.

The carefreeness and great net benefits of the end of the Cold War and American unipolarity and derived hegemony gave rise to an "End of History" mentality and narrative in the West, where Western models of liberal democracy and capitalist market economy had defeated alternative political and economic models (Fukuyama, 1992). History naturally continued and continues at the global level and in the Arctic, which appears to be cognitively challenging for Western observers.

The stark difference in experiences during and memories of the 1990s inform conflicts between Russia and the West today. For the West, the dissolution of the "Evil Empire" (the USSR) was an undisputed good. This dissolution might well be seen as a catastrophe for post-Soviet families who suffered the social, health, and safety consequences.

The Arctic has its own "End of History" narrative, seeing the liberal Circumpolar Arctic Council Arctic as a natural state and exceptional to emerging international systemic change. There seems to be little understanding that a liberal Circumpolar Arctic reflect a unipolar international system guided by a US liberal hegemon. Now the international system and order is changing, and so is the Arctic and its order.

## The Return of Russia as Great Power in the Caucasus and Eastern Europe and Effects on Arctic Order

In recent years Russia has recovered from the depth of its socioeconomic crisis in the 1990s and early 2000s following the dissolution of the USSR. During that crisis, Russia could not behave as a usual great power, imposing a sphere of influence around itself, which allowed the expansion of NATO and the European Union into the former East bloc and post-Soviet republics (Mearsheimer, 2014). That expansion was democratically mandated by these new member states, but such democratic mandates do not matter if great powers can assert different intents, as societies experienced during the Cold War from both superpowers in Central Europe, the Middle East, or Latin America and further, and which continues.

### *Russian-European Competition in the Caucasus, Black Sea, Eastern Europe and Baltic*

In 2008, Russia blocked the path of Georgia towards NATO by military action. In 2014, Russia likewise secured the Crimean Peninsula, which held strategic importance for Russia, through military action in the face of Ukrainian rapprochement to the EU. The Ukraine crisis set off an unprecedented crisis between Russia and the West, which raised speculation about similar Russian action in the Arctic as in Ukraine. Russia faced specific strategic challenges in the Black Sea region, which it could only address by hard power for lack of soft power.

## Arctic Consequences of Russian-European Eastern European Competition

The Ukraine crisis has had a significant effect on the Arctic and Arctic order through Western horizontal escalation by financial and technological sanctions against Russian Arctic oil and gas developments, which forced Russia to turn to China for funding (Farchy & Mazneva, 2017). This forced Arctic rapprochement between Russia and China has become part of the larger systemic development of power transition from West to East and its effects in the Arctic. Western horizontal escalation of the Ukraine crisis has affected the Circumpolar order through Canadian demonstrations at the Arctic Council. Arctic military security dialogues with Russia have been discontinued, making such dialogues meaningless, and now need to be reinvented.

A marginal Arctic consequence of the Ukraine crisis has been Russian counter sanctions against Icelandic and Norwegian fish exports to Russia. Russia skillfully exploited the special status of the Faroe Islands within the Kingdom of Denmark and towards the EU, as well as Faroese–EU fishing conflicts. Russia exempted the Faroe Islands from sanctions, and Faroese fish exports to Russia expanded considerably, driving economic and demographic growth in the Faroe Islands, while sowing discord in the Danish–Faroese relationship (Rigsombudsmanden på Færøerne, 2014).

## Loose Sino-American Bipolarity, US and Chinese Bounded Orders in the Arctic

The international system is entering an era of loose Sino-American bipolarity (Tunsjø, 2018), but how will this shape the Arctic? The system will be bipolar because the American and Chinese national economies stand out as much larger than other national economies. The European Union has a comparable economy in terms of size, but the EU lacks the political integration to act as third pole. The other major actor for the purposes of this analysis and discussion, Russia, has a much smaller national economy than the US, China, or the EU, and is only a pole concerning nuclear weapons. The emerging Sino-American bipolarity can be characterized as loose, since other actors, like the EU, Russia, major European countries, Japan, and India, are much stronger and have much more freedom to maneuver than secondary powers to the US and the USSR had during Cold War bipolarity and after the WWII carnage.

At the global level, the fundamental change has been the power transition from West to East with the economic return of Asia and China in particular. The international system has been dominated by Europe and North America for some centuries, which was first clear from European imperialism and colonialization of the rest of the world, world wars originating in Europe, the Cold War (also originating in Europe), and finally US unipolarity and hegemony. Russia and the USSR have historically been part of this Western center of the international system. A European–North-American-centered world is a historical anomaly. The demographic and economic center of the world is Asia. All other things being equal, economic activity reflects demography. Therefore, Asia has historically had the largest economies in the world. The world was much less interconnected before European imperialism and globalization. The interconnections of empires and current globalization mean that the relative shifts in the size of national economies today are felt much more acutely around the world, with globalization's compression of time and space (Harvey, 1989). Into the 1800s, Asia occupied more than half of the world economy, a figure that fell to less than 20% in the 1950s due to imperialism, world wars, and revolutions. Now Asia is regaining its relative share of the world economy (which is, of course, much larger than it was before the 1800s in absolute terms). This return of Asia has been driven by the strong growth in Japan after World War II followed by the flying geese of first South Korea, Singapore, Taiwan, and Hong Kong, secondly Thailand, Malaysia, and Indonesia, and finally China, Vietnam, the Philippines, etc. The economic growth of China since DENG Xiaoping's Open Door policy from 1978 is unprecedented in economic history (Asian Development Bank, 2011).

## *Sino-American International Order as Framework for Arctic Bounded Order*

The primary global security competition will be Sino-American and centered on the South China Sea in particular and East Asia in general. The US will seek to reinforce its alliances with countries like Japan, South Korea, Australia and India in a containment logic of China. Domestic American support for this engagement will be a critical variable, as was clear from the election of Donald Trump as president in 2016 and the hostility of his administration to multilateral trade agreements. Japan, South Korea, India, Australia, and other countries will each seek to balance their benefits from trade with China with security concerns. The narrow election of Joe Biden as US president in 2020 (within 100,000 votes could have swung the 2020 election for Donald Trump because of the Electoral College) shows the instability of American politics and policy.

The international order being painstakingly established by the US and China, with much pushback from the US, will be thicker than the thin US–USSR international order to control nuclear deterrence, arms control, and non-proliferation. The Sino-US order will touch upon many Arctic issues. The thicker Sino–US order will cover arms control and non-proliferation, which is actually still bipolar US–Russian. Arms control diplomacy must move from US–Russian to include China. The Arctic continues to be central for US–Russian strategic stability. It will also concern Sino-US strategic stability, primarily through US missile defense in Alaska. The US Department of Defense has expressed concerns for possible future Chinese SSBNs operating in the Arctic Ocean, but that would put them far from their support infrastructure and protection in the South China Sea (Stewart & Ali, 2019).

Another area of international order concerning the Arctic is space security and governance. Space is the ultimate high ground, dominating all other battlefields on land, sea, and in the air. There are strong drivers for the militarization of space, as evidenced by the reorganization of the US Air Force Space Command as the US Space Force (Krepon, 2017; The National Air and Space Intelligence Center, 2018; United States Air Force Space Command, 2018). This reorganization should not be mistaken for an idiosyncrasy of President Trump (Trump, 2019). Equally, the French Air Force has been renamed the Air and Space Force, and President Emmanuel Macron stated in 2019 that France will acquire the independent capability to defend itself in space (Macron, 2019). China in 2013 and India in 2019 have carried out tests of anti-satellite weapons, shooting down their own discarded satellites. The basis of space security, both defense and offense, is space situational awareness, where infrastructure in the Arctic is crucial. The US has key radars for this purpose in the Aleutians, Alaska, Thule/Greenland, and Vardø/Norway (Alby, 2013).

It is important to keep in mind, both concerning nuclear strategic stability and space security science, technology, and infrastructure in the Arctic, that it directly concerns superpower and great power security and their international order, not the Arctic per se. Although small Arctic states like the Kingdom of Denmark or Norway house this infrastructure, they have no seat at this table.

Another area of international order that concerns the Arctic at superpower and great power level is the global energy system. This system has reflected the international system. It was dominated by the US controlling the Arabian Peninsula and Persian Gulf according to the Carter Doctrine of security from the USSR and revolutionary Iran in exchange for a secure oil supply (Carter, 1981). The US Navy dominated the seas on which oil tankers moved. The USSR was a major producer for its own needs and that of its block and some exports to Western Europe (Rogers & Tynkkynen, 2021). China and emerging markets were marginal energy markets.

This energy system has shifted radically with consequences for the place of the Arctic. The US has become energy-independent and even an exporter of energy based on unconventional oil and gas. This energy independence loosens the US–Gulf security for energy relationship. China has become the world's largest energy importer, which brings China severe energy security problems from unstable supplier regions in the Middle East, Africa, and Latin America, as well as the Malacca Strait Dilemma, that its shipments can easily be interdicted by the US Navy (Zhang, 2011).

The global energy system today is characterized by an independent US, China as the largest importer, and the Kingdom of Saudi Arabia and Russia as the two major exporters competing for the Chinese market. Saudi Arabia has lower production costs, but higher political risk and further distance from China. Russian energy resources are increasingly in the Arctic, which has higher production cost, but low political risk, and proximity to China, which China is ready to pay for, for energy security reasons (Thompson, 2020). Russian Arctic energy resources play a major role in international (energy) order of superpowers and great powers, and with no seat at the table for Arctic small states and middle powers (Kobzeva & Bertelsen, 2021).

The Arctic in the global energy system points toward climate change, which greatly affects the Arctic and with feedback to the rest of the world from sea level rise, methane emissions, and albedo effects, among others. However, climate change does not originate in the Arctic (Arctic Climate Impact Assessment, 2004). Climate change can only be addressed through the land use and energy systems of the largest emitters and through UN climate diplomacy. Again, it is the international order of superpowers and great powers, with no seat at the table for Arctic small states or middle powers.

## *US and Chinese Bounded Orders in the Arctic*

The US and China will each establish their bounded regional orders for security competition, just as the US and the USSR each created their own bloc during the Cold War. The two bounded orders in the Arctic will be an American-led bounded order consisting of the Nordic and North American Arctic (Alaska, Canada, Denmark/Faroe Islands/Greenland, Finland, Norway, and Sweden). The status of the Russian Arctic will be key to the future of Arctic order. In light of Sino-American animosity and US determination to exclude China from the Nordic and North American Arctic, the Russian Arctic is the only part of the Arctic open to China – on Russian conditions.

The overall driving force in Sino-American bipolarity on the Arctic will be the combination of the US desire to keep China out of the Nordic and North American Arctic and the US/NATO/EU-Russian competition in Eastern Europe and its effects on the Sino-Russian relationship (Pompeo, 2019). The Sino-American bipolar dynamic for the Arctic will push towards the formation of a US and a Chinese bounded order in the Arctic.

## *US Arctic Bounded Order*

The US is fast reestablishing its Arctic bounded order, which in many ways is a resurrection of the US Cold War Arctic bounded order. During unipolarity, the US could step back from liberal Arctic affairs, creating a false impression of disengagement, while it remained deeply engaged at the strategic stability, missile defense, and space security level, which has been overlooked by many Arctic observers (Office of the Secretary of Defense, 2019). With the return of Russia as a great power and the rise of China, the US must recreate its bounded order in the Arctic, as it recreates bounded orders in Europe and around the world to compete with Russia and China. This bounded order also seeks to discipline and mobilize smaller allies in this competition.

The US will have a heavy strategic presence in terms of nuclear and space-related infrastructure in the Nordic and North American Arctic. The Arctic will continue to play a key role for strategic stability between the US, Russia, and increasingly China. US SSBNs may patrol in the Arctic and fast-attack submarines will threaten Russian SSBNs and any possible Chinese SSBNs.

The US may have an expanded space presence in the Arctic. There may be an increased missile defense and space situational awareness infrastructure in Alaska, such as the current Cobra Dane and Clear Air Force Station radar stations and interceptor missiles at Fort McGreely near Fairbanks. The radar station at Thule in Northwest Greenland may only increase in importance for missile defense and space situational awareness. The radar at Vardø, Eastern Finnmark, US technology operated by Norwegian defense, may also grow in importance for its closeness to the Russian Bastion around the Kola Peninsula.

Alaska has great hopes of developing an LNG sector, producing natural gas and exporting it as LNG to East Asia. Alaska has pursued this idea for decades towards Japan without result. In 2017, an agreement was made among Alaska, Sinopec, China Investment Corp, and Bank of China to possibly develop a US$43 billion LNG project (Rogers & Tynkkynen, 2021). This project sank with Sino–American animosity under the Trump Administration. It remains to be seen whether China will invest in a US-based energy supplier, which raises energy security issues for China. It also remains to be seen whether the US will tolerate such a large Chinese investment in Alaska, which is a key state for strategic stability, missile defense, and space situational awareness and power projection towards East Asia. Like Greenland, Alaska has a relatively small population and fiscal vulnerability, but naturally a state in the union.

The US–Canadian relationship within NORAD and NORTHCOM has been reinforced. The reactivation of the US 2nd Fleet covering the North Atlantic is a clear illustration of this bounded order for security competition (Burke & Davis, 2020). British Columbia has long-running aspirations for exporting LNG to East Asia, first Japan, which has never materialized. The current BC LNG export to China at 100,000 tons is marginal compared to Russian LNG exports to China (Hoong, 2019). The Sino–Canadian relationship has soured significantly, especially over the Canadian detention of the Huawei executive Meng Wanzhou at the US's request.

The Kingdom of Denmark, Iceland, and Norway are mobilized by the US to provide airfields and harbors, as well as participate in exercises close to the Russian nuclear Bastion around the Barents and Kara Seas. The US – nominally rotational – presence of P8 Poseidon anti-submarine patrol planes at Keflavik in Iceland shows the US return to the North Atlantic and Iceland. The US Navy and Air Force is much more active sailing and flying off Northern Norway and into the Barents Sea. These deployments put pressure on the Russian Northern Fleet as close as possible to its home ports on the Kola Peninsula. Small Nordic NATO allies are deeply dependent on US protection and have been put in difficult positions by these US deployments. These US deployments pressure Russia concerning its second-strike capabilities and resources for power projection south into the North Atlantic and beyond.

Norway (and Denmark) are hard-pressed to show loyalty to the US by participating in these deployments. At the same time, for Norway, participating in such deployments, which challenge Russian second strike capabilities and the foundational defense of Russia, threatens the bilateral neighborly relationship between Norway and Russia, which Norway depends on for a harmonious and practically well-functioning cross-border relationship with Russia in the Barents Sea (Nilsen, 2019a, b, 2020).

Countering China focuses on keeping Chinese investment, enterprises, and infrastructure, as well as science out of the Nordic Arctic. This exclusion is particularly clear in the Kingdom of Denmark, where the US goes to great lengths towards Denmark, the Faroe Islands, and Greenland to keep China out of the Faroe Islands and Greenland. The US will not accept Chinese presence in terms of capital or infrastructure, particularly in Greenland due to the country's unstable constitutional relationship with Denmark and very small demography and economy. The US has made this point abundantly clear to the Danish government, and the US is diplomatically and economically engaging Greenland directly. The US has equally shown to the Faroe Islands that it does not accept Chinese Huawei 5G infrastructure in the Faroe Islands, which occupies an important position in the North Atlantic. Iceland's economic and scientific relations with China have attracted much alarmist attention, but this attention illustrates the deep concern among Western observers with Chinese presence in Iceland (Higgins, 2013).

Sweden and Finland was the non-aligned buffer zone between the US and NATO and the USSR in the Nordic region from the Baltics to the Barents Sea. Swedish neutrality was always more nominal than real, with various deep collaborations with the US and NATO. Sweden and Finland are today Enhanced Opportunity Partners of NATO. They are collaborating increasingly closely for defense and with Norway in the North Calotte region. Sweden is rearming itself from reaping peace dividends of the post-Cold War period. This Swedish rearmament reflects Swedish concern with Russian reassertion of its interests in Eastern Europe, following from the Ukraine crisis. The Baltic Sea is a potential conflict area. Therefore, Sweden is reinforcing its defense of Gotland Island in the Baltic. Northern Sweden, together with Northern Finland and Norway, is part of the strategic depth for Russia around the Kola Peninsula, where Russia will seek to exclude any peer competitor from gaining a foothold. The new Swedish Arctic strategy acknowledges the strategic importance of Northern Sweden defense-wise.

Finland has historically had close economic and technological ties with the USSR and Russia. These ties are illustrated today by, for instance, the nuclear power station under construction at Hanhikivi with Russian reactor. Finland is also pursuing the project of an Arctic submarine cable from East Asia via the Northern Sea Route and Northeast Passage to Kirkenes. Such a proposal would contribute to Finland's ICT-driven growth, but it entails advanced technological collaboration with China and Russia at the strategic level. How this strategic ICT infrastructure engagement with Russia and China affects Finland's relations with the US and the EU remains to be seen.

## *The EU Between US and Sino-Russian Arctic Bounded Orders*

The European Union has at once a narrow and weak explicit Arctic actor profile in its official Arctic policy documents, and a much larger implicit Arctic actor profile in energy, ocean, and space matters (Bertelsen et al., 2020). Twenty-four percent of hydrocarbons consumed in the European Union originate from the Arctic, which largely means the Russian Arctic. The EU and Russia have a substantial energy trade and investment. Western Europe has been a significant natural gas customer of first the USSR and then Russia since the 1970s via a pipeline infrastructure crossing Eastern and Central Europe, which causes political risk to all involved parties. That infrastructure is being sought replaced by the Nordstream 1 (concluded) and 2 subsea pipelines connecting Russia and Germany directly under the Baltic Sea (Gustafson, 2020).

French Total is a very active participant in Russian natural gas production. Total holds a 20% stake in the enormous Yamal LNG project and 10% in the follow-up Arctic LNG2 project. Total is a shareholder in the Russian natural gas company Novatek and participates in developing LNG transshipment hubs on the Kola Peninsula and Kamchatka. The German oil company Wintershall DEA is also involved in Russian oil and gas production. French and German technology suppliers are active in Russian Arctic energy developments. French Technip is a key LNG technology provider for the Yamal LNG project (TechnipFMC, 2018). The author has himself seen the enormous Siemens pumps delivering oil from the Lukoil Varandey export terminal to the offshore loading point.

## *Sino-Russian Arctic Bounded Order*

The Russian Arctic appears part of the Chinese bounded order due to the Russian–Western rift made clear in the Ukraine crisis. Russia itself would prefer a multipolar world order and in the Arctic. Russia does not want to be a natural resource-province of China, to be the "Canada" of China. Russia seeks balanced European-East Asian investment in Russian Arctic energy developments and markets. The Russian-Western rift will decide whether Russia and the Russian Arctic becomes part of a Chinese bounded order (Kobzeva & Bertelsen, 2021).

There should be opportunity for the West not to push Russia further into the Chinese bounded order, as Russia prefers to stay independent of a Chinese order. Keeping Russia out of the Chinese bounded order in general, and in the Arctic specifically, will require that the West accommodates certain Russian interests contradicting the American liberal post-Cold War world order built on US unipolarity, preserves nuclear and space strategic stability, manages geopolitical competition from the Black Sea to the Baltic Sea, and that the West refrains from escalating Eastern European conflicts with Russia horizontally to the Arctic, as happened after the Ukraine crisis.

There are similarities between the diplomacy of President Richard Nixon and National Security Adviser Henry Kissinger normalizing relations with the People's Republic of China in the early 1970s and distancing the People's Republic of China from the USSR. That policy was hailed as a great strategic success for four decades, before now being criticized for strengthening a communist party state China without reforming it politically in a liberal direction (Pompeo, 2019).

Russia has consistently sought a multipolar world since this strategic interest was formulated by Foreign Minister and Prime Minister Yevgeny Primakov in the late 1990s. Primakov saw that a new US-Russian bipolarity was unattainable, US unipolarity was detrimental for Russia, and multipolarity the best attainable option for Russia (Smith, 2013). Therefore, Russia seeks a balanced relationship in the Arctic with the EU, the other Arctic states, China and other Asian states like Japan and India.

Russia's Arctic interests concern the defense and fiscal survival of the state. The Arctic is key to Russia's nuclear second-strike capability and, therefore, its ultimate line of defense. The Arctic is a natural resource base of oil, gas, and minerals, which is an essential fiscal basis for the Russian state and for private business.

Russia would prefer a balanced Arctic energy development and export heading west to Europe and east to East Asia. This balanced preference was clear from the initial composition of ownership in the Yamal LNG project, with 60% for Novatek, 20% for Total, and 20% for China National Petroleum Corporation. This composition was greatly disturbed by the fallout from the Ukraine crisis with Western financial and technological sanctions against Russia, which forced Russia to seek Chinese funding.

The upcoming Arctic LNG2 and other projects have a range of French Total, China National Petroleum Corporation, China National Offshore Oil Corporation, a Japanese consortium of Mitsui and Japan Oil, Gas and Metals National Corporation, and other investors, which shows Russia's determination to pursue multipolarity despite the bipolar pressures from a world order dominated by the US and China as by far the two largest national economies.

## Conclusion – Circumpolar Arctic Bounded Order or US and Sino-Russian Arctic Bounded Orders

There are two likely scenarios for the future Arctic under the Sino-American world order. The first is the Arctic will be divided into a US and a Chinese bounded order, with the Nordic and North American Arctic within the US bounded order and the

Russian Arctic in the Chinese bounded order. This division will result in weakened circumpolar cooperation in the Arctic Council. We already see the US working hard to keep China out of Greenland and the rest of the Nordic–Arctic region, Sino–Russian cooperation on energy and shipping, and US–Russian conflict.

The alternative Arctic order is a circumpolar Arctic, in which Russia has become part of the American bounded order. This order will require the US to divide Russia and China, as Nixon and Kissinger separated China from the Soviet bounder order after 1972. This will require the West to safeguard strategic stability, refrain from interfering in domestic Russian affairs on the environment, and human rights in general, and specifically within the Russian Arctic. China and Russia are fully aware of this logic. Russia favors a multipolar order, but the bipolar logic imposes itself from the relative economic and technological might of the US and China vis-à-vis other states, which is clear from Russian dependence on outside energy sales and investments.

In both scenarios, the Arctic will be very different from the circumpolar Arctic of the Arctic Council we have become used to since the 1990s. Circumpolar cooperation can probably be maintained especially in the fields of law of the sea, natural science research and resource exploitation. Arms control dialogue on strategic stability, nuclear weapons reduction and missile defense will be crucial for the safety of humanity. However, the environment and human rights will be downplayed in the Russian Arctic.

## References

Alby, F. (2013). 8.1. Space situational awareness systems and space traffic control. In *Safety design for space operations* (p. 412). Elsevier. https://doi.org/10.1016/B978-0-08-096921-3.00008-8

Arctic Climate Impact Assessment. (2004). *Impacts of a warming arctic: Arctic climate impact assessment* (1st ed.). Cambridge University Press.

Asian Development Bank. (2011). *Asia 2050: Realizing the asian century*. Asian Development Bank.

Bertelsen, R. G., Aronsen, K. I. R., & Nyborg, O. A. (2016). Kongeriget danmark som arktisk stat: Kommentar til redegørelsen forsvarsministeriets fremtidige opgaveløsning i arktis. *Tidsskrift for Søvæsen, 187*(2–3), 92–105.

Bertelsen, R. G., Kobzeva, M., Raspotnik, A., & Rowe, E. T. W. (2020). *Internationale forskere: EU's arktispolitik er for snæver*. Retrieved 12/04, 2020, from https://www.altinget.dk/arktis/artikel/forskere-eus-arktis-politik-er-for-snaever

Burke, S., & Davis, T. (2020). *Distinguished speakers november 2020 – Arctic continental defense: Canada and US perspectives*. Retrieved 04/23, 2021 from https://rusi-ns.ca/event/distinguished-speakers-november-2020/

Carter, J. (1981). Presidential directive/NSC-63. Persian gulf security framework (U). https://web.archive.org/web/20120915113622/http://www.jimmycarterlibrary.gov/documents/pddirectives/pd63.pdf

Case, A., & Deaton, A. (2020). *Deaths of despair and the future of capitalism*. Princeton University Press.

Christensen, S. A., & Nehring, N. (1997). *Grønland under den kolde krig: Dansk og amerikansk sikkerhedspolitik 1945–68*. Dansk Udenrikspolitisk Institut DUPI.

English, J. (2013). *Ice and water: Politics, peoples, and the Arctic Council*. Allen Lane.

Farchy, J., & Mazneva, E. (2017). *Russia wins in arctic after U.S. fails to kill giant gas project*. Retrieved 05/20, 2018, from https://www.bloomberg.com/news/articles/2017-12-14/russia-dreams-big-as-u-s-fails-to-kill-27-billion-gas-project

Fukuyama, F. (1992). *The end of history and the last man*. Free Press.

Gustafson, T. (2020). *The bridge. Natural gas in a redivided europe*. Harward University Press. https://doi.org/10.4159/9780674243842

Harvey, D. (1989). *The condition of postmodernity: An enquiry into the origins of cultural change*. Blackwell.

Heininen, L. (2013). Finland as an Arctic and European state – Finland's Northern dimension (policy). In R. Murray & A. D. Nuttall (Eds.), *International security and the Arctic: Understanding policy and governance* (pp. 294–319). Cambria Press.

Heininen, L., & Southcott, C. (Eds.). (2010). *Globalization and the circumpolar North*. University of Alaska Press.

Herz, J. H. (1950). Idealist internationalism and the security dilemma. *World Politics, 2*(2), 157–180.

Higgins, A. (2013, 03/22). Teeing off at the edge of the Arctic? A Chinese plan baffles Iceland. *The New York Times*.

Holtsmark, S. (Ed.). (2015). *Naboer i frykt og forventning: Norge og russland 1917–2014* [Neighbors in fear and anticipation: Norway and Russia 1917–2014]. Pax.

Hoong, N. W. (2019). *Canada's breakthrough LNG deal with China*. Retrieved 04/23, 2021, from https://asiatimes.com/2019/09/canadas-breakthrough-lng-deal-with-china/

Ingimundarson, V. (2011). *The rebellious ally: Iceland, the United States, and the politics of empire 1945–2006*. Republic of Letters.

Jensen, B. (2003). *Føroyar undir kalda krígnum (1945–1991)*.

Kobzeva, M., & Bertelsen, R. G. (2021). European-russian-chinese arctic energy system. In X. Li (Ed.), *China-EU relations in a new era of global transformation*. Routledge.

Krepon, M. (2017). *Op-ed | is space warfare's final frontier?* Retrieved 01/13, 2020, from https://spacenews.com/op-ed-is-space-warfares-final-frontier/

Lindsey, G. (1989). *Strategic stability in the arctic No. 241*. The International Institute for Strategic Studies.

Macron, E. (2019). *Discours d'Emmanuel Macron à l'Hôtel de Brienne* [Speech by Emmanuel Macron at l'Hôtel de Brienne]. Retrieved 12/30, 2019, from https://www.elysee.fr/emmanuel-macron/2019/07/13/discours-aux-armees-a-lhotel-de-brienne

Mearsheimer, J. J. (2014). Why the Ukraine crisis is the West's fault: The liberal delusions that provoked Putin. *Foreign Affairs, 93*(5 September/October), 77–89.

Mearsheimer, J. J. (2019). Bound to fail: The rise and fall of the liberal international order. *International Security, 43*(4), 7–50. https://doi.org/10.1162/isec_a_00342

Nilsen, T. (2019a). *B-2 stealth bomber on mission above Arctic Circle outside Norway*. Retrieved 12/15, 2019, from https://thebarentsobserver.com/en/security/2019/09/b-2-stealth-bomber-first-mission-above-arctic-circle-outside-norway

Nilsen, T. (2019b). *B-52 flights close to homeport and patrol areas for Russia's ballistic missile subs*. Retrieved 12/15, 2019, from https://thebarentsobserver.com/en/security/2019/11/us-b-52-strategic-bombers-and-norwegian-f-16s-flying-wing-wing-over-barents-sea

Nilsen, T. (2020). *U.S. sixth fleet enters the barents sea with missile defense destroyer*. https://thebarentsobserver.com/en/security/2020/05/us-sixth-fleet-enters-barents-sea

Office of the Secretary of Defense. (2019). *2019 missile defense review No. 2020*. United States Department of Defense.

Paine, S. C. M. (2021). *Why the Soviet Union lost the Cold War*. Retrieved 04/22, 2021, from https://www.clementscenter.org/press/item/2143-why-the-soviet-union-lost-the-cold-war

Petkov, B. (1995). *Trade between Iceland and the Soviet Union 1953–1993: Rise and fall of barter exchange No. 95:01*. Institute of Economic studies, University of Iceland.

Pompeo, M. R. (2019). *Remarks: Looking north: Sharpening America's Arctic Focus*. Retrieved 05/01, 2019, from https://www.state.gov/looking-north-sharpening-americas-arctic-focus/

Ramseur, D. (2017). *Melting the ice curtain: The extraordinary story of citizen diplomacy on the Russia-Alaska frontier*. University of Alaska Press.

Rigsombudsmanden på Færøerne. (2014). *Indberetning nr. 5/2014*. Rigsombudsmanden på Færøerne.

Rogers, D., & Tynkkynen, V. (2021). *Russian oil in the far North: Culture, companies, and climate.* Retrieved 04/22, 2021, from https://www.alaskaworldaffairs.org/events/russian-oil-in-the-far-north-culture-companies-and-climate-dr-doug-rogers/

Smith, M. A. (2013). Russia and multipolarity since the end of the Cold War. *East European Politics, 29*(1), 36–51.

Stewart, P., & Ali, I. (2019). *Pentagon warns on risk of Chinese submarines in Arctic.* Retrieved 06/13, 2019, from https://www.reuters.com/article/us-usa-china-military-arctic/pentagon-warns-on-risk-of-chinese-submarines-in-arctic-idUSKCN1S829H

TechnipFMC. (2018). *TechnipFMC in joint-venture with its partners JGC and chiyoda, as key players successfully contributed towards the first cargo of LNG for the megaproject yamal LNG.* Retrieved 08/06, 2020, from https://www.technipfmc.com/en/media/news/2018/04/technipfmc-in-jointventure-with-its-partners-jgc-and-chiyoda-as-key-players-successfully-contributed

The National Air and Space Intelligence Center. (2018). *Competing in space.* Retrieved 12/23, 2019, from https://media.defense.gov/2019/Jan/16/2002080386/-1/-1/1/190115-F-NV711-0002.PDF

Thompson, H. (2020). *Are we at the end of the long twentieth century?* Retrieved 12/04, 2020, from https://www.cfg.polis.cam.ac.uk/long-twentieth-century

Trump, D. J. (2019). *Space policy directive-4*. Retrieved 01/19, 2020, from https://www.spaceforce.mil/About-Us/SPD-4

Tunsjø, Ø. (2018). *The return of bipolarity in world politics: China, The United States, and geostructural realism*. Columbia University Press. Retrieved from http://search.ebscohost.com/login.aspx?direct=true&db=nlebk&AN=1708604&site=ehost-live

*AFSPC space superiority*. United States Air Force Space Command (Director). (2018). [Video/DVD] United States Space Force.

Waltz, K. N. (1959). *Man, the state, and war; a theoretical analysis*. Columbia University Press.

Waltz, K. N. (1979). *Theory of international politics* (1st ed.). McGraw-Hill.

Westad, O. A. (2005). *The global Cold War: Third world interventions and the making of our times*. Cambridge University Press. https://doi.org/10.1017/CBO9780511817991

Wormdal, B. (2015). *Spionbasen: Den ukjente historien om CIA og NSA i norge*. Pax.

Zhang, Z. (2011). China's energy security, the malacca dilemma and responses. *Energy Policy, 39*(12), 7612–7615. https://doi.org/10.1016/j.enpol.2011.09.033

**Rasmus Gjedssø Bertelsen** is Professor of Northern Studies and Barents Chair in Politics at UiT The Arctic University of Norway (Campus Tromsø). He has been guest researcher at China Nordic Arctic Research Centre in Shanghai (2016), guest professor at Kobe University (2017) and guest professor at Sorbonne University (2020–2021). He coordinates the Norwegian-Russian PhD/MA-course Society and Advanced Technology in the Arctic and the Norway-EU Science Diplomacy Network.

# Index

**A**

Abandoned projects, 309
Access, 5, 10, 14, 22, 24, 35, 47, 51, 93, 94, 98, 105–107, 129, 130, 135, 136, 141, 143, 149, 156, 230, 232, 237, 240, 242–245, 247, 281, 300, 308, 311, 339, 342, 391, 393, 397, 401, 432
Allure, 7, 130, 137–143
Alpine, 190, 192
Anthropocene, 16, 333, 335, 347, 402, 403
Arctic, 1, 21, 41, 63, 91, 113, 129, 147, 172, 190, 229, 274, 286, 297, 315, 331, 351, 370, 389, 420, 425, 443, 466
Arctic biodiversity, 223
Arctic Council, 1, 10, 14, 15, 37, 104, 111–115, 118–120, 122–125, 172–175, 180–184, 231, 238–240, 245, 247, 248, 261–263, 269, 304, 320, 322, 331, 332, 389, 394, 407–422, 427, 433, 443, 444, 450–454, 456, 458, 463, 465, 469–471, 479
Arctic gateways, 286
Arctic geopolitics, 7, 14, 92, 110, 112, 114–116, 119–124, 426, 434
Arctic infrastructure megaprojects, 299–301, 303–309
Arctic Monitoring and Assessment Programme (AMAP), 35, 111, 172–177, 179–184, 196, 222, 238, 240, 247, 409, 410, 412, 451, 453, 454
Arctic order, 15, 463–479
Arctic policy, 6, 14, 110, 113, 115, 118, 131, 136, 157, 316, 324, 371, 375, 400, 413, 416, 420, 427, 430–433, 435, 437, 477
Arctic regime, 12, 315, 316, 320–326
Arctic research, 1, 6, 64–67, 70–82, 85, 86, 114, 115, 162, 232, 268, 280, 437
Arctic shipping, 12, 118, 157, 158, 182, 316–319, 324, 325, 340, 448
Arctic states, 7, 10, 15, 16, 26–29, 105, 107, 111–115, 117, 118, 120–125, 139, 158, 229–231, 236, 238–248, 315–317, 320–322, 324, 325, 342, 343, 352, 366, 408–416, 420–422, 426, 427, 429, 432–434, 436, 439, 447–450, 452–458, 468, 473, 478

**B**

Barents, 13, 51, 66, 94, 233, 262, 286, 370, 389, 427, 467
Barents region, 13, 14, 369–371, 378, 382–384, 389–403, 437
Biomonitoring, 238
Boreal, 9, 22, 134, 175, 176, 190, 192, 193, 197–199, 211, 279, 283, 286
Borealization, 11, 216–218, 285
Bounded orders, 15, 466–468, 471–479

**C**

Central Arctic Ocean (CAO), 11, 157, 279–282, 286, 288, 290–291, 323, 421, 422, 436, 449
Chairmanship, 269, 331, 396, 412–420, 422, 448

China, 5, 7, 8, 12, 15, 69, 78, 105, 106, 117, 120, 125, 133, 136, 147–164, 172, 263, 270, 271, 274–276, 290, 305, 310, 311, 315–326, 342, 352, 371, 411, 414, 416, 421, 425, 426, 428, 429, 434, 436, 438, 439, 464–466, 471–479
Chinese economy, 150, 151, 153, 156, 159, 163, 164
Classified information, 245
Climate change, 7–9, 14, 15, 32, 34–37, 63, 104, 106, 109, 112–117, 119–125, 138, 140, 142, 143, 147, 152, 153, 159–161, 163, 171, 175, 176, 179, 181, 189, 190, 194, 195, 197, 200, 211–213, 215, 219, 220, 222, 223, 230, 245, 246, 260, 264, 274, 285–287, 298, 302, 305–307, 334, 366, 395, 399–402, 407, 408, 412, 416, 421, 425, 427, 431, 432, 443, 451–452, 458, 463, 474
Cold war, 1, 3–7, 13–15, 23, 63, 91–107, 109–114, 116, 117, 119–122, 124, 337, 345, 366, 369–384, 389, 391–393, 397, 408, 409, 425, 427, 428, 430, 437, 443, 447, 464–471, 474
Colonialism, 13, 336, 365, 366
Common interests, 7, 110, 117, 120–124, 200
Comprehensive security, 113, 117–120, 124, 469
Continuity and rupture, 26, 65–67, 142, 408, 417
Cooperation, 7, 11, 14, 15, 32, 74, 75, 91, 92, 94, 100, 101, 104–107, 109–125, 235, 239, 248, 259, 262, 270–272, 288, 289, 291, 302, 315, 317–319, 321, 324–326, 363, 370, 371, 377–380, 383, 384, 393–399, 414, 416, 421, 426, 427, 430–432, 434–439, 443–458, 463, 467–469, 479
Cultural revitalization, 366

### D

Demography, 26–29, 464, 465, 472, 476
Development, 4–8, 10–12, 14, 16, 23, 24, 29, 30, 35, 36, 41, 43, 44, 46–60, 63–70, 72, 76, 78–82, 84–86, 92, 104, 113–116, 119, 125, 129–131, 133–144, 148, 153, 155, 157, 158, 160, 163, 181, 183, 184, 194, 229–233, 240, 241, 245–247, 257–265, 268–276, 283, 297–305, 307, 309–312, 315–325, 331, 332, 335–340, 342–347, 351–366, 371, 389–403, 408, 414, 415, 420, 425–429, 433, 436, 437, 447, 448, 451, 464, 467, 469, 471, 472, 477, 478
Discussed, 3, 11, 42, 53, 57, 95, 110, 113, 115–121, 161, 163, 172–174, 200, 240, 246, 298, 300, 305, 308, 315, 364, 393, 401, 409

### E

Earth System dynamics, 4, 12, 16, 332–336, 338, 344, 347
Economic diversification, 58, 129
Encounters, 7, 22, 36, 138, 139, 142, 144, 149, 150, 155, 156, 283, 352
Environmental information, 9, 140, 231, 240, 242–244, 315
Environmental monitoring, 9, 10, 230–232, 239–245, 247, 409
European Union (EU), 14, 236, 258, 263, 274, 276, 290, 325, 397, 411, 414, 421, 425–439, 468, 470, 471, 477, 478

### F

Features of transport projects, 11, 303, 308, 310–312
Fisheries management, 100, 288, 379, 399
Fisheries research, 281, 288–291
Frontier, 1, 7, 11, 12, 44–47, 50, 58, 63, 129–144, 298, 304, 312, 351–366
Functional region, 150, 151

### G

Geopolitics, 1, 2, 5, 10–12, 114–120, 136, 425, 435
Glavsevmorput, 56, 63, 67, 76, 78–80, 82–86
Global Arctic, 2, 7, 11, 54, 113–116, 125, 147, 148, 162, 297–312
Global conventions, 15, 134, 443–458
Global flows, 7, 148, 149
Global governance, 15, 422, 444–447, 450, 456, 458
Governance, 1, 5–7, 12–15, 63, 100, 104, 107, 110, 112, 114–117, 121, 122, 125, 131, 139, 140, 268, 281, 288, 291, 322, 323, 389, 393, 394, 398–402, 411, 417, 418, 421, 422, 427, 430, 431, 435, 436, 439, 443–458, 467, 473
Governmental monitoring, 240
Great acceleration, 4, 331–333, 335, 336, 338, 339
Great power competition, 6, 7, 92, 105

# Index

Great powers, 6, 7, 14, 51, 91, 92, 94, 95, 105, 109, 112, 117–119, 122, 152, 316, 322, 325, 326, 373, 425, 426, 429, 434, 439, 463–467, 470, 471, 473, 474

Greenhouse gases, 9, 171, 178, 184, 189, 190, 195–200, 260, 267, 335, 338

## H

History, 9, 23, 24, 26, 36, 44, 47, 50–52, 63, 66, 67, 110, 112, 134, 152, 189, 190, 231, 261–263, 266, 273, 283, 285, 289, 299, 303, 307, 309, 320, 337, 340, 352, 355, 356, 358, 360, 363–366, 369–384, 390, 393, 413, 427, 428, 434, 469, 470, 472

## I

Identity, 5, 25–27, 29–32, 36, 144, 238, 323, 364, 398, 426

Implemented, 10, 11, 23, 54, 77, 86, 121, 123, 144, 181, 182, 247, 302, 305, 307–309, 311, 312, 370, 433, 437, 453, 455

Indigenous peoples, 4, 5, 7, 15, 16, 21–37, 53, 63, 109, 111–115, 117, 118, 120–122, 124, 125, 215, 223, 231, 236–238, 265, 272, 274, 280, 291, 301, 343, 344, 346, 352, 354–356, 360, 364, 372, 401, 402, 411, 427, 444, 446, 449–451, 458, 467, 469

Industrialization, 6, 9, 13, 36, 41–60, 299

Infrastructure, 4–7, 10, 11, 35, 45, 47, 56, 58, 78, 96, 105, 118, 129, 130, 135–137, 141, 144, 154, 156, 163, 233, 241, 259, 261, 271, 297–312, 318–320, 325, 340, 342, 346, 355, 397, 400–402, 429, 430, 436, 438, 468, 473, 475–477

Institutional change, 13, 393, 417, 420, 444, 458

International order, 12, 15, 463–470, 472–474

International system, 15, 323, 434, 463–466, 469–473

## K

Knowledge production, 162, 451, 455

## M

Major drivers, 44

Microplastic, 10, 230, 232, 234–235, 245, 246, 248

Mining, 4, 6, 23, 41–60, 82, 153, 155, 163, 173, 223, 232–234, 239, 265, 269, 272, 299, 303, 308, 319, 343, 344, 346, 357, 359, 360, 362, 365, 375, 399, 400, 402, 451

Mobilities, 7, 21, 57, 102, 106, 130, 143

Monitoring, 9, 81, 161, 235, 288, 396, 409, 451

Monitoring data, 9, 10, 230–232, 240, 242–245, 247, 248

## N

Norway, 2, 4, 6, 7, 10, 13, 22, 23, 27, 51, 55, 76, 91–107, 111, 117, 118, 123, 131, 135, 156, 158, 161, 172, 180, 232, 236, 238, 241, 243, 246, 261–263, 266–269, 274, 275, 288–290, 297, 302, 317, 319, 320, 342, 343, 360, 363–365, 369–384, 389–401, 411–413, 415, 418, 425, 426, 430, 433, 436–438, 447, 450, 451, 453, 456, 457, 467–469, 473–476

Norwegian defense, 6, 98, 475

## O

Outside interests, 7, 104, 365, 466

## P

Polar silk road, 105, 106, 147, 157, 163

Pollution in the Arctic, 109, 116, 121, 123, 230, 232–239, 245, 246, 409

Pops, 10, 35, 112, 230, 232, 235, 236, 238, 240, 245–248, 410, 453, 454

Post-Cold War, 7, 15, 109–125, 370, 394, 395, 464, 465, 467–470, 476, 478

Postcolonial processes, 12, 351–366

Power transition, 464, 471

## R

Regime-making, 12, 315–317, 321, 322, 324, 325

Regional architecture, 394

Regional cooperation, 15, 383, 393, 426, 438, 439, 457

Regional governance, 13, 15, 389–403, 444, 448, 456, 457

Resilience, 14, 37, 124, 223, 398, 435

Resource extraction, 36, 46, 154, 259, 339, 343, 428

Russia, 4, 10, 15, 22, 24, 26, 27, 30, 33, 36, 41–45, 47, 51, 53–55, 58, 59, 63, 66–70, 73, 85, 91, 94, 96, 104–106, 117–119, 122, 123, 131, 153, 160, 163, 172, 180, 231, 237, 240, 241, 244, 246, 260–264, 267, 269–275, 282, 288, 289, 299–302, 305, 308, 310, 312, 316–323, 325, 342, 343, 353, 354, 363–366, 369, 371–377, 380, 382, 383, 389, 391–399, 402, 412, 416, 419, 425–430, 433–439, 447, 448, 450, 454, 455, 464–466, 468, 470–471, 473–479

Russian Arctic, 15, 23, 27, 36, 43, 64, 86, 106, 118, 154, 159, 177, 238, 239, 269–271, 297, 303, 305, 312, 318, 319, 325, 366, 393, 469, 471, 474, 477, 479

## S

Sea-ice ecosystem, 9, 11, 113, 130, 211–213, 223, 279, 281, 283–285

Security, 4, 24, 57, 92, 110, 132, 156, 244, 265, 291, 318, 370, 396, 412, 425, 468

Security community, 120

Security governance, 14, 425–439

Self-determination, 12, 13, 115, 351–366, 450

Short-lived climate forcer (SLCF), 8, 171–184, 415, 452

Significant geopolitical changes, 110–114, 122

Soil organic carbon, 193

Soviet Union, 6, 11, 13, 23, 24, 48, 50, 63–87, 91–95, 97–104, 106, 111–113, 121, 258, 298, 354, 369–384, 392, 400, 409, 430

Special features, 120–123

Strategy, 6, 7, 9, 11, 14, 16, 37, 63, 86, 91–93, 95, 96, 98–104, 106, 111, 115, 118, 121, 124, 131, 149, 152, 153, 156, 157, 159, 183, 262, 263, 265–270, 302, 308, 319, 331, 342, 370, 381, 394, 397, 400, 402, 408, 409, 416, 417, 422, 427, 431, 432, 434–436, 443, 449, 453, 456, 468, 469, 476

Sustainable development, 12, 15, 16, 86, 107, 111, 123, 124, 137, 141, 239, 242, 247, 248, 269, 331–347, 394, 410, 411, 420, 431, 432, 435, 455, 467, 469

## T

Tourism, 7, 14, 16, 36, 129–144, 147, 151, 156, 269, 281, 317, 335, 343, 344, 362, 365, 399

Transport, 4, 6, 7, 11, 41, 46, 49, 53–55, 57, 60, 63, 64, 77, 78, 129, 130, 133, 135–137, 143, 155, 157, 173, 175–177, 194, 232, 233, 245, 279, 282, 283, 297, 298, 300, 302–312, 342, 353, 358, 362, 394, 395, 401, 407, 453

Trophic mismatch, 9, 214, 286

Trust-building, 13, 369–384

Trusts, 48, 300, 303, 342, 370, 374, 378, 380–383, 394, 395, 398, 434

Tundra greening, 220, 221

## W

Wildernesses, 7, 137, 138, 190

Wildlife, 9, 138, 140, 142, 143, 211–216, 237, 273, 357